Lecture Notes in Computer Science   7447

Commenced Publication in 1973
Founding and Former Series Editors:
Gerhard Goos, Juris Hartmanis, and Jan van Leeuwen

Editorial Board

David Hutchison
  Lancaster University, UK
Takeo Kanade
  Carnegie Mellon University, Pittsburgh, PA, USA
Josef Kittler
  University of Surrey, Guildford, UK
Jon M. Kleinberg
  Cornell University, Ithaca, NY, USA
Alfred Kobsa
  University of California, Irvine, CA, USA
Friedemann Mattern
  ETH Zurich, Switzerland
John C. Mitchell
  Stanford University, CA, USA
Moni Naor
  Weizmann Institute of Science, Rehovot, Israel
Oscar Nierstrasz
  University of Bern, Switzerland
C. Pandu Rangan
  Indian Institute of Technology, Madras, India
Bernhard Steffen
  TU Dortmund University, Germany
Madhu Sudan
  Microsoft Research, Cambridge, MA, USA
Demetri Terzopoulos
  University of California, Los Angeles, CA, USA
Doug Tygar
  University of California, Berkeley, CA, USA
Gerhard Weikum
  Max Planck Institute for Informatics, Saarbruecken, Germany

Stephen W. Liddle   Klaus-Dieter Schewe
A Min Tjoa   Xiaofang Zhou (Eds.)

Database and Expert Systems Applications

23rd International Conference, DEXA 2012
Vienna, Austria, September 3-6, 2012
Proceedings, Part II

 Springer

Volume Editors

Stephen W. Liddle
Brigham Young University, Marriott School
784 TNRB, Provo, UT 84602, USA
E-mail: liddle@byu.edu

Klaus-Dieter Schewe
Software Competence Center Hagenberg
Softwarepark 21, 4232 Hagenberg, Austria
E-mail: kd.schewe@scch.at

A Min Tjoa
Vienna University of Technology, Institute of Software Technology
Favoritenstraße 9-11/188, 1040 Wien, Austria
E-mail: amin@ifs.tuwien.ac.at

Xiaofang Zhou
University of Queensland
School of Information Technology and Electrical Engineering
Brisbane, QLD 4072, Australia
E-mail: zxf@uq.edu.au

ISSN 0302-9743   e-ISSN 1611-3349
ISBN 978-3-642-32596-0   e-ISBN 978-3-642-32597-7
DOI 10.1007/978-3-642-32597-7
Springer Heidelberg Dordrecht London New York

Library of Congress Control Number: 2012943836

CR Subject Classification (1998): H.2.3-4, H.2.7-8, H.2, H.3.3-5, H.4.1, H.5.3, I.2.1,
I.2.4, I.2.6, J.1, C.2

LNCS Sublibrary: SL 3 – Information Systems and Application, incl. Internet/Web
and HCI

Typesetting: Camera-ready by author, data conversion by Scientific Publishing Services, Chennai, India

Printed on acid-free paper

Springer is part of Springer Science+Business Media (www.springer.com)

Preface

This volume includes invited papers, research papers, and short papers presented at DEXA 2012, the 23rd International Conference on Database and Expert Systems Applications, held in Vienna, Austria. DEXA 2012 continued the long and successful DEXA tradition begun in 1990, bringing together a large collection of bright researchers, scientists, and practitioners from around the world to share new results in the areas of database, intelligent systems, and related advanced applications.

The call for papers resulted in the submission of 179 papers, of which 49 were accepted as regular research papers, and 37 were accepted as short papers. The authors of these papers come from 43 different countries. The papers discuss a range of topics including:

- Database query processing, in particular XML queries
- Labeling of XML documents
- Computational efficiency
- Data extraction
- Personalization, preferences, and ranking
- Security and privacy
- Database schema evaluation and evolution
- Semantic Web
- Privacy and provenance
- Data mining
- Data streaming
- Distributed systems
- Searching and query answering
- Structuring, compression and optimization
- Failure, fault analysis, and uncertainty
- Predication, extraction, and annotation
- Ranking and personalization
- Database partitioning and performance measurement
- Recommendation and prediction systems
- Business processes
- Social networking

In addition to the papers selected by the Program Committee two internationally recognized scholars delivered keynote speeches:

Georg Gottlob: DIADEM: Domains to Databases

Yamie Aït-Ameur: Stepwise Development of Formal Models for Web Services Compositions – Modelling and Property Verification

In addition to the main conference track, DEXA 2012 also included seven workshops that explored the conference theme within the context of life sciences, specific application areas, and theoretical underpinnings.

We are grateful to the hundreds of authors who submitted papers to DEXA 2012 and to our large Program Committee for the many hours they spent carefully reading and reviewing these papers. The Program Committee was also assisted by a number of external referees, and we appreciate their contributions and detailed comments.

We are thankful for the Institute of Software Technology at Vienna University of Technology for organizing DEXA 2012, and for the excellent working atmosphere provided. In particular, we recognize the efforts of the conference Organizing Committee led by the DEXA 2012 General Chair A Min Tjoa. We are gratefull to the Workshop Chairs Abdelkader Hameurlain, A Min Tjoa, and Roland R. Wagner.

Finally, we are especially grateful to Gabriela Wagner, whose professional attention to detail and skillful handling of all aspects of the Program Committee management and proceedings preparation was most helpful.

September 2012   Stephen W. Liddle
                                                                    Klaus-Dieter Schewe
                                                                       Xiaofang Zhou

Organization

Honorary Chair

Makoto Takizawa   Seikei University, Japan

General Chair

A Min Tjoa   Technical University of Vienna, Austria

Conference Program Chair

Stephen Liddle	Brigham Young University, USA
Klaus-Dieter Schewe	Software Competence Center Hagenberg and Johannes Kepler University Linz, Austria
Xiaofang Zhou	University of Queensland, Australia

Publication Chair

Vladimir Marik   Czech Technical University, Czech Republic

Program Committee

Witold Abramowicz	The Poznan University of Economics, Poland
Rafael Accorsi	University of Freiburg, Germany
Hamideh Afsarmanesh	University of Amsterdam, The Netherlands
Riccardo Albertoni	OEG, Universidad Politécnica de Madrid, Spain
Rachid Anane	Coventry University, UK
Annalisa Appice	Università degli Studi di Bari, Italy
Mustafa Atay	Winston-Salem State University, USA
James Bailey	University of Melbourne, Australia
Spiridon Bakiras	City University of New York, USA
Zhifeng Bao	National University of Singapore, Singapore
Ladjel Bellatreche	ENSMA, France
Morad Benyoucef	University of Ottawa, Canada
Catherine Berrut	Grenoble University, France
Debmalya Biswas	Nokia Research, Germany
Athman Bouguettaya	RMIT, Australia
Danielle Boulanger	MODEME,University of Lyon, France
Omar Boussaid	University of Lyon, France
Stephane Bressan	National University of Singapore, Singapore
Patrick Brezillon	University of Paris VI (UPMC), France
Yiwei Cao	RWTH Aachen University, Germany
Silvana Castano	Università degli Studi di Milano, Italy

Michal Krátký	VSB-Technical University of Ostrava, Czech Republic
Arun Kumar	IBM Research, India
Ashish Kundu	IBM T.J. Watson Research Center, Hawthorne, USA
Josef Küng	University of Linz, Austria
Kwok-Wa Lam	University of Hong Kong, Hong Kong
Nadira Lammari	CNAM, France
Gianfranco Lamperti	University of Brescia, Italy
Mong Li Lee	National University of Singapore, Singapore
Alain Toinon Leger	Orange - France Telecom R&D, France
Daniel Lemire	LICEF Research Center, Canada
Lenka Lhotska	Czech Technical University, Czech Republic
Wenxin Liang	Dalian University of Technology, China
Lipyeow Lim	University of Hawai at Manoa, USA
Tok Wang Ling	National University of Singapore, Singapore
Sebastian Link	University of Auckland, New Zealand
Volker Linnemann	University of Lübeck, Germany
Chengfei Liu	Swinburne University of Technology, Australia
Chuan-Ming Liu	National Taipei University of Technology, Taiwan
Fuyu Liu	Microsoft Corporation, USA
Hong-Cheu Liu	University of South Australia, Australia
Jorge Lloret Gazo	University of Zaragoza, Spain
Miguel Ángel López Carmona	University of Alcalá de Henares, Spain
Jiaheng Lu	Renmin University, China
Jianguo Lu	University of Windsor, Canada
Alessandra Lumini	University of Bologna, Italy
Hui Ma	Victoria University of Wellington, New Zealand
Qiang Ma	Kyoto University, Japan
Stéphane Maag	TELECOM SudParis, France
Nikos Mamoulis	University of Hong Kong, Hong Kong
Elio Masciari	ICAR-CNR, Università della Calabria, Italy
Norman May	SAP AG, Germany
Jose-Norberto Mazón	University of Alicante, Spain
Dennis McLeod	University of Southern California, USA
Brahim Medjahed	University of Michigan - Dearborn, USA
Harekrishna Misra	Institute of Rural Management Anand, India
Jose Mocito	INESC-ID/FCUL, Portugal
Riad Mokadem	IRIT, Paul Sabatier University, France
Lars Mönch	FernUniversität in Hagen, Germany
Yang-Sae Moon	Kangwon National University, Korea
Reagan Moore	University of North Carolina at Chapel Hill, USA

Franck Morvan   IRIT, Paul Sabatier University, Toulouse,
                                   France
Mirco Musolesi   University of Birmingham, UK
Ismael Navas-Delgado   University of Málaga, Spain
Wilfred Ng   University of Science and Technology,
                                   Hong Kong
Javier Nieves Acedo   Deusto University, Spain
Mourad Oussalah   University of Nantes, France
Gultekin Ozsoyoglu   Case Western Reserve University, USA
George Pallis   University of Cyprus, Cyprus
Christos Papatheodorou   Ionian University and "Athena" Research
                                   Centre, Greece
Marcin Paprzycki   Polish Academy of Sciences, Warsaw
                                   Management Academy, Poland
Oscar Pastor Lopez   Universidad Politecnica de Valencia, Spain
Jovan Pehcevski   European University, Macedonia
Reinhard Pichler   Technische Universität Wien, Austria
Clara Pizzuti   ICAR-CNR, Italy
Jaroslav Pokorny   Charles University in Prague, Czech Republic
Elaheh Pourabbas   National Research Council, Italy
Fausto Rabitti   ISTI, CNR Pisa, Italy
Claudia Raibulet   Università degli Studi di Milano-Bicocca, Italy
Isidro Ramos   Technical University of Valencia, Spain
Praveen Rao   University of Missour-KaNSAS City, USA
Rodolfo F. Resende   Federal University of Minas Gerais, Brazil
Claudia Roncancio   Grenoble University / LIG, France
Edna Ruckhaus   Universidad Simon Bolivar, Venezuela
Massimo Ruffolo   ICAR-CNR, Italy
Igor Ruiz Agúndez   Deusto University, Spain
Giovanni Maria Sacco   University of Turin, Italy
Shazia Sadiq   University of Queensland, Australia
Simonas Saltenis   Aalborg University, Denmark
Carlo Sansone   Università di Napoli "Federico II", Italy
Igor Santos Grueiro   Deusto University, Spain
N.L. Sarda   I.I.T. Bombay, India
Marinette Savonnet   University of Burgundy, France
Raimondo Schettini   Università degli Studi di Milano-Bicocca, Italy
Erich Schweighofer   University of Vienna, Austria
Florence Sedes   IRIT, Paul Sabatier University, Toulouse,
                                   France
Nazha Selmaoui   University of New Caledonia, France
Patrick Siarry   Université Paris 12 (LiSSi), France
Gheorghe Cosmin Silaghi   Babes-Bolyai University of Cluj-Napoca,
                                   Romania
Leonid Sokolinsky   South Ural State University, Russia

Bala Srinivasan   Monash University, Australia
Umberto Straccia   Italian National Research Council, Italy
Darijus Strasunskas   Strasunskas Forskning, Norway
Lena Stromback   Swedish Meteorological and Hydrological
                                   Institute, Sweden
Aixin Sun   Nanyang Technological University, Singapore
David Taniar   Monash University, Australia
Cui Tao   Mayo Clinic, USA
Maguelonne Teisseire   Irstea - TETIS, France
Sergio Tessaris   Free University of Bozen-Bolzano, Italy
Olivier Teste   IRIT, University of Toulouse, France
Stephanie Teufel   University of Fribourg, Switzerland
Jukka Teuhola   University of Turku, Finland
Taro Tezuka   University of Tsukuba, Japan
Bernhard Thalheim   Christian Albrechts Universität Kiel, Germany
J.M. Thevenin   University of Toulouse I Capitole, France
Helmut Thoma   Thoma SW-Engineering, Basel, Switzerland
A Min Tjoa   Technical University of Vienna, Austria
Vicenc Torra   IIIA-CSIC, Spain
Traian Truta   Northern Kentucky University, USA
Theodoros Tzouramanis   University of the Aegean, Greece
Marco Vieira   University of Coimbra, Portugal
Jianyong Wang   Tsinghua University, China
Junhu Wang   Griffith University, Brisbane, Australia
Qing Wang   The Australian National University, Australia
Wei Wang   University of New South Wales, Sydney,
                                   Australia
Wendy Hui Wang   Stevens Institute of Technology, USA
Andreas Wombacher   University Twente, The Netherlands
Lai Xu   Bournemouth University, UK
Ming Hour Yang   Chung Yuan Christian University, Taiwan
Xiaochun Yang   Northeastern University, China
Haruo Yokota   Tokyo Institute of Technology, Japan
Zhiwen Yu   Northwestern Polytechnical University, China
Xiao-Jun Zeng   University of Manchester, UK
Zhigang Zeng   Huazhong University of Science and
                                   Technology, China
Xiuzhen (Jenny) Zhang   RMIT University Australia, Australia
Yanchang Zhao   RDataMining.com, Australia
Yu Zheng   Microsoft Research Asia, China
Qiang Zhu   The University of Michigan, USA
Yan Zhu   Southwest Jiaotong University, Chengdu,
                                   China

External Reviewers

Hadjali Allel	ENSSAT, France
Toshiyuki Amagasa	Tsukuba University, Japan
Flora Amato	University of Naples Federico II, Italy
Abdelkrim Amirat	University of Nantes, France
Zahoua Aoussat	University of Algiers, Algeria
Radim Bača	Technical University of Ostrava, Czech Republic
Dinesh Barenkala	University of Missouri-Kansas City, USA
Riad Belkhatir	University of Nantes, France
Yiklun Cai	University of Hong Kong
Nafisa Afrin Chowdhury	University of Oregon, USA
Shumo Chu	Nanyang Technological University, Singapore
Ercument Cicek	Case Western Reserve University, USA
Camelia Constantin	UPMC, France
Ryadh Dahimene	CNAM, France
Matthew Damigos	NTUA, Greece
Franca Debole	ISTI-CNR, Italy
Saulo Domingos de Souza Pedro	Federal University of Sao Carlos, Brazil
Laurence Rodrigues do Amaral	Federal University of Uberlandia, Brazil
Andrea Esuli	ISTI-CNR, Italy
Qiong Fang	University of Science and Technology, Hong Kong
Nikolaos Fousteris	Ionian University, Greece
Filippo Furfaro	DEIS, University of Calabria, Italy
Jose Manuel Gimenez	Universidad de Alcala, Spain
Reginaldo Gotardo	Federal University of Sao Carlos, Brazil
Fernando Gutierrez	University of Oregon, USA
Zeinab Hmedeh	CNAM, France
Hai Huang	KAUST, Saudi Arabia
Lili Jiang	Lanzhou University, China
Shangpu Jiang	University of Oregon, USA
Hideyuki Kawashima	University of Tsukuba, Japan
Selma Khouri	LIAS/ENSMA, France
Christian Koncilia	University of Klagenfurt, Austria
Cyril Labbe	Université Joseph Fourier, Grenoble, France
Thuy Ngoc Le	National University of Singapore, Singapore
Fabio Leuzzi	University of Bari, Italy
Luochen Li	National University of Singapore, Singapore
Jing Li	University of Hong Kong
Xumin Liu	Rochester Institute of Technology, USA
Yifei Lu	University of New South Wales, Australia
Jia-Ning Luo	Ming Chuan University, Taiwan

Table of Contents – Part II

Ranking and Personalization

Searching I

Database Partitioning and Performance

Semantic Web

Data Mining II

Distributed Systems

Web Searching and Query Answering

Recommendation and Prediction Systems

Query Processing II

Query Processing III

Searching II

Business Processes and Social Networking

Data Security, Privacy, and Organization

Table of Contents – Part I

Data Extraction

Personalization, Preferences, and Ranking

Databases and Schemas

Privacy and Provenance

XML Queries and Labeling II

Data Streams

Structuring, Compression and Optimization

Data Mining I

Road Networks and Graph Search

Consistent Query Answering Using Relational Databases through Argumentation

Cristhian A. D. Deagustini[1,2,3], Santiago E. Fulladoza Dalibón[1,2,3],
Sebastián Gottifredi[1,3], Marcelo A. Falappa[1,3], and Guillermo R. Simari[1]

[1] Artificial Intelligence Research and Development Laboratory
Department of Computer Science and Engineering
Universidad Nacional del Sur - Alem 1253, (8000) Bahía Blanca, Buenos Aires
[2] Faculty of Administrative Sciencies
Universidad Nacional de Entre Ríos - Tavella 1424, (3200) Concordia, Entre Ríos
[3] Consejo Nacional de Investigaciones Científicas y Técnicas
{cadd,sef,sg,mfalappa,grs}@cs.uns.edu.ar

Abstract. This paper introduces a framework that integrates a reasoner based on defeasible argumentation with a large information repository backed by one or several relational databases. In our scenario, we assume that the databases involved are updated by external independent applications, possibly introducing inconsistencies in a particular database, or leading to inconsistency among the subset of databases that refer to the same data. Argumentation reasoning will contribute with the possibility of obtaining consistent answers from the information repository with the properties described. We present the *Database Integration for Defeasible Logic Programming* (DBI-DeLP) framework, which enables commonsense reasoning based on *Defeasible Logic Programming* (DeLP) by extending the system capabilities to handle large amounts of data and providing consistent answers for queries posed to it.

1 Introduction

Argumentation represents a sophisticated mechanism for the formalization of commonsense reasoning, which has found application and proven its importance in different areas of Artificial Intelligence (AI) such as legal systems, multi-agent systems, and decision support systems among others (see [1–3]). Intuitively, an argument is a coherent set of statements leading from some premises to a claim or conclusion, whose acceptance will depend on considering the relevant arguments in favor and against the argument. This process is effected through a dialectical analysis, which is formalized by some form of a proof procedure [3]. In the literature of argumentation systems, certain systems called Rule Based Argumentation Systems (RBAS) can be found [4–7]; in them, the structure of the arguments is made explicit. In the RBAS, arguments are build from a specific knowledge base of rules, usually known as a *program*. An argument in these systems will be a set of rules from which the claim of the argument can be inferred. RBAS are suitable tools to develop real-world applications that can handle the

S.W. Liddle et al. (Eds.): DEXA 2012, Part II, LNCS 7447, pp. 1–15, 2012.

inconsistency found in the data coming from real domains while providing consistent answers and human-like explanations.

However, in RBAS the arguments are build using information contained in the program; this information can be perceived as *local* information, *i. e.,* more subjective and therefore affected by the agent's particular vision of the domain. To avoid giving answers heavily influenced by fixed facts in the program, it is important to include in the reasoning new sources of supporting data from which the arguments can be build. Nevertheless, using external sources of data can introduce another problem: any massive collection of real data is prone to contain contradictory data. For example, consider a legal environment where new legal precedents are established each time a case resolution is given. A particular case can have a resolution in a first instance that declares a person guilty of a crime, but then in a second instance the sentence is changed declaring the person not guilty. In a relational database (or simply database) these two events can be stored as different records of the same table, making the database inconsistent as it says that the same person is both guilty and not guilty of the same crime, *i. e.,* it maintains conflicting data.

Conflict can also arise when data, while not by itself contradictory, still can lead to conflicting conclusions. Consider a RBAS with the following rules from a movies domain:

- If a movie has a rating below five then it is a bad movie.
- If an action movie has the actor Stallone as its star then is not a bad movie.

Suppose that the system has access to a database with film reviews that states the rating of a film as four, and there is another database showing that in the same film Stallone has the leading role. There exists a conflict between these two data sources as they can support both the conclusion of the movie being bad and not bad. This type of situations are very common and have to be addressed.

We will introduce a framework that enables argumentative reasoning over data stored in relational databases referred to as *Database Integration for Defeasible Logic Programming* (DBI-DeLP). For this, DBI-DeLP uses DeLP [7] to handle the argumentation process and feeds it with information retrieved from available domain data sources that are not expected to maintain a consistent state, but can still give consistent answers for queries posed to the system.

The paper is organized as follow: in Section 2 we briefly review DeLP, the argumentation formalism that supports DBI-DeLP, in Section 3 we show how argument supporting information can be retrieved from relational databases, and in Section 4 we present how consistent answers can be obtained from an inconsistent environment. Finally, in Section 5 we offer some conclusions and identify future lines of work.

2   Background

We begin by giving a brief summary of Defeasible Logic Programming (DeLP). DeLP combines results of Logic Programming and Defeasible Argumentation providing the capability of representing information as rules in a declarative manner, and a defeasible argumentation inference mechanism for warranting the entailed conclusions. These rules are the key element for introducing defeasibility and they will be used to represent a relation between pieces of knowledge that could be defeated after all things are considered.

A Defeasible Logic Program (de.l.p. for short) is a pair (Π, Δ) where Π is a set of strict rules and facts, and Δ is a set of defeasible rules. Facts are ground literals representing atomic information or the negation of atomic information using strong negation "$\sim$" (e. g., L, or $\sim L$). In what follows L_i are literals. Strict rules (s-rules) are denoted $L_0 \leftarrow L_1, \ldots, L_n$, and represent information that can not be refused, i. e., if Body $L_1, \ldots, L_n$ can be accepted then Head L_0 is granted. Defeasible Rules (d-rules) are denoted $L_0 \prec L_1, \ldots, L_n$. A d-rule represents tentative information that may be used if nothing could be posed against it. A d-rule $Head \prec Body$ expresses that reasons to believe in the antecedent Body give reasons to believe in the consequent Head. Since strong negation can appear in facts or in the head of rules, from a de.l.p. contradictory literals could be derived. Given a literal L, $\bar{L}$ represents the complement with respect to strong negation. Nevertheless, the set Π must be non contradictory.

We will adopt the notation of logic programming, where variable names begin with uppercase letters, and where constant and predicate names begin with lowercase letters, e. g., actor $\leftarrow$ sean_penn represents a strict rule; and good_movie(M) $\prec$ performs_in(M, arnold) represents a defeasible rule.

The last element we need to introduce is Presumptions, which are defeasible rules with an empty body. Presumptions are assumed to be true if nothing could be posed against them. In [7] an extension to DeLP that includes presumptions is presented, where an extended de.l.p. is a pair (Π, Δ) where Π is a set of strict rules and facts, and Δ is a set of defeasible rules and presumptions.

We can define how a literal will be derived from a de.l.p. A literal L is derivable from a set of facts, strict rules, presumptions and defeasible rules X, denoted by $X \mid\sim L$, iff it is derivable in the classical rule-based sense, treating strict and defeasible rules equally. These inferences are called defeasible derivations, and are computed by backward chaining applying the usual SLD inference procedure used in logic programming.

Definition 1. *(Defeasible derivation)*
Let $\mathcal{P} = (\Pi, \Delta)$ be a de.l.p and L a ground literal. A defeasible derivation of L from $\mathcal{P}$, denoted $\mathcal{P} \mid\sim L$, consists of a finite sequence $L_1, L_2, \ldots, L_n = L$ of ground literals, and each literal L_i is in the sequence because:
(a) L_i is a fact or a presumption,
(b) there exists a rule R_i in $\mathcal{P}$ (strict or defeasible) with head L_i and body $B_1, B_2, \ldots, B_k$ and every literal of the body is an element L_j of the sequence appearing before L_i $(j < i)$.

A set X is *contradictory*, denoted $X \mid\sim \perp$, iff both $X \mid\sim L$ and $X \mid\sim \sim L$ holds for some L. That is, X is contradictory (or inconsistent) if both a literal and its complement can be derived from them. For example, the sets {a, $\sim$a} and {a, b, $\sim$a $\prec$ b} are contradictory as a and $\sim$a are supported by them. We will see later how this concept can be applied to databases as well. A literal L is *consistently derivable* by X, denoted by $X\!\!\mid\!\sim^c L$, iff $X\!\!\mid\!\sim L$ and $X\!\!\not\mid\!\sim \bot$, *i. e.*, if it can be derived but not its complement. In DeLP when contradictory literals are derived, a dialectical process is used for deciding which literals are warranted. A literal L is warranted if there exists a non-defeated argument A for L. Using rules, presumptions and facts arguments can be constructed as follows.

Definition 2. *(Argument, Subargument)*
Let L be a literal and let $\mathcal{P} = (\Pi, \Delta)$ be an extended de.l.p. $\langle \mathcal{A}, L \rangle$ with $\mathcal{A} \subseteq \Delta$ is an argument *for L, iff $\Pi \cup \mathcal{A}\!\mid\!\sim^c L$ and $\mathcal{A}$ is minimal, i. e., not exists a set $A' \subseteq \mathcal{A}$ such that $\Pi \cup \mathcal{A}' \!\mid\!\sim^c L$. An argument $\langle \mathcal{B}, q \rangle$ is a* subargument *of an argument $\langle \mathcal{A}, L \rangle$, iff $\mathcal{B} \subseteq \mathcal{A}$.*

Two literals L and L_1 *disagree* regarding a de.l.p. $\mathcal{P} = (\Pi, \Delta)$, iff the set $\Pi \cup \{L, L_1\}$ is contradictory. An argument $\langle \mathcal{A}_1, L_1 \rangle$ is a *counterargument* to an argument $\langle \mathcal{A}_2, L_2 \rangle$ at a literal L, iff there is a subargument $\langle \mathcal{A}, L \rangle$ of $\langle \mathcal{A}_2, L_2 \rangle$ such that L and L_1 disagree. When this happens, we have to solve the dispute between them by using a formal comparison criterion among disagreement arguments; here we use *generalized specificity* [7, 8] to this end. According to this criterion an argument is preferred to another argument, iff the former is more *specific* than the latter. For example, $\langle \{(c\!\!-\!\!\prec a, b)\}, c \rangle$ is more specific than $\langle \{(\sim c\!\!-\!\!\prec a)\}, \sim c \rangle$, denoted by $\langle \{(c\!\!-\!\!\prec a, b)\}, c \rangle \succ \langle \{(\sim c\!\!-\!\!\prec a)\}, \sim c \rangle$, see [8] for a formal definition and further discussion.

Thus, an argument $\langle \mathcal{A}_1, L_1 \rangle$ is a *defeater* of an argument $\langle \mathcal{A}_2, L_2 \rangle$, iff there is a subargument $\langle \mathcal{A}, L \rangle$ of $\langle \mathcal{A}_2, L_2 \rangle$ such that $\langle \mathcal{A}_1, L_1 \rangle$ is a counterargument of $\langle \mathcal{A}_2, L_2 \rangle$ at literal L and either $\langle \mathcal{A}_1, L_1 \rangle \succ \langle \mathcal{A}, L \rangle$ (*proper defeat*) or $\langle \mathcal{A}_1, L_1 \rangle \not\succ \langle \mathcal{A}, L \rangle$ and $\langle \mathcal{A}, L \rangle \not\succ \langle \mathcal{A}_1, L_1 \rangle$ (*blocking defeat*).

When considering sequences of arguments, the definition of defeat is not sufficient to describe a conclusive argumentation line since it disregards the dialectical structure of argumentation, *cf.* [7]. Basically, when a query is received the dialectical proof procedure attempts to build an argument for it. Then attempts to build arguments against it (called interfering arguments), and defeaters for the interfering arguments (called supporting arguments as they *defend* the first claim), and so on. This is called an argumentation line.

In DeLP a literal L is *warranted* from a de.l.p. $\mathcal{P}$ (noted $\mathcal{P} \!\mid\!\sim_w L$), if there is an argument $\langle \mathcal{A}, L \rangle$ which is ultimately non-defeated. To decide whether $\langle \mathcal{A}, L \rangle$ is defeated or not, every argumentation line starting with $\langle \mathcal{A}, L \rangle$ has to be considered. This leads to a tree structure called dialectical tree (d-tree), denoted $\mathcal{T}_{\langle \mathcal{A}_0, L_0 \rangle}$. In a dialectical tree every node (except the root) is a defeater of its parent, and leaves are non-defeated arguments. A node is defeated if at least one of its attackers is undefeated. Again, we refer the interested reader to [7] for a deeper treatment of both argumentation lines and dialectical trees.

When we have build the dialectical tree, it is necessary to perform a *bottom-up* analysis of the tree to decide whether the argument at the root of it is defeated or not. Every leaf of the tree is marked *undefeated* and every inner node is marked *defeated* if it has at least one child node marked *undefeated*; otherwise it is marked *undefeated*. We call a literal L *warranted* in a DeLP $\mathcal{P}$, denoted by

$\mathcal{P}\!\mid\!\sim_w L$, iff there is an argument $\langle \mathcal{A}, L \rangle$ for L in $\mathcal{P}$ such that the root argument in its d-tree have all its attackers marked as defeated. Then $\langle \mathcal{A}, L \rangle$ is a *warrant* for L. We will see latter that dialectical trees (and the process to build them) will be crucial to obtain consistent answers based on inconsistent databases.

3   Supplying Information to the Argumentative Process

As mentioned earlier, DeLP is a framework that enables query resolution by an argumentative process which deals with inconsistency by considering uncertain data as defeasible information. DeLP can also resolve conflicts between conflicting data by using some appropriated preference criteria. This property is used to model commonsense reasoning similar to the one performed by humans. Nevertheless, in a production environment we often want to base our reasoning in as much information as we can gather. However, we need to have in mind that information provided by different entities can be contradictory. In Argumentation using contradictory information allows the system the construction of different arguments and counter-arguments, but we have to be careful that answers given by the system are still consistent although they are based on contradictory data. Here the non-monotonic characteristic of defeasible argumentation is useful as it allows us to have different answers at different times as knowledge that supports them is updated with new, opposite data; and still the dialectical process ensures the consistency of these answers.

Clearly, characteristics of DeLP must be combined with great amounts of dynamically updated data to be useful in any real production environment. Operationally speaking, this can not be achieved by including new data as a static part of the program, therefore we need to consider alternatives for making available the new data to the system. Two distinct problems arise in this scenario. First, we have to establish the information sources that will be used to obtain a supporting argument; secondly, we need to define an efficient way to provide the information to be used in the argumentation processes.

To address the first issue, different sources of data can be used to support arguments. We think that a good option as data sources can be relational databases, as it is probably the most popular storage system these days. Moreover, in the recent time we have witness how datasets with information from different domains have been released. Therefore it is reasonable to assume that the argumentation part of the system will need to have access to relational databases containing the information for a specific domain.

For the second problem we have to obtain from these databases the evidence to support arguments and make the data available for the argumentation process. To achieve this we present DBI-DeLP, a framework that integrates a reasoner based on Defeasible Logic Programming with a large information repository based on one or several relational databases.

Before we can show the framework itself and how it solves queries, we need to establish some definitions for elements present in the DBI-DeLP formalism. First we need to define how the records obtained from databases will be represented so they can be used by the reasoner to solve queries. As said before, we have to keep in mind that one possible problem that arises from the use of several databases is contradictions.We think that all information can be valuable in an argumentation process, so we want a database record representation strategy that allow us to reason with them without the need to clean up inconsistencies. We cannot use facts to represent database obtained information because it can lead to inconsistencies in the set of strict knowledge Π. Thus, in DBI-DeLP we introduce *operative presumptions* (OP), which are presumptions used to represent such knowledge.

Definition 3. *(Operative Presumption) Given a set of databases $D_1 \ldots D_n$, an operative presumption is a literal L_0 with the form $pred(p_1, \ldots, p_n)$ denoted "$L_0 \prec true$" such that there exists $t = \{q_1, q_2, \ldots, q_m\} \in D_k$ with $1 \le k \le n$ such that $q_i = p_i$ for all i. The set of all operative presumptions for D is called $OPset_D$. The set of all operative presumptions is $OPset = \bigcup_{i=1}^{n} OPset_{D_i}$.*

Operative presumptions provide a way to represent the information coming from external sources. Thus, we define a DBI-DeLP program as a set of strict rules, facts, defeasible rules and operative presumptions, where the set Π of strict rules and facts and the set Δ of defeasible rules are fixed parts of the program, while the set Σ of operative presumptions dynamically changes when the databases that provide them are modified. The set Σ contains all potential operative presumptions that can be obtained from information maintained in all the databases the system uses. Generally speaking, Σ can be thought as a representation in the program of all the information in the universe of databases that are available to the system. We say that a database is available if we can somehow connect and make queries to the DBMS that manages it.

Definition 4. *(DBI-DeLP Program)*
Given a set of databases $D_1 \ldots D_k$, a DBI-DeLP program (d.b.i.-de.l.p.) $\mathcal{P}$ is a triplet (Π, Δ, Σ) where (Π, Δ) is a DeLP program, and $\Sigma = \bigcup_{i=1}^{k} OPset_{D_i}$ is a set of Operative Presumptions.

Before we can obtain consistent answers for queries posed to the system based on possibly inconsistent databases, we need to establish how we can feed the argumentation process with data from these relational databases, *i. e.*, how we determine that a tuple in a database is relevant to the particular query that the system is trying to solve at a given time. One goal of our approach is the use of a massive repository of domain data formed by several relational databases, possibly provided by different entities. Therefore, as we will use external data sources that may be constantly growing, it is important to devise an efficient method to extract information relevant to a particular query, instead of constantly encoding new data in the program. Different approaches have been developed over the years regarding information extraction and consistent query answering from possibly inconsistent databases. In [9] inconsistencies are solved by means of *database repair*, that is, explicit addition, modification or suppression of conflicting tuples in databases. A different focus is used in [10, 11], where modifications to queries are made so consistent answers can be returned (although they also use repairs as auxiliary objects to obtain them). Despite its differences, both approaches center in obtain consistent data from the databases (or its repairs), and then give them as answers. But, keeping in mind that what needs to be consistently answered is the query posed to the system and not the SQL query sent to databases, we think that another possibility is to extract from the databases **all** relevant data (contradictory or not) and then work our way around them. So, in our strategy we will not try to obtain consistent information from the databases nor there is the need to modify them to maintain their consistency. Instead, inconsistencies in data obtained will be handled by the dialectical process in DBI-DeLP. This will allow us to retrieve every possible support data for building arguments, regardless of the possible inconsistencies among them. For instance, suppose that we have a database with a tuple indicating that *tux* is a bird, and another one saying that it is a penguin. A priori these database tuples are not contradictory. Also suppose that we know that almost every bird fly , and that penguins does not fly . If we have the query *fly(tux)?* then those tuples became conflictive, as they support complementary conclusions, but still we will retrieve both and let the argumentation process in charge of the problem.

To make a record in a database available to the argumentation process, we need to do two main things. First, we need to determine that a certain database (and which tables and fields in it) may contain important data to the query the system is solving. Secondly, we need to retrieve that data and make it available to the process. In a nutshell, we use the elements in the literal that the dialectical procedure is trying to warrant to determine which available databases are related to it, and then we extract relevant information from these databases using the pertinent SQL queries. Finally, we transform all retrieved results into operative presumptions so they become available to the argumentation process.

To start the process of obtaining from databases relevant data to a particular query, we first identify the literal in the query that is being solved. In section 2 we have outlined how DeLP constructs arguments to solve queries by a backward chaining process. That is, when DeLP is trying to build an argument for a literal L it searches its knowledge base looking for strict rules, facts, presumptions and defeasible rules that has L as Head, and then it tries to prove the literals in the Body . We will call these literals in the Body part of a rule *Function Objectives (FO)*, as they will be the next *targets* of the inference procedure.

Definition 5. *(Function Objective (FO))*
Given a strict rule $L_0 \leftarrow L_1, \ldots, L_n$, or a defeasible rule $L_0 \prec L_1, \ldots, L_n$, we call Function Objective to each literal L_i $(1 \leq i \leq n)$. The set of every possible operative presumption is called FOset.

For example, if DeLP tries to prove recommended(commando, mike) and finds a rule recommended(Film, User) $\prec$ genre(Film, Genre), likes(User, Genre), high_rating(Film); then genre(commando, Genre), likes(mike, Genre) and high_rating(commando) are Function Objectives. As we can see, all FO are first order predicates in the form $func(p_1, \ldots, p_n)$, where *func* is the predicate's functor and $p_1, \ldots, p_n$ is a list of the predicate's parameters (*e. g.*, for genre(commando, Genre) its functor is genre while commando and Genre are its parameters).

The SLD procedure will try to prove every Function Objective using all rules, facts and presumptions in the DBI-DeLP program; but for the purpose of this section we will focus on how presumptions can be obtained from databases. To do this, a search for presumptions is launched (see condition (a) of Def. 1), in order to retrieve from the databases, if exists, some information offering support to the literal. For this we have to identify our data sources, that is which databases (and which tables and fields in them) are expected to have useful data for the FO. We will say that a data source (actually, tables and fields in a database) is pertinent to a FO if the data source has the potential to support the FO, *i. e.*, we *may* use some data stored in those table and fields to prove the literal.

Definition 6. *(Data Source Pertinency)*
Let O be a FO func$(p_1, \ldots, p_m)$ *, T be a set of m {table, field} pairs,* A_i *be the set of all possible values for* p_i *and* B_i *be the set of all possible values for {table, field}*$_i$*. We say that T is pertinent to O iff* $B_i \subseteq A_i$ *for all* $1 \leq i \leq m$

Thus, the FO-DataSource pertinency relation states that for a particular data source there are certain fields in tables that corresponds to a FO. For example, if we have an available database set up like the one shown in Fig. 1, the set {[movies, title], [actors, name]} is pertinent to the FO performs_in(movie, actor) because by looking in the title field of table movie and the name field of the actors table we can find out if a certain actor had been cast in a given film. Actually, we will have to look in those fields but in the SQL JOIN of the three presented tables, but for the purpose of this paper we will not focus on this.

Example Database

Movie	
id	title
1	Commando
2	Rambo
3	Terminator
4	Demolition Man

Movie-Actor		
movie_id	actor_id	performance
1	2	Great
2	1	Ok
3	2	Great
4	1	Ok
4	3	Regular
4	4	Great

Actor	
id	name
1	Stallone
2	Schwarzenegger
3	Bullock
4	Snipes

Fig. 1. Database with data about films's casts

If a database is pertinent to the FO, we can use it to support the FO, *i. e.*, we can obtain the necessary operative presumptions from it. For example performs_in("Commando", "Stallone")? is proven to be true. Notice that the pertinency relation does not guarantee that a given database will in fact prove the literal. For instance, we cannot use the database in Fig. 1 to prove performs_in("The One", "Jet Li"), as there is no tuple in the database indicating that Jet Li performs in the movie The One. Instead, the pertinency relation indicates that the data source is a candidate as it may have the required data.

Nevertheless, pertinency of a data source to a FO is necessary to obtain the correspondent operative presumption, but may not be sufficient. Sometimes we may want to subject the retrieval of a result from the databases to certain conditions (besides values given in the FO). For instance, lets say that we will consider the FO good_performance("Demolition Man", Actor)?, trying to retrieve from

the database all actors that have had a good performance in the movie Demolition Man. If we establish that only the Ok or Great performances can be
considered as good performances, then good_performance("Demolition Man",
"Stallone")? is true, but good_performance("Demolition Man", "Bullock")? is
not, as database state that Sandra Bullock has a regular performance in the
film. So, although {[movies, title], [actors, name]} is pertinent to the FO, the
tuple for Sandra Bullock should not be retrieved. So, sometimes we will restrict
which tuples should be retrieved from databases by stating some *conditions* that
need to hold to consider a certain tuple a valid result. Basically, a condition for a
FO over an available database D is the sets of values that certain fields in a table
from D *must* have to enable the retrieval of tuples in that mentioned database.

Definition 7. *(Condition)*
*Let D be a database, $T_1, T_2, \ldots, T_n$ tables in D, F_i a field in T_i, V_i be a set of
values valid for F_i, and FO be a Function Objective. Then a condition for FO
over D is the triplet (FO, D, $\{(T_1, F_1, V_1), (T_2, F_2, V_2), \ldots, (T_n, F_n, V_n)\}$).*

So, for the FO good_performance("Demolition Man", Actor) in the last example,
the condition (good_performance, Example Database, { (Movie-Actor table, Performance field, { "Ok", "Great" })}) is established. We will only use the functor of
the FO in the condition as a condition is valid for all FO that shares the functor,
e. g., the conditions imposed to the FO good_performance("Demolition Man",
Actor) and the FO good_performance("Rambo", "Stallone") are the same.

So, we have defined all we need to determine when a database tuple have to
be retrieved to make it available to the argumentation process. Now we need to
define the presumption retrieval function used to retrieve these tuples. Generally
speaking, the goal of the function is to feed the argumentation process with
relevant data obtained from the databases that are available to the system. As
expected, relevancy of data in a database is settled by the combination of the
pertinency of the data source and conditions established for the FO.

Definition 8. *(Presumption Retrieval Function (PRF))*
*Let FOset be the set of all the Function Objectives, OPset be the set of all the op-
erative presumptions, FO be a function objective $func(p_1, \ldots, p_m)$, C be a condition (FO, D, $\{(T_1, F_1, V_1), (T_2, F_2, V_2), \ldots, (T_n, F_n, V_n)\}$) for FO over an available
database D. The presumption retrieval function (PRF) $PRF : FOset \mapsto OPset$
is such that PRF(FO) = S where*

1. *$func(q_1, \ldots, q_m) \prec true$ in S iff exists a tuple t $\{q_1, q_2, \ldots, q_m, r_1, r_2, \ldots, r_n\}$
   in the database D such that*
   - *if $ground(p_i)$, then $q_i = p_i$ for all $1 \le i \le m$*
   - *and $r_i \in V_i$ with r_i being the value stored in field F_i of table T_i for all
     $1 \le i \le n$.*
2. *the set of m {table, field} pairs for t is pertinent to $func(p_1, \ldots, p_m)$*
3. *S is maximal: there is no set S' of OP such that S is a proper subset of S'
   which satisfies conditions (1) and (2).*

So, the PRF will retrieve **all** database tuples which values are equal to the correspondent grounded values in the FO and that fulfill all conditions imposed to the FO. For example, given the data shown in Fig. 1, for the FO good_performance("Demolition Man", Actor), if conditions imposed are the same ones that had been stated before, the PRF will retrieve the tuples ("Demolition Man", "Stallone", "Ok") and ("Demolition Man", "Snipes", "Great"). So, the resulting set S is {good_performance("Demolition Man", "Stallone") —≺ true, good_performance("Demolition Man", "Snipes") —≺ true}, unifying the non grounded parameter Actor with both "Stallone" and "Snipes". As we can see, good_performance("Rambo", "Stallone") —≺ true is not included in S as the tuple ("Rambo","Stallone", "Ok") does not match the value required for the grounded parameter. Also, good_performance("Demolition Man", "Bullock") is not included as the tuple ("Demolition Man", "Bullock", "Regular") does not fulfill the restriction imposed, as stated before. As a remark, notice that operative presumptions built by the function will not include values for conditions, as they does not belong to the literal in which the OP are based.

4   Consistent Query Answering Based on Multiple Data Sources

In Section 3 the method used by DBI-DeLP to obtain argument supporting information is presented. As stated before, in our approach multiple repositories will be used as data sources for the argumentative process, and is safe to assume that inconsistencies will appear in a massive collection of data, both in the sense of opposite data or data that can lead to conflicting conclusions. In this section, we will introduce the strategies used to solve these clashes to give useful answers, but first we need to define what is a query in our framework. We have already seen queries in past sections, *e. g.*, *fly(tux)?*. Intuitively, a query in DBI-DeLP is a yes/no question we pose to the system. Thus, queries DBI-DeLP will solve are not the ones like *Give me all stocks certain company has bought* but ones like *Should I buy stocks from that company?*. That is why we do not need to retrieve consistent information as long as we can still give a consistent answer. Queries are expressed as a first order predicate that the system will try to prove, *i. e.*, will try to build arguments for and against it and then consider them to give a final answer. Thus, a query (or more precisely the process involved in its answering) is dynamic: it begins when the question arrives and ends when the system gives an answer. Syntactically, we distinguish queries from regular predicates by adding the question mark at the end of queries.

Definition 9. *(Query Warranting)*
Let L be a queried literal in the form $func(p_1, \ldots, p_m)$ where $ground(p_i)$ for all $1 \leq i \leq m$, Σ be the set of all operative presumptions for the available databases and $\Sigma' \subseteq \Sigma$ be the set of all operative presumptions obtained by the PRF for all Functions Objectives in arguments included in the dialectical tree $\mathcal{T}_{\langle \mathcal{A}, L \rangle}$ built for L. Let $\mathcal{P} = (\Pi, \Delta, \Sigma)$ and $\mathcal{P}' = (\Pi, \Delta, \Sigma')$ be two d.b.i.-de.l.p. We say that L is warranted from $\mathcal{P}$ iff $\mathcal{P}' \hspace{1pt}\vdash_w L$.

Remark: The process of answering a query is considered as part of an atomic transaction in the sense used in database theory.

So, as stated by this last definition, to give an answer the system combines the knowledge represented in strict rules, facts and defeasible rules with data obtained from the available databases so arguments can be builded and the warranting procedure can be carried out. Basically, the process is the same as the introduced in Section 2: given a queried literal L we build an argument in favor of it, then interfering and supporting arguments to this first argument, and we analyze them using the dialectical tree structure. Nevertheless, there is one obvious difference: when arguments are build we encounter several FO that may be proved by searching for supporting data in available databases in the way described in Section 3. Thus, we do not use the entire set Σ to answer a query. Instead, we use a subset of Σ formed only by presumptions needed to build arguments in the d-tree of the queried literal.

Regarding disagreement and defeat relations between arguments in DBI-DeLP, we will use the notions of proper and blocking defeaters defined just like in [7]. Nevertheless, to determine which argument prevails when two arguments disagree in DBI-DeLP we will use a combination of the *generalized specificity* comparison criterion introduced earlier with an additional comparison criterion that will be used to define the preference of data coming from some available databases over others. Thus, we will give a notion of preference between arguments that reflects this difference from DeLP. We begin by defining a comparison criterion for operative presumptions, and then we will show how it will be used by the framework.

Definition 10. *(Operative presumptions preference criterion) Let DB_1 and DB_2 be two available databases. Let OP_{DB_1} and OP_{DB_2} be operative presumptions retrieved from DB_1 and DB_2, respectively. Let ">" be a preference relation explicitly defined between DB_1 and DB_2. We say that OP_{DB_1} is preferred than OP_{DB_2} (denoted $OP_{DB_1} \triangleright OP_{DB_2}$) iff $DB_1 > DB_2$.*

Notice that we do not define which ">" stands for, so we can choose whatever criteria we want to determine which databases are preferred.

So, basically we say that operative presumptions obtained from sources in which we trust more are preferred in arguments. Although this distinction is important, we still think that arguments as a whole are obviously a more strong structure than the components that built them, as arguments may contain more information than data obtained from databases. So, when comparing two arguments in DBI-DeLP, we give the *generalized specificity* criterion more importance than the *operative presumptions preference* criterion. To do this, when two arguments disagree, we will only use the *operative presumptions preference* criterion if the *generalized specificity* criterion cannot solve the disagreement.

Definition 11. *(Argument Preference) Let $\langle \mathcal{A}_1, L_1 \rangle$ and $\langle \mathcal{A}_2, L_2 \rangle$ be arguments. We say that $\langle \mathcal{A}_1, L_1 \rangle$ is preferred than $\langle \mathcal{A}_2, L_2 \rangle$ iff*

(a) $\langle \mathcal{A}_1, L_1 \rangle$ is more specific than $\langle \mathcal{A}_2, L_2 \rangle$ or

(b) $\langle \mathcal{A}_2, L_2 \rangle$ *is not more specific than* $\langle \mathcal{A}_1, L_1 \rangle$ *and exists an* OP_{A_1} *in* $\langle \mathcal{A}_1, L_1 \rangle$ *and an* OP_{A_2} *in* $\langle \mathcal{A}_2, L_2 \rangle$ *such that* $OP_{A_1} \rhd OP_{A_2}$ *and not exist an* OP'_{A_2} *in* $\langle \mathcal{A}_2, L_2 \rangle$ *and an* OP'_{A_1} *in* $\langle \mathcal{A}_1, L_1 \rangle$ *such that* $OP'_{A_2} \rhd OP'_{A_1}$.

Remark: We use the *generalized specificity* criterion to compare arguments as a whole as it is the more widely used criterion in DeLP and its extensions to compare arguments, but clearly this criterion can be replaced by any other, *e. g.*, priority among rules, without changing anything else in the framework.

So, now that we have defined a defeat relation between arguments in DBI-DeLP, we can build dialectical trees for queries posed to the system and give answers to them. It is important to note that any answer given by DBI-DeLP will fulfill the consistency requirement, although no consistency restrictions are imposed to the Σ set. To demonstrate that, we show that the set of warranted queries inferred by a *d.b.i-de.l.p.* is non-contradictory, *i. e.*, a *d.b.i-de.l.p.* cannot warrant a literal and its complement simultaneously.

Proposition 1. *Given a d.b.i-de.l.p.* $\mathcal{P} = (\Pi, \Delta, \Sigma)$ *where* Π *is consistent, the set* $W = \{Q_i \mid \mathcal{P}\hspace{-2pt}\mid\sim_w Q_i\}$ *is non-contradictory.*

Proof. Let W be the set of warranted queries from $\mathcal{P}$ and $L_1 \in W$. Thus, there exists an argument $\langle \mathcal{A}_1, L_1 \rangle$ for L_1 which is undefeated in a dialectical process. Suppose $L_2 \in W$, and L_1 and L_2 are complementary ($L_1 = \bar{L}_2$). As L_2 is warranted there exists an argument $\langle \mathcal{A}_2, L_2 \rangle$ for L_2, which is undefeated in the argumentation process. As L_1 and L_2 are contradictory $\langle \mathcal{A}_1, L_1 \rangle$ is a counter-argument for $\langle \mathcal{A}_2, L_2 \rangle$ and vice versa. Depending on the comparison criterion used, $\langle \mathcal{A}_2, L_2 \rangle$ can be a proper defeater of $\langle \mathcal{A}_1, L_1 \rangle$ (or viceversa), or $\langle \mathcal{A}_2, L_2 \rangle$ and $\langle \mathcal{A}_1, L_1 \rangle$ are blocking defeaters. Since $\langle \mathcal{A}_2, L_2 \rangle$ is undefeated and is a defeater (blocking or proper) for $\langle \mathcal{A}_1, L_1 \rangle$, L_1 is not warranted from $\mathcal{P}$, which contradicts our original hypothesis. $\square$

To briefly illustrate the process we introduce a reduced example in which we show how we can use inconsistent databases to support that argumentative process and still obtain consistent answers. For space reason, we will show builded arguments without focusing on how operative presumptions that supports them are obtained. Consider the d.b.i.-de.l.p and conditions to FOs shown in Fig. 2 and databases shown Fig. 3 in that contains information about the movies domain, as a recommender system may have, so it can advice to the user whether or not to watch a certain film.

Suppose that we want to answer the query *recommended("Just go with it", "ted")?*. To solve the query, the DeLP Core starts the argumentation process by trying to build an argument supporting the claim indicating that "Just Go With It" is a film that can be recommended to the user Ted.

The process starts using recommended(Film, User) —≺likes_film(User, Film2), share_genre(Film, Film2, Genre), share_actor(Film, Film2, Actor),Film $\neq$ Film2. Based on the fact likes_film("Ted", "Click") and information in the Movies Database, it builds an argument in favor of the recommendation, as Click is a film the user Ted likes that shares actor and genre with Just Go With It.

$$\Pi = \left\{ \begin{array}{c} child\_restricted(Film) \leftarrow has\_violence(Film). \\ fanatic("ted", "Adam Sandler"). \\ fanatic("ted", "Jennifer Aniston"). \\ likes\_film("ted", "Click") \end{array} \right\}$$

$$\Delta = \left\{ \begin{array}{c} recommended(Film, User) \prec \\ likes\_film(User, Film2). share\_genre(Film, Film2, Genre), \\ share\_actor(Film, Film2, Actor), Film \neq Film2. \\ \sim recommended(Film, User) \prec \\ lead\_role(Actor, Film), poor\_performance(Actor, Film). \\ share\_genre(Film, Film2, Genre) \prec \\ genre(Film, Genre), genre(Film2, Genre). \\ share\_actor(Film, Film2, Actor) \prec \\ performs\_in(Film, Actor), performs\_in(Film2, Actor). \end{array} \right\}$$

Conditions given to Function Objectives				
Predicate	Database	Table	Field	Value
lead_role	Critics DB	Movie-Actor	role	leading
poor_performance	Critics DB	Movie-Actor	performance	Poor
poor_performance	Forum Users DB	ActorFilm	review	Bad
~ poor_performance	Critics DB	Movie-Actor	performance	Great
~ poor_performance	Forum Users DB	ActorFilm	review	Good

Fig. 2. The *d.b.i.-de.l.p.* for the example

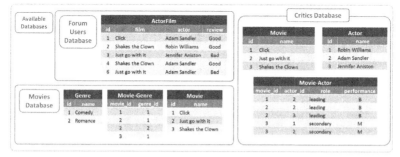

Fig. 3. Available databases

The dialectical process will try to build a counter-argument, *i. e.*, it will try to prove ~recommended('Just go with it', 'ted'), using ~recommended(Film, User) ⊰ lead_role(Actor, Film), poor_performance(Actor, Film). Both lead_role and poor_performance are proved using data in the available databases. For the FO lead_role the PRF retrieves for example lead_role("Adam Sandler", "Just go with it"). As for the poor_performance predicate both the Forum Users and the Critics databases are listed as data sources for it. In the Forum Users database we can see that a user has stated that Adam Sandler had a bad performance in the film, and as this is the value indicating a poor performance in the Condition for this FO then poor_performance("Adam Sandler", "Just go with it") is proved. Then an argument for ~recommended("Just go with it", "ted")is built, and it defeats the first argument as is more specific.

The process continues by trying to defeat ~recommended("Just go with it", "ted"). After considering all possibilities the dialectical process build an argument for ~poor_performance("Adam Sandler", "Just go with it") based on the presumption supported by the Great critic's review stored in the Critics Database, attacking ~recommended("Just go with it", "ted") in its sub-argument, *i. e.*, here we have a conflict between operative presumptions. To solve this conflict we use the operative presumption preference criterion. Suppose that Critics Database > Forum Users Database, *i. e.*, we prefer a critics' opinion, then ~poor_performance("Adam Sandler", "Just go with it") ▷ poor_performance("Adam Sandler", "Just go with it"), thus making ∼recommended("Just go with it", "ted") a defeated argument.

Finally, the dialectical tree is build and marked, establishing that the root argument recommended("Just go with it", "ted") is warranted, as all its attackers are defeated, thus the server sends YES as the answer to the query *recommended("Just go with it", "ted")?*, indicating that the film Just Go With It can be recommended to the user Ted as it shares genre and actor with another film that he likes, in this case the film Click. In Figure 4 the builded dialectical tree is shown. It can be seen that a consistent, reasoned answer to the received query is build although inconsistent data is used in the process.

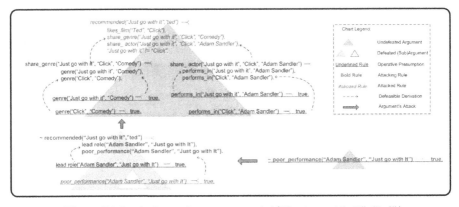

Fig. 4. Dialectical tree for *recommended("Just go with it", "ted")*

5   Conclusions and Related Work

We have introduced an approach to commonsense reasoning using information stored in relational databases; their use to store information from which arguments are build is more efficient than the explicit codification of facts in a program. In this manner, other systems may feed the data repository without the need of complex interfaces, making the system's knowledge evolution easier. Nevertheless, the use of externally updated data sources has at least a drawback: since no control over added data is effected, inconsistencies can appear. We have shown how Defeasible Argumentation is a suitable tool to handle inconsistencies, allowing a consistent query answering based on potentially conflicting data without the need for modifications neither on databases nor on queries sent to them. We have presented a framework that extracts query related data (possibly inconsistent) from databases and answers the query through an argumentative process. As for future work, we have identified several lines of research such as: semantic-level translation between predicates and database schemas, reasoning over database schemas, novel mechanisms for resolving non-ground queries, and the addition (learning) of new rules based on obtained data (rule mining). Also, we will study different presumption comparison criteria.

To the best of our knowledge, there has been no work integrating relational database technologies with defeasible argumentation systems in the manner presented here. However, in Section 3 we have already introduced some work regarding information extraction and consistent query answering from possibly inconsistent database [9–11] and discussed the differences between these approaches and ours. In addition, in [12] the problem of using defeasible reasoning in a massive data repository is addressed. In that work we can see how defeasible reasoning is a suitable tool to solve inconsistencies, in the same manner that we have shown here. Nevertheless, instead of using databases to look for information to support conclusions (and solving the possible inconsistencies), they use the Web as repository, more specifically the Semantic Web.

In a more general way, in [13] a view of the inconsistency issue is given; they state, in the same manner as we did, that inconsistency is not always a bad thing, even can be a desirable thing for certain systems, as long as we can somehow handle it and give sound answers. Moreover, they mention databases as an example of possible inconsistency holders, and defeasible argumentation as a possible solution to it.

Reference

1. Bench-Capon, T.J.M., Dunne, P.E.: Argumentation in artificial intelligence. Artif. Intell. 171, 619–641 (2007)
2. Besnard, P., Hunter, A.: Elements of Argumentation. The MIT Press (2008)
3. Rahwan, I., Simari, G.R.: Argumentation in Artificial Intelligence. Springer (2009)
4. Prakken, H., Sartor, G.: Argument-based extended logic programming with defeasible priorities. Journal of Applied Non-Classical Logics 7(1) (1997)
5. Dung, P., Kowalski, R., Toni, F.: Dialectic proof procedures for assumption-based, admissible argumentation. Artificial Intelligence 170(2), 114–159 (2006)
6. Amgoud, L., Kaci, S.: An Argumentation Framework for Merging Conflicting Knowledge Bases: The Prioritized Case. In: Godo, L. (ed.) ECSQARU 2005. LNCS (LNAI), vol. 3571, pp. 527–538. Springer, Heidelberg (2005)
7. García, A.J., Simari, G.R.: Defeasible logic programming an argumentative approach. TPLP, 95–138 (2004)
8. Stolzenburg, F., García, A.J., Chesñevar, C.I., Simari, G.R.: Computing generalized specificity. Journal of Applied Non-Classical Logics 13(1), 87–113 (2003)
9. Santos, E., Martins, J.P., Galhardas, H.: An argumentation-based approach to database repair. In: 19th European Conf. on AI (ECAI), pp. 125–130 (2010)
10. Celle, A., Bertossi, L.: Querying Inconsistent Databases: Algorithms and Implementation. In: Palamidessi, C., Moniz Pereira, L., Lloyd, J.W., Dahl, V., Furbach, U., Kerber, M., Lau, K.-K., Sagiv, Y., Stuckey, P.J. (eds.) CL 2000. LNCS (LNAI), vol. 1861, pp. 942–956. Springer, Heidelberg (2000)
11. Bertossi, L.: Consistent query answering in databases. SIGMOD Record 35(2), 68–76 (2006)
12. Bassiliades, N., Antoniou, G., Vlahavas, I.P.: A defeasible logic reasoner for the semantic web. Int. J. Semantic Web Inf. Syst. 2(1), 1–41 (2006)
13. Bertossi, L., Hunter, A., Schaub, T.: Introduction to Inconsistency Tolerance. In: Bertossi, L., Hunter, A., Schaub, T. (eds.) Inconsistency Tolerance. LNCS, vol. 3300, pp. 1–14. Springer, Heidelberg (2005)

Analytics-Driven Lossless Data Compression for Rapid In-situ Indexing, Storing, and Querying

John Jenkins[1,2], Isha Arkatkar[1,2], Sriram Lakshminarasimhan[1,2], Neil Shah[1,2], Eric R. Schendel[1,2], Stephane Ethier[3], Choong-Seock Chang[3], Jacqueline H. Chen[4], Hemanth Kolla[4], Scott Klasky[2], Robert Ross[5], and Nagiza F. Samatova[1,2,⋆]

[1] North Carolina State University, NC 27695, USA
[2] Oak Ridge National Laboratory, TN 37831, USA
[3] Princeton Plasma Physics Laboratory, Princeton, NJ 08543, USA
[4] Sandia National Laboratory, Livermore, CA 94551, USA
[5] Argonne National Laboratory, Argonne, IL 60439, USA
samatova@csc.ncsu.edu

Abstract. The analysis of scientific simulations is highly data-intensive and is becoming an increasingly important challenge. Peta-scale data sets require the use of light-weight query-driven analysis methods, as opposed to heavy-weight schemes that optimize for speed at the expense of size. This paper is an attempt in the direction of query processing over losslessly compressed scientific data. We propose a co-designed double-precision compression and indexing methodology for range queries by performing unique-value-based binning on the most significant bytes of double precision data (sign, exponent, and most significant mantissa bits), and inverting the resulting metadata to produce an inverted index over a reduced data representation. Without the inverted index, our method matches or improves compression ratios over both general-purpose and floating-point compression utilities. The inverted index is light-weight, and the overall storage requirement for both reduced column and index is less than 135%, whereas existing DBMS technologies can require 200-400%. As a proof-of-concept, we evaluate univariate range queries that additionally return column values, a critical component of data analytics, against state-of-the-art bitmap indexing technology, showing multi-fold query performance improvements.

1 Introduction

Increasingly complex simulation models, capable of using high-end computing architectures, are being used to simulate dynamics of various scientific processes with a high degree of precision. However, coupled with this opportunity to augment knowledge and understanding of the highly complex processes being studied are the challenges of conducting exploratory data analysis and knowledge discovery. Specifically, data size on the tera- and peta-scale is becoming a limiting factor in understanding the phenomena latent in these datasets, especially in a post-processing context.

Due to massive dataset sizes, full context analysis is a crucial bottleneck in the knowledge discovery pipeline, being restrained by the limits of computer memory and

---

⋆ Corresponding author.

S.W. Liddle et al. (Eds.): DEXA 2012, Part II, LNCS 7447, pp. 16–30, 2012.

I/O bandwidth. Most commonly, the applications that such data exploration processes are characteristic of are interactive and require close to real-time I/O rates for full data exploration. However, I/O access rates are too slow to support efficient random disk access in real-time for large-scale data sets, necessitating new approaches geared towards reducing the I/O pressure of extreme-scale data analytics.

A *knowledge priors* approach to data analytics is promising in restricting data to smaller and more practical sizes. Often times, scientists have some prior knowledge about the regions of interest in their data. For example, fusion scientists aiming to understand plasma turbulence might formulate analyses questions involving correlations of turbulence intensities in different radial zones ($0.1 < \psi < 0.15; 0.3 < \psi < 0.35; 0.5 < \psi < 0.55; 0.7 < \psi < 0.75; 0.9 < \psi < 0.95$). Likewise, climate scientists aiming to understand factors contributing to natural disasters might limit their search to particular regions or perhaps only a single region.

Formulating queries on scientific simulation data constrained on variables of interest is an important way to select interesting or anomalous features from large-scale scientific datasets. Traditional database query semantics can effectively be used for formulating such queries. This allows us to leverage a great deal of work done in the database community on query processing. The indexing techniques used in traditional database systems, such as $B-$trees [7] or bitmap indexes [20], have been used extensively in the literature. However, while indexing is a blessing for fast and efficient query processing, it is arguably a curse in terms of storage; the index size is often 100-300% of the original column size for high-cardinality data (such as double-precision data), which is a huge bottleneck for storage-bound extreme-scale applications.

A number of bitmap index compression techniques have been introduced to reduce the size of the bitmap index while keeping fast query retrieval possible. In particular, Word Aligned Hybrid (WAH) [13] bitmap compression is used in FASTBIT [20], a state-of-the-art scientific database technology with fast query processing capabilities. Overall, the storage footprint used in FASTBIT for a high-cardinality column and its corresponding index is around 200% of the original size, which still becomes prohibitive for extreme-scale data sets. Furthermore, these indexing schemes are optimized for returning the record ID, or region index in the context of spatio-temporal data sets. However, for data analytics, the *actual values* of the variables associated with these points are equally important.

Therefore, we present a co-designed data reduction and indexing methodology for double-precision datasets, optimized for query-driven data analytics. We believe that a tight cohesion between the methods allows us to optimize storage requirements while at the same time facilitating both fast indexing at simulation-time and range query processing with value retrieval, desirable features for data analytics. Our focus in particular is on write-once, read-many (WORM) datasets utilizing double-precision floating-point variables, representing large-scale, high-fidelity simulation runs that are subsequently analyzed by numerous application scientists in multiple (often global) contexts. A few examples of such data are in the particle-based fusion simulation GTS [17] and in the direct numerical combustion simulation S3D [6], each of which are comprised of primarily double-precision, high-cardinality variables ($\approx 100\%$ unique for GTS, $\approx 50\%$ unique for S3D).

To be more specific, our paper makes the following contributions:

- We present a lossless compression methodology for floating-point (single and double-precision) columns that can be utilized for indexing and range query processing, utilizing unique-value encoding of the most significant bytes. Our lossless compression reduces the size of a number of high-entropy, double-precision scientific datasets by at least 15%. Compared to lossless compression techniques like FPC [4], optimized for double-precision data, we report superior average compression ratios.
- Using our lossless compression method, we optimize range query evaluation including value retrieval by binning the column data by the distinct significant byte metadata, integrating efficient compressed-data organization and decompression of retrieved results. Compared to state-of-the-art techniques like FASTBIT [20], we provide comparable or better performance on range queries retrieving record IDs. For range queries additionally retrieving variable values, we achieve a performance improvement of one-to-two orders of magnitude.
- For query processing, we utilize an inverted index that uses approximately 50% space with respect to the original column size. Considering both the compressed column data and index, our method has a smaller storage footprint compared to other database indexing schemes.

2 Background

Search and query processing operations on traditional database systems like Oracle, MySQL, and DB2 involve the use of indexing techniques that are usually variants of either bitmap indexes or $B-$trees. While these techniques are effective in speeding up query response times, they come at the cost of a heavy-weight index management scheme. Indexing with $B-$trees [7], which tends to be more suitable for transactional databases that require frequent updates, is observed to consume storage that is three to four times the size of the raw column data for high-cardinality attributes. Scientific data, which is typically read (or append) only, have been shown to be better served with bitmap-based indexing techniques [16, 20], providing faster response times with lower index storage overhead.

While there are numerous technologies that use variants of bitmap indexing, we primarily focus on FASTBIT [20], a state-of-the-art bitmap indexing scheme, that is used by a number of scientific applications for answering range queries. FASTBIT employs a Word-Aligned-Hybrid (WAH) compression scheme based on run-length encoding, which decreases the index storage requirement and allows FASTBIT to perform logical operations efficiently on the compressed index and compute partial results by scanning the index. For those records that cannot be evaluated with the index alone, FASTBIT resorts to performing a read of the raw data, in what is called *candidate checks*. Unfortunately, the bitmap index created is sensitive to the distribution and cardinality of the input data, taking anywhere from 30 to 300% of the raw column size. The space can partly be reduced through techniques such as *precision binning*, at the cost of disturbing the distribution of values along the bins.

On the other side of the coin, data compression methods within databases have been widely studied [9, 12, 18]. For example, the column-oriented database C-Store [2] uses null compression (elimination of zeroes), dictionary encoding, and run-length encoding for effective data reduction of attributes organized contiguously, as opposed to the traditional row-store organization. MonetDB, on the other hand, uses the patched frame of reference (PFOR) algorithm and variants, which promotes extremely fast decompression speeds for query processing [23]. While these methods have limited use on double-precision data due to high-entropy significant bits, our work does share similarity with the dictionary encoding method, in that we compress floating-point data through identifying unique values and assigning them reduced bitwise representations. However, we perform this on only the most significant few bytes of the double-precision data, as opposed to the full dataset as in C-Store, and discard the representation entirely when using the inverted index for our query processing methodology.

As mentioned, many general-purpose and specialized compression methodologies fail to provide high compression ratios on double-precision data. Part of the reason for this is that floating-point scientific data is notoriously difficult to compress due to high entropy significands, of which floating-point data is primarily composed of (23 of 32 bits for single precision and 52 of 64 bits for double-precision). Much work has been done to build compressors for these kinds of data, mostly based on difference coding. Algorithms such as FPC [4] and fpzip [15] use *predictors* like the Lorenzo predictor [10], FCM [21] and DFCM [8] to compress. Given an input stream of double-precision values, the predictors use the previously seen values to predict the next value in the stream, and rather than attempt to compress the double values themselves, the compression algorithm uses a measure of error between the predicted and actual value, typically as an XOR operation.

Our methodology is based on treating the most significant bytes of double-precision data differently than the least significant bytes. Isenburg *et al.* use the same underlying concept in a prediction-based compression utility, which partitions the sign, exponent, and significand bits of the prediction error, followed by compression of each component [11]. Unlike their method, our method must maintain the approximability of the floating point datasets by treating the most significant bytes as a single component (sign, exponent, and the most significant significand bits), enabling efficient index generation and range query processing over the compressed data.

3 Method

3.1 System Overview

As mentioned, the goal of this paper is to facilitate query-driven analysis of large-scale scientific simulation data with storage-bound requirements. There are two stages where we focus our design to achieve this goal: first, while simulation data is being generated and in memory, or as a post-processing step, we can process and reorganize a double-precision dataset to compress the data. Second, we can modify the new organization of data to optimize query processing on the preprocessed data. For this purpose, we introduce two components in the scientific knowledge discovery pipeline, the *lossless compressor* and *query engine*.

3.2 Compression

Scientific simulations use predominantly double-precision floating-point variables, so the remainder of the paper will focus on compression and query processing for these variables, though our method can be applied to variables of different precision. The underlying representation of these variables, using the IEEE 754 floating-point standard [1], is a primary driver of our compression and querying methodology, so we briefly review it here. The standard encodes floating point values using three components: a *sign* bit, a *significand* field, and an *exponent* field. 64-bit double-precision values use one sign bit, 11 exponent bits, and 52 significand bits. Given the sign bit s, the unsigned integral representation of the exponent field e, and each significand bit m_i (most to least significant), the resulting value encoded by a double-precision variable is:

$$\text{value} = (-1)^s \times 2^{e-1023} \times (1 + \sum_{i=1}^{52}(m_i 2^{-i})). \tag{1}$$

Note that, all other components being equal, a difference of one in the exponent fields of two double-precision variables leads to a 2x difference in the represented values.

Our key observation for the compression process is that there is similarity with respect to orders of magnitude in our target datasets. For instance, in a simulation grid, adjacent grid values are unlikely to differ in orders of magnitude, except perhaps along simulation-specific phenomenon boundaries. Furthermore, the encoding naturally lends itself to accurate approximation given the exponent components. Hence, we base our compression and query processing methodology on the commonality in the sign and exponent field of double-precision datasets.

Figure 1 gives an overview of the compression process, developed under the assumption of similar exponent components and with the intention of applying to range query processing. For an N-element *partition*, or compression stream of maximum bounded size, we split the $8N$-byte double-precision column stream into two components: a kN-byte *high-order byte stream* consisting of the most significant k bytes of each value, and the remaining $(8-k)N$-byte *low-order byte stream* consisting of the remaining significant bytes. Using the observation of highly similar sign and exponent values, we identify the unique high-order bytes and discard redundant values. Let n be the number of unique high-order byte patterns. We define a *bin* to be a set of low-order bytes with equivalent high-order bytes, with bin edges $B_1, B_2, \ldots, B_n$ corresponding to the sorted unique patterns. The low-order bytes are reorganized into their respective bins, and a record ID (RID) to bin mapping M is generated to maintain the original organization, using a bitmap with $\lceil \log(n) \rceil$ bits per identifier. The unique high-order bytes, M, and optionally the low-order bytes are then compressed using the general purpose compressor bzip2. We do not consider using more complex algorithms, such as prediction-based compressors, in this paper. We feel that the use of a general-purpose compression algorithm provides a solid baseline of performance that applications can improve on, given additional application-specific knowledge of dataset characteristics.

Three data structures are produced as the result of the compression process: (1) the compression metadata, defining the high-order byte values and file offsets of each bin, (2) the compressed RID-to-bin mapping M, and (3) the bin-organized low-order bytes.

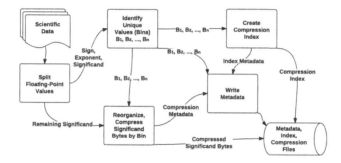

Fig. 1. Various stages of the compression methodology, described in Section 3.2. The bitmap index is used for compression, while the inverted index is used in query processing.

The value of k should be chosen with two goals in mind: to cause the cumulative number of distinct high-order bytes to stabilize with an increasing stream size, and to maximize the redundancy of the patterns (for compression) while encoding the entirety of the sign and exponent components (for future query processing). For scientific floating point data, we found $k = 2$ to be the most effective; it covers the sign bit, all exponent bits, and the first four significand bits of double-precision values (approximately two significant figures in base 10 scientific notation). This makes sense, as higher degrees of precision in scientific data tend toward high-entropy values. To verify our choice of k for this paper, Figure 2 shows the number of distinct high-order bytes recorded as a data stream is processed. For both $k = 2$ and 3, a relatively small cardinality is seen relative to the number of points processed, with the distinct values quickly reaching a (near) maximum.

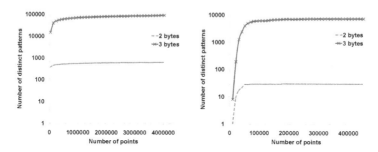

Fig. 2. Cumulative growth of the number of distinct higher order 2-byte and 3-byte pattern for increasing data size

Recall that the metadata consists of unique high-order bytes as well as their respective file offsets to the low-order byte payload. Hence, the metadata size is directly proportional to the number of unique high-order bytes. As shown in Figure 2, for two of the scientific datasets, the size of metadata is less than 0.1% of the dataset for $k = 2$, due to the small number of distinct patterns. For $k = 3$, however, the number of distinct patterns increases by a factor of 100 due to the addition of the higher-entropy significand bits. This increases the metadata size similarly, while additionally increasing the size of the RID to bin mapping logarithmically. Thus, we use $k = 2$ in this paper. Given the trends in Figure 2, we expect random sampling to be sufficient to determine a good value of k for double-precision datasets.

3.3   Query Processing: Index Generation

The compression methodology presented in Section 3.2 is, as will be shown, effective at improving the compression ratio of many scientific datasets, but is not optimized for query processing. If a range query is performed using our compression index, the entire RID-to-bin mapping M would need to be traversed to map the binned data back to RIDs. Thus, at the cost of additional storage, we optimize for range queries by using an inverted index M^{-1} which maps each bin to a list of RIDs sharing the same high-order bytes, creating a *bin-based value-to-RID* mapping. Figure 3 illustrates the index used in compression compared to the inverted index. This organization is advantageous for range query processing, because we now access the RIDs by bin, the same as accessing the low-order bytes. The organization is disadvantageous because of the increased space, both for the index itself as well as the additional metadata, such as file offsets, needed to access the new index. This means, for a partition of N elements, approximately $N\log(N)$ bits is needed to store the index, with marginally additional space to store metadata such as the number of elements within each bin. Bounding the maximum partition size to 32GB of double-precision data ensures that each RID in the inverted index needs no more than four bytes, making the index size less than 50% of the raw column size, or lower for smaller partitions. As a simple example, a partition size of 2GB of double-precision data requires 28 bits for each RID, translating to an index size of 43.75% of the raw column size. This is assuming, of course, that the partition is completely filled. Furthermore, we do not consider compression of the inverted indexes, a well-studied topic [19, 22] that we hope to integrate into our method in the future.

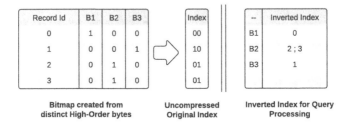

Fig. 3. Building an inverted index for query processing from the index used in compression

3.4   Query Processing: File Layout

The data used by the query processing engine is split into three components: a metadata file, an index file, and a compression file, each corresponding to its purpose described in the previous sections. The metadata file is shown in Figure 4.

The metadata file contains partition information, including file offsets for each partition and bin, the number and bounds (high-order bytes) of bins, and the number of values per bin per partition. The index file and the compression file contain the RIDs and compressed low-order bytes, respectively. A single scan of the metadata file is necessary for query processing and is small enough to be held in memory to optimize future queries. In our experimentation, however, we do not consider this possibility.

```
<N number of partitions>
<Metadata offset for partition t>  (0 ≤ t < N)
<Index offset for partition t>  (0 ≤ t < N)
<Compression offset for partition t>  (0 ≤ t < N)
(Repeat for 0 ≤ t < N)
<P number of elements in partition t>
<B number of bins>
<Number of elements in bin b>  (0 ≤ b < B)
<Bin bound b>  (0 ≤ b < B)
<Compression offset b>  (0 ≤ b < B)
(End Repeat)
```

Fig. 4. Metadata file format

3.5 Query Processing: Range Queries

The processing of range queries is based on two characteristics of our compression/indexing process: data arranged per-bin (low-order bytes and inverted index) are organized on disk in increasing order of high-order bytes, and bin edges (the high-order bytes) provide a lower bound on the values of RIDs within each bin by treating the high-order bytes as a truncated double-precision value.

The query evaluation process is shown in Figure 5. Given a variable constraint $[v_1, v_2)$, the metadata file shown in Figure 4 is traversed to obtain the necessary high-order bytes and bin file-offsets. Using the high-order bytes as a lower-bound for values within a bin, the boundary bins B_x and B_y are obtained using a binary search. Then, a single seek per-partition is needed in the index and low-order bytes files to fetch the data corresponding to the range of bins $B_x, B_{x+1}, \ldots, B_y$, taking advantage of the bin organization in file. The column data corresponding to the low-order bytes are reconstructed and only the data in boundary bins are filtered against the query bounds.

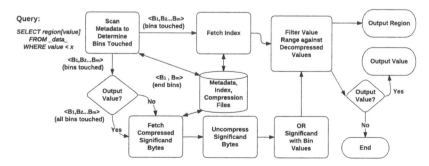

Fig. 5. Query processing methodology, taking into account metadata, index, and compression data fetching and aggregating

In the case of queries requesting only RIDs, not all of the low-order bytes need to be fetched and reconstructed. Only the bins at each boundary need be checked against the query constraints, as all remaining bins are guaranteed to fit within the query bounds.

4 Results and Discussions

4.1 Experimental Setup

We performed our experiments on the Lens cluster, dedicated to high-end visualization and data analysis, at Oak Ridge National Laboratory. Each node in the cluster is made up of four quad-core 2.3 GHz AMD Opteron processors and is equipped with 64GB of memory. In the following figures and tables, we refer to our methodology as CDI, corresponding to the Compressed representation of both the column Data and Index. All experiments were run with data located on the Lustre filesystem. For the indexing and query processing experiments, we compare against WAH encoding within the FASTBIT software. To avoid database-related overheads such as concurrency control, transaction support, etc. and provide a fair comparison between technologies, we wrote a minimal query driver for FASTBIT using only the necessary indexing and querying functions provided in the FASTBIT API. Furthermore, for fairness of comparison, we use the same partition size of 2GB for both our method and FASTBIT.

4.2 Datasets

To evaluate our compression, indexing, and query processing performance, we use a collection of double precision datasets from various sources. The majority of the datasets (*msg*, *num*, and *obs*) are publicly available and discussed by Burtscher and Ratanaworabhan [5]. We additionally use timeslice data for numerous variables generated by the GTS [17], FLASH [3], S3D [6], and XGC-1 [14] simulations.

In particular, we used the following two scientific simulation datasets to evaluate our query performance in terms of value centric queries and region centric queries: 1) GTS [17], a particle-based simulation for studying plasma microturbulence in the core of magnetically confined fusion plasmas of toroidal devices, and 2) S3D [6], a first-principles-based direct numerical simulation (DNS) of reacting flows which aids the modeling and design of combustion devices.

4.3 Query Processing

Index Generation

We evaluate the performance of our index generation methodology with respect to both computational efficiency as well as storage efficiency. Table 1 shows the results that we obtained from these experiments. Without low-order byte compression, our indexing operates an order of magnitude or more faster than FASTBIT when not considering I/O, and requires storage smaller than that of all tested configurations for FASTBIT for 17 of the 24 datasets tested. With low-order byte compression, our method performs roughly two to three times faster, while having a smaller storage footprint on 19 of

Table 1. Query index generation throughput and storage footprint. C: CDI. C_b: CDI with bin compression. F_D: FASTBIT with default configuration (10^5 bins). $F_{2,3}$: FASTBIT with bin boundaries at two/three significant digits.

Dataset	Index Gen. (MB/sec)						Storage (data+index) Req. (%)				
	In-situ			Post-processing							
	C	C_b	F_3	C	C_b	F_3	C	C_b	F_2	F_3	F_D
msg_bt	180	21	9	71	18	7	125.01	**119.36**	152.05	178.13	192.58
msg_lu	187	21	9	72	17	7	125.01	**124.44**	162.63	197.86	201.55
msg_sp	205	20	10	79	18	8	125.01	**124.01**	126.24	157.04	197.67
msg_sppm	191	37	13	77	30	11	125.03	**59.60**	114.75	116.75	125.32
msg_sweep3d	204	22	9	72	19	8	125.02	**96.62**	148.39	187.49	200.86
num_brain	215	20	9	86	17	7	125.00	124.50	**122.93**	191.54	202.31
num_comet	153	17	6	81	15	5	125.04	**116.20**	150.32	193.07	196.06
num_control	164	21	6	78	18	5	125.03	**124.05**	154.83	199.63	200.89
num_plasma	184	62	9	48	32	8	125.02	**51.44**	126.15	189.31	197.56
obs_error	222	30	10	35	22	8	125.00	**94.90**	130.34	167.63	176.93
obs_info	207	37	8	15	19	7	125.05	**75.06**	117.53	181.31	219.32
obs_spitzer	213	20	10	89	17	8	125.01	**94.37**	138.29	195.90	198.31
obs_temp	187	21	7	43	15	6	125.03	**125.03**	174.65	200.11	209.95
gts_phi_l	164	21	7	50	16	5	125.04	**125.04**	181.49	199.42	208.79
gts_phi_nl	169	21	7	62	16	5	125.04	**125.04**	183.64	199.70	208.85
gts_chkp_zeon	168	21	5	42	13	4	125.10	**125.10**	176.35	198.87	220.36
gts_chkp_zion	175	21	5	57	14	5	125.11	**125.11**	166.08	194.58	220.00
gts_potential	143	20	15	62	17	14	125.00	**125.00**	184.01	197.95	199.85
xgc_iphase	133	22	8	65	19	7	125.00	**105.33**	168.28	172.33	176.91
s3d_temp	223	19	17	70	16	12	125.00	123.28	**117.17**	135.41	202.25
s3d_vvel	186	20	9	76	17	8	125.01	**125.01**	168.89	194.96	202.12
flash_velx	209	21	11	92	18	10	125.00	125.00	**123.76**	157.18	195.68
flash_vely	217	21	12	91	18	9	125.00	125.00	**112.30**	137.32	193.07
flash_gamc	219	17	**20**	89	15	14	125.00	121.37	**100.40**	102.14	198.11

the 24 datasets. We attribute these gains to the less computationally intensive unique value encoding method as well as the data reduction enabled by data reorganization and redundant value removal.

End-to-End Query Performance Evaluation

For an end-to-end performance comparison, we perform queries under a number of scenarios, using the GTS potential (gts_potential) and S3D temperature (s3d_temp) variables. We look at two types of range queries: those that output record IDs given constraints on variables, which we will refer to as "region-centric" queries, named for the use-case of retrieving "regions of interest" from datasets arranged on a spatial grid structure, and those that additionally output the values of the variables, which we will refer to as "value-centric" queries. We compare each of these query types against FAST-BIT, which is specifically optimized for range queries.

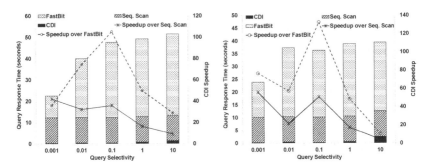

Fig. 6. Comparison of speedup of our method over FASTBIT and sequential scans, for value-centric queries when the query selectivity is varied from 0.001% to 10.0%. The left plot is for GTS potential, while the right plot is for S3D temperature.

Value-Centric Queries. Figure 6 shows the speedup in value-based query response time using our method, compared to FASTBIT's default and precision-based indexing, with varying query selectivity. By query selectivity, we refer to the percentage of the raw dataset returned by a query. For two scientific application variables S3D velocity and GTS potential, we provide a speedup of greater than a factor of 28. Due to the clustering of the data, a very small number of I/O seek operations are needed by our method as opposed to FASTBIT. The reason that sequential scan performs better than FASTBIT in this context is that, in parallel file systems such as Lustre, seeks are a very high-latency operation; FASTBIT resorts to seeking per item, while sequential scan reads all data in a single read.

For value-centric queries, not much difference is observed in the response time by FASTBIT using precision binning and default binning. This is because, in both cases, random disk access dominates processing time. While FASTBIT has a very fast CPU processing time for each query, the I/O time spent on random file access dominates the overall query response time.

The speedup observed increases from a factor of 35 for 0.001% selectivity to 105 for 0.1% selectivity. Here the performance improvement is due to a significantly lower number of seeks. On decreasing query selectivity, FASTBIT can fetch more consecutive blocks of file from disk, thus reducing I/O seek time. The I/O read time contributes to most of the query response time. Thus, the speedup comes down for 10% selectivity to a factor of about 28.

Region-Centric Queries. Figure 7 shows region query response time with varying number of hits (records returned) for our method compared to FASTBIT with precision and default binning. For region-centric queries, only the points falling within misaligned bins need to be evaluated. For FASTBIT, the type of binning used plays a definitive role in determining the time taken to respond to region queries. In the case of precision binning for FASTBIT, it can answer queries involving three decimal point precision by going through the set of bitmap indexes alone. It need not seek to the disk if the range specified in the query involves less than three decimal points. On the other hand, the default binning option needs to perform raw data access to evaluate edge bins.

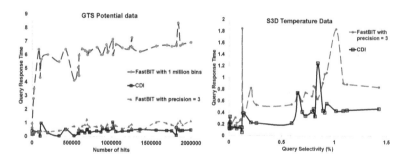

Fig. 7. Comparison of response return by FASTBIT against our method for region-centric queries with varying number of query hits

The performance of our method is better than the precision binning in many cases, but both methods see instances of lower performance. This is caused by partitioning methods that split on a fixed, rather than arbitrary, precision, causing lower degree of regularity between the bins. This happens when misaligned bins happen to be those with the largest number of points contained in them. In these cases, there is a higher false positive rate, causing it to be slower than FASTBIT, though FASTBIT is seen to have similar issues when using the precision-binning option.

4.4 Performance Analysis

Figure 8 shows the breakup of overall query processing time into I/O and compute components, corresponding to index/bin loading and processing, respectively. The dataset tested on is S3D using the velocity variable. I/O is the dominant cost of query processing, while the application of the query constraints and data transformations is a low but not insignificant component. We believe multithreading or asynchronous I/O would be able to hide most of the compute costs by interleaving it with the more costly I/O operations.

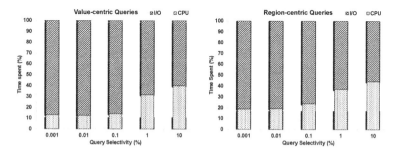

Fig. 8. Comparison of computation and I/O time distribution for our method for different query types of varying selectivity, on the S3D temperature variable. In comparison, FASTBIT spends over 90% of the time on I/O.

4.5 Compression

To analyze the performance of our lossless data compression scheme, we compared the compression ratios obtained with our method (without the inverted index) to those obtained by other standard lossless compression utilities, as well as more recent floating-point compressors. Out of the datasets tested, our method performed better than all of the other compressors tested (gzip, fpzip [15], bzip2, and FPC [5]) on 18 of 24. FPC gave superior performance compared to our method on two of the 27 datasets, while fpzip gave better performance on the remaining four. Overall, our method was consistent in yielding comparable or better compression ratios than the other compressors, providing evidence of strong compression ratios in other application datasets.

Table 2. Compression ratio and CDI storage components. CDI$_b$: CDI with bin compression.

Dataset	Compression Ratio					Storage Requirement (%)		
	gzip	fpzip	bzip2	FPC	CDI$_b$	Data	Index	Metadata
msg_bt	1.12	1.20	1.09	1.29	**1.40**	69.35	1.86	≈0.00
msg_lu	1.05	1.13	1.01	1.17	**1.30**	74.42	1.97	0.01
msg_sp	1.10	1.11	1.06	1.26	**1.33**	73.98	1.11	≈0.00
msg_sppm	7.41	3.25	7.09	5.30	**8.87**	9.58	1.66	0.02
msg_sweep3d	1.09	1.33	1.32	**3.09**	2.11	46.60	0.67	0.02
num_brain	1.06	1.25	1.06	1.16	**1.28**	74.50	3.39	≈0.00
num_comet	1.16	1.27	1.17	1.16	**1.34**	66.16	8.16	0.03
num_control	1.05	1.12	1.03	1.05	**1.15**	74.02	12.22	0.02
num_plasma	1.77	1.06	6.17	15.05	**80.67**	1.40	1.04	0.03
obs_error	1.44	1.37	1.36	**3.60**	2.59	44.90	5.88	≈0.00
obs_info	1.14	1.06	1.22	2.27	**3.52**	24.97	3.36	0.04
obs_spitzer	1.23	1.07	1.78	1.03	**1.90**	44.36	8.05	≈0.00
obs_temp	1.03	1.09	1.03	1.02	**1.13**	75.00	12.70	0.03
gts_phi_l	1.04	1.18	1.02	1.07	**1.19**	75.00	8.56	0.03
gts_phi_nl	1.04	1.17	1.01	1.07	**1.19**	75.00	9.2	0.03
gts_chkp_zeon	1.04	1.09	1.02	1.01	**1.17**	75.00	10.04	0.10
gts_chkp_zion	1.04	1.10	1.02	1.02	**1.18**	75.00	9.6	0.11
gts_potential	1.04	1.15	1.01	1.06	**1.18**	75.00	9.60	≈0.00
xgc_iphase	1.36	1.53	1.37	1.36	**1.58**	55.33	7.56	≈0.00
s3d_temp	1.18	**1.46**	1.15	1.34	1.35	73.38	0.77	≈0.00
s3d_vvel	1.04	1.24	1.02	1.15	**1.27**	75.00	3.74	≈0.00
flash_velx	1.11	**1.34**	1.08	1.26	1.32	75.00	0.81	≈0.00
flash_vely	1.13	**1.43**	1.09	1.29	1.32	75.00	0.80	≈0.00
flash_gamc	1.28	**1.62**	1.28	1.53	1.40	71.37	0.06	≈0.00

To justify our superior performance on most of the datasets, we argue that the bin-based compression of the data generally allows a much greater exploitation of existing compression algorithms than the normal distribution of scientific data that was passed to the other compressors. The reorganization of the data allowed gzip and bzip2's algorithms to be utilized as best as possible, causing the data to be reduced significantly because of the splitting of the low-entropy and high-entropy sections of the data.

As evidenced by the small compressed index and metadata sizes, the reorganization is a low-overhead operation with respect to storage. We attribute the better performance of FPC and fpzip on some of the datasets to the encoding of data dependency which the FCM [21], DFCM [8], and Lorenzo [10] predictors used by FPC and fpzip were able to capture in their predictions.

5   Conclusion

As the size of scientific datasets in various disciplines continues to grow, new methods to store and analyze the datasets must be developed, as I/O capabilities are not growing as fast, and new technologies, such as SSDs are not currently able to achieve the storage density and cost-efficiency of traditional mechanical disk drives. Successful methods of mitigating this growing gap must involve data reduction in all stages of the knowledge discovery pipeline, including storage of raw data as well as analytics metadata. We believe our effort at compression, indexing, and query processing of scientific data represents a step in the right direction, allowing both efficient lossless compression of double-precision data for accuracy-sensitive applications as well as efficient query processing on variable constraints, all with less space and I/O requirements than other database technologies.

Acknowledgements. We would like to thank ORNLs leadership class computing facility, OLCF, for the use of their resources. This work was supported in part by the U.S. Department of Energy, Office of Science and the U.S. National Science Foundation (DE-1240682, DE-1028746). Oak Ridge National Laboratory is managed by UT-Battelle for the LLC U.S. D.O.E. under contract no. DEAC05-00OR22725.

References

1. IEEE standard for floating-point arithmetic. IEEE Standard 754-2008 (2008)
2. Abadi, D., Madden, S., Ferreira, M.: Integrating compression and execution in column-oriented database systems. In: Proceedings of the 2006 ACM SIGMOD International Conference on Management of Data, SIGMOD 2006, pp. 671–682. ACM, New York (2006)
3. Fryxell, B., Olson, K., Ricker, P., Timmes, F.X., Zingale, M., Lamb, D.Q., MacNeice, P., Rosner, R., Truran, J.W., Tufo, H.: FLASH: An adaptive mesh hydrodynamics code for modeling astrophysical thermonuclear flashes. The Astrophysical Journal Supplement Series 131, 273–334 (2000)
4. Burtscher, M., Ratanaworabhan, P.: High throughput compression of double-precision floating-point data. In: IEEE Data Compression Conference, pp. 293–302 (2007)
5. Burtscher, M., Ratanaworabhan, P.: FPC: A high-speed compressor for double-precision floating-point data. IEEE Transactions on Computers 58, 18–31 (2009)
6. Chen, J.H., Choudhary, A., Supinski, B., DeVries, M., Hawkes, S.K.E.R., Liao, W., Ma, K., Mellor-Crummey, J., Podhorszki, N., Sankaran, S.S.R., Yoo, C.: Terascale direct numerical simulations of turbulent combustion using S3D. Comp. Sci. and Discovery 2(1)
7. Comer, D.: The ubiquitous B-Tree. ACM Comput. Surv. 11, 121–137 (1979)
8. Goeman, B., Vandierendonck, H., Bosschere, K.D.: Differential FCM: Increasing value prediction accuracy by improving table usage efficiency. In: Seventh International Symposium on High Performance Computer Architecture, pp. 207–216 (2001)

9. Graefe, G., Shapiro, L.: Data compression and database performance. In: Proceedings of the 1991 Symposium on Applied Computing, pp. 22–27 (April 1991)

10. Ibarria, L., Lindstrom, P., Rossignac, J., Szymczak, A.: Out-of-core compression and decompression of large n-dimensional scalar fields. Computer Graphics Forum 22, 343–348 (2003)

11. Isenburg, M., Lindstrom, P., Snoeyink, J.: Lossless compression of predicted floating-point geometry. Computer-Aided Design 37(8), 869–877 (2005); CAD 2004 Special Issue: Modelling and Geometry Representations for CAD

12. Iyer, B.R., Wilhite, D.: Data compression support in databases. In: Proceedings of the 20th International Conference on Very Large Data Bases, VLDB 1994, pp. 695–704. Morgan Kaufmann Publishers Inc., San Francisco (1994)

13. Wu, K., Ahern, S., Bethel, E.W., Chen, J., Childs, H., Cormier-Michel, E., Geddes, C., Gu, J., Hagen, H., Hamann, B., Koegler, W., Lauret, J., Meredith, J., Messmer, P., Otoo, E., Perevoztchikov, V., Poskanzer, A., Prabhat, Rubel, O., Shoshani, A., Sim, A., Stockinger, K., Weber, G., Zhang, W.-M.: FastBit: interactively searching massive data. Journal of Physics: Conference Series 180(1), 012053 (2009)

14. Ku, S., Chang, C., Diamond, P.: Full-f gyrokinetic particle simulation of centrally heated global ITG turbulence from magnetic axis to edge pedestal top in a realistic Tokamak geometry. Nuclear Fusion 49(11), 115021 (2009)

15. Lindstrom, P., Isenburg, M.: Fast and efficient compression of floating-point data. IEEE Transactions on Visualization and Computer Graphics 12, 1245–1250 (2006)

16. Sinha, R.R., Winslett, M.: Multi-resolution bitmap indexes for scientific data. ACM Trans. Database Syst. 32 (August 2007)

17. Wang, W.X., Lin, Z., Tang, W.M., Lee, W.W., Ethier, S., Lewandowski, J.L.V., Rewoldt, G., Hahm, T.S., Manickam, J.: Gyro-kinetic simulation of global turbulent transport properties in Tokamak experiments. Physics of Plasmas 13(9), 092505 (2006)

18. Westmann, T., Kossmann, D., Helmer, S., Moerkotte, G.: The implementation and performance of compressed databases. SIGMOD Rec. 29(3), 55–67 (2000)

19. Witten, I.H., Moffat, A., Bell, T.C.: Managing Gigabytes: Compressing and Indexing Documents and Images, 2nd edn. Morgan Kaufmann (1999)

20. Wu, K.: Fastbit: an efficient indexing technology for accelerating data-intensive science. Journal of Physics: Conference Series 16, 556 (2005)

21. Yiannakis, S., Smith, J.E.: The predictability of data values. In: Proceedings of the 30th Annual ACM/IEEE International Symposium on Microarchitecture, MICRO 30, pp. 248–258. IEEE Computer Society, Washington, DC (1997)

22. Zobel, J., Moffat, A.: Inverted files for text search engines. ACM Computing Surveys 38(2) (July 2006)

23. Zukowski, M., Heman, S., Nes, N., Boncz, P.: Super-scalar ram-cpu cache compression. In: Proceedings of the 22nd International Conference on Data Engineering, ICDE 2006. IEEE Computer Society, Washington, DC (2006)

Prediction of Web User Behavior by Discovering Temporal Relational Rules from Web Log Data

Xiuming Yu[1], Meijing Li[1], Incheon Paik[2], and Keun Ho Ryu[1,2]

[1] Database / Bioinformatics Laboratory, Chungbuk National University, South Korea
{yuxiuming,mjlee,khryu}@dblab.chungbuk.ac.kr
[2] The University of Aizu, Aizu-Wakamatsu, Fukushima, Japan
paikic@u-aizu.ac.jp

Abstract. The Web has become a very popular and interactive medium in our lives. With the rapid development and proliferation of e-commerce and Web-based information systems, web mining has become an essential tool for discovering specific information on the Web. There are a lot of previous web mining techniques have been proposed. In this paper, an approach of temporal interval relational rule mining is applied to discover knowledge from web log data. Comparing our proposed approach and previous web mining techniques, the attribute of timestamp in web log data is considered in our approach. Firstly, temporal intervals of accessing web pages are formed by folding over a periodicity. And then discovery of relational rules is performed based on constraint of these temporal intervals. In the experiment, we analyze the result of relational rules and the effect of important parameters used in the mining approach.

Keywords: the Web, e-commerce, web log data, temporal interval, relational rules.

1 Introduction

Recently, the Web has become an important channel for conducting business transactions. With the rapid development of web technology, the Web has become an important and favorite platform for distributing and acquiring information. The data collected automatically by the Web and application of web servers represents the navigational behavior of web users. Web mining is a technology to discover and extract useful information from web log data. Because of the tremendous growth of information sources, increased interest of various research communities, and recent interest in e-commerce, the area of web mining has become vast and more interesting.

There are so many techniques of web, but most of previous studies have a common problem in that all those studies are limited without considering the temporal interval of web log data during the mining task, which may miss some of the time series characteristics.

Temporal interval relational rule mining is a new approach to discover the relationships among transactions; it involves temporal interval data. The advantage of temporal interval data mining is presented in [1]. In this paper, we apply the approach of

S.W. Liddle et al. (Eds.): DEXA 2012, Part II, LNCS 7447, pp. 31–38, 2012.
© Springer-Verlag Berlin Heidelberg 2012

temporal interval relation rule mining to web usage mining. Infrequent events can be discarded after obtaining a large event set and uniform event set, and relational rules are generated from a generalized database that only includes the uniform events, which can reduce the scale of data set to perform an effective task.

The reminder of this paper is organized as follows. In section 2, we present the detail of the temporal relational rule mining from a web log via an example. In section 3, we run several experiments to show the result of the approach of temporal relational rule mining from web log data, and we discuss the effect of important parameters used in the mining approach. Finally, we summarize our work, present conclusions, and discuss future work in section 4.

2   The Process of Mining Task

This section presents the process of mining temporal interval relation rules from web log data. The process mainly includes data preprocessing and relation rules discovery.

2.1 Data Preprocessing

Generally, the data applied to the access patterns mining task is not fit to be used directly. The web log data should be converted into general temporal interval data. An operation of preprocessing is essential in order to avoid noise, outliers, and missing values. We introduce a process of preprocessing with an example.

The aim of preprocessing is to obtain clean data that satisfies the format of relation rules mining task. An experimental data set is selected from the web log file that comes from an institution's official web site, which it is used as experimental data in paper [2].

We remove the data with wrong statue numbers that start with the numbers 4 or 5 and the irrelevant attributes for clarity. There are many attributes in one record of web log file, such as IP address, user id, time, method, URL, and so on. In this paper, we need the attributes of IP Address, Time, and URL; thus, the rest of attributes of Method, Status, and Size need to be discarded.

The URLs always come with long strings, which make it difficult to distinguish the requested URLs of web log data in thousands of records, the URLs can be transformed into code numbers for simplicity.

There are also lots of noise data in the web log files caused by the breaking of loading web page, resulting in inability to obtain the URL of the page. These records also need to be removed.

2.2 Relational Rules Mining

After preprocessing of data is complete, we get more formal data. Different from previous approaches, the proposed approach contains three main steps: getting a large event set, getting temporal interval data, and getting access patterns using the approach of discovering temporal relational rules.

Getting Large Event Set. A Large Event Set (LES) is an event set that contains the events that satisfy a user specified minimum support threshold. The events in LES represent the transactions or objects with large proportion in the entire data set. In this paper, a web log file denotes a data set, and one web page is defined as an event and LES denotes the set of web pages that are accessed by web users with enough frequency over a period of time. An important definition for generating LES is user session. Here the user session time is defined as one hour for simplicity. Then, the example data is grouped by one hour for each web user. According to the experimental data, candidate event types are extracted and their supports are calculated. To calculate the support count for each candidate, we need to count the visit times that are accessed by different web users. Finally, a user specified Minimum Support threshold for Large Event (MSLE) must be defined. MSLE denotes a kind of abstract level that is a degree of generalization. The support value will be determined by the proportion of web users accessing times of web pages. Selecting MSLE is very important; if it is low, then we can get a detailed event. If it is high, then we can get general events. In this example, MSLE is defined as 75%. In other words, if a web page is accessed by greater than 75% web users, then this web page can be denoted as a large event. After dealing with the entire example data, LES is obtained as shown in Table 1.

Table 1. Large event set

Event Type	Support
2	3
4	3
6	3
7	4

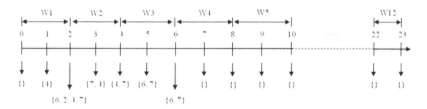

Fig. 1. Temporal interval graph

Getting Uniform Event Set. The sequences that are only included in the above large event types table are expressed with temporal interval. As we know, events within an element do not have to occur at simultaneously. Window Size (WS) can be defined to specify the maximum allowed time difference between the latest and earliest occurrences of events in any element of a sequential pattern. In this paper, the value of WS is defined as 2. There are 12 windows generated in 24 temporal intervals according to access time, and the events are signed into the time line. So, the event sequences with temporal intervals are as shown in Fig. 1. A Uniform Event Set (UES) means that the set of events has occurred continuously. Uniform events in a sequence can be summarized into one event with a temporal interval. A criterion for judging whether the event is uniform or non-uniform is needed; it is a Minimum Support threshold for Uniform Event (MSUE), and we define MSUE as a user-defined value of frequency of event that occurs in intervals, as was previously done in [3]. In this example, the frequency of each event in LES is calculated, and the candidates of uniform events are then obtained. Then, the uniform event set <{4}, {6}, {7}> is obtained from the candidate of uniform events, which satisfies the given value of MSUE 25%.

Getting Generalized Database. In order to discover a candidate temporal interval relation, the Generalized Database (GD) has to be generated. A generalized event with the parameters of VS and VE means that an event has occurred continuously in the temporal interval [VS, VE]. According to the event types in the uniform event type set and the original data in the sorted database, the generalized database is created as shown in Table 2. There, VS is the start time of web user who accessed the web site (sequence in event type) that are only included in the uniform event type set. And VE is the last time of accessing the web site that is also contained in the uniform event type set.

Table 2. Generalized database

IP Address	VS	VE	Event Type
82.117.202.158	1	4	4
82.117.202.158	2	5	6
82.117.202.158	3	5	7
82.208.207.41	1	2	4
82.208.207.41	2	5	7
82.208.255.125	2	5	6
82.208.255.125	5	5	7
83.136.179.11	1	3	4
83.136.179.11	2	6	6
83.136.179.11	4	6	7

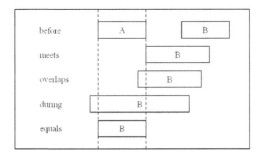

Fig. 2. Relations between events

Getting Relational Rules. Any two intervals of the events have one relation. There are many types of relations between two events, such as before, meets, overlaps, starts, during, and so on. In this paper, we only consider the relations of before, meets, overlaps, during, and equals, as is shown in Fig. 2. Then, temporal interval relation rules are generated from the generalized database as follows: first, we obtain all the candidate relation rules by comparing the time interval of each two event types. For example, for the first web user "82.117.202.158" in Table IV, the generalized event set contains three events <4, 6, 7>; we compare   with the time interval VS and VE of each two events <4, 6>, <4, 7> and <6, 7>. There are three relation rules, such as 4 [overlaps] 6, 4 [overlaps] 7, and 7 [during] 6. Following this step, we can obtain all candidate relation rules. Then, after generating the candidate relation rules and calculating the supports of each candidate relation rule, a suitable value of user-defined Minimum Support threshold for Relation Rule (MSRR) is defined. In the example, we define the value of MSRR as 50%; it means that there are four web users. If there are at least two web users that satisfy a relation rule, then it is a frequent relation rule. After discarding the infrequent relation rules, the result of temporal interval relation rule is as shown in Table 3.

Table 3. Relation rules

Relation Rules	Support
4 [overlaps] 6	50%
7 [during] 6	75%

3   Experiments and Results

This section presents several experiments based on the approach of temporal interval relational rule mining. The first experiment shows the real application of temporal interval relational rule mining for web mining. The other experiments show the effect of parameters of our approach, such as MSLE, MSUE, and MSRR.

3.1 Data Set

The first experimental data records the accessing information of web site (http://www.vtsns.edu.rs), requests to the institution's official website on November 16, 2009 which is referenced in the paper [4]. The other experimental data come from NASA Web log files (http://ita.ee.lbl.gov/html/contrib/NASA-HTTP.html) which is referenced in the paper [5].

3.2 Result and Discussion

Analysis of Result. In this section, we show the results that are obtained by applying the approach of discovering temporal relational rule mining to the above web log file. An experiment is performed with the above data after data preprocessing. At the step of getting a large event set, 7% is defined as the value of MSLE. In the step of getting uniform event set, 8% is defined as the value of MSUE; an event that is not a very frequently uniform set can be obtained, and 50% is defined as the value of MSRR. Then, we get the result of temporal interval relation rules as shown in Table 4. From the result of the experiment, we can obtain the hidden information in the web log data. For example, for the rule "6 [before] 18", we can determine that 55.6% of the web users who visit web site 18 (oglasna.php) always visit web site 6 (is-pit_raspored_god.php) first. According to this discovery, the web site designer can improve the layout of web page; for instance, he can modify web page 6 by adding a hyperlink in web page 6 to web page 18.

The application of our approach can be used to find the relationship between web-sites. For example it can be used to find the relational merchandise on web sites. When a customer finishes his online shopping, it will show the next popular mer-chandises that are also popular for the persons who purchased the same merchandise.

Table 4. Experimental result

Relation Rules	Support
6 [before] 7	88.9%
6 [before] 8	77.8%
6 [before] 18	55.6%
7 [meets] 8	54.4%
7 [before] 18	54.4%
8 [during] 18	66.7%
18 [before] 7	88.9%
30 [before] 8	66.7%

Effect of Parameters. Previous studies have shown the approach of access patterns mining using the approach of temporal relation rule mining. To discuss the effect of parameters used in the mining task, we perform some experiments.

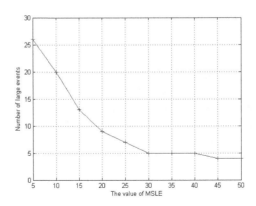

Fig. 3. Effect of parameter of MSLE

Effect of MSLE. The process of getting a large event set aims at extracting the events that satisfy a user defined MSLE. It can discard the infrequent events to reduce the size of experimental database for reducing the search space and time and maintaining the accuracy of the whole process of mining task. We compare the numbers of large events when the values of MSLEs are changed. The experimental result is shown in Fig. 3. We can see that the smaller the MSLE, the more generalized the LES. There always exists a value of MSLE, and from the value, the number of large events will not change, or will change very little. This value is always selected to be used as the value of MSLE in the experiment.

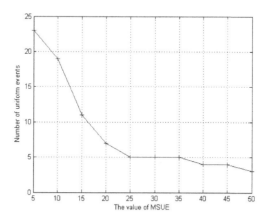

Fig. 4. Effect of parameter of MSUE

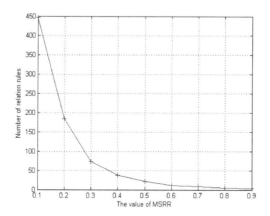

Fig. 5. Effect of parameter of MSRR

Effect of MSUE. In the process of getting a uniform event set, the parameter of MSUE is used to judge whether the event is uniform or non-uniform. This step can discard the events that do not occur continuously to avoid the situation that a web user accesses a web page repeatedly in a short period of time. We perform the experiment by using the records from a processed database that only contains the events in LES. In this experiment, we compare the numbers of uniform events when the values of

MSUEs are changed. The experimental result is shown in Fig. 4. We can see that greater MSUE, the more uniform UES. There always exists a value of MSUE, from the value, the number of large events will not change, or it will change very little. This value is always selected to be used as the value of MSUE in the experiment.

Effect of MSRR. In the process of getting relation rules, the parameter of MSRR is used to judge whether the candidate relation rules are frequent. This step can discard the infrequent rules that do not satisfy MSRR to exploit for the efficient discovery of relation rules. We perform the experiment by using the candidate relation rules that contain 450 rules. In this experiment, we compare the numbers of relation rules and the values of MSRRs. The experimental result is shown in Fig. 5. We can see that greater MSRR, the more efficient the relation rules. In other words, if specific rules need to be obtained, then the value of MSRR should be greater.

4   Conclusion and Future Research

In this paper, we presented the application of mining access patterns via discovering the temporal interval relation rules in web log data. In the process of mining task, a novel approach of data preprocessing and relation rules mining method are introduced: removing irrelevant data and discarding irrelevant attributes and missing value data are used to obtain cleaning data; Getting large event set, uniform event set, and relation rules is the main steps of discovery of temporal interval relation rules. This approach can help web site designers to improve the layout of web pages and make the web pages more comfortable for web users. We can also uncover some hidden information about the relationships between web pages via the result of mining task.

In future works, we will find a more efficient algorithm for access patterns mining, and more efficient way for preprocessing the data in web log files.

Acknowledgments. This work was supported by the National Research Foundation of Korea (NRF) grant funded by the Korea government (MEST) (No. 2012-0000478).

References

1. Lee, Y.J., Lee, J.W., Chai, D.J., Hwang, B.H., Ryu, K.H.: Mining temporal interval relational rules from temporal data. The Journal of Systems and Software, 155–167 (2009)
2. Dimitrijević, M., Bošnjak, Z., Subotica, S.: Discovering Interesting Association Rules in the Web Log Usage Data. Interdisciplinary Journal of Information, Knowledge, and Management 5, 191–207 (2010)
3. Pray, K.A., Ruiz, C.: Mining Expressive Temporal Associations from Complex Data. In: Perner, P., Imiya, A. (eds.) MLDM 2005. LNCS (LNAI), vol. 3587, pp. 384–394. Springer, Heidelberg (2005)
4. Maja, D., Zita, B.: Discovering Interesting Association Rules in the Web Log Usage Data. Interdisciplinary Journal of Information, Knowledge, and Management 5 (2010)
5. Chordia, B.S., Adhiya, K.P.: Grouping Web Access Sequences Using Sequence Aliment Method. Indian Journal of Computer Science and Engineering 2(3), 308–314 (2010)

A Hybrid Approach to Text Categorization Applied to Semantic Annotation

José Luis Navarro-Galindo[1], José Samos[1], and M. José Muñoz-Alférez[2]

[1] Department of Computer Languages and Systems, Computing School, Universidad de Granada, C/ Periodista Daniel Saucedo Aranda s/n, 18071 Granada, Spain
[2] Department of Physiology, Pharmacy School, Universidad de Granada, Campus de Cartuja, 18071 Granada, Spain
jlnavar@correo.ugr.es, {jsamos,malferez}@ugr.es

Abstract. In this paper, a hybrid approach is presented as a new technique for text categorization, based on machine learning techniques such as Vector-Space Model combined with n-grams. Given a specified content this technique takes care of choosing from different categories the one that best matches it. FLERSA is an annotation tool for web content where this technique is being used. The hybrid approach provides to FLERSA the capability for automatically define semantic annotations, determining the concepts that the content of a web document deals with.

Keywords: semantic annotation, vector space model, n-gram, FLERSA, information retrieval, semantic web.

1 Introduction

One of the main issues to be resolved in order to progress towards the Semantic Web is how to convert existing and new unstructured web content that can be understood by humans into semantically-enriched structured content that can be understood by machines.

The semantic markup of web documents is the first step towards adapting unstructured web content to the Semantic Web. Annotation consists of assigning a note to a specific portion of text. The note assigned contains semantic information in the form of metadata in order to establish a link between a reference ontology [5] and the specific part of text which is being marked-up.

FLERSA (Flexible Range Semantic Annotation) is our annotation tool for web content. The tool has been developed over a CMS and allows both manual [6] and automated annotation of web content [7].

The main contribution of this paper is to present our hybrid approach for text categorization applied to create automated semantic annotations on web documents, and to illustrate its performance in FLERSA, a tool in which this approach has been implemented. The paper begins with the theoretical basis of our approach; then, the paper undertakes a detailed study of the hybrid approach proposed; finally, the paper ends with the approach evaluation in the physiology area, conclusions and bibliography.

S.W. Liddle et al. (Eds.): DEXA 2012, Part II, LNCS 7447, pp. 39–47, 2012.
© Springer-Verlag Berlin Heidelberg 2012

2   The Hybrid Approach

This section presents our proposal for text categorization based on theoretical principles such as Vector-Space Model combined with n-grams. First, the theoretical basis of the study is summarized. Next, our hybrid approach is presented, demonstrating its performance through an illustrative example. Finally, how FLERSA implements and benefits from the approach is shown.

2.1   Theoretical Basis

In the **Vector-Space Model** [8], each document in a collection is considered to be a weight vector in a space of T dimensions where T is the number of different terms that appear in the collection.

$$D_i = (d_{i1}, d_{i2}, d_{i3}, ..., d_{it}). \tag{1}$$

Equation 1 represents a document in a Vector-Space of T dimensions. d_{ij} is the weight of the j-nth term for the document D_i.

The Vector-Space Model, when calculating the weight of a term in a document, takes into account the following aspects:

- The frequency of occurrence in the document, *tf* [4]. The most repeated words in a document are, in principle, more relevant than those less used.
- The number of documents in the collection in which the term appears, *df* [3]. The most common terms in the collection will be less relevant than the rarest.
- The length of the document, to ensure that all documents behave similarly regardless of their length. In other words, there is no relationship between relevance and length.

The Vector-Space Model works as follows:

- Equation 2 is calculated for each term belonging to the documents in the collection. When working with retrieval information systems, the same procedure is needed for queries.

$$w_{i,j} = tf_{i,j} * log\left(\frac{D}{df_j}\right). \tag{2}$$

  where,
  - $tf_{i,j}$ is the number of times *j-nth* terms occurs in the *i-nth* document.
  - df_j is the number of documents containing the j-nth term.
  - D is the number of documents in the collection.
- Given a query, it is possible to measure the similarity between the query vector and vectors belonging to the document collection using the cosine equation 3.

$$Sim(Q, D_i) = \frac{\sum_i w_{Q,j} w_{i,j}}{\sqrt{\sum_j w_{Q,j}^2}\sqrt{\sum_i w_{i,j}^2}}. \tag{3}$$

This model is well known and commonly used for categorization of text given that systems based on it can be easily trained. However, it has several limitations:

1. It is very calculation-intensive. From the computational standpoint it is very slow, requiring a lot of processing time.
2. Each time a new term is added into the term space, recalculation of all vectors is needed.
3. The model presents false negative/positive matches when working with documents with similar content.

In computational linguistics context [1], an **n-gram** is a contiguous sequence of n items from a given sequence of text or speech. These items can be phonemes, syllables, letters or words according to the application.

In the present work, the n-gram concept refers to a contiguous sequence of alphabetic characters separated from other sequences by a space or a punctuation mark. An n-gram, therefore, matches with what is commonly called "word" in its orthographic dimension.

The letter "n" in "n-gram" is a mathematical symbol that is used to evolve the whole series of natural numbers, i.e. numbers from one (1, 2, 3 ...). The three types of n-grams that are considered in the present work are: the "monograms", also called "unigrams" or "1-grams', and refer to individual words (e.g. "white"); the "bigrams' or "2-grams" which are strings of two words (e.g. "white blood"); finally, the "trigrams" or "3-grams" which are unified formations of three words (for example, "white blood cells").

2.2   Our Hybrid Approach

When the FLERSA tool was being developed, the automated semantic annotation process needed a machine learning technique to determine the concepts that the content of a web document deals with. We needed to obtain an effective categorization technique which could be used for the automated creation of semantic annotation. Our research derived from this need, and led to the creation of our approach.

The majority of machine learning techniques we studied, including the Vector-Space Model, rely on statistical calculations of individual lexical frequencies (i.e., single word counts) to estimate the concept that a document deals with, on the assumption that such lexical statistics are sufficiently representative of informative content. These kinds of estimations assume words occur independently from each other, ignoring the compositional semantics of language and causing several problems, such as ambiguity in understanding textual information, misinterpreting the original informative intent, and limiting the semantic scope of text. These problems reduce the accurate estimation of concepts that web documents are involved in, and hence decrease the effectiveness of automated annotation process.

Our hybrid approach proposes the use of lexical statistics, such as the Vector-Space Model, combined and enriched with n-grams, with the aim of enhancing the estimation of the concept that a document deals with, avoiding inherit problems from the Vector-Space Model, explained before, that cause erroneous results like false negatives and false positives.

In the Vector-Space Model a relationship exists between the frequency and informative content of terms. Based on this, we propose to use monograms, bigrams and trigrams as terms when calculating the Vector-Space Model equations. Monograms are made up of single words, bigrams by words taken in pairs and trigrams by words taken three at a time. Web document statistics are modelled using n-grams, which are contiguous sequences of parts of a sentence, extracted from text. This proposal takes into account the semantic interrelationship between words that make up sentences, in order to enhance the effectiveness of concept estimations.

Table 1. Our Hybrid Approach Representation Scheme

Term	$W_{C1i}\ W_{C2i}\ ...\ W_{CNi}$	W_{Qi}
$n-gram_1$		
$n-gram_2$	CORPUS	QUERY
...	WEIGHTS	WEIGHTS
$n-gram_N$		

The statistical scheme proposed by our hybrid model (Vector-Space Model combined with n-grams) is shown graphically in table 1. Some considerations:

- Weights are calculated following the equation 2
- Column W_{Ci} shows the weights of concept "C" for i-nth term
- Column W_{Qi} shows the weights of query for i-nth term

When building a hybrid approach model applied to a specific area, a Corpus of texts is needed for each concept to be modeled. Then, n-grams (monograms, bigrams and trigrams) are extracted from the Corpus texts and weights are calculated according to equation 2. Finally, the similarity equation 3 is calculated for each concept-query pair. Similarity analysis is done separately for monograms, bigrams and trigrams and their values are weighted according to equation 4 shown below.

$$
\begin{aligned}
Sim_{Global}(Q, C_i) = &\alpha \cdot Sim_{Tri}(Q, C_i) + \beta \cdot Sim_{Bi}(Q, C_i) \\
&+ \gamma \cdot Sim_{Mono}(Q, C_i), \text{ where } \alpha + \beta + \gamma = 1.
\end{aligned} \tag{4}
$$

The determination of the most appropriated weights (α, β and γ) for equation 4 is beyond of the scope of this paper. The weights should be experimentally adapted depending on the knowledge area where they are used.

2.3 The Hybrid Approach Implementation

In spite of the disadvantages presented by the Vector-Space Model, it suits the automated semantic process perfectly, as well as the task of determining the concept that deals with a text fragment that is being marked-up.

We realised that in the semantic annotation context, concepts are used instead of categories, and there are annotations (text fragments) rather than documents, however, the categorization problem remains the same in essence. The FLERSA tool's capacity to resolve the disadvantages of the Vector Model is addressed as follows:

1. **Calculation-intensive.** Since the calculation of equation 3 is expensive in terms of computational time, the FLERSA tool pre-computes the denominator element ($\sqrt{\sum_i w_{Ci,j}^2}$) which is independent of the query vector, providing adequate response times for users of the automated annotation system.

2. **Term changes.** The knowledge worker is responsible for providing a basic Corpus from which to begin the training process. For each concept the system works with, the Corpus supplies at least one text (in the form of manual annotation) that contains words (used as terms) closely related with the concept that they evoke. Weights are computed for training terms and, gradually, new terms can be incorporated into the Corpus as the web content and semantic annotations grow. Term changes occur infrequently. Recalculation of all vectors is only necessary when the knowledge worker validates new annotations suitable for inclusion in the Corpus. The recalculation process is executed explicitly by the knowledge worker from a back-end application. The FLERSA tool follows a "pay as you go" approach [2]. The longer the knowledge worker spends on "fine tuning" the system, the better the results of the hybrid approach will be.

3. **False matches.** The Vector-Space Model does not specify which elements of a document must be used as terms. That is to say, use of words is not required and other possibilities are admitted, such as n-grams. The FLERSA tool uses n-grams (sequence of words) for ensuring low error rates when determining the concept a text fragment deals with.

Figure 1 illustrates the overall data flow for the hybrid approach to text categorization. The square boxes represent states and the oval boxes actions. As can be seen, the automated semantic annotation process consists of 4 steps which are described below:

- **Step 1**: A new input text arrives at the system for classification.
- **Step 2**: The input text n-gram frequency profile is computed following equation 2. The system has pre-computed profiles for each concept modelled in the Corpus annotations.
- **Step 3**: The system compares the input text profile against the pre-computed profiles of each of the concepts modelled in the system. An easily-calculated distance measure following equation 3 is used. Similarities of monograms, bigrams and trigrams are calculated separately, so a weighted value of similarity values is needed, as in equation 4.

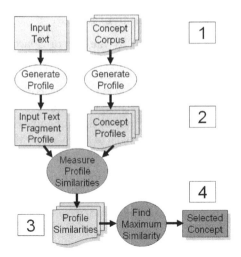

Fig. 1. Dataflow for text categorization

– **Step 4**: The system classifies the text fragment as belonging to the concept having the highest similarity. It also considers the possibility of not classifying the input text when its similarity value is less than a threshold value, so text fragments belonging to foreign categories are assigned no category. When a text fragment is satisfactorily classified as passing the threshold value, a semantic annotation is created in which categorization information is included.

3   Evaluation of the FLERSA Tool

The aim of this section is to evaluate the quality and effectiveness of the automated semantic annotation process. An evaluation environment has been built for this purpose; it is made up of a domain ontology, a set of test articles and two Corpus.

3.1   Environment

A complete content infrastructure has been created that enables testing of the component functionality: an ontology for physiology, two sets of articles for training the automated semantic annotation system and 52 example articles. The environment is accessible at http://www.scms.es/physiology, using user and password: "demo".

The ontology shown in Figure 2 works as a taxonomy from which concepts of the test environment are sorted. It is used to establish links between concepts and semantic annotations according to the issue that they deal with. The high level concept *"human_physiology"* was created in order to provide the framework for

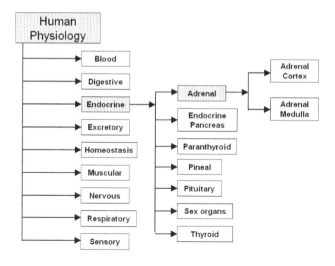

Fig. 2. Physiology taxonomy

working with global automated annotations. Nine subconcepts associated with human systems have been considered, namely: *blood, digestive, endocrine, excretory, homeostasis, muscular, nervous, respiratory and sensory.* They have been used for determining the human system that is discussed in a web document.

For local automated annotations, glands belonging to the human endocrine system have been considered. There are nine subconcepts associated with glands under consideration: *adrenal, adrenal cortex, adrenal medulla, endocrine pancreas, parathyroid, pineal, pituitary, thyroid and sex organs.* These specialized concepts have been used for determining the issue that deals with a specific paragraph within a web document during the local automated annotations process.

Regarding the set of articles for training, specific articles were sized (about 20K characters each) for each of the concepts (categories) detailed above. These articles work as a Corpus, from which the n-gram frequency profiles were computed to represent each of the concepts. The global Corpus is used for automated annotations of the entire text of a web document. It is made up of selected information collected by the knowledge worker that will be used for determining the system that is to be discussed in a web document. The local Corpus was also developed by the knowledge worker and it is used for generating automated annotations at paragraph level on web documents.

Finally, a set of test articles was made up of 52 articles that deal with different human diseases. The evaluation process was intended to check if the hybrid approach was able to determine the human system discussed in these articles, at both global and local level. Additionally, foreign articles dealing with foreign concepts such as wines and pizza types were included, in order to test how the system worked with information from other scopes.

3.2   Results

The weights used for equation 4 were $\alpha = 0.57$, $\beta = 0.28$ and $\gamma = 0.15$. The same values were used for automated annotations, at both global and local level. A threshold value was also used so that a semantic annotation would be created

Table 2. Summary of automated annotation results

Concept	Type	Chars	1-gram	2-grams	3-grams	Correct	False-	False+
Adrenal cortex	Local	3811	286	522	580	100,00%	0,00%	0,00%
Adrenal gland	Local	3138	188	325	349	100,00%	0,00%	0,00%
Adrenal medulla	Local	2234	202	330	357	100,00%	0,00%	0,00%
Endocrine pancreas	Local	10756	613	1280	1457	90,57%	9,43%	0,00%
Parathyroid gland	Local	2817	248	410	451	100,00%	0,00%	0,00%
Pineal gland	Local	6004	408	624	669	80,00%	0,00%	20,00%
Pituitary gland	Local	5121	365	740	881	92,31%	0,00%	7,69%
Sex organs	Local	1788	241	352	361	25,00%	75,00%	0,00%
Thyroid gland	Local	16278	905	1793	2009	100,00%	0,00%	0,00%
Blood system	Global	17540	793	1841	2079	57,14%	0,00%	42,86%
Digestive system	Global	24726	941	2360	2669	75,00%	0,00%	25,00%
Endocrine system	Global	18882	721	1723	2017	85,71%	14,29%	0,00%
Excretory system	Global	18359	719	1687	1939	60,00%	0,00%	40,00%
Homeostasis	Global	12007	582	1192	1329	0,00%	0,00%	100,00%
Muscular system	Global	8063	427	771	877	50,00%	50,00%	0,00%
Nervous system	Global	25996	1052	2500	2837	50,00%	0,00%	50,00%
Respiratory system	Global	13486	605	1324	1495	75,00%	0,00%	25,00%
Sensory system	Global	20848	918	2107	2380	100,00%	0,00%	0,00%
Average		**11769**	**567**	**1215**	**1374**	**87,69%**	**6,15%**	**6,15%**

only when the similarity value passed the threshold value. The threshold was set to 0.07 for local automated annotations and was set to 0.1 for global automated annotations.

The tests were conducted from the 52 test articles; 195 different automated semantic annotations were taken: 143 at the local level and 52 at a global level.

The shown data in table 2 summarizes several kinds of information that were obtained during the evaluation process: column 1 shows the different concepts that were considered, column 2 indicates the concept scope, columns from 3 to 6 show statistical information about the size and number of n-grams that make up the Corpus, columns from 7 to 9 are percentages of correct annotations and false negatives/positives, respectively.

Regarding the results obtained, initially the automated annotations system worked quite well although some errors took place. As the Corpus was updated, and validated annotations incorporated into it, the overall effectiveness of the automated annotation system was increased. We are satisfied with the results obtained. Although we are in the early stages, the success-rate hits of the hybrid approach in automated semantic annotation is encouraging. When working with the hybrid approach technique, it is worth highlighting the importance of knowledge worker on the training process in order to obtain satisfactory results.

4   Conclusions and Future Work

In this paper, a hybrid approach has been presented, in which the Vector-Space Model is combined with n-grams to provide a more effective text categorization technique.

In our approach, the Vector-Space Model has been enriched with n-grams in order to avoid problems such as ambiguity in understanding textual information, misinterpreting the original informative intent, and limiting the semantic scope of texts.

The hybrid approach is suitable for use in information retrieval systems, and it has been implemented in an annotation tool called FLERSA. It is used in the automated annotation process for determining the concepts that a web document deals with. A test environment was prepared and an evaluation process was undertaken. The results obtained demonstrate that using n-grams avoids inherit problems from Vector-Space Model that cause wrong results.

We are currently still working in FLERSA, refining the automated annotation process based on the hybrid approach presented in this paper. As regards future work, the next step will be to continue the experimentation in specific domains such as educational or medical environments.

References

1. Cavnar, W.B., Trenkle, J.M.: N-gram Based Text Categorization. In: 3rd Annual Symposium on Document Analysis and Information Retrieval, pp. 161–175 (1994)
2. Halevy, A., Franklin, M., Maier, D.: Principles of Dataspace Systems. In: Proceedings of the Twenty-Fifth ACM SIGMOD-SIGACT-SIGART Symposium on Principles of Database Systems, PODS 2006, pp. 1–9. ACM, New York (2006)
3. Jones, K.S.: A statistical interpretation of term specificity and its application in retrieval. Journal of Documentation 28, 11–21 (1972)
4. Luhn, H.P.: The Automatic Creation of Literature Abstracts. IBM Journal of Research and Development 2(2), 159–165 (1958)
5. Maedche, A., Motik, B., Stojanovic, L., Studer, R., Volz, R.: Ontologies for Enterprise Knowledge Management. IEEE Intelligent Systems 18(2), 26–33 (2003)
6. Navarro-Galindo, J.L., Jiménez, J.S.: Flexible Range Semantic Annotations Based on RDFa. In: MacKinnon, L.M. (ed.) BNCOD 2010. LNCS, vol. 6121, pp. 122–126. Springer, Heidelberg (2012)
7. Navarro-Galindo, J.L., Samos, J.: Manual and Automatic Semantic Annotation of Web Documents: The FLERSA Tool. In: 12th International Conference on Information Integration and Web-based Applications and Services, vol. 1, pp. 540–547 (November 2010)
8. Salton, G., Lesk, M.E.: The SMART automatic document retrieval systems - an illustration. Commun. ACM 8, 391–398 (1965)

An Unsupervised Framework
for Topological Relations Extraction
from Geographic Documents

Corrado Loglisci[1], Dino Ienco[2], Mathieu Roche[3],
Maguelonne Teisseire[2], and Donato Malerba[1]

[1] Dipartimento di Informatica, Università degli Studi di Bari "Aldo Moro", Bari, Italy
[2] UMR Tetis, IRSTEA, France
[3] LIRMM, Universite' Montpellier 2, France

Abstract. In this paper, we face the problem of extracting spatial relationships from geographical entities mentioned in textual documents. This is part of a research project which aims at geo-referencing document contents, hence making the realization of a Geographical Information Retrieval system possible. The driving factor of this research is the huge amount of Web documents which mention geographic places and relate them spatially. Several approaches have been proposed for the extraction of spatial relationships. However, they all assume the availability of either a large set of manually annotated documents or complex hand-crafted rules. In both cases, a rather tedious and time-consuming activity is required by domain experts. We propose an alternative approach based on the combined use of both a spatial ontology, which defines the topological relationships (classes) to be identified within text, and a nearest-prototype classifier, which helps to recognize instances of the topological relationships. This approach is unsupervised, so it does not need annotated data. Moreover, it is based on an ontology, which prevents the hand-crafting of *ad hoc* rules. Experimental results on real datasets show the viability of this approach.

1 Introduction

For decades, the spatial databases of Geographic Information Systems (GIS) have been considered the main collector of geographic information. With the advent of the Web, this is becoming less and less true. Nowadays, much geographical information resides in unstructured format in web resources, such as Web pages [8], and there is an emerging trend to query Web search engines by specifying geographic entities. This poses new challenges to GIS scientists since geographical knowledge is no longer well structured as it used to be in traditional GIS, it is often incomplete and fuzzy in nature, and its semantics is more difficult to express due to the ambiguity of the natural language representation. In recent years, particular attention has been paid to the design of Geographic Information Retrieval (GIR) systems which augments information retrieval with geographic metadata and enables spatial queries expressed in natural language. The task is

S.W. Liddle et al. (Eds.): DEXA 2012, Part II, LNCS 7447, pp. 48–55, 2012.

very demanding and the greatest challenge is to bridge the semantic gap between the formal models used in GIS technology and the informal human's description of the spatial information available in textual documents. Documents can be analyzed to recognize geographic information (e.g., place names and their attributes) which can be used to populate a GIS [11]. A more comprehensive view of the spatial information contained in a document goes beyond the extraction of geographic places and their properties, considering also the information on the spatial arrangement (or spatial relationships) holding over geographic entities. The extraction of spatial relationships from documents can thus play a prominent role to improve modern GIS and to design advanced functionalities such as, spatial querying, spatial analysis, and spatial reasoning [2].

Works on the extraction of spatial relationships have concentrated basically on two main research lines. In the first one, spatial relationships to be extracted are structured in *ad hoc* templates which can be filled with the spatial terms present in the text [5]. These templates are manually encoded and identify only specific relationships, which often do not correspond to well-established topological, distance, and directional relationships [10], and are therefore of limited use. In the second research line, the focus is on the recognition of topological, distance, and directional relationships by means of models learned from annotated documents [12].

A common and brittle assumption of these methods [5,10,12] is the availability of a large set of relations previously recognized and extracted from manually annotated documents. Actually, this contrasts with the more common situation in which we have an abundance of unannotated documents and any manual intervention could be expensive especially because of the inherent complexity of the activity of annotation of relations. Indeed, the identification of a relation within a text requires first the recognition of the objects (in the case of geographic documents, place names) involved in the relations and then the recognition of relations of interest by discriminating them from those uninteresting which can equally be present in the document.

In this paper we propose an alternative approach to extract relationships between geographic places. Specifically, we focus on the extraction of topological relationships between two regions [3]. The proposed approach is based on the combined use of both a spatial ontology, which prevents the hand-crafting of *ad hoc* rules, and a nearest-prototype classifier [9], which helps to automatically recognize instances of the topological relationships within text. The spatial ontology is exploited to determine the prototypes of the topological relationships, which, in the classification task, correspond to representatives of the classes. Differently from existing works, we follow an unsupervised approach since neither classified data (annotated geographic documents) nor classes of relationships are supposed to be available.

The paper is structured as follows. In Section 2 we describe our approach to document pre-processing, use of spatial ontologies and nearest-prototype technique. A case study with real geographic documents is presented in Section 3. Finally, conclusions are drawn.

2   Topological Relation Extraction

Before describing our proposal, we introduce some useful terms. Names of geographic places within text are called *geo-entities* in accordance with the usual practice of considering words with a proper noun as named entities. Geo-entities concern specific *spatial objects* endowed of a locational property defined on a georeferenced system. Given two spatial objects, their locational properties implicitly define their spatial relations. In this work, we focus on *topological relations* which are invariant under topological transformations, such as rotation, translation, and scaling. Their semantics is precisely defined by the nine-intersection model [3] and depends on the physical representation (e.g., punctual, linear and areal) of the involved spatial objects. For areal objects considered in this work, only eight topological relations are possible, namely { *disjoint, covers, covered by, contains, inside, meet, overlap, equal}*. *Spatial ontologies* describe and formalize knowledge on spatial and geographical concepts. They are used to support reasoning in various application domains where space plays a central role.

The extraction of topological relations between two regions is realized through a framework which includes both pre-processing and nearest-prototype classification techniques (Figure 1). The pre-processing techniques extract dependency relations between two geo-entities. These correspond to grammatical relations which connect two geo-entities within text and, in the structure of a discourse, can express spatial relations between the two geo-entities. Extracted dependencies are classified by means of some prototypes which are obtained from a spatial ontology. Classification is based on the textual dissimilarity between the extracted dependencies and the prototypes of the topological relationships.

A toy example follows. Consider the sentence "The Thames flows through London". The component *Geographic document pre-processing* produces the dependency relation which involve the recognized geo-entities "Thames", "London". The dependency is built with the grammatical relations which involve the words present in the sentence and the two geo-entities. By exploiting a spatial ontology, the component *Determination of prototypes* produces a text-based representative of each topological relationship { *disjoint, covers, covered by, contains, inside, meet, overlap, equal}*. Finally, the component *Nearest-prototype technique* classifies the dependency between "Thames" and "London" with one of the eight admissible topological relationships. The class is determined as the topological relationship closest to the dependency relation.

2.1   Geographic Document Pre-processing

The geographic document pre-processing component pipelines some natural language processing tools available in Stanford CoreNLP[1]. It returns the dependency relations in the documents together with lexical, syntactic and domain-specific annotations for each word occurring in a dependency relation. The pipeline is so composed:

---

[1] http://nlp.stanford.edu/index.shtml

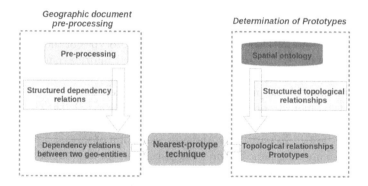

Fig. 1. Overview of the proposed unsupervised framework

Segmentation. Each document is first split into linguistic units (tokens and words) and then segmented into sequences of words, namely sentences.

POS tagging. Morpho-syntactic categories are automatically assigned to the words in the sentences.

Lemmatization. Words denoting grammatical variants are converted to normal or root forms.

Geo-Entity Recognition. The names of geographic places are identified by means of gazetteers automatically created with the geographic names database available in Geonames.[2]

Dependency Graph Extraction. For each sentence, a graph-based structure is generated with the annotations of POS tagging. Nodes correspond to words while edges correspond to grammatical relations between two words. A grammatical relation is represented as a triple composed of the type of relation, word which defines the relation and word object of the relation. A path (afterwards *dependency path*, DP) which links two geo-entities (words) denotes a dependency relation. Since a sentence (and therefore its associated graph) can report multiple geo-entities, several DPs can be extracted from a sentence. Nevertheless, there is only one DP connecting two geo-entities. It can express a topological relation between the two geo-entities and the prototype-based classification aims to recognize this.

2.2   Prototype-Based Classification

In prototype-based classification, prototypes of the classes are generated by either selecting representatives from the classified data or generating new representatives ad-hoc [1]. In the problem at hand, the unavailability of classified data constrains us to construct the prototypes of topological relationships, so we propose to leverage the spatial knowledge acquired in spatial ontologies to define these prototypes. At this aim, we consider the SUMO ontology [7] because, besides offering a formal definition of spatial relationships, it also presents an

---

[2] http://geonames.org/, retrieved on February, 2012.

unstructured and textual description of these relationships based on the linguistic ontology Wordnet. We exploit this textual information to determine a structured and computable representation of the prototypes of the relationships. Only one prototype is constructed from the textual information associated to one topological relationship present in SUMO. This representation is operatively produced by means of the pipeline described in Section 2.

Following the blueprint of prototype-based classification [9], the class to assign to each DP is determined as the topological relationship whose prototype is the closest one to the DP. In particular, the representation of DPs and prototypes in terms of textual and linguistic features suggests us to use a semantic dissimilarity measure. Therefore, the class of each DP is identified by the prototype which has the smallest dissimilarity measure with the DP.

3   Case Study

A case study was conducted on a corpus of geographic documents populated with the Featured Articles wiki-pages of the category "Geography and places"[3]. It comprises one hundred and twenty-three wiki-pages (retrieved on January 2012) that totally include 1088 sentences and 3631 geo-entities (9.98 sentences per document). Two datasets of DPs were obtained, one (denoted as DPnc) produced by the pipeline illustrated in the section 2.1, the other one (denoted as DPwc) produced by processing the DPs of DPnc with a technique of Co-reference Resolution. Such a technique solves the references to geo-entities reported with the pronouns or common nouns. This allows us to consider DPs in the classification process which otherwise would have been discarded. As decision function for the nearest-prototype technique, we consider the Lin [6] and Jiang & Conrath [4] semantic dissimilarity measures which can exploit the $is - a$ hierarchy arrangement in Wordnet.

Case Study Design. In the SUMO ontology, thirty spatial relationships are reported, nineteen of which have a natural language description. These include the eight topological relationships. A preliminary inspection revealed that the natural language description is not even coherent with the hierarchical organization of the ontology. Indeed, the semantic dissimilarities computed on the topological relationships revealed that the relationships having the same father can be actually associated to dissimilar DPs and, conversely, similar DPs can be classified as relationships positioned far apart in the ontology. Therefore, we derived five groups containing very similar relationships, and we manually assigned one topological relationship to each group. The assignment is reported in Table 1: $disjoint, equal, overlaps$ are associated to no relationship in the ontology ($distance, orientation$ are not topological).

Evaluation Scheme. Evaluation was conducted on the results of nearest-prototype technique, namely the topological relationships found in documents,

---

[3] http://en.wikipedia.org/wiki/Wikipedia:Featured_articles#Geography_
and_places

Table 1. The grouped spatial relationships and the corresponding topological ones

Groups in the ontology	Topological relationships
meetsSpatially, fills	*meet*
partiallyFills	*covers*
partiallyFills	*coveredBy*
propertPart, part, located	*contains*
propertPart, part, located	*inside*
distance	−
orientation	−
−	*disjoint*
−	*equal*
−	*overlap*

with respect to ground truth. In particular, we evaluated the performances by means of *Precision* and *Recall* along two main perspectives: the used dissimilarity measures and the topological relations to be recognized.

Ground truth was defined in terms of qualitative topological relationships by exploiting the geographic quantitative information available in the geographic database Geonames (such as latitude, longitude coordinates). We combined it to derive the likely topological relationship between the geo-entities involved in each extracted DP.

Results. The classification process is performed on the two datasets described above. The dataset DPnc (without co-references) amounts to 6650 DPs (6.11 DPs per sentence) while DPwc (with co-references) to 7954 (7.31 DPs per sentence). We report only the evaluation for the topological relationships that Geonames database allows to identify, as discussed above. In Table 2, we have *% relationships in Ground Truth* as the frequency of the topological relationships (fifth column) in the ground truth and *% classified in DPs* as the frequency of the DPs classified out of those in the ground truth.

The first consideration on the results concerns the difference of the number of relationships existing in DPnc with respect to DPwc. Indeed, the co-reference analysis leads to increase the number of relationships of kind *meet* (32.74) and *contains* (24.71), while the occurrences of *inside* decrease proportionally in a set greater than DPs, namely DPwc. This recurs also in the classification results. We have a greater set of the classified DPs in correspondence of a greater set of ground truth for Lin measure, and a smaller set of DPs (43) with a smaller set of labels (39,56) for J&C. Correspondingly, the values of F-score are encouraging considering that the approach works without any supervision on documents: in the case of DPnc, we have values of average F-score with Lin measure higher than those with J&C. This is mainly due to the differences of values produced by the measures and, therefore, to their different computation. Indeed, the J&C measure can return two types of values: *i)* negative values, which are difficult to handle as dissimilarities, and *ii)* values included in a very wide range, which can lead dissimilar DPs to be labeled with the same topological relationship. Instead, the Lin measure returns a scaled value which allows to attribute equal importance to similar DPs with respect to the same topological relationship.

Table 2. Evaluation of the classified DPs

Dataset	% classified in DPs	% relations in ground truth	Dissimilarity measure	Topological relations	Evaluation			
					Precision	Recall	F-score	avg F-score
DPnc	33,3	22,8	Lin	meet	0,581	0,585	0,582	
	48,12	24,15		contains	0,865	0,597	0,706	0,665
	48,12	42,45		inside	0,865	0,597	0,706	
	29,29	22,8	J&C	meet	0,520	0,525	0,522	
	44,69	24,15		contains	0,900	0,529	0,666	0,618
	44,69	42,45		inside	0,900	0,529	0,666	
DPwc	34,56	32,74	Lin	meet	0,5840	0,632	0,607	
	54,59	24,61		contains	0,8330	0,706	0,764	0,712
	54,59	39,56		inside	0,8330	0,706	0,764	
	32,86	32,74	J&C	meet	0,5310	0,645	0,582	
	43	24,61		contains	0,7956	0,4625	0,585	0,584
	43	39,56		inside	0,7956	0,4625	0,585	

This different behaviour of the measures seems to justify also the different avg F-score between the two datasets where the Lin measure outperforms the J&C one.

Another consideration can be done on the performances obtained with the same measure in the two datasets: the avg F-score with the Lin measure increases from DPnc to DPwc while that with J&C has opposite behaviour. Indeed, considering additional DPs (namely, DPwc) improves the results only for the *meet* relationship while decreases those for the other two relationships. This can be due also to possible errors originated in the co-reference resolution which, anyway, can introduce relationships which involve erroneously geo-entities. An interesting aspect is the performance exhibited with respect to the % classified DPs (second column): we have high F-score (0.706, 0.764) in correspondence of the greater set of the classified DPs (those obtained with the Lin measure) while, when the framework recognizes less topological relationships (e.g. 29.29), the performance decreases. This can be indicative of the fact that, with huge sets of DPs, the framework could exhibit very good performances. A final consideration can be done on the performances with respect to the single relationships. In each experiment (dataset-measure), we have the better F-score in correspondence of *contains* and *inside*. This can be seemingly attributed to the fact that the group of spatial relationships assigned to *contains* and *inside* (Table 1) enumerates three relationships against two in the case of *meet*. Indeed, an higher number of spatial relationships could lead a group to have an higher number of textual features (which represent a prototype) with the probability to have an higher similarity with DPs. To mitigate this bias, which could be more evident with an unbalanced ontology, we normalized the values of the dissimilarity by the number of such features.

4   Conclusions

In this work we investigated the problem of extracting spatial relations between geo-entities. Our proposal follows an unsupervised approach which, differently from the most of alternative methods, disregards the too restrictive assumption of the availability of training annotated documents. The framework integrates a spatial ontology into a nearest-prototype classifier to label relationships expressed in text. The case study reported quite good performances as evaluated with a (although not complete) ground truth and underline the influence of knowledge acquired in the ontology on the final results: enriching the description of the spatial relationships of the ontology or insert new relationships could be beneficial and provide a more accurate recognition of topological relationships.

Acknowledgments. This work is in partial fulfillment of the research objectives of the PRIN 2009 Project Learning Techniques in Relational Domains and Their Applications funded by the Italian Ministry of University and Research (MIUR).

References

1. Bezdek, J.C., Kuncheva, L.: Nearest prototype classifier designs: An experimental study. Int. J. Intell. Syst. 16(12), 1445–1473 (2001)
2. Chen, J., Zhao, R.L.: Spatial relations in gis: a survey on its key issues and research progress. Acta Geodaetica et Cartographica Sinica 28(2), 95–102 (1999)
3. Egenhofer, M.J., Franzosa, R.D.: Point set topological relations. International Journal of Geographical Information Systems 5(2), 161–174 (1991)
4. Jiang, J., Conrath, D.: In: Proc. of the Int'l. Conf. on Research in Computational Linguistics, pp. 19–33 (1997)
5. KordJamshidi, P., van Otterlo, M., Moens, M.-F.: Spatial role labeling: Towards extraction of spatial relations from natural language. TSLP 8(3), 4 (2011)
6. Lin, D.: An information-theoretic definition of similarity. In: Proceedings of the 15th International Conference on Machine Learning, pp. 296–304. Morgan Kaufmann (1998)
7. Niles, I., Pease, A.: Towards a standard upper ontology. In: FOIS, pp. 2–9 (2001)
8. Palkowsky, B., MetaCarta, I.: A new approach to information discovery-geography really does matter. In: Proceedings of the SPE Annual Technical Conference and Exhibition (2005)
9. Reed, S.K.: Pattern recognition and categorization. Cognitive Psychology 3, 393–407 (1972)
10. Worboys, M.F.: A generic model for planar geographical objects. International Journal of Geographical Information Science 6(5), 353–372 (1992)
11. Zhang, C., Zhang, X., Jiang, W., Shen, Q., Zhang, S.: Rule-Based Extraction of Spatial Relations in Natural Language Text. In: 2009 Int. Conf. on Computational Intelligence and Software Engineering, pp. 1–4. IEEE (2009)
12. Zhang, X., Zhang, C., Du, C., Zhu, S.: Svm based extraction of spatial relations in text. In: ICSDM, pp. 529–533. IEEE (2011)

Combination of Machine-Learning Algorithms for Fault Prediction in High-Precision Foundries

Javier Nieves, Igor Santos, and Pablo G. Bringas

S^{3}Lab, DeustoTech Computing
University of Deusto
Bilbao, Spain
{jnieves,isantos,pablo.garcia.bringas}@deusto.es

Abstract. Foundry is one of the activities that has contributed to evolve the society, however, the manufacturing process is carried out in the same manner as it was many years ago. Therefore, several defects may appear in castings when the production process is already finished. One of the most difficult defect to detect is the *microshrinkage*: tiny porosities that appear inside the casting. Another important aspect that foundries have to control are the attributes that measure the faculty of the casting to withstand several loads and tensions, also called *mechanical properties*. Both cases need specialised staff and expensive machines to test the castings and, in the second one, also, destructive inspections that render the casting invalid. The solution is to model the foundry process to apply machine learning techniques to foresee what is the state of the casting before its production. In this paper we extend our previous research and we propose a general method to foresee all the defects via building a meta-classifier combining different methods and without the need for selecting the best algorithm for each defect or available data. Finally, we compare the obtained results showing that the new approach allows us to obtain better results, in terms of accuracy and error rates, for foretelling microshrinkages and the value of mechanical properties.

Keywords: fault prediction, machine learning, meta-classification, process optimization.

1 Introduction

The manufacturing process is an important part of the current society. Thanks to it, consumers can have different products and services. Within the manufacturing process, the casting production or the foundry process is considered as one of the main factors that influences the development of the world economy. Thousands of castings are created in foundries composing complex systems. In fact, the actual capacity of the casting production of the world, which is higher than 60 million metric tones per year, is strongly diversified.

Due to current trends, it is really easy to produce castings and suddenly discover that every single one is faulty. The techniques for the assurance of failure-free foundry processes are exhaustive production control and diverse simulation

S.W. Liddle et al. (Eds.): DEXA 2012, Part II, LNCS 7447, pp. 56–70, 2012.
© Springer-Verlag Berlin Heidelberg 2012

techniques [1] but they are extremely expensive and only achieve good results in an *a posteriori* fashion. These methods are also still incapable of preventing two of the most difficult targets to detect in ductile iron castings, i.e., *the microshrinkage* and *the mechanical properties.* The first one, also called secondary contraction, consists in tiny porosities that appear inside the casting when it is cooling down. For the second one: mechanical properties, we have selected the *ultimate tensile strength* that is the force, which a casting can withstand until it breaks, in other words, it is the maximum stress any material can withstand when subjected to tension.

The problem of foreseeing the apparition of both flaws is very difficult to solve [2–5] due to the following reasons: (i) a huge amount of data, not prioritised or categorised in any way, is required to be managed, (ii) it is very hard to find cause-effect relationships between the variables of the system, and (iii) the human knowledge used in this task usually tends to be subjective, incomplete and not subjected to any test. One way to solve this problem is the employment of machine learning methods.

Currently, *machine-learning* classifiers have been applied in domains alike with outstanding results, for instance, for fault diagnosis [6], malware detection [7] or for cancer diagnosis [8]. Machine learning is being used increasingly in the field of metallurgy in several aspects such us classifying foundry pieces [9], optimising casting parameters [10], detecting causes of casting defects [11] amongst other related problems [12]. We have also applied these ideas and, we tested several machine-learning classifiers [2, 4, 5, 13–15] to identify which is the best classifier to predict microshrinkages and the ultimate tensile strength.

These classifiers, used as a stand-alone solution, are capable to predict several defects. But this process has some shortcomings such as: (i) we cannot be completely sure that the selected classifier is the best one to generalise the manufacturing process, (ii) the learning algorithms employed for creating some of the machine learning classifiers only find a local maximum and, hence, the final result is not optimal and (iii) by using a single classifier, we should generate a classifier close to the process nature (linear or non-linear). Combination of different classifiers can solve these problems. Firstly, it is more safe if we use all the classifiers instead of selecting one. Secondly, by combining different sub-optimal classifiers, we can approximate their behaviour to the optimal one. Finally, in this combination process we are able to select several classifiers building a linear meta-classifier (all of them are linear classifiers), non-linear (all of them are non-linear classifier) or hybrid (classifiers belong to both classes).

Against this background, we present here the first approach that employs a meta-classification technique, specifically, methods that allow us to combine several machine learning classifiers for categorising castings and to foresee microshrinkages and the ultimate tensile strength. These methods are able to learn from labelled data to build accurate classifiers that are going to share its knowledge under some rules. We propose the adoption of this method for the detection of microshrinkages and the ultimate tensile strength using features extracted from the foundry production parameters as we did before [2, 4, 5, 13–15].

2   High Precision Foundry

2.1   Foundry Process

A foundry is a factory where metal is melted and poured into containers specially shaped to produce objects such as wheels and bars. In other words, the factory in which metals are melt. Despite the fact that the process seems to be simply, the whole process become complex due to the hard conditions in which is developed. In this research we focus on foundries which produce castings that are close to the final product shape, i.e., 'near-net shape' components. To obtain the final casting, metals, in our case iron metals, have to pass through several stages in which raw materials are transformed. The most important stages are the following [16]:

- **Pattern making.** In this step, moulds (exteriors) or cores (interiors) are produced in wood, metal or resin in order to be used to create the sand moulds in which the castings are made.
- **Sand mould and core making.** The sand mould is the most widely extended method for ferrous castings. Sand is mixed with clay and water or other chemical binders. Next, the specialised machines create the two halves of the mould and join them together to provide a container in which the metals are poured into.
- **Metal melting.** In this process, raw materials are melt and mixed. Molten metal is prepared in a furnace and depending on the choice of the furnace, the quality, the quantity and the throughput of the melt change.
- **Casting and separation.** Once the mixture is made, the molten material is poured onto the sand mould. It can be done using various types of ladles or, in high volume foundries that generate small castings, automated pouring furnaces. Later, the metal begins to cool. This step is one of the most important because the majority of the defects can appear during this phase. Finally, when the casting has been cooled enough to maintain the shape, the casting is separated from the sand. The removed sand is recovered for further uses.
- **Removal of runners and risers.** Some parts of the casting that had been used to help in the previous processes are then removed. They can be detached by knocking off, sawing or cutting.
- **Finishing.** To finish the whole process some actions are usually performed, e.g., cleaning the residual sand, heat treatment and rectification of defects by welding.

As aforementioned, to detect faulty castings and in order to know the behaviour of the casting to withstand several forces and loads, several tests are done when the casting is finished. The complexity of carrying out this process before doing it, i.e., using *ex-ante* methods, stems from the huge amount of variables to monitor along the whole foundry process and, therefore, the way in which these variables influence the final design of a casting. Consequently, we have simplified the manufacturing and the main variables to control in order to foresee the faulty castings, and also features of the casting, can be classified into the following categories: (i) *metal-related* and (ii) *mould-related*.

- **Metal-related variables**
  - *Composition:* Type of treatment, inoculation and charges [17].
  - *Nucleation potential and melt quality:* Obtained by means of a thermal analysis program [18].
  - *Pouring:* Duration of the pouring process and temperature.
- **Mould-related variables**
  - *Sand:* Type of additives used, sand-specific features and carrying out of previous test or not.
  - *Moulding:* Machine used and moulding parameters.

Generally, the dimension and geometry of the casting also play a very important role in this practice and, thus, we included several variables to control these two features. In addition, we took into account other parameters regarding the configuration of each machine working in the manufacturing process [19]. Finally, we can represent the castings with 24 different variables [2].

2.2 Microshrinkages

An irregularity in the casting is called a casting defect. When a defect appears, the casting must be corrected or, in the worst case, rejected. There are several defects that might arise along the foundry process and affect the metal [16].

Michroshrinkages is a kind of defect that usually appears during the cooling phase of the metal but it cannot be noticed until the production is finished. Particularly, this flaw consists of a form of filamentary shrinkage in which the cavities are very small but large in number and can be distributed over a significant area of the casting, i.e., a minuscule internal porosities or cavities. The reason of its apparition is that metals are less dense as a liquid than as a solid. And during the solidification process, the density of the metal increases while the volume decreases in parallel. In this way, diminutive, microscopically undetectable interdendritic voids may appear leading to a reduction of the castings hardness and, in the cases of high precision foundries (where the casting is a part of a very sensitive piece), this defect renders the piece useless [20].

The existing tests to detect microshrinkages use non-destructive inspections. The most widely techniques are the analysis via X-ray and ultrasound emissions. Unfortunately, both require suitable devices, specialised staff and quite a long time to analyse all the parts. Moreover, every test has to be done once the casting is done. Therefore, post-production inspection is not an economical alternative to the pre-production detection of microshrinkages.

Although we have already obtained overall significant results through a supervised machine-learning-based approach predicting those imperfections [2, 13, 14], these approaches require to test several classifiers and identify which classifier fits to the foundry process. Moreover, if the research is not developed in an exhaustive manner or the selected learning methods only detect local maximums, we may select a non-optimal classifier to foresee microshrinkages. In addition, as we show in our previous research, we do not select always the same classifier with the same configuration to predict every defect.

2.3   Mechanical Properties

When the foundry process is accomplished, the final casting is a part of a more complex system that will be subject to several forces (loads). During the design step, engineers calculate these forces and how the material deforms or breaks as a function of applied load, time or other conditions. And later, after the whole process, they select some specimens to test their actual behaviour. Therefore, it is important to recognise how mechanical properties influence iron castings [18]. Specifically, the most important mechanical properties of foundry materials are the following ones [21]: strength (there are many kinds of strength such as ultimate strength and ultimate tensile strength), hardness, toughness, resilience, elasticity, plasticity, brittleness, ductility and malleability.

To assure the performance of castings, there are common or standard procedures for measuring the mechanical properties of the materials in a laboratory. Unfortunately, the only way to know how the castings withstand the forces and loads and take measurements of the behaviour is employing destructive inspections. In addition, this complex process, like in microshrinkages tests, requires suitable devices, specialised staff and quite a long time to analyse the materials.

Regarding the ultimate tensile strength, on which we focus here on, its checking method is performed as follows. First, a scientist prepares a testing specimen from the original casting. Second, the specimen is placed on the tensile testing machine. Finally, the machine pulls the sample from both ends and measures the force required to pull the specimen apart and how much the sample stretches before breaking.

Moreover, the main variables to control in order to predict the mechanical properties of metals are the composition [17], the size of the casting, the cooling speed and thermal treatment [18]. In this way, the system should take into account all these variables to issue a prediction on those mechanical properties. Hence, our machine-learning models are composed of about 25 variables.

We developed several researches applying machine-learning-based classifiers with the aim of predicting these features [4, 5, 13, 15]. By carrying out this approach, foundries can reduce the cost of their quality tests because the destruction of the casting is no longer required. In our research, we obtained significant results that prove the plausibility of this technique. Nevertheless, as happened with microshrinkages, we cannot assure that the optimal classifier is included in the tested classifiers.

3   Combining Machine-Learning Classifiers

Classifiers by themselves are able to obtain good results, but we cannot ensure that a specific classifier is perfectly suitable for the prediction of every defect in the foundry process. To solve this problem, several studies have been developed to combine classifiers [22]. These techniques seek to obtain a better classification decision despite of incorporating a higher degree of complexity to the process.

From a statistical point of view [23], assuming a labelled data set $\mathbf{Z}$ and the n number of different classifiers with relatively good performance making predictions for $\mathbf{Z}$, we can select one of them to solve classification problems, but there is a risk of not choosing the proper one. Therefore, the safest option is to use all of them and make an 'average' of their outputs. The resulting classifier is not necessarily better but will decrease or eliminate the risk induced because of the use of non appropriate classifiers.

From a computational point of view [22], some supervised machine-learning algorithms, in their learning phase, generate models based on local maximum solutions. Thus, an aggregation of classifiers is much closer to the optimal classifier than only one of them.

Similarly, the foundry process itself can be categorised into linear or nonlinear. By using these combination methods, we are capable of designing a collective intelligence system for classification which incorporates both linear and nonlinear classifiers.

The combination methods we used to develop the experiments are detailed below.

3.1   By Vote

The democracy for classifying elements is one of the oldest strategies for decision making. Extending the electoral theory, other methods can allow the combination of classifiers [24]:

- **Majority Voting Rule.** Assuming that the labelled outputs of classifiers are given as c-dimensional binary vectors $[d_{i,1}, ..., d_{i,c}]^T \in \{0,1\}^c, i = 1, ..., L$ where $d_{i,j} = 1$ if the classifier D_i categorises $\mathbf{x}$ in ω_j, or 0 otherwise. The plurality of the votes results in a set of classification for the class ω_k such as $\sum_{i=1}^{L} d_{i,k} = \max_{j=1}^{c} \sum_{i=1}^{L} d_{i,j}$. Regarding the problem of ties, these are solved arbitrarily.
- **Product Rule.** This second method takes into account the probabilities [24]. Thus, for the Product Rule, $p(x_1, ..., x_R|\omega_k)$ represents the joint probability distribution of the measurements taken from the classifiers. We assume that these representations are statistically independent. By including the Bayesian decision theory[24], the method assigns $Z \rightarrow \omega_j$ if $P^{-(R-1)}(\omega_j) \prod_{i=1}^{R} P(\omega_j|x_i) = \max_{k=1}^{m} P^{-(R-1)}(\omega_k) \prod_{i=1}^{R} P(\omega_k|x_i)$. The decision rule quantifies the probability of a hypothesis by combining the *a posteriori* probabilities generated by the classifiers. Indeed, this fusion rule is really hard because it may inhibit one of the outputs when the probability is close to 0.
- **Average Rule.** To obtain the Average Rule, we must start generating the Sum Rule to subsequently make a division employing the number of base

classifiers, R, as denominator [24]. Assuming that the *a posteriori* probabilities computed for each classifier are not derived from *a priori* probabilities, we obtain the Sum Rule in which we assign $Z \to \omega_j$ if $(1 - R)P(\omega_j) + \sum_{i=1}^{R} P(\omega_j|x_i) = \max_{k=1}^{m}[(1 - R)P(\omega_k) + \sum_{i=1}^{R} P(\omega_k|x_i)]$

- **Max Rule.** We start with the Sum Rule and obviate the product of *a posteriori* probabilities and assuming prior equalities, the method assigns $Z \to \omega_j$ if $\max_{i=1}^{R} P(\omega_k|x_i) = \max_{k=1}^{m} \max_{i=1}^{R} P(\omega_k|x_i)$

- **Min Rule.** For the Min Rule, starting with the Product Rule and obviating the product of *a posteriori* probabilities and assuming prior equalities, we will assign $Z \to \omega_j$ if $med_{i=1}^{R} P(\omega_j|x_i) = \max_{k=1}^{m} med_{i=1}^{R} P(\omega_k|x_i)$

3.2   Grading

The base classifiers are all the classifiers that we want to combine through the *grading* method [25] and these have been evaluated using k-fold cross-validation [26] ensuring that each of the instances has been employed for the learning phase of each classifier.

Formally, let p_{ikl} as the calculated class probability for each base classifier k for the class l and the instance i. To simplify the equations, we write P_{ikl} to refer to the vector $(p_{IK1}, p_{IK2}, ..., p_{ikn_l})$ of all probabilities for the instance i and the classifier k. In addition, the prediction of the base classifier k for i is the class L, p_{ikL}, is calculated by the maximum likelihood, in other words, $c_{ik} = argmax_l\{p_{ikl}\}$.

Moreover, *grading* builds n_c training datasets, one for each base classifier k, adding the predictions g_{ik} to the original data set as the new class. $prMeta_{ik}$ is the probability calculated by the meta-classifier of k that the base classifier k is going to correctly foresee the instance i. Regarding this information, the final estimated probability for the class l and the instance i, if there is, at least, one meta-classifier which indicates that its classifier is going to foresee the result in a correct manner (i.e., $prMeta_{ik} > 0.5$), is calculated as $prGrading_{il} = \sum\{prMeta_{ik}|c_{ik} = l \wedge prMeta_{ik} > 0.5\}$.

Therefore, the classification step is as follows [25]. First, each base classifier makes a prediction for the instance you want to foresee. Second, meta-classifiers qualify the result obtained by the base classifiers for the instance we are trying to classify. And, finally, the classification is derived using only the positive results. Conflicts (i.e., multiple classifiers with different predictions have got a correct result) can be solved using the *by vote* method or employing the estimated confidence for the base classifier.

3.3   Stacking

The *stacking* method [27] is another manner of combining classifiers that tries to improve the union based on cross-validation method.

Hence, we use several classifiers or generalisers. To learn these classifiers, we select a set of r partitions, each one divides θ (the training set) into two sets, usually disjoint. We label the set of partitions as θ_{ij}, where $1 \leq i \leq r$ and $j \in \{1, 2\}$. Then, we define the space, in which these classifiers are, as the level 0 space. The classifiers use the original data set θ for the learning step.

Then, for each r_i partition of θ, $\{\theta_{i1}, \theta_{i2}\}$, we generate a set of k numbers. Typically, this k numbers can be: (i) the assumptions made by the original classifier or generaliser, (ii) the input component θ_{i2} or (iii) the vector in the input space which connects the component θ_{i2} to its θ_{i1} nearest neighbour. Subsequently, we take each group of k numbers as input component in a second space, *level 1 space*. Due to we have r partitions of θ, there are r points in the space of level 1. These points are known as the *reduced* or *level 1* training set for the level 1 classifiers.

For the classification process, firstly, we carry out a question to the classifiers in level 0 (original classifiers). Secondly, once we get the answer from all of them, we apply the transformations of k numbers that produce the input data set for the level 1 (this is the results transformation step). Thirdly, level 1 classifiers will derive the solution. And finally, the response is transformed back into the level 0 space to provide the final result. The whole process is known as *stacked generalisation* and can be more complex adding multiple stacking levels.

To apply the *stacking* method, we miss a set of rules such as (i) which classifiers should be selected at level 0, (ii) which ones at level 1 and which k numbers ought to be employed to generate the level 1 space [27].

3.4   Multischeme

This method is a meta-classification method implemented by Weka [28] which allows the combination of classifiers in a simple manner. This method employs a combination rule based on the results obtained by the cross-validation and the error rate measured as the mean square error from several classifiers.

The cross-validation is a simple way of mapping a classifier G with a set of training data θ and estimating the error rate of G when θ is generalised. More rigorously, multischeme method is performed by calculating the mean for each instance i and the error rate of G achieved predicting the output target related to the input data set θ_{i2} and when the learning phase is performed using the rest of θ, θ_{i1}. The error estimation is made through cross-validation as follows: $CV(G, \theta) \equiv (\sum_i [G(\theta_{i1}, \text{input of } \theta_{i2})(\text{output of } \theta_{i2})]^2)/m$

By using this measure, multischeme method is able to determine which classifier has to be taken into account to make the most precise classification.

4   Experimental Results

To prove our hypothesis, we have collected data from a real foundry specialised in safety and precision components for the automotive industry, principally in disk-brake support with a production over $45,000$ tons a year. These experiments are focused exclusively on the prediction of the aforementioned targets:

(i) microshrinkages and (ii) the ultimate tensile strength. Note that, as we have already mentioned, the only way to examine both objectives is after the production is done.

Moreover, pieces flawed with a microshrinkage or an invalid ultimate tensile strength must be rejected because of the very restrictive quality standards (which is an imposed practice by the automotive industry). Therefore, regarding the acceptance/rejection criterion of the studied models, we defined several risk levels that resembles the one applied by the final requirements of the customer.

For microshrinkages, we defined the following 4 levels of risks: *Risk 0* (there is no microshrinkages), *Risk 1* (the probability of being microshrinkages is really low), *Risk 2* (there are some possibilities that the casting is flawed with a miscroshrinkage) and *Risk 3* (It is sure that the casting has a microshrinkage). In these experiments, the machine-learning classifiers have been built with the aforementioned 24 variables. We have worked with two different references (i.e., type of pieces) and, in order to test the accuracy of the predictions, with the results of the non-destructive X-ray and ultrasound inspections from 951 castings (note that each reference may involve several castings or pieces) performed in beforehand.

For the ultimate tensile strength, we have defined two risk levels: Risk 0 (more than 370 MPa[1]) and Risk 1 (less than 370 MPa). In these experiments, the machine-learning models have been built with the aforementioned 24 variables. We have worked with 11 different references and, in order to test the accuracy of the predictions, we have used as input data the results of the destructive inspection from 889 castings performed in beforehand. In spite of the fact that in our previous research we have examined this dataset with diverse sizes [4, 15], currently, we are interested in the accuracy level with the full original dataset since the foundry, we are collaborating with, always works with the whole dataset.

Specifically, we have conducted the next methodology in order to evaluate properly the combination of classifiers:

- **Cross-validation:** We have performed a *k-fold cross-validation* [26] with $k = 10$. In this way, our dataset is 10 times split into 10 different sets of learning (90% of the total dataset) and testing (10% of the total data).
- **Learning the model:** We have made the learning phase of each algorithm with each training dataset, applying different parameters or learning algorithms depending on the model. More accurately, we have use the same set of models that in our previous work [2, 4, 5, 13–15]: *Bayesian networks* (with K2, Hill Climber, Tree Augmented Naïve (TAN) as structural learning algorithms and Naïve Bayes), *K-Nearest Neighbour* (with values for k between 1 and 5), *Support Vector Machines* (with polynomial, normalised polynomial, radial basis function (RBF) and Pearson VII function-based kernels), Decision Trees (using random forests with different amount of trees (n),s $n = 50$, $n = 100$, $n = 150$, $n = 200$, $n = 250$, $n = 300$ and $n = 350$, and a J48 tree) and *Artificial Neural Networks* (specifically a MultiLayer Perceptron).

---

[1] MegaPascal, unit of pressure.

- **Learning the combination of the classifiers:** Once the machine-learning classifiers were built, we teach the different combination methods using the aforementioned models. More accurately, we tested the following combination methods:
  - *By vote:* There are several ways to combine the results of the classifiers by vote. For these experiments we have used *the majority vote rule* [22], *the product rule* [24], *the average rule* [24], *the max rule* [24] and *the min rule* [24].
  - *Grading:* This method needs a first level classifiers that have to assure that the predictions achieved by the original classifiers are correct. In this way, we have performed our experiments using the following first level classifiers: *a Naïve Bayes, Tree Augmented Naïve* and a *K-Nearest Neighbour* where $1 \leq k \leq 5$.
  - *Stacking:* For combining the original classifiers, *stacking* creates two separate spaces, in the first space, there are the original classifiers, and in the second one, there are several classifiers that derive the final result accordingly to the results achieved by the previous ones. To create the second space we have tested the following classifiers: *a Naïve Bayes, Tree Augmented Naïve*, a *K-Nearest Neighbour* with $k = 1$, $k = 2$, $k = 3$, $k = 4$ and $k = 5$; and a J48 decision tree.
  - *Multischeme:* This method combines the results using the cross-validation outputs and the error rates from the original classifiers. Thus, in this research we have used *multicheme* as it is.
- **Testing the model:** For each combination method, we have evaluated the percentage of correctly classified instances and the area under the Receiver Operating Characteristic (ROC) area that establishes the relation between false negatives and false positives [29]. We have decided to use the ROC area due to the realization that simple classification accuracy is often a poor metric for measuring the performance [30].

After applying the aforementioned methodology, we have obtained the following results. In order to facilitate the readability, we have divided the results by the classification target.

4.1 Microshrinkage

As we mentioned before, we have evaluated the meta-classifiers in terms of prediction accuracy and the area under the ROC curve. In this way, Table 1 illustrates the obtained results in terms of prediction accuracy. Using the full original dataset of 951 evidences, we can achieve a 94.47% of accuracy level. *Stacking built through a Tree Augmented Naïve* outperformed the rest of combination methods. On the one hand, the *stacking* method seems to be the best combination method due to the first three meta-classifiers was built using this technique. In addition, these three meta-classifiers achieved a similar result, more than a 94% of accuracy. The deviation between the majority of the meta-classifiers is really small (only 2.29 units). On the other hand, there are three meta-classifiers

Table 1. Results in terms of accuracy and AUC predicting microshrinkages

Combination Method	Accuracy (%s)	AUC
Stacking (TAN)	94.47	0.9873
Stacking (Naïve Bayes)	94.12	0.9659
Stacking (KNN k=5)	94.12	0.9657
Grading (KNN k=5)	93.94	0.9077
MultiScheme	93.94	0.9820
Grading (KNN k=4)	93.93	0.9073
Grading (KNN k=3)	93.87	0.9062
Grading (J48)	93.86	0.9059
By Vote (Majority Voting Rule)	93.85	0.9071
Grading (Naïve Bayes)	93.81	0.9053
Grading (TAN)	93.78	0.9049
Stacking (KNN k=3)	93.72	0.9528
Stacking (KNN k=4)	93.66	0.9600
Grading (KNN k=2)	93.49	0.9017
Grading (KNN k=1)	93.48	0.9013
By Vote (Average Rule)	93.33	0.9820
Stacking (J48)	93.26	0.9145
Stacking (KNN k=2)	92.46	0.9420
Stacking (KNN k=1)	92.18	0.9021
By Vote (Max Rule)	73.29	0.9538
By Vote (Product Rule)	68.17	0.9076
By Vote (Min Rule)	68.17	0.9076

that obtained an accuracy under the 75%. Those classifiers are based on *By Vote Rule*. Except for Majority Voting and Average Rule, this method is not adequate for the foundry process.

Notwithstanding, despite some Stacking meta-classifiers have achieved better accuracy levels than the Grading, Grading-based classifiers could achieve good results. Surprisingly, *MultiSheme*, one of the simplest method, is ranked in the fifth position. Hence, using this method (with an accuracy of 93.94%) or Majority Voting Rule (with an accuracy of 93.85%), other really simply method for combining classifiers, we can reduce the computational complexity while we maintain good results.

Table 1 also shows the area under the ROC curve (AUC). In this way, the obtained results in terms of AUC are similar to the ones of prediction accuracy. The *Stacking built through a Tree Augmented Naïve* also outperformed the rest of algorithms. More accurately, ROC analysis provides tools to select possible optimal models and to discard the suboptimal ones [29]. Therefore, if the results are closer to a value of 1 than to a value of 0, the classifier achieves a better performance because it obtains less amount of false positives and false negatives. In summary, although all of them accomplish acceptable values (they exceed the line of no-discrimination, in other words, more than a 0.90), Stacking based on a TAN classifier outshine the other classifiers achieving a 0.9873.

As it is shown in [23], these methods approximate the results achieved in our previous research using single classifiers [2, 13, 14]. Actually, there is one of them that overtakes the single classifiers. Therefore, we can conclude that the combination of classifiers can obtain (i) a good generalisation of the process improving the accuracy and reducing the error rates; and (ii) reduction of the problem for selecting one single classifier.

4.2   Mechanical Properties

We have measured the same parameters for testing the combination methods in the prediction of the ultimate tensile strength, more particularly, the accuracy and the area under the ROC curve. Thus, table 2 shows the achieved results. In this case, the meta-classifier with the best performance was the *Grading method (using a Tree Augmented Naïve)*. In terms of accuracy, the classifier obtained an 86.63%. As we can deduct, the combination of classifiers also depends on the defect to be foreseen. For microshrinkages, the *stacking* method built through a TAN was the better, however, for the ultimate tensile strength, it is the worst.

Table 2. Results in terms of accuracy and AUC predicting ultimate tensile strength

Combination Method	Accuracy (%s)	AUC
Grading (TAN)	86.63	0.8101
MultiScheme	86.37	0.9171
Grading (J48)	86.34	0.8077
By Vote (Majority Voting Rule)	85.95	0.8005
Grading (Naïve Bayes)	85.91	0.7990
By Vote (Average Rule)	85.86	0.9146
Stacking (TAN)	85.73	0.9098
Grading (KNN k=5)	85.71	0.7977
Stacking (Naïve Bayes)	85.48	0.9094
Stacking (KNN k=5)	85.33	0.8778
Grading (KNN k=4)	85.23	0.7929
Grading (KNN k=3)	85.15	0.7922
Grading (KNN k=1)	85.05	0.7940
By Vote (Max Rule)	84.92	0.8968
Stacking (J48)	84.79	0.7996
Grading (KNN k=2)	84.77	0.7884
Stacking (KNN k=3)	84.17	0.8527
Stacking (KNN k=4)	83.69	0.8665
Stacking (KNN k=1)	81.43	0.7676
Stacking (KNN k=2)	79.57	0.8258
By Vote (Product Rule)	69.80	0.5994
By Vote (Min Rule)	69.80	0.5994

The *MultiScheme* and the *Grading* methods follow closely the best meta-classifier. For the second one, its first level classifiers are J48 decision trees. In terms of accuracy, the difference between the three methods does not exceed 0.29 units. Surprisingly, as in the microshrinkages, two very simple methods, such as MultiScheme and the Majority Voting Rule, are among the best meta-classifiers. Similarly, results illustrate us that the Product Rule and Min Rule are not able to make good predictions of the ultimate tensile strength. Moreover, although the classifiers are not ranked in the same position that in the microshrinkages, the overall behaviour is the same. All of them got very similar results, while three of them departed from the general behaviour.

Regarding the error rates, all of the classifiers obtained very similar results, but the best classifier is not the best dealing with false positives. Nevertheless, the second one, the MultiSchema, is the best one. Thus, and because of the fact that the success rate is approximately the same, this could be the choice for predicting this feature. Another interesting aspect that we saw is that the best-performing classifier for microshrinkages problem is one of the best classifiers in terms of area under the ROC curve. Notwithstanding, the difference in accuracy with the best classifier is 0.9 units, hence, this method could be used for foreseeing both defects.

In this second experiment, the behaviour of a single classifier could not be improved. In our previous work [4, 5, 13, 15] we reach an accuracy of 86.84% using *Random Forests with 250 trees*. However, the difference is pretty small, hence, we can confirm that combining methods can approximate the results of the best classifier [23]. In addition, we do not care if the process is lineal or non-linear due to this method allow us to create a model that includes both type of classifiers. And finally, although it does not exceed the single classifiers, this method reduces the error rates, hence, we consider it as the best way for predicting the ultimate tensile strength.

5   Conclusions

On the one hand, microshrinkages are tiny porosities that appear when the casting is cooling down. On the other hand, ultimate tensile strength is the capacity of a metal to resist deformation when subject to a certain load. The prediction of the apparition of microshrinkages and the value of ultimate tensile strength renders as the hardest issues in foundry production, due to many different circumstances and variables that are involved in the casting process and determine it.

Our previous research [2, 4, 5, 13–15] pioneers the application of Artificial Intelligence to the prediction of these two features. Specifically in this paper, we have extended that model to the prediction via the combination of stand-alone classifiers. The majority of them behave well, but stacking built with a *Tree Augmented Naïve* for microshrinkages and grading also built with a TAN outperform the rest of the meta-classifiers. Moreover, the achieved results, in the case of microshrinkages, improve the classification done with a single classifier. On the other hand, as it is shown in [22], for the ultimate tensile strength, the meta-classifier approximates the previous results and reduces the error rates.

In addition, as we noticed in our previous work [2, 4, 15], there are some irregularities in the data that may alter the outcome rendering it not as effective as it should. More accurately, these inconsistencies appear because the data acquisition is performed in a manual fashion.

Accordingly, future work will be focused on four main directions. First, we plan to extend our analysis to the prediction of other defects in order to develop a global system of incident analysis. Second, we plan to integrate this meta-classifier, which will work as a black box combining all partial results to predict any defect, into a Model Predictive Control system in order to allow an hybrid prediction model. Third, we plan to employ some techniques (e.g., Bayesian compression) to give more relevance to the newer evidences than to the older ones. The main objective is to develop a new method to quickly adapt the machine learning classifiers included in this meta-classifier. And, finally, we plan to test a preprocessing step to reduce the irregularities in the data.

Reference

1. Sertucha, J., Loizaga, A., Suárez, R.: Improvement opportunities for simulation tools. In: Proceedings of the 16th European Conference and Exhibition on Digital Simulation for Virtual Engineering (2006) (invited talk)
2. Santos, I., Nieves, J., Penya, Y.K., Bringas, P.G.: Optimising Machine-Learning-Based Fault Prediction in Foundry Production. In: Omatu, S., Rocha, M.P., Bravo, J., Fernández, F., Corchado, E., Bustillo, A., Corchado, J.M. (eds.) IWANN 2009, Part II. LNCS, vol. 5518, pp. 554–561. Springer, Heidelberg (2009)
3. Santos, I., Nieves, J., Penya, Y.K., Bringas, P.G.: Towards noise and error reduction on foundry data gathering processes. In: Proceedings of the International Symposium on Industrial Electronics, ISIE (2010) (in press)
4. Nieves, J., Santos, I., Penya, Y.K., Rojas, S., Salazar, M., Bringas, P.G.: Mechanical properties prediction in high-precision foundry production. In: Proceedings of the 7th IEEE International Conference on Industrial Informatics, INDIN 2009, pp. 31–36 (2009)
5. Nieves, J., Santos, I., Penya, Y.K., Brezo, F., Bringas, P.G.: Enhanced Foundry Production Control. In: Bringas, P.G., Hameurlain, A., Quirchmayr, G. (eds.) DEXA 2010, Part I. LNCS, vol. 6261, pp. 213–220. Springer, Heidelberg (2010)
6. Yang, J., Zhanga, Y., Zhu, Y.: Intelligent fault diagnosis of rolling element bearing based on svms and fractal dimension. Mechanical Systems and Signal Processing 1, 2012–2024 (2007)
7. Santos, I., Penya, Y.K., Devesa, J., Bringas, P.G.: N-grams-based file signatures for malware detection. In: Proceedings of the 11th International Conference on Enterprise Information Systems (ICEIS), vol. AIDSS, pp. 317–320 (2009)
8. Lisboa, P., Taktak, A.: The use of artificial neural networks in decision support in cancer: a systematic review. Neural Networks 19(4), 408–415 (2006)
9. Lazaro, A., Serrano, I., Oria, J., de Miguel, C.: Ultrasonic sensing classification of foundry pieces applying neuralnetworks. In: 5th International Workshop on Advanced Motion Control, pp. 653–658 (1998)
10. Zhang, P., Xu, Z., Du, F.: Optimizing casting parameters of ingot based on neural network and genetic algorithm. In: ICNC 2008: Proceedings of the 2008 Fourth International Conference on Natural Computation, pp. 545–548. IEEE Computer Society, Washington, DC (2008)
11. Perzyk, M., Kochanski, A.: Detection of causes of casting defects assisted by artificial neural networks. Proceedings of the I MECH E Part B Journal of Engineering Manufacture 217 (2003)
12. Sourmail, T., Bhadeshia, H., MacKay, D.: Neural network model of creep strength of austenitic stainless steels. Materials Science and Technology 18(6), 655–663 (2002)
13. Santos, I., Nieves, J., Bringas, P.G., Penya, Y.K.: Machine-learning-based defect prediction in high-precision foundry production. In: Becker, L.M. (ed.) Structural Steel and Castings: Shapes and Standards, Properties and Applications, pp. 259–276. Nova Publishers (2010)
14. Santos, I., Nieves, J., Bringas, P.G.: Enhancing fault prediction on automatic foundry processes. In: World Automation Congress (WAC), pp. 1–6. IEEE (2010)
15. Santos, I., Nieves, J., Penya, Y.K., Bringas, P.G.: Machine-learning-based mechanical properties prediction in foundry production. In: Proceedings of ICROS-SICE International Joint Conference (ICCAS-SICE), pp. 4536–4541 (2009)

16. Kalpakjian, S., Schmid, S.: Manufacturing engineering and technology. Pearson Pentice Hall (2005)
17. Carrasquilla, J.F., Ríos, R.: A fracture mechanics study of nodular iron. Revista de Metalurgía 35(5), 279–291 (1999)
18. Gonzaga-Cinco, R., Fernández-Carrasquilla, J.: Mecanical properties dependency on chemical composition of spheroidal graphite cast iron. Revista de Metalurgia 42, 91–102 (2006)
19. Sertucha, J., Suárez, R., Legazpi, J., Gacetabeitia, P.: Influence of moulding conditions and mould characteristics on the contraction defects appearance in ductile iron castings. Revista de Metalurgia 43(2), 188–195 (2007)
20. Margaria, T.: Inoculation alloy against micro-shrinkage cracking for treating cast iron castings, November 13 (2003); WO Patent WO/2003/093,514
21. Lung, C.W., March, N.H.: Mechanical Properties of Metals: Atomistic and Fractal Continuum Approaches. World Scientific Pub. Co. Inc. (July 1992)
22. Kuncheva, L.I.: Combining Pattern Classifiers: Methods and Algorithms. Wiley-Interscience (2004)
23. Dietterich, T.G.: Ensemble Methods in Machine Learning. In: Kittler, J., Roli, F. (eds.) MCS 2000. LNCS, vol. 1857, pp. 1–15. Springer, Heidelberg (2000)
24. Kittler, J., Hatef, M., Duin, R., Matas, J.: On combining classifiers. IEEE Transactions on Pattern Analysis and Machine Intelligence 20(3), 226–239 (1998)
25. Seewald, A.K., Fürnkranz, J.: An Evaluation of Grading Classifiers. In: Hoffmann, F., Adams, N., Fisher, D., Guimarães, G., Hand, D.J. (eds.) IDA 2001. LNCS, vol. 2189, pp. 115–124. Springer, Heidelberg (2001)
26. Kohavi, R.: A study of cross-validation and bootstrap for accuracy estimation and model selection. In: International Joint Conference on Artificial Intelligence, vol. 14, pp. 1137–1145 (1995)
27. Wolpert, D.: Stacked generalization. Neural Networks 5(2), 241–259 (1992)
28. Garner, S.: Weka: The Waikato environment for knowledge analysis. In: Proceedings of the New Zealand Computer Science Research Students Conference, pp. 57–64 (1995)
29. Singh, Y., Kaur, A., Malhotra, R.: Comparative analysis of regression and machine learning methods for predicting fault proneness models. International Journal of Computer Applications in Technology 35(2), 183–193 (2009)
30. Provost, F., Fawcett, T.: Analysis and visualization of classifier performance: Comparison under imprecise class and cost distributions. In: Proceedings of the Third International Conference on Knowledge Discovery and Data Mining, pp. 43–48. Amer. Assn. for Artificial (1997)

A Framework for Conditioning Uncertain Relational Data

Ruiming Tang[1], Reynold Cheng[2], Huayu Wu[3], and Stéphane Bressan[1]

[1] National University of Singapore
{tangruiming,steph}@nus.edu.sg
[2] Department of Computer Science, The University of Hong Kong
ckcheng@cs.hku.hk
[3] Institute for Infocomm Research, Singapore
huwu@i2r.a-star.edu.sg

Abstract. We propose a framework for representing conditioned probabilistic relational data. In this framework the existence of tuples in possible worlds is determined by Boolean expressions composed from elementary events. The probability of a possible world is computed from the probabilities associated with these elementary events. In addition, a set of global constraints conditions the database. Conditioning is the formalization of the process of adding knowledge to a database. Some worlds may be impossible given the constraints and the probabilities of possible worlds are accordingly re-defined. The new constraints can come from the observation of the existence or non-existence of a tuple, from the knowledge of a specific rule, such as the existence of an exclusive set of tuples, or from the knowledge of a general rule, such as a functional dependency. We are therefore interested in computing a concise representation of the possible worlds and their respective probabilities after the addition of new constraints, namely an equivalent probabilistic database instance without constraints after conditioning. We devise and present a general algorithm for this computation. Unfortunately, the general problem involves the simplification of general Boolean expressions and is NP-hard. We therefore identify specific practical families of constraints for which we devise and present efficient algorithms.

1   Introduction

The exponential growth of the volume of digital data available for analysis is both a blessing and a curse. It is a blessing for unprecedented applications can be devised that exploit this wealth. It is a curse for the challenges paused by the integration and cleaning of these data are obstacles to their effective usage.

We consider such challenges as modeling, integrating and cleaning economic and scientific data ([4,8]). We consider the collected data is uncertain and the adjunction of knowledge and observations as a "conditioning process".

Traditional data models, under the strict closed world assumption, presume the exactitude and certainty of the information stored. Modern applications, such as the ones we consider here, need to process uncertain data coming from

S.W. Liddle et al. (Eds.): DEXA 2012, Part II, LNCS 7447, pp. 71–87, 2012.
© Springer-Verlag Berlin Heidelberg 2012

diverse and autonomous sources (we do not discuss here the orthogonal issue of the source of uncertainty). This requirement compels new data models able to manage uncertainty. Probabilistic data models are candidates for such solutions. A probabilistic database instance defines a collection of database instances, called *possible worlds*, in an underlying traditional data model: relational, semistructured or else. Each possible world is associated with a probability. The probability may be interpreted as the chance of the instance to be an actual one.

The probabilities of possible worlds are derived from probabilities associated with data units, which can be tuples, attribute values in relational databases, and elements in XML databases, depending on the underlying probabilistic data model. Each probability value can be interpreted as the probability of the data unit to belong to the actual instance. A language of probabilistic events (*events* for short) and expressions can be used to define dependencies among data units.

The addition of knowledge to a probabilistic database instance is called *conditioning*. Conditioning removes possible worlds that do not satisfy the given constraint, and thus possibly reduces uncertainty. The computation of the remaining possible worlds, of their probabilities and of the respective probabilities of tuples after conditioning is one of the practical challenges.

The contributions of this paper are:

1. We propose a model for conditioned probabilistic relational databases. For the sake of rigors and generality, we devise a model that natively caters for constraints rather than treating them as add-ons. We devise a tuple-level model as we consider tuples to be the units of observation. We prove that, for every *consistent* conditioned probabilistic database, there exists an *equivalent* probabilistic database on the same sample space without constraints. We define the necessary notions of consistency and equivalence.
2. We adapt an algorithm from [9] for our proposed model, to compute the new equivalent probabilistic database in general case. This computation can leverage Bayes' Theorem since the sample space is unchanged. Unfortunately, the general problem is NP-hard. Fortunately, there are some valuable and practical classes of constraints for which we can devise efficient algorithms.
3. We identify two such classes, observation constraints and X-tuple constraints, and present the corresponding efficient algorithms. An observation constraint is the knowledge of the truth value of an event. In the simplest case, it is the knowledge of the existence or non-existence of a tuple. An X-tuple constraint is the knowledge of the mutual exclusiveness of the existence of some tuples. There are two cases of X-tuple constraints: X-tuple with "maybe" semantics that considers that none of the candidate tuples could exist and X-tuple without "maybe" semantics that requires exactly one of the candidate tuples to exist. Observation constraints and X-tuple constraints naturally stem from integration, lineage and data cleaning applications. The usage of these classes of constraints in such applications are discussed in [1] and [4], for instance.

Let us consider an example of a probabilistic database of port names and their container throughput as collected from various websites. Table 1[1] contains

---

[1] TEU stands for Twenty-foot Equivalent Unit. It is a measure of container capacity.

excerpts of information about the container throughput of ports (e.g., the port of Hong Kong). The uncertainty [2] of the collected information is represented by the Boolean expressions and probabilities associated with the Boolean variables in these expressions. The Boolean expression associated with a tuple, as shown in column exp of Table 1, is interpreted as the condition for the tuple to exist in the actual instance of the database. The probability associated with a Boolean variable, as shown in Table 2, is to be interpreted as the probability of the variable to be true. It induces the probability of a tuple to exist in the actual instance of the database.

Table 1. Table R

tid	port	year	throughput	exp
t_1	Hong Kong	2007	23.998 million TEUs	e_1
t_2	Hong Kong	2008	24.494 million TEUs	e_2
t_3	Hong Kong	2009	21.040 million TEUs	e_3
t_4	Hong Kong	2009	20.9 million TEUs	e_4

Table 2. probability of variables

Boolean event	probability
e_1	1/2
e_2	3/5
e_3	1/2
e_4	3/5

Let us use the additional knowledge that the container throughput values are unique in the same year of the same port. Conditioning the probabilistic database with this constraint should have the effect to filter those possible worlds that do not satisfy this constraint. Table 3 illustrates four possible worlds that do not satisfy this constraint. The conditioned probabilistic database is obtained by removing these four possible worlds.

Table 3. Four possible worlds which do not satisfy the constraint

tid	port	year	throughput
t_3	Hong Kong	2009	21.040 million TEUs
t_4	Hong Kong	2009	20.9 million TEUs

tid	port	year	throughput
t_1	Hong Kong	2007	23.998 million TEUs
t_3	Hong Kong	2009	21.040 million TEUs
t_4	Hong Kong	2009	20.9 million TEUs

tid	port	year	throughput
t_2	Hong Kong	2008	24.494 million TEUs
t_3	Hong Kong	2009	21.040 million TEUs
t_4	Hong Kong	2009	20.9 million TEUs

tid	port	year	throughput
t_1	Hong Kong	2007	23.998 million TEUs
t_2	Hong Kong	2008	24.494 million TEUs
t_3	Hong Kong	2009	21.040 million TEUs
t_4	Hong Kong	2009	20.9 million TEUs

The probabilities of the remaining possible worlds in the conditioned database instance are conditional probabilities in the same sample space. They are computed using Bayes' Theorem. Let pwd denote a possible world of the original probabilistic database. Let $p(pwd)$ denote the probability of the possible world pwd in the original probabilistic database. Let $p'(pwd)$ denote the probability of the possible world pwd in the conditioned probabilistic database.

---

[2] Uncertainty could be quantified in various ways, for instance using ranks, reliability and trust or simply multiplicity of the sources. This issue is orthogonal to the contributions of this paper.

The Bayesian equation tells us that $p(pwd \wedge C) = p(pwd|C) \times p(C)$. We know that $p'(pwd) = p(pwd|C)$. Therefore $p'(pwd) = \frac{p(pwd \wedge C)}{p(C)}$. We can now compute the respective probabilities of the remaining possible worlds.

In the example, let us consider the remaining possible world in which t_1, t_2 and t_3 exist. The probability of this possible world in the original database is $p(pwd) = p(e_1) \times p(e_2) \times p(e_3) \times (1 - p(e_4)) = \frac{3}{50}$. The probability of this possible world in the conditioned database is $p'(pwd) = \frac{p(pwd \wedge C)}{p(C)} = \frac{\frac{3}{50}}{\frac{7}{10}} = \frac{3}{35}$.

In the example, an observation constraint that ascertains the existence of the tuple t_1 states that $e_1 = true$. The authors of [4] define cleaning as the addition of such observation constraints. In the example, an X-tuple without "maybe" semantics indicates that only one of t_3, t_4 exists, namely that t_3, t_4 form one X-tuple, is denoted by the Boolean expression $(e_3 \wedge \neg e_4) \vee (\neg e_3 \wedge e_4)$. Similarly, an X-Tuple with "maybe" semantics indicates that at most one of t_3, t_4 exists, is denoted by $(e_3 \wedge \neg e_4) \vee (\neg e_3 \wedge e_4) \vee (\neg e_3 \wedge \neg e_4)$.

The rest of this paper is organized as follows. Section 2 presents a rapid overview of related work on probabilistic data models and their applications. In Section 3, we present our probabilistic data model and define the conditioning problem. In Section 4, we give general solutions and present practical instances of the general problem for which efficient solution exists and devise such solutions, namely, the observation and X-tuple constraints. We evaluate the performance of the algorithms proposed in Section 5. Finally, we conclude our work in Section 6.

2   Related Work

2.1   Probabilistic Relational Models

We are interested in probabilistic relational models. We are not aware of relation-level models. Such models may indeed not be practical. The authors of [7,10,5,1] propose or use tuple-level models. The authors of [2,9] propose or use attribute-level models. The choice of a model at the appropriate granularity depends on the application.

The model of [10] is the first tuple-level probabilistic relational model. In this model the probability of a tuple is the joint probability of its attributes. Tuples in one probabilistic instance are mutually exclusive and the sum of their probabilities is one. This model is suitable when an instance stores a unique object. It requires several instances for several objects. The model of [5] overcomes this limitation by using keys to differentiate different objects in the same instance. Objects need not exist in some possible world and therefore the sum of the corresponding probabilities may be less than 1. The model is similar to the model of [1]. A group of tuples representing one object is called an "X-tuple". We later refer to the corresponding dependencies as an X-tuple constraint.

The model of [7] associates each tuple with a basic event or a Boolean expression. Events are associated with probabilities. The sample space is the space of events that are true. Possible worlds and their probabilities are defined by the probabilities of events and event expressions to be true. The authors of [6] presents an algebra for this model.

The model in [3] is the first proposed attribute-level probabilistic relational model. Attribute values are associated with probabilities. The authors of [2] propose a 1st Normal Form representation of the attribute-level probabilistic relations themselves (by means of a binary model).

As we have seen, existing work on probabilistic relational data model are either tuple-level models or attribute-level models. We propose a tuple-level model with additional constraints integrated as first class citizens. The model we present caters for constraints rather than treating them as add-ons, as other tuple-level models can be adapted to (e.g. [7]).

2.2   Conditioning

The authors of [9] propose an approach to conditioning probabilistic relation instances. They use the attribute-level model of [2]. They adapt algorithms and heuristics for Boolean validity checking and simplification to solve the general NP-hard conditioning problem. As the authors claim, [9] seems to be the only existing work on conditioning probabilistic relational databases.

Our work differentiates itself from [9] in three aspects. Firstly, although the model we are proposing is fit to tackle the "conditioning problem" as defined in [9], it can also be used to address other problems. This is possible because we have defined constraints as first class citizens of the model. For instance, we can address problems such as the equivalence of conditioned databases or the reverse problem of finding a conditioned database from a given set of possible worlds. Secondly, we do not only study the conditioning problem in the general case (for which it is proved to be an NP-hard problem) but we also identify some practical classes of constraints for which we can devise efficient algorithms. Lastly, they do conditioning on an attribute-level model. We propose a tuple-level model for the sake of applications that we are interested in.

3   Probabilistic Data Model

In this section, we propose a model for conditioned probabilistic relational databases. For the sake of rigor and generality, we devise a tuple-level model that natively caters for constraints rather than treating them as add-ons.

3.1   Probabilistic Relation

A probabilistic database instance is a collection of probabilistic relation instances. A probabilistic relation instance represents a set of traditional relation instances, called possible worlds, together with a probability for each of these instances to be actual. A probabilistic relation instance is represented as a set of tuples (a traditional relation instance), each of which is associated with a Boolean expression whose truth value determines the presence or absence of the tuple in the actual instance and whose probability to be true or false is defined by probabilities associated with the events of which it is composed. In addition, a set of complex events define integrity *constraints* that instances must verify.

Definition 1. *Let E be a set of symbols called* events (e). *A complex event* (ce) *is a well formed formula of propositional logic in which events are propositions.*

$$ce = e|ce \lor ce|ce \land ce|ce \rightarrow ce|\neg ce$$

$C(E)$ is the set of complex events formed with the events in E. An interpretation of a complex event is a function from E to $\{true, false\}$.

Definition 2. *A probabilistic relation instance is a quintuple $< R, E, f, C, p >$: R is a traditional relation instance, E is a set of events, f is a function from R to $C(E)$, C is a subset of $C(E)$, and p is a function from E to $[0,1]$.*

Definition 3. *The probability of a complex event c, noted $p(c)$, is as follows.*

$$p(c) = \sum_{I \in \mathcal{M}(c)} ((\prod_{I(e)=true} p(e)) \times (\prod_{I(e)=false} (1 - p(e))))$$

where $\mathcal{M}(c)$ is the set of models of c.

Figure 1 presents the uncertain information about port mentioned in Section 1, in our probabilistic relation model. R (the first part of the probabilistic relation $\mathcal{R}$ in Figure 1) is the one in Table 1, but without exp column.

$$\mathcal{R} = \begin{cases} R \\ E = \{e_1, e_2, e_3, e_4\} \\ f(t_1) = e_1, f(t_2) = e_2, f(t_3) = e_3, f(t_4) = e_4 \\ C = \emptyset \\ p(e_1) = \frac{1}{2}, p_1(e_2) = \frac{3}{5}, p(e_3) = \frac{1}{2}, p(e_4) = \frac{3}{5} \end{cases}$$

Fig. 1. Example of a conditioned probabilistic relation instance

3.2   Possible Worlds

Informally and in short, a possible world is a traditional relation instance such that the complex events associated with the tuples in the possible world are true, the complex events associated with the tuples not in the possible world are false, and the constraints are true. The probability of a possible world is the probability to find such a model given the probabilities of individual events, under the condition that the constraints are held.

Definition 4. *Let $\mathcal{R} =< R, E, f, C, p >$ be a probabilistic relation instance. R' is a possible world of $\mathcal{R}$ if and only if there exists a model of the following formula F with a non zero probability.*

$$F = (\bigwedge_{t_i \in R'} f(t_i)) \land (\bigwedge_{t_i \notin R'} \neg f(t_i)) \land (\bigwedge_{c \in C} c) \quad and \quad p(F) \neq 0$$

where $\mathcal{M}(R')$ is the set of models of the above formula.

We call $p_{\mathcal{R}}(R')$ the probability, $p(F|C)$, of the possible world R' in the probabilistic relation instance $\mathcal{R}$. We call $\mathfrak{P}(\mathcal{R})$ the set of possible worlds of $\mathcal{R}$.

Note that a possible world R' is a subset of R. Note that R' can be empty.

Definition 5. *Let $\mathcal{R} =< R, E, f, C, p >$ be a probabilistic relation instance. We say that $\mathcal{R}$ is **consistent** (resp. **inconsistent**) if and only if there exists a possible world (resp. there does not exist a possible world) of $\mathcal{R}$, i.e. $\mathfrak{P}(\mathcal{R}) \neq \emptyset$ (resp. $\mathfrak{P}(\mathcal{R}) = \emptyset$).*

Theorem 1. *Let $\mathcal{R} =< R, E, f, C, p >$ be a probabilistic relation instance. $\mathcal{R}$ is inconsistent iff C is always evaluated to be false, i.e. $\mathcal{M}(C) = \emptyset$, where $\mathcal{M}(C)$ is the set of models of C.*

Theorem 2. *Let $\mathcal{R}_1 =< R, E, f, C, p >$ and $\mathcal{R}_2 =< R, E, f, \emptyset, p >$ be two probabilistic relation instances. If R' is a possible world of $\mathcal{R}_1$ then it is a possible world of $\mathcal{R}_2$.*

The proofs to Theorem 1 and 2 can be found in our technical report [12].

3.3   World Equivalence

We now define an equivalent relation on one probabilistic relation instance. Two probabilistic relation instances are *world equivalent* if they have the same possible worlds with the same probabilities.

Definition 6. *Let $\mathcal{R}_1 =< R, E_1, f_1, C_1, p_1 >$ and $\mathcal{R}_2 =< R, E_2, f_2, C_2, p_2 >$ be two probabilistic relation instances. We say that $\mathcal{R}_1$ and $\mathcal{R}_2$ are **world equivalent** if and only if:*

$$\forall R' \subset R \, ((R' \in \mathfrak{P}(\mathcal{R}_1)) \leftrightarrow (R' \in \mathfrak{P}(\mathcal{R}_2))) \qquad and$$

$$\forall R' \in \mathfrak{P}(\mathcal{R}_1)(p_{\mathcal{R}_1}(R') = p_{\mathcal{R}_2}(R'))$$

We write $\mathcal{R}_1 \equiv_w \mathcal{R}_2$.

Theorem 3. *Let $\mathcal{R}_1 =< R, E_1, f_1, C, p_1 >$ be a probabilistic relation instance. If $\mathcal{R}_1$ is consistent then there exists a probabilistic relation $\mathcal{R}_2 =< R, E_2, f_2, \emptyset, p_2 >$ such that $\mathcal{R}_2 \equiv_w \mathcal{R}_1$.*

Proof. This proof by construction follows Poole's independent choice logic [11].

For each tuple $t_i \in \mathcal{R}_1$, take the completed disjunctive normal form of $f_1(t_i) \wedge C$. Use d_{ij} to represent each conjunct of $f_1(t_i) \wedge C$, i.e. $f_1(t_i) \wedge C = \bigvee d_{ij}$.

For each distinct d_{ij}, create a new event e_k, using a mapping function(i.e. $r(d_{ij}) = e_k$), and associate e_k with the probability $p_2(e_k) = \frac{p(d_{ij})}{p(C)}$. Here we model the possible worlds as unique mutually exclusive events.

We construct mapping function f_2 for $\mathcal{R}_2$ as: $f_2(t_i) = \bigvee_{\bigvee d_{ij}=f_1(t_i) \wedge C, e_k=r(d_{ij})} e_k$. f_2 makes sure $\mathcal{R}_2$ has the same set of possible worlds as $\mathcal{R}_1$.

Based on the constructing f_2, $p(f_2(t_i)) = \frac{p(\vee d_{ij})}{p(C)} = \frac{p(f_1(t_i) \wedge C)}{p(C)} = p(f_1(t_i)|C)$. This equation guarantees that $\mathcal{R}_2$ and $\mathcal{R}_1$ associate the same possible world with the same probability value.

Note that the new created events e_k are mutually exclusive. We can use a set of independent events w_k to represent e_k, as: $e_1 = w_1, e_k = \neg w_1 \wedge ... \wedge \neg w_{k-1} \wedge w_k (k = 2, ..., n-1), e_n = \neg w_1 \wedge ... \wedge \neg w_{n-1}$. One can easily compute the probability $p_2(w_k)$ by knowing $p_2(e_k)$. We can also easily replace e_k by independent events w_k, in f_2.

By the construction above, we get $\mathcal{R}_2$ such that $\mathcal{R}_2 \equiv_w \mathcal{R}_1$.   □

3.4   Conditioning and the Conditioning Problem

Conditioning a probabilistic relation consists in adding constraints. For example conditioning $\mathcal{R}_1 =< R, E_1, f_1, C_1, p_1 >$ with a set of constraints C is creating the probabilistic relation instance $\mathcal{R}_C =< R, E_1, f_1, C_1 \cup C, p_1 >$. Notice that R, E_1, f_1 and p_1 being unchanged, the underlying sample space is the same.

The conditioning problem consists in finding a world equivalent probabilistic relation instance with no constraint, given one probabilistic relation instance with constraints. Following the example above, given $\mathcal{R}_C =< R, E_1, f_1, C_1 \cup C, p_1 >$, the conditioning problem is finding $\mathcal{R}_2 =< R, E_2, f_2, \emptyset, p_2 >$ such that $\mathcal{R}_2 \equiv_w \mathcal{R}_C$. Notice that R is unchanged.

The existence and form of an equivalent probabilistic relation with no constraint is a direct consequence of Theorem 3, provided the conditioning yields a consistent instance.

Corollary 1. *Let $\mathcal{R}_1 =< R, E_1, f_1, C_1, p_1 >$ be a probabilistic relation instance. Let C be a set of constraints. $\mathcal{R}_C =< R, E_1, f_1, C_1 \cup C, p_1 >$ is the result of conditioning $\mathcal{R}_1$ with C. If $\mathcal{R}_C$ is consistent then the conditioning problem given $\mathcal{R}_C$ always has a solution (i.e. there exists $\mathcal{R}_2 =< R, E_2, f_2, \emptyset, p_2 >$ such that $\mathcal{R}_2 \equiv_w \mathcal{R}_C$).*

4   Algorithms

In this section, we propose different conditioning algorithms for different cases, including the general case of constraints and two special classes of constraints, i.e., observation constraints and X-tuple constraints. We will show that under the special cases, the complexity of solving the conditioning problem can be linear in the size of the probabilistic database, in contrast with NP-hard of the conditioning problem of the general case (as stated in [9]). For each case, before enforcing constraints, we need to make sure that the conditioned database is consistent. We start from the general case.

4.1   General Case

In the general case, the complex events associated with tuples, as well as the constraint C, are general elements of $C(E)$. In order words, there are no other requirements on f values and C.

Algorithm 1. Conditioning problem algorithm with a general constraint

---

Data: $\mathcal{R}_1 =< R, E_1, f_1, C, p_1 >$ (C is a general constraint)
Result: $\mathcal{R}_2 =< R, E_2, f_2, \emptyset, p_2 >$ such that $\mathcal{R}_2 \equiv_w \mathcal{R}_1$

1 Phase 1: encoding;
2 **for** *each tuple $t_i \in R$* **do**
3 $\quad$ normalize $f_1(t_i)$ to its completed disjunctive normal form;
4 $\quad$ construct a new R_{new} by duplicating t_i, with its f_1 value being one conjunct clause;
5 normalize C to the completed disjunctive normal form;
6 construct $\mathfrak{C}$ by adding all the conjunct clauses of C;
7 **for** *each $t_i \in R_{new}$* **do**
8 $\quad$ **if** $f_1(t_i) \notin \mathfrak{C}$ **then**
9 $\quad\quad$ remove t_i;
10 Phase 2: computing f_2 and p_2;
11 **for** *each $t_i \in R_{new}$* **do**
12 $\quad$ $f_2(t_i) = update(f_1(t_i))$;
13 $\quad$ **for** *each event e_i' in $f_2(t_i)$* **do**
14 $\quad\quad$ $p_2(e_i') = p(e_i|(e_1 \wedge ... \wedge e_{i-1} \wedge C))$;
15 $\quad\quad$ //e_i is the original basic event of e_i' before updating;
16 **if** *any two e_i', e_j' have the same original event $\wedge$ $p_2(e_i') = p_2(e_j')$* **then**
17 $\quad$ eliminate one of them, replace the eliminated event anywhere by the kept one;

Checking Consistency. According to Theorem 1, checking consistency of the conditioned database is to check whether there exists a model of its constraint C. In the general case, the constraint C can be an arbitrary logical expression. Thus actually it is an SAT problem, which is an NP-complete problem.

Conditioning. Algorithm 1 is an algorithm adapted from [9] in order to solve the conditioning problem with a general constraint. The general idea of Algorithm 1 follows the construction in the proof of Theorem 3. Algorithm 1 computes the possible worlds (of the conditioned probabilistic database) in which each tuple exists, in the phase 1 (line 1 to line 9). The algorithm computes f_2 and p_2 values using Bayesian equation, in the phase 2 (line 10 to line 17). In this phase, it provides some optimizations to minimize the number of events used during the conditioning process, thus it avoids to compute probabilities of unnecessary events. However, since the conditioning problem is an NP-hard problem in general, no matter how to optimize the algorithm, it is not practical to make it scalable to database size theoretically.

4.2 Special Case 1: Positive and Negative Observations

A positive (resp. negative) observation is the knowledge of one event being true (resp. false). After building a probabilistic database, one may get some information about the existence of some tuples or the truth value of some events.

This kind of observation is used in some research, as [4]. In [4], the authors build a probabilistic database with the information describing prices of items. There are several possible prices for each item, but only one of them is true. Later, they clean this uncertain information by observing the actual price of some items. Thus, their cleaning operations are adding observation constraints on their probabilistic database.

Checking Consistency. A probabilistic relation with a set of observation constraints can be consistent or inconsistent. A set of observation constraints form a conjunction of positive and negative events, i.e., $C=p_1 \wedge p_2 \wedge \ldots \wedge p_m \wedge \neg q_1 \wedge \neg q_2 \wedge \ldots \wedge \neg q_n$. The approach to check consistency is to check whether there exists the positive and negative form of the same event in C. If there exist e_i and $\neg e_i$ in C, C is always evaluated to be false (because C is a conjunction). In this case, the probabilistic relation is inconsistent. By first sorting the positive and negative events, checking consistency can be checked in $O(nlogn)$, where n is the number of events in C. .

Algorithm 2. Conditioning problem algorithm with a set of observation constraints

---

Data: $\mathcal{R}_1 =< R, E_1, f_1, C, p_1 >$ (C is a set of observations)
Result: $\mathcal{R}_2 =< R, E_2, f_2, \varnothing, p_2 >$ such that $\mathcal{R}_2 \equiv_w \mathcal{R}_1$

1 **for** *each* $e_i \in C$ **do**
2     $p(C) \times = p_1(e_i);$ //only needed when computing p(C);
3     $p_2(e_i) = 1;$
4 **for** *each* $\neg e_i \in C$ **do**
5     $p(C) \times = (1 - p_1(e_i));$ //only needed when computing p(C);
6     $p_2(e_i) = 0;$

Conditioning. Algorithm 2 is the algorithm solving the conditioning problem with a set of observation constraints. The basic idea is to update the probabilities of the positive events in the constraints to 1, and to update the probabilities of the negative events in the constraints to 0. As we can see, the only component updated in Algorithm 2 is p_1, while other parts remain the same. The time complexity of Algorithm 2 is linear in the size of the constraint. In the worst case the time complexity is $O(n)$, where n is the number of tuples in the probabilistic relation. The two lines (line 2 and 5) are needed only when we compute $p(C)$.

A Running Example. Assume that we have one probabilistic relation $\mathcal{R}$ (shown in Section 1), with a set of observation constraints $C = e_1 \wedge \neg e_3$. The equivalent probabilistic relation without constraints after conditioning should be $< R, E, f, \varnothing, p_2 >$, with $p_2(e_1) = 1, p_2(e_2) = \frac{3}{5}, p_2(e_3) = 0, p_2(e_4) = \frac{3}{5}$.

4.3 Special Case 2: X-tuple without "Maybe" Semantics

In this section, for $\mathcal{R} =< R, E, f, C, p >$, we consider the special case that each f value is one unique event.

Definition 7. *An **X-tuple** is a set of tuples among which every tuple is mutually exclusive to others. An **X-tuple** without "maybe" semantics means within the X-tuple, one tuple must be chosen. The independence between different X-Tuples are assumed.*

Many research works (such as [1,4]) use X-tuple model as their probabilistic data model. Thus, their probabilistic data model can be viewed as the conditioning result of adding a set of X-tuple constraints. This section considers about the X-tuple without "maybe" semantics that requires exactly one of the candidate tuples to exist. The next section considers about the X-tuple with "maybe" semantics that considers none of the candidate tuples could exist.

Assume we have k X-tuples: $X_1, X_2, ..., X_k$, and each X-tuple X_i has n_i candidate tuples: $t_{(i,1)}, ..., t_{(i,n_i)}$. The constraint of an X-tuple X_i (with n_i tuples) without "maybe" semantics can be expressed as:

$$C_i = \bigvee_{j=1,...,n_i} (f(t_{(i,j)}) \wedge (\bigwedge_{r=1,...,n_i \wedge r \neq j} \neg f(t_{(i,r)})))$$

Thus a set of X-tuples without "maybe" semantics constraints telling that there are k X-tuples present as:

$$C = \bigwedge_{i=1,...,k} C_i$$

For example, in the probabilistic relation $\mathcal{R}$ (shown in Section 3), if we have a set of X-tuple without "maybe" semantics constraints telling that there are two X-tuples: $X_1 = \{t_1, t_2\}, X_2 = \{t_3, t_4\}$, the set of constraints C present as:

$$C = ((e_1 \wedge \neg e_2) \vee (\neg e_1 \wedge e_2)) \wedge ((e_3 \wedge \neg e_4) \vee (\neg e_3 \vee e_4))$$

Checking Consistency. In this special case, one probabilistic relation with a set of X-tuple without "maybe" semantics constraints, with each f value being one unique event, is always consistent. According to Definition 5 and Theorem 1, in order to check consistency, we need to check whether C is always evaluated to be false or not. From the above example, if we transform the constraint C into its disjunctive normal form, we can see that each event involved in C appears only once in each conjunct. Thus the case of $e_i \wedge \neg e_i$ never happens, which gives chances to C to be true. In this case, one probabilistic relation is always consistent.

Conditioning

Definition 8. *A **compact encoding** of an X-tuple without "maybe" semantics X_i, which has n_i tuples, is:*

$$f(t_{(i,j)}) = \begin{cases} e_{i_j}, j=1 \\ \neg e_{i_1} \wedge ... \wedge \neg e_{i_{j-1}} \wedge e_{i_j}, j=2, ..., n_i - 1 \\ \neg e_{i_1} \wedge ... \wedge \neg e_{i_{j-1}}, j=n_i \end{cases}$$

where $t_{(i,j)}$ represents the j-th tuple of X_i, $j=1$, ..., n_i, and all e_{i_j} are new events which have never been used.

Theorem 4. *The compact encoding can express the X-tuple semantics.*

The proof to this theorem can be found in [12].

The independence is assumed among X-tuples, thus when we compute the probability of one tuple $t_{(i,j)}$ existing in the equivalent probabilistic database without constraints, we only need to consider the X-tuple X_i containing $t_{(i,j)}$.

Theorem 5. $\mathcal{R}_1 =< R, E_1, f_1, C, p_1 >$ *is one probabilistic relation. C is a set of X-tuple without "maybe" semantics constraints claiming that there exist k X-tuples: $X_1, X_2, ..., X_k$, while each X-tuple X_i has n_i candidate tuples: $t_{(i,1)}, ..., t_{(i,n_i)}$. The probability of one tuple $t_{(i,j)}$ existing in the equivalent conditioned probabilistic database without constraints, $\mathcal{R}_2 =< R, E_2, f_2, \varnothing, p_2 >$, can be computed as:*

$$p(f_2(t_{(i,j)})) = \frac{dif_{ij}}{XDif_i}$$

where $dif_{ij} = \frac{p(f_1(t_{(i,j)}))}{1-p(f_1(t_{(i,j)}))}$, and $XDif_i = \sum_{j=1}^{n_i} dif_{ij}$.

Proof. Since independence is assumed among X-tuples, we only need to consider the sub-constraint C_i which is the constraint claiming X-tuple X_i. According to the semantics of an X-tuple constraint, we can compute the probability of C_i as:

$$p(C_i) = \sum_{j=1}^{n_i} p(f_1(t_{(i,j)}) \wedge \bigwedge_{r=1,...,n_i \wedge r \neq j} \neg f_1(t_{(i,r)}))$$

$$= \sum_{j=1}^{n_i} (p(f_1(t_{(i,j)})) \times \prod_{r=1,...,j-1,j+1,...n_i} (1 - p(f_1(t_{(i,r)}))))$$

$$= \sum_{j=1}^{n_i} ((\prod_{r=1}^{n_i} (1 - p(f_1(t_{(i,r)})))) \times \frac{p(f_1(t_{(i,j)}))}{1 - p(f_1(t_{(i,j)}))})$$

$$= (\prod_{r=1}^{n_i} (1 - p(f_1(t_{(i,r)})))) \times XDif_i$$

We can easily get:

$$p(f_1(t_{(i,j)}) \wedge C_i) = (\prod_{r=1}^{n_i} (1 - p(f_1(t_{(i,r)})))) \times \frac{p(f_1(t_{(i,j)}))}{1 - p(f_1(t_{(i,j)}))}$$

$$= (\prod_{r=1}^{n_i} (1 - p(f_1(t_{(i,r)})))) \times dif_{ij}$$

Thus we can induce:

$$p(f_2(t_{(i,j)})) = p(f_1(t_{(i,j)})|C_i) = \frac{p(f_1(t_{(i,j)}) \wedge C_i)}{p(C_i)} = \frac{dif_{ij}}{XDif_i} \qquad \Box$$

Using Theorem 5, we can save lots of computations. When computing $p(f_2(t_{(i,j)}))$, we only need to compute dif_{ij} and $XDif_i$ of X-tuple X_i.

Actually, we can also simplify the computation of probability of the constraint C, which is a set of X-tuple constraints, using the theorem below.

Theorem 6. $R_1 =< R, E_1, f_1, C, p_1 >$ is one probabilistic relation. C is a set of X-tuple without "maybe" semantics constraints claiming there exist k X-tuples: $X_1, X_2, ..., X_k$, while each X-tuple X_i has n_i candidate tuples: $t_{(i,1)}, ..., t_{(i,n_i)}$. The probability of C can be computed as:

$$p(C) = Base \times Dif$$

where $Base = \prod_{i=1}^{k} XBase_i$, $XBase_i = \prod_{r=1}^{n_i}(1-p(f_1(t_{(i,r)})))$, $Dif = \prod_{i=1}^{k} XDif_i$. $XDif_i$ is defined in Theorem 5.

Proof. Since X_i, X_j are independent, C_i, C_j are independent as well. We can compute the probability of C, as (using Theorem 5):

$$p(C) = \prod_{i=1}^{k} p(C_i) = \prod_{i=1}^{k}((\prod_{r=1}^{n_i}(1 - p(f_1(t_{(i,r)})))) \times XDif_i)$$

$$= \prod_{i=1}^{k}(Xbase_i \times XDif_i) = (\prod_{i=1}^{k} Xbase_i) \times (\prod_{i=1}^{k} XDif_i) = Base \times Dif \qquad \Box$$

Algorithm 3 is the algorithm solving the conditioning problem for a set of X-tuple without "maybe" semantics constraints. Algorithm 3 is built based on Theorem 4, 5 and 6. The encoding phase (line 1 to line 4) is based on Theorem 4. The computing phase (line 5 to line 17) comes from Theorem 5 and 6. Actually, we do not need to compute $p(C)$ (line 9 and line 11) if we only aim to solve the conditioning problem. However, the probability of C can be computed by Algorithm 3, without affecting the time complexity. Obviously, the complexity of Algorithm 3 is linear in the size of X-tuple constraints. In worst case, it is $O(n)$, where n is the size of the probabilistic relation.

A Running Example. Assume that we have one probabilistic relation $\mathcal{R}$ (shown in Section 1), with a set of X-tuple constraints $C = ((e_1 \wedge \neg e_2) \vee (\neg e_1 \wedge e_2)) \wedge ((e_3 \wedge \neg e_4) \vee (\neg e_3 \vee e_4))$ (it tells t_1, t_2 is one X-tuple, and t_3, t_4 is another X-tuple). The equivalent probabilistic relation without constraints after conditioning should be $< R, E_2, f_2, \emptyset, p_2 >$

According to the compact encoding, we have f_2 function:

$$f_2(t_1) = e_1', f_2(t_2) = \neg e_1', f_2(t_3) = e_2', f_2(t_4) = \neg e_2',$$

and $E_2 = \{e_1', e_2'\}$.

Algorithm 3. Conditioning problem algorithm with a set of X-Tuple without "maybe" semantics constraints

---

Data: $\mathcal{R}_1 = <R, E_1, f_1, C, p_1>$ (C is a set of X-tuple constraints)
Result: $\mathcal{R}_2 = <R, E_2, f_2, \varnothing, p_2>$ such that $\mathcal{R}_2 \equiv_w \mathcal{R}_1$

---

1 Phase 1: encoding;//only needed when doing conditioning;
2 **for** *each X-tuple X_i* **do**
3   **for** *each tuple $t_{(i,j)}$ of X_i* **do**
4   update $f_1(t_{(i,j)})$ to $f_2(t_{(i,j)})$ by encoding in a compact manner;
5 Phase 2: computing;
6 **for** *each X-tuple X_i* **do**
7     $XBase_i = 1; XDif_i = 0$;
8   **for** *each tuple t_{ij} of X_i* **do**
9        $base_{ij} = 1 - p(f_1(t_{(i,j)}))$; $XBase_i \times = base_{ij}$;//only needed when computing p(C);
10        $dif_{ij} = \frac{p(f_1(t_{(i,j)}))}{1-p(f_1(t_{(i,j)}))}$; $XDif_i + = dif_{ij}$;
11 $Base = \prod_{i=1}^{k} XBase_i$;//only needed when computing p(C);
12 $Dif = \prod_{i=1}^{k} XDif_i$;
13 $p(C) = Base \times Dif$;
14 **for** *each X-tuple X_i* **do**
15   **for** *each tuple $t_{(i,j)}$ of X_i* **do**
16   set up an equation $p(f_2(t_{(i,j)})) = \frac{dif_{ij}}{XDif_i}$;
17 solve the equations, and get the p_2 value for all new variables in f_2;

Next, compute the p_2 function.

$$Base = (1 - p_1(e_1)) \times (1 - p_1(e_2)) \times (1 - p_1(e_3)) \times (1 - p_1(e_4)) = \frac{1}{25}$$

$$XDif_1 = \frac{p_1(e_1)}{1-p_1(e_1)} + \frac{p_1(e_2)}{1-p_1(e_2)} = \frac{5}{2}, XDif_2 = \frac{p_1(e_3)}{1-p_1(e_3)} + \frac{p_1(e_4)}{1-p_1(e_4)} = \frac{5}{2}$$

$$XDif = XDif_1 \times XDif_2 = \frac{25}{4}$$

$$p(C) = Base \times XDif = \frac{1}{25} \times \frac{25}{4} = \frac{1}{4}$$

$$p(f_2(t_1)) = \frac{\frac{p_1(e_1)}{1-p_1(e_1)}}{\frac{p_1(e_1)}{1-p_1(e_1)} + \frac{p_1(e_2)}{1-p_1(e_2)}} = \frac{2}{5}, p(f_2(t_2)) = \frac{\frac{p_1(e_2)}{1-p_1(e_2)}}{\frac{p_1(e_1)}{1-p_1(e_1)} + \frac{p_1(e_2)}{1-p_1(e_2)}} = \frac{3}{5}$$

$$p(f_2(t_3)) = \frac{\frac{p_1(e_3)}{1-p_1(e_3)}}{\frac{p_1(e_3)}{1-p_1(e_3)} + \frac{p_1(e_4)}{1-p_1(e_4)}} = \frac{2}{5}, p(f_2(t_4)) = \frac{\frac{p_1(e_4)}{1-p_1(e_4)}}{\frac{p_1(e_3)}{1-p_1(e_3)} + \frac{p_1(e_4)}{1-p_1(e_4)}} = \frac{3}{5}$$

Thus we have p_2 function:

$$p_2(e_1') = \frac{2}{5}, p_2(e_2') = \frac{2}{5}.$$

4.4   Special Case 3: X-tuple with "Maybe" Semantics

In this section, we study the case of X-tuple with "maybe" semantics constraint that considers none of the candidate tuples could exist. X-tuple with "maybe" semantics has a slightly different compact encoding approach.

Assume we have k X-tuples: $X_1, X_2, ..., X_k$, and each X-tuple X_i has n_i candidate tuples: $t_{(i,1)}, ..., t_{(i,n_i)}$. The constraint of an X-tuple X_i (with n_i tuples) with "maybe" semantics can be expressed as:

$$C_i = (\bigvee_{j=1,...,n_i} (f(t_{(i,j)}) \wedge (\bigwedge_{r=1,...,n_i \wedge r \neq j} \neg f(t_{(i,r)})))) \vee (\wedge_{r=1,...,n_i} \neg f(t_{(i,r)}))$$

Thus a set of X-tuples with "maybe" semantics constraints telling that there are k X-tuples present as:

$$C = \bigwedge_{i=1,...,k} C_i$$

Definition 9. *A compact encoding of an X-tuple with "maybe" semantics X_i, which has n_i tuples, is:*

$$f(t_{(i,j)}) = \begin{cases} e_{i_j}, j=1 \\ \neg e_{i_1} \wedge ... \wedge \neg e_{i_{j-1}} \wedge e_{i_j}, j=2, ..., n_i \end{cases}$$

where $t_{(i,j)}$ represents the j-th tuple of X_i, j=1, ..., n_i.

The algorithm solving the conditioning problem for a set of X-tuple with "maybe" semantics constraints is almost the same as Algorithm 3, except that we need to insert one line between line 10 and 11: "$XDif_i+ = 1;$".

5   Performance Evaluation

We mainly show the advantage of our approach in handling special classes of constraints, i.e., achieving scalable performance. Since the first special case is trivial, we only test the performance of the last two special cases.

All algorithms were implemented in Java and performed by a 2.83GHz CPU with 3.00GB RAM. We generated $300k$ tuples as our synthetic data set. f value

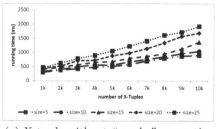 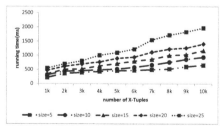

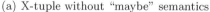

(a) X-tuple without "maybe" semantics   (b) X-tuple with "maybe" semantics

Fig. 2. Running time varying number of X-tuples and size of X-Tuples

of each tuple is one unique event, and the probability of each event is randomly assigned between [0,1]. Every X-tuple contains the same number of candidate tuples. We use two parameters to control the complexity of X-tuple constraint: the number of X-tuples and the number of candidate tuples in each X-tuple.

Figure 2 shows the running time by differentiating the cases of X-tuples with different number of candidate tuples, and varying the number of X-tuples. Figure 2(a) and 2(b) show the running time of solving the conditioning problem with a set of X-tuple without and with "maybe" semantics constraints respectively. The number of X-tuples varies from $1k$ to $10k$. We performed five runs, each of which adopts different number of candidate tuples in X-tuples, from 5 to 25.

We can observe that for both cases, when we fix the number of X-tuples, the running time increases as the number of candidate tuples in each X-tuple increases; when we fix the number of candidate tuples in each X-tuple, the running time increases as the number of X-tuples increases. The reason is the system takes more effort to compute the probability of the constraint, and to conditioning the probabilistic relation. Also, the time of each run is scalable to the number of X-tuple, and thus to the relation size. It verifies our previous analysis.

6   Conclusion

In this paper, we formalize a tuple-level probabilistic relational data model with additional constraints integrated as first class citizens. The model gives the flexibility to incorporate knowledge and observations in the form of constraints which are specified at any time. This process is called "conditioning". Furthermore, we propose and devise algorithms for the compact representation and the computation of possible worlds and their probabilities after conditioning. We also identify valuable and practical classes of constraints (i.e., observation constraints and X-tuple constraints), devise efficient algorithms for them, and demonstrate that solving the conditioning problem under them can be performed in linear time.

Acknowledgment. Stéphane Bressan was partially funded by the A*Star SERC project "Hippocratic Data Stream Cloud for Secure, Privacy preserving Data Analytics Services" 102 158 0037, NUS Ref: R-702-000-005-305. Reynold Cheng was supported by the Research Grants Council of Hong Kong (GRF Project 711309E).

References

1. Agrawal, P., Benjelloun, O., Sarma, A.D., Hayworth, C., Nabar, S., Sugihara, T., Widom, J.: Trio: A system for data, uncertainty, and lineage. In: VLDB (2006)
2. Antova, L., Jansen, T., Koch, C., Olteanu, D.: Fast and simple relational processing of uncertain data. In: ICDE, pp. 983–992 (2008)
3. Barbará, D., Garcia-Molina, H., Porter, D.: The management of probabilistic data. IEEE Trans. on Knowl. and Data Eng. 4, 487–502 (1992)

4. Cheng, R., Chen, J., Xie, X.: Cleaning uncertain data with quality guarantees. PVLDB 1(1), 722–735 (2008)
5. Dey, D., Sarkar, S.: Psql: A query language for probabilistic relational data. Data Knowl. Eng. 28(1), 107–120 (1998)
6. Fink, R., Olteanu, D., Rath, S.: Providing support for full relational algebra in probabilistic databases. In: ICDE, pp. 315–326 (2011)
7. Fuhr, N., Rölleke, T.: A probabilistic relational algebra for the integration of information retrieval and database systems. ACM Trans. Inf. Syst. 15(1) (1997)
8. Halevy, A.Y., Rajaraman, A., Ordille, J.J.: Data integration: The teenage years. In: VLDB, pp. 9–16 (2006)
9. Koch, C., Olteanu, D.: Conditioning probabilistic databases. PVLDB (2008)
10. Pittarelli, M.: An algebra for probabilistic databases. IEEE Trans. Knowl. Data Eng. 6(2), 293–303 (1994)
11. Poole, D.: Probabilistic horn abduction and bayesian networks. Artif. Intell. 64(1), 81–129 (1993)
12. Tang, R., et al.: A framework for conditioning uncertain relational data, http://www.comp.nus.edu.sg/~tang1987/trb1-12-conditioning.pdf

Cause Analysis of New Incidents
by Using Failure Knowledge Database

Yuki Awano[1], Qiang Ma[2], and Masatoshi Yoshikawa[2]

[1] Undergraduate School of Informatics and Mathematical Science,
Faculty of Engineering, Kyoto University Yoshida-Honmachi,
Sakyo, Kyoto, 606–8501 Japan
[2] Graduate School of Informatics, Kyoto University Yoshida-Honmachi,
Sakyo, Kyoto, 606–8501 Japan
`awano@db.soc.i.kyoto-u.ac.jp`, `{qiang,yoshikawa}@i.kyoto-u.ac.jp`

Abstract. Root cause analysis of failed projects and incidents is an important and necessary step to working out measures for preventing their recurrences. In this paper, to better analyze the causes of failed projects and incidents, we propose a novel topic-document-cause(TDC) model that reveals the corresponding relationships among topics, documents, and causes. We use the JST failure knowledge base to construct a TDC model with machine learning methods such as LDA and perceptron. The experimental results show that our approach performed better at discovering the causes of failures for projects and incidents.

1 Introduction

When we watch the news or read a newspaper, we can find a variety of accidents reported on such as failure at launching a new rocket, defects in electronic devices, collapsed buildings, air planes crashes, etc. When we look into the background of these accidents, some have the same causes. Why do we make similar accidents again and again? How can we prevent the recurrence of these accidents? Preventing the recurrences of these accidents is important. To do this, we think that analyzing the root cause of these accidents and sharing these causes with others are important. Thus, we study on methods for supporting root cause analysis.

Many corporations and organizations construct knowledge bases to avoid the recurrence of accidents. In some databases, they record not only the descriptions of accidents but also the backgrounds and effects of them. In this paper, we call these databases "failure knowledge databases."

Failure knowledge databases are useful resources for supporting root cause analysis. However most of these knowledge bases are closed and not open to public access. Fortunately, there exist public databases that are provided by governments and organizations. For example, there are the JST[1] failure knowledge database [1] and a database on management crisis of startup companies [2].

---

[1] Japan Science and Technology Agency `http://www.jst.go.jp/`

S.W. Liddle et al. (Eds.): DEXA 2012, Part II, LNCS 7447, pp. 88–102, 2012.
© Springer-Verlag Berlin Heidelberg 2012

However, they are not used effectively despite the fact that it costs much to build them. JST says that one of the reasons is that the knowledge of an accident is not conveyed properly from a creator of the information to users. To solve this problem, JST thinks that revealing the structure of an accident is important. JST proposed a failure knowledge database that contains the structure of an accident, called a "scenario." However, this database only provides keyword based search function for existing accidents and cannot tackle new incidents in contemporary society.

We propose methods for extrapolating the causes of new incidents by using an existing failure knowledge database. Our methods can assist users in understanding incidents and constructing a failure knowledge database.

The idea is very simple: similar causes trigger similar results. When a new incident occurs, we think it is helpful to extract the possible causes quickly by searching for similar incidents and concluding their causes from the failure knowledge databases.

Table 1. Incidents similar in words but different in behaviors and situations

e1	Title	Fatigue breakdown of gears of overhead traveling cranes with hoist.
	Overview	While lifting up a heavy object with a double-rail overhead traveling crane, the crane suddenly stopped and became inoperative. A gear in the hoisting machine was broken at the base.
e2	Title	Backlash of gears was too big and caused intense noise.
	Overview	A optical fiber base material synthesis device was designed and put to use. Noise of the device was larger than 80 decibels. A commercially available shaft was used to spin the base material of optical fiber (Rpm in the hundreds). However, the class of gears was JIS 4. This generated a big backlash and caused intense noise. To solve this problem, the material of one gear was changed to fiber-reinforced plastic(FRP).

The challenge is how to find similar incidents recorded in the knowledge database. The conventional methods, such as tf-idf based approach is considerable but not a good choice. For example, let's consider the two incidents, e1 and e2, in the JST failure knowledge database which are shown in Table 1. Both incidents are categorized in the "machinery" category. In this category average similarity between two incidents is 0.0198 according to the vector space model and cosine similarity with the tf-idf term weight. The incident similarity between $e1$ and $e2$ is 0.189, which is larger than the average similarities between two incidents. This means that e1 and e2 are similar incidents. However, these two incidents are different in terms of behavior and situation. As shown in Table 1, $e1$ is an incident in which the gears of a crane were destroyed by the fatigue of materials. $e2$ is an incident in which gears generated intense noise because the material of the gears was not suitable. While these two incidents have the same keyword "gear", they are different. This shows that tf-idf depends heavily on the words used to describe incidents. tf-idf does not work well when comparing two incidents in terms of their behaviors and situations.

Here, documents mean the descriptions of a project or incident, and records means the documents that are stored in the database.

As one of the solutions, we propose methods that compute the similarities between documents through topics generated by LDA [3]. One of the notable features of our methods is that we focus on behaviors and situations of incidents than surface text (tf-idf based keyword appearance). We expect that topics can represent the behaviors and situations in incidents. We will explain the details of the intervention of topics and evaluate the effects of topics later. The major contributions can be summarized as follows.

1. **Models for analyzing the causes of new accidents**
   In Section 3.2, we propose models that can represent the relations between causes, documents, and topics. We can extrapolate the causes of a new incident by considering multiple similar incidents by using our models.
2. **Revealing the underlying relations between incidents from the descriptions of circumstances and behaviors**
   In Section 4.1, we explain our methods to reveal the underlying relations between incidents through topics generated by LDA. In Section 5, we evaluates its effectivity.
3. **Revealing the relations between incidents and causes**
   In Section 3.2, we explain how we discover relations between incidents and causes by using a perceptron from a dataset. In Section 5, we evaluates the effectivity of the perceptron used to understand these relations.

2   Related Work

2.1   JST Failure Knowledge Database

The JST failure knowledge database was designed to share experiences and lessons obtained by analyzing failure incidents. The database launched on March 23, 2005, and the service was closed on March 31, 2011. At present, it is made public on the website of the Hatamura Institute for the Advancement of Technology [1]. Records in the JST failure knowledge database are stored under 16 categories, such as Machinery, Chemistry, and Construction. There are about 1000 incidents in the database. Each record contains case name, overview, sequence, cause, background, and other columns. An example is shown in Table 2.

The JST failure knowledge database is intended to communicate failure knowledge. The main concept of this database is the structure of failure knowledge called "scenario." A scenario is a sequence of causes, actions and results. The elements of a scenario are defined in a cause mandala, as shown in Figure 1. In the database, a scenario is given to each incident. We focus on the cause part of a scenario. By using this part, we can extrapolate the causes of a new given document.

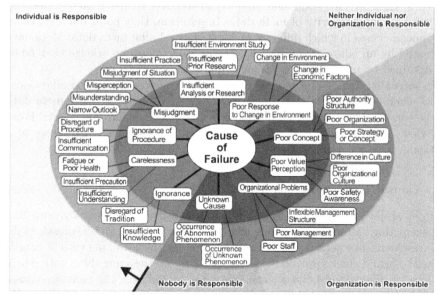

Fig. 1. Cause mandala

Table 2. Example of Record

Case Name	Loss of the deep sea remote operated vehicle (ROV), Kaikou
Date	May 29, 2003
Place	Kochi, Japan
Location	About 130 km South East Off the Muroto cape
Overview	Following a successful long period survey mission for acquisition of data about the Nankai Earthquake in the Nankai trough,...
Sequence	May 29th 09:30 The KAIKOU was launched into the water from the deep sea research ship KAIREI at the underwater operation area (fig 1)...
Cause	1. The damage to the secondary cable a. The strength of the aramid fiber around the opening of the sheath near the detaining harness was...
Background	The KAIKOU was the only deep sea survey system existing in the world able to explore the deepest part of the ocean,...
Scenario	Misjudgment, Misperception, Usage, Operation/Use, Failure, Degradation, Loss to Organization, Economic Loss

2.2 The Adequacy of Meta Data

Ito et al.[4] examined the implicit relations between incidents in the database without using meta data (scenarios). They stated that meta data is created from a single editor's point of view and some meta data is intentionally adjusted.

In their paper, they calculated the similarities of sequential and associational structures between incidents by using the co-occurrence of words. It is clearly shown in their investigation that they could discover relations that cannot be retrieved by the similarity of meta data. In addition, they point out that there are opposite cases in which different incidents have similar meta data. Meta data is created by an editor intending to communicate failure knowledge that he or she obtained by understanding an incident.

To solve this single point of view problem, we consider multiple similar documents when we extrapolate the causes of a new incident. Inadequate meta data can be corrected by using other adequate meta data of similar documents. From this view point, our method not only can help a user to find the causes of a particular incident, but also can improve the meta data in the database.

2.3   Causal Extraction and Search

Aono et al. [5] proposed a method for building a causal network by searching and extracting causes from documents on the Web by using clue expressions. In their research, they get the causes of a given incident by searching with keywords related to it and clue expressions. After that, they do the same thing with newly retrieved incidents and build a causal network. In addition, they analyze causal networks by examining their partial structure.

Nojima et al. [6] proposed a method for extracting a cause and result pair from news articles by using clue expressions and structure of news article. They also proposed a hypothesis that states that the headlines of news articles are the results of a written incident. By using this hypothesis, their method can extract a cause and result pair, which is not written in one sentence. When they use the clue expressions, their method requires that a cause and result pair appears in the same sentence or the same document.

Ishii et al. [7] proposed a topic-event cause model and an incremental method for causal network construction. In their method, events are represented as SVO tuple. They showed in the results of experiments that their method could detect similar event vertices better than the baseline method did, which uses a bag of words as a representation of an event. One of the notable features of their work is to construct causal network incrementally to improve the freshness of knowledge base.

To extrapolate the causes of a new incident, our method uses the relations between incidents. Here, incidents are composed of multiple sentences. Thus, we cannot use their method solve to our problem. The possibility is that we can use it to get the causal relations in an incident or between incidents that have unique names. Our future work is to be able to use those relations as detailed information on the cause of a new incident.

3   Document-Cause Model

We propose a document-cause model that represents the relations between documents and causes in order to extrapolate the causes of a new incident.

3.1   Input and Output

The input of our method is the unstructured text representation of incidents in a dataset, structured scenarios, and unstructured text of a new incident as a query. The output is the estimated causes of a new incident.

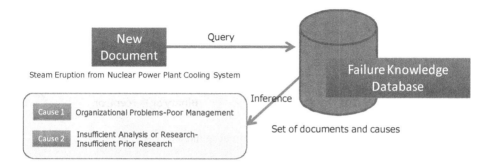

Fig. 2. Overview of Our Approach

There are many kinds of failure knowledge databases. In this paper, we use the JST failure knowledge database as the dataset. We also used the case name and overview of a record in the database as the document.

3.2   Document-Cause Model(DC Model)

The document-cause model is a network model that is built from a dataset and represents the relations between document and causes. We extrapolate the causes of new incidents by comparing incidents in the network. The document represents the behaviors and situations of an incident. In the JST failure knowledge database, for each document, there are a case name and an overview description. The cause is an element that is defined in a cause mandala. In the database, the cause part of a scenario corresponds to it.

Here, a query document describes a new incident. In our method, we compare a query document with not only just the most similar document but also with multiple similar documents. Thus, we expect that this method can consider the relations between documents and causes more precisely than a method considering only the most similar document. This model is composed of a document-cause layer. In this paper, we express weighted edges in a network by using a matrix. We regard no edge as an edge with a weight of 0. Suppose n is the number of documents. m is the number of causes, and the relations between documents and causes are represented by $n \times m$ matrix A.

Construction of Document-Cause Layer. Without using supervised data, we cannot extrapolate the relations between causes and documents in the real world. Thus, we construct a document-cause layer by analyzing existing datasets with two ways.

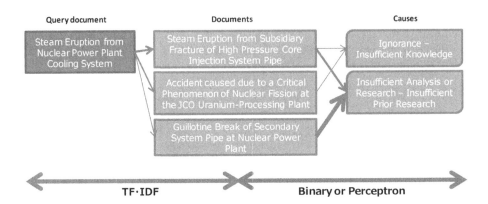

Fig. 3. Document-cause model

1. **Binary** When a cause is assigned in a dataset, we create an edge with a weight of 1.
2. **Perceptron** We use binary as the initial state. The system learns the weights between incidents and causes by using perceptron with the standard delta rule.

In the former, the weight of an edge be 0 or 1. The latter lets the weight of each edge range from 0 to 1. Edges in perceptron cases can represent more sensitive relations between incidents and causes, and we expect that the performance would be better in perceptron case. We evaluate this in Section 5.

Causes of a New Incident. Suppose the tf-idf based similarities between a query document and existing documents are $\mathbf{d_q}(d_{qi} = Sim(d_q, d_i))$, a cause vector of d_q is $\mathbf{c_q}$, and the weight of c_i in d_q is c_{qi}.

$$\mathbf{c_q} = A\mathbf{d_q} \tag{1}$$

Intuitively, top K results with higher c_q scores will return as the failure causes of a new incident (d_q). For example, return the top three causes as extrapolated causes. In this paper, we return causes that are weighted more than a certain threshold as extrapolated causes.

When a new incident happened, we think that the user has understood the incident. We think that the user wants to understand the incident in terms of not only what happened but also the potential causes. Thus, our methods provide the possible causes for help. In our method, a user is provided with a list of the possible causes that have been extracted without containing perspectives or personal views. This can solve the problem that registered meta data is created from only a single editor's point of view.

4   Topic-Document-Cause Model

We also propose a topic-document-cause model which is an advanced version of the topic-document model (In Section 3.2).

4.1   Topic-Document-Cause Model (TDC Model)

In the TDC model, we calculate the similarities between a query document and existing documents through the intervention of topics, which are collections of words and are generated by LDA, which is a topic model. In LDA, it is assumed that a document is a mixture of topics. Through the intervention, we can get the underlying similarities between incidents. The relations between incidents and causes are the same as in the DC model.

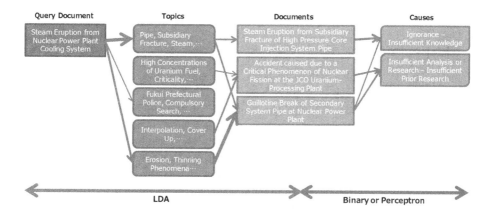

Fig. 4. Structure of the topic-document-cause model

Topics are collections of words in documents. Each word in the collection has probability to the topic. Each topic has probability to documents. These probabilities are sampled by LDA.

The TDC model is a network model that has two layers: a topic-document layer and a document-cause layer. In each layer, there are edges that represent the relations between two ends. In the topic-document layer, the edges between topics and documents indicate the relations between them. Weights show how strong the relation is. If there is a strong edge, we assume that the topic is highly related to the connected document. In the document-cause layer, an edge represents relations between documents and causes. When a weight is high, it means that the document is highly related to the cause.

Suppose that n is the number of documents, m is the number of topics, and l is the number of causes. The relations between topics and documents can be described with a $n \times m$ matrix A. The relations between documents and causes can be described with a $l \times m$ matrix B.

Construction of Layers. In the topic-document layer, we calculate the relations between documents and topics by using LDA. In our method, we expect that topics represent the behaviors and situations that caused accidents. Through the intervention of topics, we get underlying similarities.

The LDA requires the number of topics for analysis. The number of topics should be sufficient enough so that each topic can represent the behaviors and situations that caused accidents. We discuss this issue in the evaluation section (Section 5). The document-cause layer is built in the same way in the DC model.

Causes of a New Incident. Suppose that the probabilities of topics to a query document are a vector $\mathbf{t_q}$, a cause vector of d_q is $\mathbf{c_q}$, and the weight of c_i in d_q is c_{qi}.

$$\mathbf{c_q} = BA\mathbf{t_q} \tag{2}$$

We extrapolate the causes of a new incident in the same way in the DC model.

5   Experiments

5.1   Overview

We collected 1175 records from the web-site of the JST failure knowledge database. We used the case name, url, overview, and scenario in the record as the document for the causal analysis. A combined string of a case name and overview was the unstructured text representation of an incident.

Incidents in the database are categorized under 16 categories (e.g., *Machinery*, *Chemical*, and *Construction*). In our experiment, we evaluated our method in each category. We assume that the category of a new incident is given in advance.

We divided each category into a learning set and a test set. The test set is the first ten incidents in the original set. The remaining incidents are the learning set. We built models by using the learning set. After that, we evaluated the models by using the test set. We used the cause part of a scenario as the cause of an incident. We evaluated extrapolated causes with precision, recall, and f-measure.

As preliminary experiment, We varied the number of topics of TDC model in the machinery category, and the results are shown in Table 3.

Table 3. Number of Topics

method(number of topics)	precision	recall	f-measure
TDC-B(25)	0.173	0.383	0.236
TDC-B(50)	0.182	0.383	0.245
TDC-B(100)	0.195	0.383	0.257
TDC-B(500)	0.182	0.383	0.246
TDC-P(25)	0.198	0.333	0.247
TDC-P(50)	0.230	0.383	0.280
TDC-P(100)	0.208	0.283	0.239
TDC-P(500)	0.145	0.167	0.154

We could not find difference between the numbers of topics. Thus we use 50 as the number of topics. The threshold for whether a cause is returned or not is 0.1. When a query document is given, we search the most similar document by using vector space model with tf-idf term weight. We return the causes of the most similar document as extrapolated causes. We use this naive method as a baseline.

5.2 Experimental Results and Consideration

Under this condition, we evaluated five methods (Table 4), and precision, recall, and f-measure are shown in Tables 5, 6, and 7. Here, precision, recall, and f-measure are computed as follows. Relevant results are causes which are recorded in the scenario part of the knowledge base.

$$precision = \frac{\text{number of extrapolated causes which are relevant}}{\text{number of extrapolated causee}} \tag{3}$$

$$recall = \frac{\text{number of extrapolated causes which are relevant}}{\text{number of relevant causes}} \tag{4}$$

$$f - measure = \frac{2 \times precision \times recall}{precision + recall} \tag{5}$$

Table 4. Abbreviation of methods

TF-IDF	Baseline
DC-B	DC model (Binary)
DC-P	DC model (Perceptron)
TDC-B	TDC model (Binary)
TDC-P	TDC model (Perceptron)

The f-measure showed that our proposed models, the DC and TDC models, worked better than the baseline method did. In particular, recall with our methods was much better than that of the baseline one. The precisions were almost the same in all five methods. This means that our methods helped users to find causes that they could not find.

An example of our method being applied to an incident titled "Explosion and fire of a gas holder at an iron factory" is shown in Table 8. For DC-B and baseline, while the precision was the same value at 0.333, the recall of baseline was 0.5, and that of DC-B was 1.0. For the baseline, a user may not have noticed "Poor Safety Awareness".

This can solve the problem in which a meta data is created from one point of view. An editor will be given multiple causes, and he or she will consider causes that he or she could not find by themselves. Improving recall is really important when we consider the use case of our method.

Table 5. Precision

Category	TF-IDF	DC-B	DC-P	TDC-B	TDC-P
Other	0.233	0.242	0.254	0.287	0.302
Chemistry	0.250	0.185	0.148	0.200	0.196
Machinery	0.000	0.150	0.133	0.182	0.230
Metals	0.267	0.224	0.197	0.266	0.258
Construction	0.275	0.244	0.232	0.249	0.220
Nuclear Power	0.300	0.190	0.187	0.207	0.192
Aerospace	0.317	0.160	0.185	0.183	0.175
Natural Disasters	0.292	0.202	0.175	0.196	0.196
Motor Vehicles	0.133	0.205	0.199	0.233	0.220
Food	0.200	0.200	0.203	0.207	0.207
Oil	0.167	0.150	0.138	0.157	0.157
Petrochemistry	0.467	0.240	0.261	0.302	0.302
Shipping and Maritime	0.217	0.137	0.134	0.158	0.138
Railways	0.275	0.179	0.176	0.206	0.192
Electrical, Electronic, and IT	0.300	0.210	0.239	0.240	0.237
Electric Power and Gas	0.200	0.144	0.130	0.158	0.170

Table 6. Recall

Category	TF-IDF	DC-B	DC-P	TDC-B	TDC-P
Other	0.134	0.469	0.516	0.478	0.503
Chemistry	0.233	0.550	0.317	0.683	0.650
Machinery	0.000	0.333	0.167	0.383	0.383
Metals	0.350	0.650	0.600	0.567	0.617
Construction	0.433	0.775	0.783	0.775	0.725
Nuclear Power	0.300	0.483	0.483	0.467	0.500
Aerospace	0.253	0.540	0.610	0.540	0.540
Natural Disasters	0.192	0.483	0.500	0.533	0.567
Motor Vehicles	0.167	0.542	0.542	0.517	0.517
Food	0.233	0.417	0.417	0.267	0.267
Oil	0.267	0.500	0.433	0.500	0.500
Petrochemistry	0.417	0.600	0.625	0.775	0.775
Shipping and Maritime	0.300	0.450	0.450	0.500	0.500
Railways	0.258	0.717	0.717	0.717	0.717
Electrical, Electronic, and IT	0.200	0.500	0.567	0.467	0.467
Electric Power and Gas	0.153	0.357	0.357	0.353	0.373

Table 7. F-measure

Category	TF-IDF	DC-B	DC-P	TDC-B	TDC-P
Other	0.150	0.286	0.313	0.329	0.349
Chemistry	0.237	0.274	0.200	0.307	0.299
Machinery	0.000	0.205	0.142	0.245	0.280
Metals	0.287	0.322	0.290	0.352	0.354
Construction	0.323	0.363	0.348	0.364	0.329
Nuclear Power	0.283	0.256	0.251	0.268	0.259
Aerospace	0.275	0.239	0.274	0.257	0.251
Natural Disasters	0.199	0.255	0.248	0.276	0.282
Motor Vehicles	0.130	0.277	0.270	0.293	0.282
Food	0.207	0.244	0.249	0.231	0.231
Oil	0.197	0.226	0.203	0.234	0.234
Petrochemistry	0.420	0.335	0.358	0.429	0.429
Shipping and Maritime	0.243	0.206	0.203	0.239	0.215
Railways	0.230	0.271	0.269	0.302	0.289
Electrical, Electronic, and IT	0.223	0.288	0.330	0.305	0.300
Electric Power and Gas	0.140	0.182	0.169	0.187	0.202

Table 8. Explosion and fire of a gas holder at an iron factory

Method	Extrapolated causes
TF-IDF	Carelessness-Insufficient Precaution Organization Problems-Poor Management Insufficient Analysis of Research-Insufficient Practice
Our method(DC-B)	Organization Problems-Poor Management Poor Value Perception-Poor Safety Awareness Carelessness-Insufficient Precaution Insufficient Analysis of Research-Insufficient Practice Carelessness-Insufficient Understanding Misjudgement-Misjudgement of Situation
Truth	Poor Value Perception-Poor Safety Awareness Carelessness-Insufficient Precaution

TDC Model and DC Model. Table 7 shows that the TDC model did not improve the results better than did the DC model. One of the considerable reasons is that LDA could not represent behaviors and situations well.

In the machinery category, by analyzing our model, we get the most related topics for causes, shown in Table 9. A topic is regarded as highly related to multiple causes. When we consider the number of topics was 50 in our experiment, some of the topics had dominant relations to causes, and this would be one reason why the TDC model had not worked well. We will show more details of this problem later (Section 5.2).

Now, we focus on the quality of the topics. We expected that topics represent behaviors and situations. However, topics became a set of words that represent more detailed categories such as "teeth steel gear stainless," and "chain

Table 9. Topics that are most related to causes

Cause	Topic number
Organizational Problems-Inflexible Management Structure	6
Ignorance-Insufficient Knowledge	8
Carelessness-Insufficient Precaution	27
Insufficient Analysis or Research-Insufficient Practice	10
Poor Value Perception-Poor Safety Awareness	17
Poor Value Perception-Poor Organizational Culture	0
Organizational Problem-Inflexible Management Structure	6
Ignorance of Procedure-Insufficient Communication	46
Ignorance of Procedure-Disregard of Procedure	38
Misjudgment-Misunderstanding	7
Insufficient Analysis or Research-Insufficient Practice	48
Poor Response to Change in Environment-Change in Economic Factors	0
Poor Concept-Poor Strategy or Concept	0
Misjudgment-Misunderstanding	0
Poor Response to Change in Environment-Change in Environment	48
Insufficient Analysis or Research-Insufficient Environment Study	43
Misjudgment-Narrow Outlook	11
Unknown Cause-Occurrence of Abnormal Phenomenon	9
Unknown Cause-Occurrence of Unknown Phenomenon	26
Carelessness-Insufficient Understanding	27
Organizational Problems	33
Poor Concept-Poor Organization	33
Misjudgment-Misjudgment of Situation	9

glass driving light fiber." When a topic did represent a more detailed category, documents that had similar topics meant documents which were put in similar categories. The results are far from what we wanted. This is the same way the tf-idf works. This is why it is difficult to find difference between the DC model and TDC model.

To improve this, we are going to filter words by using word classes or frequencies. A topic is defined as a bag of words. We think that the order of words is important for representing behaviors and situations. A new representation model for incidents that focuses on behaviors and situations is necessary. This is our future work.

Perceptron and Binary. Perceptron based methods did not achieve better results than binary based ones. One of the considerable reason is the bias of dataset we used. As shown in Table 10, in our dataset, cause distribution is biased.

"Insufficient Analysis or Research-Insufficient Prior Research" and "Insufficient Analysis or Research-Insufficient Practice" had 115 incidents and 64 incidents, respectively. These numbers are much larger than those of other categories. This bias in the database also explains why perceptron does not work

Table 10. Distribution of causes

Cause	Count
Organizational Problems-Inflexible Management Structure	7
Ignorance-Insufficient Knowledge	48
Carelessness-Insufficient Precaution	33
Insufficient Analysis or Research-Insufficient Practice	64
Poor Value Perception-Poor Safety Awareness	11
Poor Value Perception-Poor Organizational Culture	1
Organizational Problems-Inflexible Management Structure	5
Ignorance of Procedure-Insufficient Communication	2
Ignorance of Procedure-Disregard of Procedure	5
Misjudgment-Misunderstanding	19
Insufficient Analysis or Research-Insufficient Prior Research	115
Poor Response to Change in Environment-Change in Economic Factors	1
Poor Concept-Poor Strategy or Concept	2
Misjudgment-Misperception	1
Poor Response to Change in Environment-Change in Environment	25
Insufficient Analysis or Research-Insufficient Environment Study	4
Misjudgment-Narrow Outlook	4
Unknown Cause-Occurrence of Abnormal Phenomenon	1
Unknown Cause-Occurrence of Unknown Phenomenon	2
Carelessness-Insufficient Understanding	19
Organizational Problems-Poor Staff	2
Poor Concept-Poor Organization	1
Misjudgment-Misjudgment of Situation	1

better than binary. When a dataset has a strong bias, the binary method can represent characteristics of the dataset correctly. This means that the initial state in the perceptron has learnt enough. We should consider this characteristic of this dataset more carefully in future work.

6   Conclusion

In this paper, we proposed methods that can extrapolate the causes of a new incident by using the JST failure knowledge database. We also carried out experiments to verify these methods. As shown by the results, our methods performed better than the baseline method, which returns the causes of the most similar incident to a query as extrapolated causes. An incident is a sequence of behaviors and situations, and each behavior and situation has a structure. From the experimental results, it is obvious that a bag of words is not sufficient for representing the behaviors and situations of incidents. We think that the order of words is important for representing behaviors and situations. A new representation model that represents behaviors and situations is necessary. Other future work includes 1) to compare our methods with the existing CBR[2] methods 2) to

---

[2] Case-based reasoning

consider the characteristics of the JST failure knowledge database. When we use the JST failure knowledge database, the characteristic in the database should be properly considered. It is one of our future work.

Acknowledgement. This research is partly supported by scientific research grants (No. 20300042, No. 20300036) made available by MEXT, Japan.

References

1. Failure knowledge database (in Japanese), `http://www.sozogaku.com/fkd/`
2. Failure Knowledge Database of Startups (in Japanese),
   `http://www.meti.go.jp/policy/newbusiness/kikidatabase/`
3. David, M., Andrew, Y., Michael, I.: Latent Dirichlet Allocation. Journal of Machine Learning Research 3, 993–1022 (2003)
4. Ito, K., Horiguchi, Y., Nakanishi, H., Nakanishi, H.: Structural Analysis of Latent Similarity among Errors by Applying Text-Mining Methods to Documents of Failure Cases. In: 23rd Fuzzy System Symposium, pp. 527–532 (2007)
5. Aono, H., Ohta, M.: Construction of a Causal Network and Acquisition of Causal Knowledges by Searching Factors. In: The 2nd Forum on Data Engineering and Information Management B9-1 (2010) (in Japanese)
6. Nojima, Y., Ma, Q., Yoshikawa, M.: Causal Relation Mining by Utilizing Contents and Structural Features of News Articles. In: The 3rd Forum on Data Engineering and Information Management F4-4 (2011) (in Japanese)
7. Ishii, H., Ma, Q., Yoshikawa, M.: Incremental Construction of Causal Network from News Articles. Journal of Information Processing 20(1), 207–215 (2012)

Modeling and Querying Context-Aware Personal Information Spaces

Rania Khéfifi[1], Pascal Poizat[1,2], and Fatiha Saïs[1]

[1] LRI, CNRS and Paris Sud University
[2] Evry Val d'Essonne University
{rania.khefifi,pascal.poizat,fatiha.sais}@lri.fr

Abstract. Personal information management (PIM) is the practice and analysis of the activities performed by people to acquire, organize, maintain, and retrieve information for everyday use. It is characterized by its heterogeneity, its dispersion and its redundancy. In this paper, our reflection focused on the development of a model that will be user-based. It help him create his data model depending on his needs.The meta-model that we propose allows users creating personal information and organizing them according to different points of view (ontologies) and different contexts. Contextual queries are defined to allow users to retrieve their personal information depending on context valuation.

Keywords: Personal Information Management, Ontologies, Contextual data.

1 Introduction

Personal information management (PIM) is the practice and analysis of the activities performed by people to acquire, organize, maintain, and retrieve information for everyday use. PIM is a growing area of interest because, everyone is looking for better use of our limited personal resources of time, money and energy. There exist several personal information management system [2–4]. MEMEX (Memory Extender), that allows automatic creation of references between library objects and establishing links between pieces of information. SEMEX [3] offers users a flexible framework with two main goals: browsing personal information by semantically meaningful associations and enrich personal information space to increase users' productivity, it uses data annotation, similarity computation from ontology-based framework. PIMO [4] is a framework for representing personal information model. It is a framework using multiple ontologies to represent concepts and documents using by persons in their daily activities. The increasingly big amount of personal information (e.g., mails, contacts, appointments) managed by a user is characterized by its heterogeneity, its dispersion and its redundancy. Processing this amount of information is difficult and not obvious. Indeed, personal information is often managed by very specific and autonomous tools which deal with specific kinds of user data (*e.g. iCal* to manage appointments and meetings, *thunderbird* and *outlook* to manage e-mails). Despite the existence of various applications,

S.W. Liddle et al. (Eds.): DEXA 2012, Part II, LNCS 7447, pp. 103–110, 2012.

there are no connections and no links between the pieces of personal information managed by the user applications. Indeed, many data present in one application can relate and concern other applications which leads to the redundancy of personal information. PIM users may create objects with the same name by referring to different entities and may create different objects with different names by referring them to the same entity. This kind of behavior may conduct to the semantic heterogeneity of personal information which makes the task of integration and interoperability between personal information more difficult. This is the reason why in this work we have developed an ontology based approach to deal with semantic heterogeneity as much as possible. Our system that we present in this paper, offers users the ability to organize their personal information in a way to ensure their semantic consistency, their reusability and their transparent querying. It is based on a meta-model allowing the user to create its own personal data model using different existing domain ontologies to describe and organize its personal information. We have studied the problem of designing model to represent data with context and selecting the most appropriate data according to a context asked in the query [1]. Martinenghi and Torlone in [5] have proposed a logical model as a foundation of context-aware database management system, and they have used many context to describe their data. Furthermore, user' personal information can be described according to different points of view thanks to the semantic Web technologies (*e.g.* OWL, RDF, SPARQL) allowing a richer descriptions of personal data.

2   Preliminaries

We present in this section the semantic Web materials on which our approach is based: OWL ontology model (http://www.w3.org/TR/owl2-profiles/), RDF data model and SPARQL query language.

OWL ontology. We consider an OWL ontology $\mathcal{O}$ that is defined as a tuple $\mathcal{O} = \{\mathcal{C}, \mathcal{P}, \mathcal{R}\}$ where $\mathcal{C}$ is a set of classes, $\mathcal{P}$ is a set of properties and $\mathcal{R}$ is a set of relations (subsumption, partOf, equivalence, disjointness).
Figure 2, gives an example of two ontologies O1 and O2. Each represents a part of the set of classes $\mathcal{C}$, partially ordered by the subsumption relation (pictured by '→'). The properties are pictured by bubbles linked to their corresponding domain and range.

RDF data model. RDF (Resource Description Framework) is a language which allows to describe semantically data in a understandable way for machines. Using RDF, data is represented as a set of triples expressed in the N-TRIPLE syntax as: <*subject property object*>. RDF language allows to use *reification* mechanism to add new elements to the descriptions of the RDF declarations, like data author, creation date, etc. For the reification semantics, see (http://www.w3.org/TR/rdf-mt/#Reif).

3   Personal Information Modelling

In this section, we will present the main components of the personal information representation model that we have designed. In Figure 1, we show the the the three levels of this model: *(i) Point-of-View level, (ii) Personal Information Type Level,* and *(iii) Personal Information Instance level.* The *Personal Information*

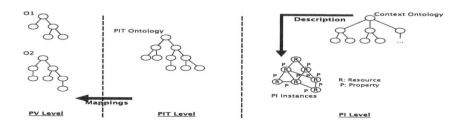

Fig. 1. Personal Information Model Architecture

Type (PIT) level represents the model of a structured vocabulary that the user aims to use to describe his/here personal information. The *Personal Information Instance (PI) level* represents the personal information that are instances of the PIT model. A set of mappings is added between the user PIT model and the ontologies that are considered in the *Point-of-View (PV) level.* As it is shown in Figure 1, personal information can also be represented according to different contexts, e.g. geographical, social and temporal (for a survey on context-aware systems, see [1]). Indeed, personal information is context-aware, which means that the concrete values of the properties may variate in function of the considered context. For example, a person may have a personal address and a professional address.

3.1   Point-of-View Level

We consider that a point of view is an ontology. It is composed of a set of classes (concepts) representing a specific domain. We allow declaring disjunction constraints between classes of a same point-of-view as well as between classes of different points-of-view.

3.2   Personal Information Type Level

This level is dedicated to represent the *Personal Information Types* (PITs) defined by the user and represented in an OWL ontology. A PIT is defined by a class for which a set of properties P is associated. We define a set of mappings between PIT classes (resp. properties) and Point-of-View classes (resp. properties) in order to exploit existing knowledge that are declared in PV ontologies.

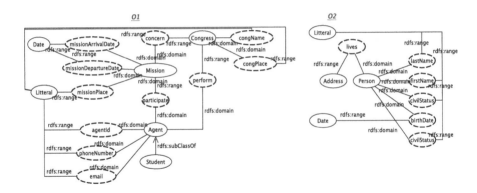

Fig. 2. Sample of some point-of-view ontologies

Definition 1 –*Personal Information Type (PIT)*. *A PIT is defined by a tuple* $(c, \mathcal{P}, \mathcal{C}, M_p)$ *where:*

- c *is the class that defines the PIT. A label is associated to the class.*
- P *is a set of properties associated by the user to the class c from a set of candidate properties of the PV ontologies.*
- $\mathcal{C}$ *is the set of classes that are involved in the PIT property definition. These classes should not be pairwise disjoint, w.r.t. the disjunction constraints that are declared between classes of the different ontologies.*
- $M_p = \{(c.p_1 \equiv o_1 : c_2.p_2), (c.p_1 \preceq o_2 : c_3.p_3), \ldots, (c.p_j \equiv o_k : c_n.p_m.)\}$ *is the set of mappings between the properties P of the PIT and properties of the PV ontologies. These mappings correspond to a subset of the whole set of mappings $\mathcal{M}_P$ between the PIT model and the PV model.*

Example 1 *Let there two ontologies: a* work *ontology O_1 and a* family *ontology O_2. The Person concept is described by the set of properties {lastName, firstName, e-mail, job} and by the set of properties {lastName, firstName, birthDate, civilStatus,numberOfChildren} in the ontology O_2. In spite of the existence of two different descriptions in ontologies O_1 and O_2, the user can choose to create a new concept (e.g. Friend) represented by a new PIT having as name Friend. The new PIT friend can be described by a subset of properties derived from the two sets of properties of the concepts $O_1 : Employee$ and $O_2 : person$, (e.g. describe Friend by name and $firstName$ derived from $O_1 : Employee$, and civilStatus derived from $O_2 : Person$). A set of mappings is then generated between the user PIT ontology and the PV ontologies O_1 and O_2:*

$$\mathcal{M}_p = \{(Friend.firstName \equiv O_1 : Employee.firstName), (Friend.lastName \equiv O_1 : Employee.name), (Friend.email \equiv O_1 : Employee.email), (Friend.civilStatus \equiv O_2 : Person.civilStatus)\}.$$

3.3   Personal Information Instance Level

Once the different PITs needed by the user are defined, the personal information (*e.g.* contacts, mails, publications, appointments) can be attached to these PITs and then instantiate the different properties describing these PITs. As it is shown in Figure 1, our personal information instances are context aware. Hence, instead of having a value for each mono-valued property we consider a set triples *(property, context, value)*.

Definition 2 *–**Personal Information(PI)** is an instance of a PIT which refers to a user information. Each PI is also represented by a subset of the property instances which describe its corresponding PIT. We consider a PI as defined by a tuple $(T, Reference, \mathcal{P}_{inst})$, where:*

- *T, represents the PIT that is instantiated by the PI.*
- *$Reference$, is URI which is an identifier used to represent the Personal Information instance in the system.*
- *$\mathcal{P}_{inst}$, is a set of tuples $(property, value, context)$ which instantiates the PIT "T" properties.*

We consider that the context values are declared in a given context ontology. In this work we will focus our study on the geographical context of personal information. We consider that the geographical ontology is composed of a set of classes that are organized in a hierarchy thanks to the *partOf* relation. We assume that one class cannot be *partOf* more than one class. We assume also the declaration of disjointness constraints and equivalence constraints between the geographical context classes. In order to represent in RDF the context values associated to the PI instances (*e.g.* a person leaving partly in France and in UK may have two Social welfare numbers (SSN)), we need to describe RDF statement, representing PI, using an other RDF statement. This is possible thanks to the reification mechanism. Then, Personal information are represented in RDF as a set of statements. Each statement instantiates one PIT property by assigning to the statement a URI *uriRef* as follows:

```
<uriRef rdf:type rdf:Statement>  <uriRef rdf:subject subj>
<uriRef rdf:predicate property> <uriRef rdf:object value>
<uriRef :contextGO context>
```

where, *subj* defines the PI URI, *property* expresses the URI of the property, *value* represents the value given to the instance of property and *context* is the value of the context, in our case is the geographical context.

An example of RDF representation of "phoneNumber" property from Friend#001 is done as follows:

<t1 **rdf:type** rdf:Statement>	<t2 **rdf:type** rdf:Statement>
<t1 **rdf:subject** Friend#001>	<t2 **rdf:subject** Friend#001>
<t1 **rdf:predicate** O2:phoneNumber>	<t2 **rdf:predicate** O2:phoneNumber>
<t1 **rdf:object** "phone11">	<t2 **rdf:object** "phone12">
<t1 **:contextGO** France>	<t2 **:contextGO** England>

4   Context-Based SPARQL Querying

It is important to provide the user a query interface with his PIM in a transparent way. This is possible, thanks to the PIT ontology that is used to describe personal information. The user simply expresses his queries using the vocabulary defined in the PIT ontology. Furthermore, in order to exploit the contextual values, the user may express his queries by specifying the desired context value. We have defined three kinds of contextual-queries:

1. *Strict-context query*: a query where the user asks for answers having the same context than the one given in the query;
2. *Disjoint-context query*: a query where the user asks for answers having a context that is distinct from the one given in the query;
3. *Similar-context query*: a query where the user asks for answers having a context that is similar to the one given in the query.

4.1 Strict Query Execution Algorithm (SQE)

We perform our queries by giving as parameters a context and a function between contexts that we seek to have. The function between contexts belongs to this operator set $\{=, \neq, \sim\}$. The algorithm returns all tuples whose valid context of these values in context. The disadvantage of these results is that sometimes the data is represented in a more generic and that the fact that data is not asked in the context when there are other data in a context that is more generic context requested by the query is near (the notion of nearness is processed by the notion of hierarchical similarity detailed in subsection 4.3).
$$\|q^{rc}\| = \bigcup_{c' \in Valid(\mathcal{O}, r, c)} (\|f_1(q, c')\|^{\text{CSPARQL}}) \times \delta_1(r, c, c') \text{ where:}$$

- $Valid(\mathcal{O}, r, c)$ is a function that determines the list of contexts similar to a given context c according to the function r based on an ontology $\mathcal{O}$.
- $\delta_1(r, c, c')$ relevance degree that allows to determine the relevance of the returned value according to the query context.(if $c = c'$ then $\delta_1 = 1$, if $c \neq c'$ then $\delta_1 = 1$, if $c \sim c'$ then $\delta_1 = \Re(c, c')$, $\Re$ is a function using similarity measure that compute relevance degree (see subsection 4.3)).
- f_1: takes as parameter a user query "q" and a context "c'" and returns all answers where values have the context "c'".

4.2 Flexible Query Execution Algorithm (FQE)

Our solution consist to return to the same tuple of values having a similar context to context requested by the petition and to evaluate these values by a degree of relevance calculated by reference to the context of the query and the ontology of context. For example, if the user requests a phone number and an address of a contact that can be valid in *France*, and we assume that the data of the user's contacts are valid in different contexts: taking the example of a contact where his phone number is given in *Europe* and the address is given in *Paris*. Using Algorithm 1, this tuple will not be returned by the query. Indeed, in algorithm 1 we consider two constraints *(i)* the contexts of values belonging to the same tuple must be the same *(ii)* the context of the tuple must verify the function $Valid(\mathcal{O}, r, c)$. In Algorithm 2, we remove the first constraint to allow obtaining more flexible tuples, which can be composed by several values in different contexts. The algorithm execution of this query is given as follows:
$$\|q^{rc}\|^{\mathcal{O}} = h_1(g_1(\|f_2(q)\|^{\text{CSPARQL}}, \mathcal{O}, r, c), \epsilon)$$

- f_2: function takes as parameter only user query and returns all answers satisfying query.

- g_1: is a function allowing to select answers resulting from f_1, according to context "c" and satisfying "r" function and compute relevance score for each answer.
- h_1: is the final selection where the function select answer that have a relevance degree greater than ϵ.

4.3   Relevance Degree Computation between Contexts

In case where there are no answers for the user query, because of the absence of user context in the personal data, the user query is rewritten into a set of queries where the user context c is replaced by all the contexts c' that are similar to. To compute this similarity between two contexts, we exploit the hierarchical structure of our context ontology. We are choose in this work to use Wu and Palmer measure [6], but we can also use other one. The similarity computation principle of this measure is: Given an ontology $\mathcal{O}$ composed of a set of nodes and a root node R. A and B represent two ontology elements of which the we aim to compute the similarity. The similarity score is computed as function of the distances: (i) $N1$ respectively $N2$ which represents the distance of A from their LCS (Least Common Subsumer), respectively, the distance of B from their LCS, and (ii) the distance N which computes the distance between the LCS of A and B and the hierarchy root R. The similarity measure of Wu and Palmer is computed by the following formula: $Sim_{WP}(A, B) = \frac{2 \times N}{N1+N2+2 \times N}$. The relevance degree $\delta(A, B)$ between two classes A and B is obtained as follows: $\delta(A, B) = 0 \ if \ \forall A \not\equiv B, \ 1 \ if \ \forall A \equiv B \ \vee \ (B \leq A) \vee \ (A \leq B) \ or \ Sim_{WP}(A, B)$

5   Experimentation

To evaluate the efficiency of our approach, we have performed our system on real datasets. Our dataset is composed of a set of user contacts which is described by name, address, birth date, phone number, ..., and each one of those properties is validate in different geographical contexts, the size of our test file is 22177 tuples. The aim of our first test is to present the usability of context and the importance of the different function that we have defined to execute queries (*Equal, Similar, Disjoint*). Figure 3(a) presents queries results, we have execute query in three cases: (i) without context $Q1$,(ii) context is "$= Lyon$" $Q2$ and (ii) context is "$= France$" $Q3$. We observe that $Q1$ returns a lot of tuples unlike in the case of $Q2$ and $Q3$, we observe that using context decrease the number of selected tuples because context is considered as filter applied to the query. The second test consists on comparing SQE and FQE algorithms. We have applied the two algorithm on queries with two different contexts and using equal and similar function between context. The first query returns about 12 answers and the second about 7000 answers. The number of answers in each case is shown in figure 3(b). In figure 3(b), we observe that when the context function is equality we found the same number of selected tuple using SQE and FQE and in the case of similarity context function, we obtain a large number of answers for FQE comparing tothe case of SQE algorithm. This is due to the fact that SQE

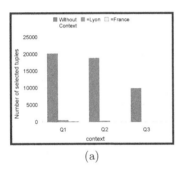
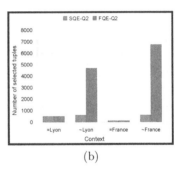

(a)   (b)

Fig. 3. (a) Results with and without context (b) SQE vs FQE: number of answers

algorithm require the same context for all values of the same answer, while for *FQE* we can have different context on the same answer. The conclusion of this experimentations is the use of one or other of the algorithm depends on the user needs and on the nature of its application.

6   Conclusion and Future Work

In this article, we have presented our personal information management system which is context-aware and ontology-based. Our meta model is defined to deal with several problems: (i) heterogeneity, (ii) redundancy and (iii) dispersion. It exploits ontologies to deal with semantic heterogeneity and to take benefits from the intensive work on knowledge representation and reasoning in this domain. The proposed contextual querying allows the system to give the user more relevant answers. Our model applies to multiple-user application, for example help the user to fulfill his form of an automatic and fast way using there stored data. In the future, we plan to continue the work over personal information querying, the automatic creation of e-governance process described in the form of workflows using contextual queries.

References

1. Baldauf, M., Dustdar, S., Rosenberg, F.: A survey on context-aware systems. International Journal of Ad Hoc and Ubiquitous Computing 2(4), 263–277 (2007)
2. Bush, V.: As we may think. The Atlantic Monthly 176(1), 101–108 (1945)
3. Dong, X., Halevy, A.: A platform for personal information management and integration. In: Proceedings of VLDB 2005 PhD Workshop, p. 26. Citeseer (2005)
4. Leo, S., Ludger, V.E., Andreas, D.: Pimo - a framework for representing personal information models. In: Proceedings of I-Semantics, pp. 270–277 (2007)
5. Martinenghi, D., Torlone, R.: Querying Context-Aware Databases. In: Andreasen, T., Yager, R.R., Bulskov, H., Christiansen, H., Larsen, H.L. (eds.) FQAS 2009. LNCS, vol. 5822, pp. 76–87. Springer, Heidelberg (2009)
6. Wu, Z., Palmer, M.: Verbs semantics and lexical selection. In: 32nd Annual Meeting of the Association for Computational Linguistics, pp. 133–138 (1994)

Ontology-Based Recommendation Algorithms for Personalized Education

Amir Bahmani[1], Sahra Sedigh[2], and Ali Hurson[1]

[1] Department of Computer Science,
[2] Department of Electrical and Computer Engineering
Missouri University of Science and Technology
Rolla, MO, USA
{abww4,sedighs,hurson}@mst.edu

Abstract. This paper presents recommendation algorithms that personalize course and curriculum content for individual students, within the broader scope of Pervasive Cyberinfrastructure for Personalizing Learning and Instructional Support (PERCEPOLIS). The context considered in making recommendations includes the academic background, interests, and computing environment of the student, as well as past recommendations made to students with similar profiles. Context provision, interpretation, and management are the services that facilitate consideration of this information. Context modeling is through a two-level hierarchy of generic and domain ontologies, respectively; reducing the reasoning search space. Imprecise query support increases the flexibility of the recommendation engine, by allowing interpretation of context provided in terms equivalent, but not necessarily identical to database access terms of the system. The relevance of the recommendations is increased by using both individual and collaborative filtering. Correct operation of the algorithms has been verified through prototyping.

1 Introduction

The growing ubiquity of mobile devices and the rapid growth of wireless networks have facilitated "anytime, anywhere" access to information repositories [11]. Moreover, the increasingly seamless integration of computing devices with daily life has brought about the "pervasive computing" paradigm - the key technology enabling this integration. By the same token, "pervasive learning," where transparent computing is leveraged to enhance education, has emerged as an application [10].

One of the critical functionalities of pervasive computing environments in general, and pervasive learning in particular, is service discovery. In pervasive environments, service discovery mechanisms must overcome intrinsic hardware, software, and networking heterogeneity [16] and support learners with a sufficient selection of services that meet their current requirements; i.e., context awareness. Applications in pervasive and ubiquitous computing utilize context to capture

S.W. Liddle et al. (Eds.): DEXA 2012, Part II, LNCS 7447, pp. 111–120, 2012.

differences and adapt the application's behaviors to the operating environment [15]. Context-aware computing is defined in Ref. [6] as the use of context in software applications that adapt their behavior based on the discovered contexts.

In our work, knowledge of the operating context (including the profile of the operator) is used to personalize courses and curricula for individual students, based on their academic profile, interests, and learning style. This paper describes the methodology that brings context-awareness to Pervasive Cyberinfrastructure for Personalized Learning and Instructional Support (PERCEPOLIS) to create an adaptive learning environment that facilitates resource sharing, collaboration, and personalized learning. PERCEPOLIS utilizes pervasive computing through the application of intelligent software agents to facilitate modular course development and offering, blended learning, and support of networked curricula [3].

Modularity, blended learning, and support of networked curricula are three essential attributes of PERCEPOLIS. Modularity increases the resolution of the curriculum and allows for finer-grained personalization of learning artifacts and associated data collection. Blended learning allows class time to be used for active learning, interactive problem-solving, and reflective instructional tasks, rather than for traditional lectures. In networked curricula, different courses form a cohesive and interconnected whole, and learning in one area reinforces and supports learning in others.

Fig. 1 depicts the personalization of a computer architecture course (CS 388) within the CS curriculum. Beginning from the top of the figure, modularity can be seen in successively decomposing the curriculum into courses, subject topics, and finally modules. The connections between these entities illustrate the networked curriculum. For each module, alternatives can be developed that differ from each other in level of difficulty, use of multimedia, appropriateness for out-of-class study (in blended learning), and other features. PERCEPOLIS recommends one of these alternatives, based on context that includes the student profile, his or her access environment, and the content of each module.

Pervasive learning overcomes the limitations of traditional passive, lecture-based classroom platform by allowing a student to peruse learning resources outside of class. Personalization strives to ensure that the learning resources recommended to each student are the most appropriate for him or her. To determine a personalized course trajectory for each student, PERCEPOLIS requires a complex recommender system that leverages computational intelligence to recommend learning artifacts and resources; such as books, hyperlinks, courses, and modules based on each student's profile and recommendations made to students with similar profiles [3].

The remainder of this paper is organized as follows. Section 2 provides a brief survey of related research as advanced in the literature. The major components of the proposed context-aware system are introduced in Sect. 3. The workflow and prototype of the proposed model are given in Sections 4 and 5, respectively. The paper closes with concluding remarks that includes avenues considered for extensions to this research.

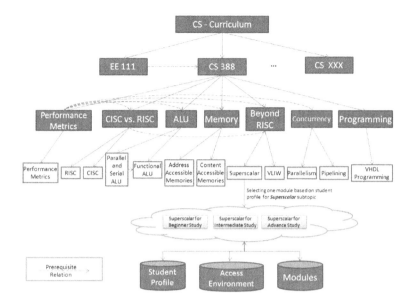

Fig. 1. Personalization of a computer architecture course (CS 388)

2   Related Literature

Context-awareness is a core attribute of PERCEPOLIS, and in large part responsible for the intelligence of the system. The underlying data structures play a significant role in the representation and exchange of context information. Key-value pair, markup-scheme, graphical, object-oriented, logic-based, and ontology-based models are data structures whose use for context representation is described and evaluated in [14]. Ontology-based models are cited as showing the greatest utility in pervasive computing, due their formal expressiveness and the fact that they facilitate reasoning [4]

Ontologies provide a controlled vocabulary of concepts, each with explicitly-defined and machine-processable semantics. By defining shared and common domain theories, ontologies allow humans and machines to communicate concisely, supporting exchange of not only syntax, but also semantics. Inherent to the use of ontologies is a trade-off between expressiveness and complexity of reasoning [5, 8]. Abundant literature exists on context-aware frameworks that utilize ontologies as the underlying context model [11], [8] and [2].

The C-CAST context management architecture was proposed in [11]. It supports mobile context-based services by decoupling provisioning and consumption. The decoupling is considered in the model to enhance scalability in terms of physical distribution and context reasoning complexity. The system is built based on three basic functional entities: the context consumer (CxC), context broker (CxB), and context provider (CxP).

A context-aware framework (CAF) developed for the HYDRA middleware project is described in [2]. The intent of HYDRA was to develop middleware to support an intelligent networked embedded system based on a service-oriented architecture. The CAF in HYDRA underpins the context-aware applications and services, while being domain-agnostic and adaptable. The CAF contains two core components: the data acquisition component and the context manager.

The novelty of our approach is threefold. Firstly, the suggested context-aware system captures and integrates capabilities of existing models. Secondly, it utilizes the summary schema model (SSM) to overcome heterogeneity and provide transparency. Thirdly, the proposed recommendation algorithm not only leverages both content-based and collaborative filtering techniques, but it can also consider a student's physical capability and adapt the content of recommended objects accordingly. The proposed system is discussed in detail in Sect. 3.

2.1   Intelligent Software Agents

A *software agent* is a computer program that acts autonomously on behalf of a person or organization. Agents can be particularly beneficial in pervasive learning environments, as they can assist in transparently managing information overload [13]. Leveraging pervasive computing and communications at various levels through the use of agent-based middleware is a defining feature of PERCEPO-LIS. A number of existing personalized learning systems, enumerated below and reported in [9] and [10], similarly employ multi-agent systems.

ISABEL [9] uses four types of agents: 1) device agents that monitor and profile each student's access device, 2) student agents that construct a complete profile of each student's interests, 3) tutor agents that interact with and identify similarities among a group of student agents characterized by a specific domain of interest, and 4) teacher agents that are associated with and manage the learning artifacts of an e-learning suite.

The pervasive learning infrastructure reported in Ref. [10] uses four types of agents. Location-aware learner agents are created for each learner logged in within a specific coverage area. Each learner agent uses the learner's preferences or previous behavior to populate the student model used by the infrastructure for storing and updating relevant information about learners. Connection agents are responsible for managing the connection between the mobile devices and the agent platform. Service agents are available for each service provided by the infrastructure. Finally, resource agents are responsible for managing resources.

PERCEPOLIS uses agents similarly. As articulated in Sect. 4, an agent is created for each instructor/advisor, course, and student; respectively. These agents communicate with each other to determine a recommendation.

3   Context-Awareness in PERCEPOLIS

In this section, we propose a context-aware system that introduces new features while maintaining the benefits of existing context-aware frameworks. As depicted

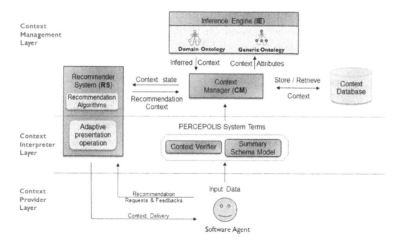

Fig. 2. Context-awareness in PERCEPOLIS

in Fig. 2, the proposed context-aware system includes three main layers: 1) the context provider layer, 2) the context interpreter layer, and 3) the context management layer.

Two types of contextual information - explicit and inferred - are captured in these layers and utilized in PERCEPOLIS. *Explicit contextual information* is provided directly by the learner or institution by completing surveys. This information can be classified into five categories:

1. Learner profile: Includes academic records (list of courses and modules passed, grades, GPA, target degree, major, etc.) and personal profile (location, disabilities, interests, needs and skills).
2. Course profile: Includes information such as department ID, mandatory topics elective topics, and default modules (identified as core content by the instructor of a course).
3. Topic profile: Describes and enumerates attributes of a topic.
4. Subtopic profile: Describes and enumerates attributes of a subtopic.
5. Module profile: Includes information such as prerequisites, contents (by topic and learning artifact), nominal level of difficulty (beginner, intermediate or advanced), and developer of module.
6. Instructor profile: Includes courses taught, skills, research interests, etc.

Inferred contextual information falls into one of two categories:

1. Learner tacit profile: Includes learning style; the learner's infrastructure (device, operating system, networking); access records; tacit skills, e.g., passing a certain module may enable a new skill; skill level; tacit interests, e.g., passing a certain module with a high grade may reflect the learner's interest in that topic.

2. Module tacit profile: Includes the implicit level of difficulty (inferred from grades) and the audience (based on frequency of use in specific courses or learners who have taken the module).

3.1 Context Provider Layer

Student and instructor agents in this layer are responsible for capturing contexts and forwarding the captured information to the layer above - the context interpreter layer. PERCEPOLIS software agents and their tasks are described in Sect. 4.

3.2 Context Interpreter Layer

Heterogeneity, which complicates transparency and scalability of computing systems, is an intrinsic characteristic of pervasive environments [16]. In PERCE-POLIS, two functions of the context interpreter layer address these challenges: 1) context verification and 2) adaptive presentation.

One of the main impediments to transparency is the difference between terminology used by developers and users, respectively. As an example, in describing a topic, the system or module developer may have used the term "*arithmetics*," but the user query may specify "*mathematics*." Unless translation takes place, a null response will be returned to the user - despite the existence of learning artifacts relevant to mathematics. In our system, the context verifier utilizes the *summary schema model* (SSM) for reconciling terminology. The SSM creates a taxonomy/ontology based on the linguistic meaning and relationships among terms.

3.3 Context Management Layer

The information resulting from context interpretation is forwarded to the context management layer, which houses the PERCEPOLIS ontology and recommendation algorithms. In the interest of scalability and efficiency, we utilize a two-tier scheme that includes generic and domain ontologies, respectively.

The filtering techniques used by recommender systems fall into one of two categories - *content-based* or *collaborative* filtering [15]. Content-based filtering techniques focus solely on identifying resources based on the profile of an individual or artifact. In contrast, collaborative filtering techniques take a peer group approach - recommendations are made based on similarities among individuals or artifacts. [3].

To prioritize and recommend the most appropriate learning artifacts; e.g., courses or topics; the recommender system must find associations among the learning artifacts and the preferences, skills, and background of a student. Ontologies can be used to represent and facilitate later identification of such relationships. We utilize the ACM Computing Classification System [1] as the common ontology in PERCEPOLIS. Fig. 3 includes a portion of the ACM CCS that decomposes hardware into six constituent topics.

To place increased emphasis on the relationships between modules and student preferences, we defined two parameters: *Relevancy Weight* (RW) and *Preference Weight* (PW). RW quantifies the relevance of a module to particular topic; PW quantifies the interest of a student in a particular topic. For instance, in Fig. 3, the RW between the module *"ALU for intermediate study"* and the topic *"Arithmetic and Logic Structures"* is 0.3. The PW between the same topic and the student in question is 0.6.

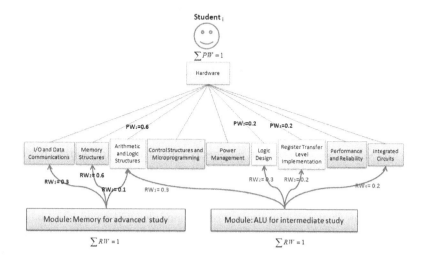

Fig. 3. Relevancy and preference weights

One task of context management is to create an individual preference tree (IPT) for each student. IPTs are created based on domain ontology and explicit and implicit contextual information. In Fig. 3, the arcs labeled with "PW" values and the nodes they connect represent the IPT of student$_i$.

Both content-based and collaborative filtering require quantification of the similarity of one or more entities. To this end, we utilize Pearson correlation, which quantifies the extent to which two variables linearly relate with each other [12], [7].

4   Workflow of the Recommendation Scheme

PERCEPOLIS recognizes three sets of entities as comprising the educational environment: i) the set of instructors/advisors, I; ii) the set of students, S; and iii) the set of courses, C. Each course $c \in C$ is a collection of interrelated mandatory and elective modules. Each of I, S, and C, respectively; is represented by a community of software agents that communicate and negotiate with each other to determine the best trajectory for each student through a course or curriculum.

More specifically, each student has an agent, S_A, which is responsible for acquiring from the context provider layer information related to a student's profile; e.g., student ID, name, department, and explicit interests/skills. S_A is subsequently responsible for sending this information to the context interpreter layer. With the help of the context verifier and SSM, the context interpreter layer interprets the input data, reconciling terminology when needed. The resulting context information is sent to the context management layer. As mentioned previously, the use of the generic ontology results minimizes the reasoning search space. For instance, if a student needs a home department course, instead of searching among all courses offered by the university, the search space is reduced to the student's department. The Inference Engine (IE) will check the provided context information against the generic ontology; finally all captured context information will be stored in Context Database.

S_A passes recommendation requests (made by a user through the user interface) to the recommender system (RS), which in turn identifies the appropriate artifacts. The personalization process depicted in Fig. 4 begins after the RS receives a recommendation request. The RS contacts the Context Manager (CM) to request the information considered in personalization, which includes IPTs, department rules, a list of courses, and course relevancy weights. IPTs are created by the CM based on student preferences and topic and module relevancy weights that are assigned by module developers.

Finally, after the student selects the desired subtopics, the RS identifies the most appropriate modules, based on student's background and interests and other context information. The content of the modules will be adapted based on the functionality and capability of the student's device using adaptive presentation . For instance, if a student's device is the Samsung SGH A107, then the IE will infer that the cell phone does not support video services, and hence will avoid recommendation of artifacts that include video.

5   PERCEPOLIS Prototype

As of publication of this paper, we have completed the first PERCEPOLIS prototype - including the proposed context-aware system. The prototype and profile databases have been implemented in Java SE 6 and MySQL 5.5.8, respectively. The prototype allows for operation in one of two modes: 1) *administrator*, and 2) *student*. Seven entities can be managed (searched/added/updated/deleted) in administrator mode: 1) words, 2) modules, 3) subtopics 4) topics, 5) courses, 6) SSM, and 7) ontology.

In student mode the system can carry out the following tasks: 1) Display the *Most Recent History*: where students can track the modules recommended by the system, as well as their own final selections; 2) *Update Profile*: where students can provide their interests as explicit context; 3) *Find Appropriate Courses*: where the system recommends the most appropriate courses based on students' academic profiles and departmental rules; and 4) Find appropriate modules: where, after students select their desired courses from among acceptable/recommended courses, the system helps students find the best trajectory of topics/subtopics/modules for each course and allows the students to select their desired subtopics from among acceptable, elective subtopics.

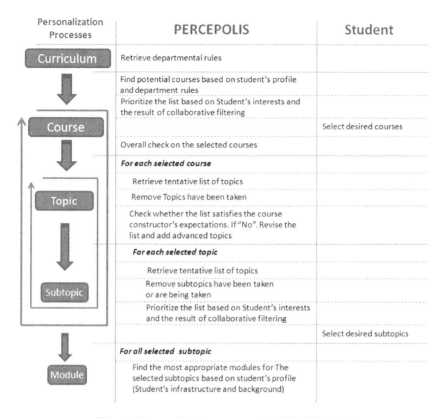

Fig. 4. Personalization process in PERCEPOLIS

6 Conclusion

In this paper, within the scope of PERCEPOLIS, we have presented a new layered context-aware system and its functionalities. The proposed model is composed of three layers: 1) the context provider layer, 2) the context interpreter layer, and 3) the context management layer. The novelty of our model is threefold. Firstly, the suggested context-aware system captures and integrates capabilities of existing models. Secondly, it utilizes SSM to overcome heterogeneity and provide transparency. Thirdly, the proposed recommendation algorithm leverages both content-based and collaborative filtering techniques and adapts the content of recommended artifacts based on a student's physical capability.

References

1. Association for Computing Machinery. The ACM Computing Classification System (1998 version), http://www.acm.org/about/class/1998 (retrieved June 2012)
2. Badii, A., Crouch, M., Lallah, C.: A context-awareness framework for intelligent networked embedded systems. In: Proceedings of the International Conference on Advances in Human-Oriented and Personalized Mechanisms, Technologies, and Services, pp. 105–110 (2010)

3. Bahmani, A., Sedigh, S., Hurson, A.R.: Context-aware recommendation algorithms for the percepolis personalized education platform. In: Proceedings of the Frontiers in Education Conference (FIE), pp. F4E–1–F4E–6 (2011)
4. Baldauf, M., Dustdar, S., Rosenberg, F.: A survey on context-aware systems. International Journal of Ad Hoc and Ubiquitous Computing 2, 263–277 (2007)
5. Bettini, C., Brdiczka, O., Henricksen, K., Indulska, J., Nicklas, D., Ranganathan, A., Riboni, D.: A survey of context modelling and reasoning techniques. Pervasive and Mobile Computing 6(2), 161–180 (2010)
6. Coppola, P., Della Mea, V., Gaspero, L., Lomuscio, R., Mischis, D., Mizzaro, S., Nazzi, E., Scagnetto, I., Vassena, L.: AI techniques in a context-aware ubiquitous environment. In: Hassanien, A.-E., Abawajy, J.H., Abraham, A., Hagras, H. (eds.) Pervasive Computing. Computer Communications and Networks, pp. 157–180. Springer, London (2010)
7. Dowdy, S., Wearden, S., Chilko, D.M.: Statistics for Research. John Wiley and Sons (2004)
8. Ejigu, D., Scuturici, M., Brunie, L.: An ontology-based approach to context modeling and reasoning in pervasive computing. In: Proceedings of the Fifth IEEE International Conference on Pervasive Computing and Communications Workshops, pp. 14–19 (2007)
9. Garruzzo, S., Rosaci, D., Sarne, G.: Isabel: A multi agent e-learning system that supports multiple devices. In: Proceedings of the 2007 IEEE/WIC/ACM International Conference on Intelligent Agent Technology, pp. 485–488 (2007)
10. Graf, S., MacCallum, K., Liu, T., Chang, M., Wen, D., Tan, Q., Dron, J., Lin, F., Chen, N., McGreal, R., Kinshuk, N.: An infrastructure for developing pervasive learning environments. In: Proceedings of the IEEE International Conference on Pervasive Computing and Communications, pp. 389–394 (2008)
11. Knappmeyer, M., Baker, N., Liaquat, S., Tönjes, R.: A Context Provisioning Framework to Support Pervasive and Ubiquitous Applications. In: Barnaghi, P., Moessner, K., Presser, M., Meissner, S. (eds.) EuroSSC 2009. LNCS, vol. 5741, pp. 93–106. Springer, Heidelberg (2009)
12. Resnick, P., Iacovou, N., Suchak, M., Bergstrom, P., Riedl, J.: Grouplens: an open architecture for collaborative filtering of netnews. In: Proceedings of the 1994 ACM Conference on Computer Supported Cooperative Work, pp. 175–186 (1994)
13. Sakthiyavathi, K., Palanivel, K.: A generic architecture for agent based e-learning system. In: Proceedings of the International Conference on Intelligent Agent Multi-Agent Systems, pp. 1–5 (2009)
14. Strang, T., Linnhoff-Popien, C.: A context modeling survey. In: Proceeings of the Workshop on Advanced Context Modelling, Reasoning and Management - The Sixth International Conference on Ubiquitous Computing (2004)
15. Xu, C., Cheung, S.C., Chan, W.K., Ye, C.: Partial constraint checking for context consistency in pervasive computing. ACM Transactions on Software Engineering and Methodology 19, 9:1–9:61 (2010)
16. Xu, K., Zhu, M., Zhang, D., Gu, T.: Context-aware content filtering & presentation for pervasive & mobile information systems. In: Proceedings of the 1st International Conference on Ambient Media and Systems, pp. 20:1–20:8 (2008)

Towards Quantitative Constraints Ranking in Data Clustering

Eya Ben Ahmed, Ahlem Nabli, and Faïez Gargouri

University of Sfax
eya.benahmed@gmail.com,
ahlem.nabli@fsegs.rnu.tn,
faiez.gargouri@isimsf.rnu.tn

Abstract. Expressing the expert expectations in form of boolean constraints during the data clustering process seems to be a promising issue to improve the quality of the generated clusters. However, in some real problems, an explosion in the volume of the processed data and their related constraints overwhelm the expert. In this paper, we aim to explicitly formulate the expert preferences on supervising the clustering mechanism through injecting their degree of interest on constraints using scoring functions. Therefore, we introduce our algorithm $\mathcal{SHAQAR}$ for quantitative ranking of constraints during the data clustering. An intensive experimental evaluation, carried out on OLAP query logs collected from a financial data warehouse, showed that $\mathcal{SHAQAR}$ outperforms the pioneer algorithms, $i.e.$ Klein et al's algorithm.

Keywords: semi-supervised clustering, quantitative ranking, numerical constraint, score function, data warehouse, OLAP query log.

1 Introduction

Data mining has been broadly addressed for the last decades. Particularly, the clustering technique is playing an elementary role in real-world application domains where massive amounts of unlabeled data exist, $e.g.$ unfiltered anomalies or unspecified handwriting or unclassified emails. Labeling unfiltered objects requires a qualified expertise in the application domain. For that reason, human experts opt to describe some objects using labels. Therefore, the data mining community is increasingly turning to semi-supervised learning, which considers both, labeled and unlabeled data.

In this context, we stress on semi-supervised clustering where several merging operations of objects are available and the expert may efficiently select which objects should or should not be merged through enforcing constraints in merging mechanism.

In this paper, we propose $\mathcal{SHAQAR}$ a semi-supervised hierarchical active clustering algorithm which prevents the use of the boolean constraints, $i.e.$ *cannot-link constraints* used to avoid objects with different labels from being merged together; and *must-link constraints* enforcing similarly labeled objects to be gathered in a common cluster.

S.W. Liddle et al. (Eds.): DEXA 2012, Part II, LNCS 7447, pp. 121–128, 2012.
© Springer-Verlag Berlin Heidelberg 2012

In several clustering processes, an explosion in the size of handled data and their related constraints overwhelmed the user. Hence, ranking constraints during the clustering process comes as a great way for soliciting user expectations.

In semi-supervised clustering, we distinguish two pools of approaches investigating the constraints ranking :

The qualitative approach expressing a relative formulation of constraints ranking, such as a user prefers merging x and y over x and z. Such a formulation is natural for humans [5];

The quantitative approach describing an absolute formulation of constraints ranking, for example a user selects merging x and y than x and z to a lesser degree of interest. Such a formulation allows both of total and partial ordering of constraints. Preferences in constraints are specified using scoring functions that associate a numeric score with every constraint.

In this paper, we develop the quantitative approach of constraints ranking during the clustering process. Characterized by its highlight accuracy, its leading features are:

- *Score-based constraints ranking*: the expert expresses the rank for any constraint by assigning a numeric score between 0 and 1, or rejecting it, or obviously affirming indifference. By default, indifference is considered;
- *Combination of ranked constraints*: we introduce original combined operators;
- *Independence of various constraints preferences* : the rank of any constraint may be altered without affecting any score of an unrelated constraint.

This paper is structured as follows: in the next section, we introduce our method for quantitative ranking of constraints in semi-supervised clustering. Section 3 reports experimental results of performance evaluation of our proposed method. Section 4 discusses related work in semi-supervised clustering. Finally, we draw the conclusion and present some future work in section 5.

2 $\mathcal{SHAQAR}$: A Novel Method For Quantitative Constraints Ranking in Data Clustering

2.1   Formal Quantitative Constraints Ranking Background

We detail, in this subsection, our key concepts that will be of use in the remainder.

Definition 1. Constraint cr
Let C_i and C_j be two clusters belonging to the set C of clusters. Such a relationship between C_i and C_j is called **constraint** *and noted as $cr=(C_i,C_j)$.*

Definition 2. Score-based Constraint cr
For any constraint cr_i, a constraint's rank is expressed using the **score function**, *denoted by score(cr_i). This score function is defined on the interval [0, 1].*

Definition 3. Quantitative Ranking $\mathcal{QR} = (CR, >_{\mathcal{QR}})$

Given a set CR of constraints, a ranking QR is a strict partial order $\mathcal{QR} = (CR, >_{\mathcal{QR}})$, where $>_{\mathcal{QR}} \subseteq dom(CR) \times dom(CR)$.

We can conclude the following properties of the relation $>_{\mathcal{QR}}$:

- **Irreflexivity**:$\forall cr_1; cr_1 \not>_{QR} cr_1$;
- **Asymmetry**: $\forall cr_1, cr_2; cr_1 >_{QR} cr_2 \Rightarrow cr_1 \not>_{QR} cr_2$;
- **Transitivity**: $\forall cr_1, cr_2, cr_3; (cr_1 >_{QR} cr_2) \wedge (cr_2 >_{QR} cr_3) \Rightarrow cr_1 >_{QR} cr_3$.

Both of atomic and combined ranked constraints are useful. In the following, we present the related settings.

Atomic Operators

We introduce two atomic operators, namely **Lowest** and **Highest** constraint.

Definition 4. Lowest$(>_{\mathcal{QR}})$

cr_k *is called the* **Lowest** *constraint, if* $\forall cr_i \in CR, score(cr_i) >_{QR} score(cr_k)$.

Definition 5. Highest$(>_{\mathcal{QR}})$

cr_k *is called the* **Highest** *constraint, if* $\forall cr_i \in CR, score(cr_k) >_{QR} score(cr_i)$.

Combined Operators

The combined operators are expressed using (i) *Pareto constraint* articulating an equality of constraints ranking; and (ii) *Prioritized constraint* indicating that given constraint is more important than another.

Definition 6. *Pareto constraint:* $cr_1 \otimes cr_2$

If cr_1 and cr_2 are equally important then $\mathcal{QR} = > cr_1 \otimes cr_2$.

Definition 7. *Prioritized constraint:* $cr_1 \ominus cr_2$

If cr_1 is more important than cr_2 then $\mathcal{QR} = > cr_1 \ominus cr_2$.

2.2 $\mathcal{SHAQAR}$ Method

Despite the accuracy of semi-supervised hierarchical algorithms, they show serious limitations when several constraints have to be enforced. Such a case is so widespread, particularly when the volume of data is massive.

To overcome this drawback, $\mathcal{SHAQAR}$ method is developed, a new hierarchical method within the expert investigation through numerical scoring of constraints. In this part of our paper, we present the notations used within $\mathcal{SHAQAR}$. After that, we focus on the processing steps of our algorithm.

Notations. The following notations will be used in the remainder:
- O : the set of objects;
- x_i : data point;
- $\mathcal{C}$: the set of generated clusters;
- C_i : the cluster;
- CR : the set of constraints;
- $>_{\mathcal{QR}}$: the set of quantitative ranked constraints;
- cr_i : the i constraint;
- α_i : the value of the score function associated to the cr_i constraint;
- cr_g : the most important constraint having the greatest function score.

$\mathcal{SHAQAR}$ **Algorithm.** The processing steps of our algorithm can be summarized as follows:

1. **Initialization**: The $\mathcal{SHAQAR}$ algorithm is initialized when each data point is associated to a separate cluster;

2. **Computing similarity matrix**: The similarity between clusters is computed using the **Jaccard distance** [14]. For two clusters C_i and C_j, this metric is defined by the following equation:

$$J(C_i, C_j) = \frac{|C_i \bigcap C_j|}{|C_i \bigcup C_j|} \qquad (1)$$

3. **Estimation of the best merge**: The expert dynamically provides a quantitative ranking of constraints related to eventual merging operations. The **Highest** operator is applied to determinate the most important constraint having the greatest score ;

4. **Merge**: The couple of the highly scored clusters is merged. The number of current clusters is decremented ;

5. **Update**: The similarity matrix is updated and the process is sequentially repeated until the number of generated clusters reaches the expert specified number.

Algorithm 1. $\mathcal{SHAQAR}$: Semi-supervised Hierarchical Active clustering based on QuAntitaive Ranking constraints

---

Data: O, $>\mathcal{QR}$
Result: $\mathcal{C}$

begin
    1. Initialize clusters :
    Each data point x_i is placed in its own cluster C_i;
    repeat
        2. Compute the similarity matrix ;
        3. Estimate the best merge:
        foreach $cr_i \in CR$ **do**
             score$(cr_i) = \alpha_i$;
        3.1. $cr_g \leftarrow$ *Highest*$(>\mathcal{QR})$;
        // Function *Highest* outputs the most important constraint ;
        3.2. $(C_i, C_j) \leftarrow cr_g$;
        4. Merge :
        the two closets clusters C_i and C_j from the current clusters to get cluster C;
        5. Update :
        5.1. Remove and from current clusters;
        5.2. Add cluster C to current clusters;
    until *Number of clusters converges* ;
    Return $\mathcal{C}$;
end

---

3   Experimental Evaluation

We have conducted experiments to evaluate the accuracy of our proposed method $\mathcal{SHAQAR}$, developed in Java, and compared it to the well known classifiers.

3.1   Framework

The architecture used for testing is an Intel Core 2 Duo 3 GHz with 2 GB of main memory. All tests fall within a real financial data warehouse created to assist the financial decision maker in his potential investments. The collected data is significantly analyzed using the OLAP technology. Such OLAP queries are so complex and may return massive amounts of data.

Therefore, a groupization of the various data warehousing analysts to draw their common preferences seems crucial [3]. Certainly, several axis of groupization are defined [4]: (i) the *function exerted*: we assume that analysts working in the same position have similar preferences, (ii) the *granted responsibilities* to accomplish defined goals: in fact two portfolio managers can not assume the same responsibilities, (iii) the *source of group identification*: it is explicit when an analyst specifies to which group he belongs otherwise this task may be implicitly performed, (iv) the *dynamic identification* of groups: the detected groups will be updated or will remain static.

The most discriminating criterion for groups identification is probably the historical analysis of these financial actors where the different analytical queries are stored. Indeed, in our experiments, the generated log files are gathered according to each criterion through applying our $\mathcal{SHAQAR}$ method. Besides, the similarity between these clusters is measured through the extension of the Jaccard metric in the multidimensional framework as defined by the following equation.

$$J(C_i, C_j) = \frac{C_{Queries(C_i, C_j)}}{\sum Queries(C_i) + \sum Queries(C_j) - C_{Queries(C_i, C_j)}}. \quad (2)$$

With $C_{Queries(C_i, C_j)}$: Number of common queries in two clusters (*i.e.* log files) C_i of analyst i and C_j of analyst j,
$\sum Queries(C_i)$: Sum of all existing queries in the cluster (*i.e.* log file) C_i.

3.2   Evaluation Criteria

To evaluate the quality of generated clusters, we compare our method to Induction Decision Tree (ID3) classification method [13] using Weka platform 3.6.5 edition [1]. Three main evaluation criteria are used during our carried out experiments: (i) The True Positive rate (TP) expressing the proportion of positive cases correctly classified; (ii) the False Positive rate (FP) measuring the proportion of negatives cases that were incorrectly identified as positive; (iii)The receiver operating characteristic (ROC) is the relationship between the rate of TP and FP [12].

3.3   Results and Discussion

It was worth the effort to experiment our $\mathcal{SHAQAR}$ algorithm and compare its behavior to Klein et *al.*'s method already used for data warehousing analysts

---

[1] http://www.cs.waikato.ac.nz/~ml/weka/

groupization [4]. Figure 1 shows the effect of TP and FP variation by comparing the ROC (*Receiver Operating Characteristic*) curve of $\mathcal{SHAQAR}$ illustrated by boxes *vs.* that of Klein et *al.* 's algorithm depicted by linepoints with respect to ID3 classification for the four identified groupization criteria: : the *function exerted*, the *granted responsibilities*, the *source of group identification* and the *dynamic identification* of groups. It is important to mention that $\mathcal{SHAQAR}$ outperforms the Klein et *al.*'s algorithm because only the most important constraints are accurately enforced during the clustering process. It is the same for the four studied criteria. Hence, it is worth noting that introducing quantitatively ranked constraints may improve the quality of generated clusters which will wholly satisfy the expert expectations. Such constraints ranking model suits the data warehouse analysts groupization issue.

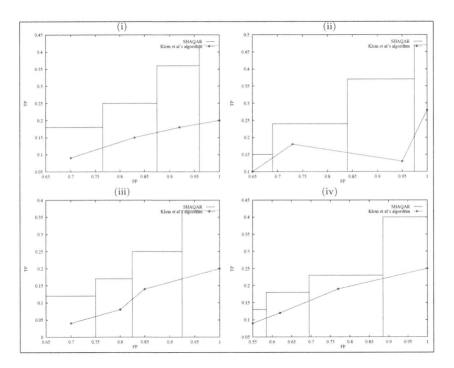

Fig. 1. $\mathcal{SHAQAR}$ performance *vs.* Klein et *al.*'s algorithm with respect to ID3 classification according to the (i) **function-based groupization**, (ii) **responsibilities-based groupization**, (iii) **Source-based of groupization**, (iv) **dynamicity-based groupization**.

4   Related Work

There is a large body of researches related to semi-supervised clustering. In this section, we focus on the most relevant approaches. Klein et *al.* [10] introduced a method for inducing spatial effects of pairwise constraints in an active learning scheme which radically reduces the number of mandatory constraints to achieve a specified accuracy. Both of (*Must-link* and *Cannot-link*) constraints are applied in clustering strategy. Basu et *al.* [2] proposed an original method for actively choosing informative pairwise constraints to get enhanced clustering performance. This strategy suits both of large datasets, and high dimensional data. Exclusively focused on the first level of hierarchical clustering algorithm, Kestler et *al.* [9] uses constraints with divisive hierarchical clustering. Such a proposed method is applied in DNA microarray analysis. Bade et *al.* [1] proposed a semi-supervised hierarchical clustering approach, coupled with a biased cluster extraction process. Such an integrated approach is used to personalize hierarchical structure from a document collection. In fact, the labeling is applied during the post-processing step. Bohm and Plant [6] introduced HISSCLU a hierarchical, density based method for semi-supervised clustering which prevents the use of explicit constraints. HISSCLU expands the clusters starting at all labeled objects at the same time. During the expansion, class labels are given to the unlabeled most coherent objects to the cluster structure. Davidson and Ravi [7] presented an agglomerative hierarchical clustering using constraints and demonstrated the enhancement in the clustering accuracy. Dubey et *al.* [8] proposed a cluster-level semi-supervision model for inter-active clustering where the user intervenes through updating cluster descriptions and assignment of data items to clusters. Nogueira et *al.* [11] proposed a new active semi-supervised hierarchical clustering method. It is founded on an innovative concept of merge confidence in clustering process. When there is lower confidence in a cluster merge, the user can be asked to provide a cluster-level constraint. Ben Ahmed et *al.* [5] introduced a qualitative ranking of constraints involved during the hierarchical clustering process.

However, such a qualitative approach is generic and does not express accurately the expert preferences. Therefore, we aim in this paper to explicitly formulate the expert preferences on supervising the clustering mechanism through injecting their degree of interest on constraints using scoring functions.

5   Conclusion

This paper introduces, $\mathcal{SHAQAR}$, a method for semi-supervised clustering within quantitative ranking of enforced constraints. To this end, we apply scoring functions to associate degree of interests to constraints. Compared to qualitative approaches, such numerical ranking of constraints is highly accurate and strictly reflects the expert expectations. Carried out experiments outlined that $\mathcal{SHAQAR}$ is highly competitive in terms of accuracy compared to popular semi-supervised methods.

Future work will include: (i) the investigation of new metric in order to evaluate the quality of derived clusters and perfectly assist the expert in this data clustering process and (ii) the learning of the scoring functions based on expert preferences.

References

1. Bade, K., Hermkes, M., Nürnberger, A.: User Oriented Hierarchical Information Organization and Retrieval. In: Kok, J.N., Koronacki, J., Lopez de Mantaras, R., Matwin, S., Mladenič, D., Skowron, A. (eds.) ECML 2007. LNCS (LNAI), vol. 4701, pp. 518–526. Springer, Heidelberg (2007)
2. Basu, S., Banerjee, A., Mooney, R.J.: Active semi-supervision for pairwise constrained clustering. In: Proceedings of the SIAM International Conference on Data Mining (SDM 2004), pp. 333–344 (2004)
3. Ben Ahmed, E., Nabli, A., Gargouri, F.: Building MultiView Analyst Profile From Multidimensional Query Logs: From Consensual to Conflicting Preferences. The International Journal of Computer Science Issues (IJCSI 2012) 9(1), 124–131 (2012)
4. Ben Ahmed, E., Nabli, A., Gargouri, F.: Performing Groupization in Data Warehouses: Which Discriminating Criterion to Select? In: Bouma, G., Ittoo, A., Métais, E., Wortmann, H. (eds.) NLDB 2012. LNCS, vol. 7337, pp. 234–240. Springer, Heidelberg (2012)
5. Ben Ahmed, E., Nabli, A., Gargouri, F.: $\mathcal{SHACUN}$: Semi-supervised Hierarchical Active Clustering Based on Ranking Constraints. In: Perner, P. (ed.) ICDM 2012. LNCS, vol. 7377, pp. 194–208. Springer, Heidelberg (2012)
6. Bohm, C., Plant, C.: Hissclu: A hierarchical density-based method for semi-supervised clustering. In: Proceedings of the International Conference on Extending Database Technology (EDBT 2008), New York, USA, pp. 440–451 (2008)
7. Davidson, I., Ravi, S.S.: Using instance-level constraints in agglomerative hierarchical clustering: theoretical and empirical results. Data Mining and Knowledge Discovery 18(2), 257–282 (2009)
8. Dubey, A., Bhattacharya, I., Godbole, S.: A Cluster-Level Semi-supervision Model for Interactive Clustering. In: Balcázar, J.L., Bonchi, F., Gionis, A., Sebag, M. (eds.) ECML PKDD 2010. LNCS, vol. 6321, pp. 409–424. Springer, Heidelberg (2010)
9. Kestler, H.A., Kraus, J.M., Palm, G., Schwenker, F.: On the Effects of Constraints in Semi-supervised Hierarchical Clustering. In: Schwenker, F., Marinai, S. (eds.) ANNPR 2006. LNCS (LNAI), vol. 4087, pp. 57–66. Springer, Heidelberg (2006)
10. Klein, D., Kamvar, S.D., Manning, C.D.: From instance-level constraints to space-level constraints: making the most of prior knowledge in data clustering. In: Proceedings of the International Conference on Machine Learning (ICML 2002), pp. 307–314. Springer, San Francisco (2002)
11. Nogueira, B.M., Jorge, A.M., Rezende, S.O.: Hierarchical confidence-based active clustering. In: Proceedings of the Symposium on Applied Computing, pp. 535–537 (2012)
12. Provost, F., Fawcett, T.: The case against accuracy estimation for comparing induction algorithm. In: Proceedings of the International Conference on Machine Learning, Madison, Wisconsin, USA, pp. 445–453 (1998)
13. Quinlan, J.R.: Induction of decision trees. Machine Learning, 81–106 (1986)
14. Tan, P.-N., Steinbach, M., Kumar, V.: Introduction to Data Mining. Addison-Wesley, Boston (2005)

A Topic-Oriented Analysis of Information Diffusion in a Blogosphere

Kyu-Hwang Kang[1], Seung-Hwan Lim[1], Sang-Wook Kim[1,*], Min-Hee Jang[1], and Byeong-Soo Jeong[2]

[1] Department of Computer Science and Engineering,
Hanyang University, South Korea
{kkhfiles,jesuslove,wook,zzmini}@agape.hanyang.ac.kr
[2] Department of Computer Science and Engineering,
Kyung Hee University, South Korea
jeong@khu.ac.kr

Abstract. A blogosphere is a representative online social network established through blog users and their relationships. Understanding information diffusion is very important in developing successful business strategies for a blogosphere. In this paper, we discuss how to predict information diffusion in a blogosphere. Documents diffused over a blogosphere deal with information on *different topics* in reality. However, previous studies of information diffusion did not consider the information topic in analysis, which leads to low accuracy in predictions. In this paper, we propose a *topic-oriented model* to accurately predict information diffusion in a blogosphere. We also define four primary factors associated with topic-oriented diffusion, and propose a method to assign a diffusion probability between blog users for *each topic* by using regression analysis based on these four factors. Finally, we show the effectiveness of the proposed model through a series of experiments.

Keywords: Blogosphere, information diffusion, topic-oriented analysis.

1 Introduction

On-line social networks and communities are becoming increasingly popular as places to exchange information and get to know other people. A *blog* is a personal web page where people can post their opinions and thoughts in a form of a document on-line [1][2][3][4][5]. A *blogosphere* is a prime example of an on-line social community in which users (blog users) can establish relationships with other blog users while exchanging their opinions and information. A blogosphere can be represented by a graph with nodes denoting blogs and edges the relationship between blogs. In this paper, we call it *a blog network*.

A blog user is often provided with a functionality that enables her to keep track of the blogs of her interest, which makes it easy and convenient for her to visit those blogs. Such functionality is called *bookmark*, *blogroll*, or *neighbor*.

---

[*] Corresponding author.

S.W. Liddle et al. (Eds.): DEXA 2012, Part II, LNCS 7447, pp. 129–140, 2012.

A blog user performs actions on a document in someone else's blog, such as *read, comment, trackback,* and *scrap.* Unique to blog sites in Korea, *scrap* is an action of copying someone else's document to one's own blog. The difference is that trackback creates a new document with a link to the original document, whereas scrap just copies someone else's document to one's own blog. Trackback and scrap actions can be viewed as a way of reproducing the original document. Such activities provide the framework for information diffusion in a blogosphere. In this paper, we discuss how to predict the degree and pathway of information diffusion in a blog network.

Information can be classified into several topics in the blogosphere, such as 'travel,' 'football,' 'food,' 'cars,' and so on. However, most studies which analyze information diffusion did not consider the topics of information. Such an analysis incurs a problem in correctly predicting future information diffusion. In this paper, we propose a topic-oriented model for analyzing information diffusion more accurately.

The authors of [6] proposed a method to compose a blog network while assigning a diffusion probability to each edge by considering previous information diffusion history. A simple way to consider topic-oriented information diffusion is to compose separate blog networks according to topics using the method described in [6] and subsequently analyze the information diffusion according to a target topic. We can correctly predict information diffusion to a degree even with such a simple model since we can assume that future information diffusion is more probable if past information diffusion history exists. We call this simple model for topic-oriented analysis the *Naïve approach.*

The Naïve approach assigns a diffusion probability to each edge only when information diffusion on a specific topic exists between users. Thus, we cannot predict information diffusion on a specific topic between two users if they have no prior diffusion history on that topic. However, in the real world, blog users may visit some blogs more frequently than others, which may affect information diffusion even though they may not propagate the documents of a specific topic. In this regard, the Naïve approach is not appropriate for predicting information diffusion.

In this paper, we propose a new analysis model to more accurately predict future information diffusion on a given topic. First, we define four factors that affect diffusion probability values obtained from analyzing the relationships among blog users. We also describe our information diffusion model based on diffusion probability, which is calculated based on these four factors using regression analysis. Our proposed model can be used to predict future information diffusion between blog users more accurately than existing methods since it assigns a diffusion probability between blog users even when previous propagation history does not exist. We show the superiority of our approach via extensive experiments.

This paper is organized as follows. In Section 2, we briefly describe previous work related to information diffusion in a blog network. We explain our approach to analyzing topic-oriented information diffusion in Section 3. We verify the effectiveness of our approach by analyzing real blog data in Section 4, and finally conclude the paper in Section 5.

2 Related Work

2.1 Analyzing Inforamtion Diffusion

Existing models for analyzing information diffusion in social networks include the *linear threshold model* [8], the *independent cascade model* [9], and the *general cascade model* [7]. These studies introduced the phenomenon that nodes in a blog network may influence one another, and therefore a node influenced by another node may have characteristics similar to those of the influencing node. They called this phenomenon *assimilation* and proposed some criteria for defining assimilation.

The *linear threshold model* [8] designates a threshold value for each node and assigns a weight to the relationship between nodes. When a specific node's accumulated influence received from surrounding nodes is greater than its threshold value, it is regarded as assimilated by those surrounding nodes. However, the linear threshold model is inappropriate to be applied to a blog network. This is because, while information diffusion occurs via *independent relationships* among blog users in the blogosphere, the linear threshold model calculates the total influence of a node by summing the weighted influence from its neighboring nodes.

The *independent cascade model* [9] assigns a probability value to the relationship between nodes, and assimilation is decided based on this value. We call this assimilation probability (or propagation probability) between nodes. The reasoning behind this model is that a blog user diffuses a certain document in the blogosphere not because of the influence from her/his neighbors, but because of the influence from a single blog user to whom the document belongs. Therefore, the independent cascade model is appropriate for explaining information diffusion in the blogosphere.

The *general cascade model* [7] generalizes the characteristics of the *linear threshold model* and the *independent cascade model*. Thus, it is appropriate for explaining the information diffusion phenomenon, which has the characteristics of both models. However, it is also not appropriate for explaining information diffusion occurring in the blogosphere.

2.2 Calculating Diffusion Probability

The linear threshold model, the independent cascade model, and the general cascade model assume that a threshold value for information diffusion or diffusion probability in the blogosphere is given. Thus, in order to apply the independent cascade model for informtion diffusion analysis in the blogosphere, diffusion probability values must be assigned between nodes (i.e., blog users). In this section, we introduce the Naïve approach for calculating the diffusion probability.

Given two blog users, A and B, the diffusion probability of A to B is calculated by counting the number of documents diffused from user A to user B. However, in order to prevent user A's ineligible documents from reducing the diffusion probability, we define *effective diffusion* as the degree to which blog users perform actions on each blog and calculate the diffusion probability according to this value.

Equation 1 shows how the degree of effective diffusion on each document is calculated. $score(d)$ represents the degree of diffusion on a document d, A_k implies one of the actions (read, scrap, trackback) that blog users can perform on a document, and also $freq(d, A_k)$ indicates the frequency with which other blog users perform A_k action on a document d. W_{A_k} is the weight value for A_k action. Thus, $score(d)$ is the sum of the frequency that other blog users perform each action on a document d multiplied by the weight corresponding to the action.

$$score(d) = \sum_{A_k \in Action} W_{A_k} * freq(d, A_k) \tag{1}$$

To determine the probability that blog user B propagates blog user A's document, $P_{A \to B}$, we use the ratio of the total of *effective diffusion* values of all blog user A's documents and the total of *effective diffusion* of the blog user A's documents propagated to blog user B. The *effective diffusion* value is then normalized as follow. We first calculate *effective diffusion* values for all the documents and assign 1 to the top $n\%$ of documents, and then the remaining documents' effective values are adjusted between 0 and 1 by dividing their *effective diffusion* values by the smallest effective diffusion value among the top $n\%$ of documents.

Equation 2 is used to calculate the probability value, $P_{A \to B}$, using the normalized *effective diffusion* value. $Norm\_score(d)$ is the normalized *effective diffusion* value of document d and $D_{A \to B}$ is a set of documents propagated from user A to user B. In this case, the normalized *effective diffusion* value of every document in $D_{A \to B}$ should be 1, regardless of the number of actions performed on that document because every document in $D_{A \to B}$ has already been propagated from user A to user B.

$$P_{A \to B} = \frac{|D_{A \to B}|}{\sum_{d \in D_A} Norm\_score(d)} \tag{2}$$

3   The Proposed Method

3.1   Terminology

Table 1 summarizes the terminology and symbols used in this paper. U_A represents individual user A, D_A is a set of documents (documents) owned by user A, $D_{A \to B}$ is a set of user A's documents that are propagated from D_A to user B, T_k is topic k belonging to a set of total topics T, and D_{A,T_k} is a set of documents related to topic T_k within D_A. N_{A,T_k} denotes the neighbors of U_A for topic T_k with a set of users who propagate D_{A,T_k}. $InfRatio_{A \to B,T_k}$ is the ratio of the degree of influence that U_A has on U_B in terms of topic T_k to the number of documents in D_{A,T_k} that have been propagated to U_B. Int_{B,T_k} is the interest, U_B has in T_k, expressed by the number of documents in D_B that satisfy T_k.

Table 1. Terminology and Symbols

term	description
U_A	Blog user A
D_A	A set of documents owned by user A
$D_{A\to B}$	A set of documents propagated from D_A to user B
T_k	Topic k
T	A set of topics $= \{T_1, T_2, \cdots\}$
D_{A,T_k}	A set of documents related to topic T_k within D_A
$D_{A\to B,T_k}$	A set of documents propagated from D_{A,T_k} to U_B
N_{A,T_k}	A set of users who propagate D_{A,T_k}
$InfRatio_{A\to B,T_k}$	Ratio of the influence U_A has on U_B in terms of topic T_k to the number of documents D_{A,T_k} propagated to U_B
Int_{B,T_k}	Degree of interest U_B has in T_k

3.2 Basic Idea

Our approach is two-fold: 1) establish the edges of a blog network, and 2) assign the information diffusion probability on each edge considering the document's topic.

First, a blog network is created according to the document's topic as in the Naïve approach. However, contrary to the Naïve approach, we establish an edge between blog users in case information diffusion occurs even though no propagation history exists between blog users on the topic in question. For example, if U_A owns documents dealing with T_k and U_B propagates D_A regardless of the document's topic (i.e., $|D_{A,T_k}| \geq 1$ and $|D_{A\to B,T}| \geq 1$), then U_B can potentially propagate D_{A,T_k} and establish edge $A\to B$ on a blog network of topic T_k. This is based on the fact that U_B can propagate documents of topic T_k from U_A if U_B has propagated any documents from U_A regardless of the topic. Thus, we eliminate the problem of overlooking any possibility of information diffusion when no previous propagation history is present.

Second, we assign a topic-oriented diffusion probability value on each edge. We identified the following four factors to affect the possibility that information diffusion will occur between blog users: 1) the degree of propagation for documents of topic T_k between U_A and U_B; 2) the degree of propagation for documents not related to topic T_k between U_A and U_B; 3) the degree of influence that U_A has on topic T_k over a blogosphere; and 4) the degree of interest that U_B has in topic T_k.

Since the possibility of information diffusion is affected by these factors, the relationship between the diffusion probability and factors can be represented as shown in Equation 3. In Equation 3, $P_{A\to B,T_k}$ is the probability that U_B propagates U_A's documents related to topic T_k and $arg1, arg2, arg3$, and $arg4$ are the corresponding factors identified above.

$$P_{A\to B,T_k} = f(arg1, arg2, arg3, arg4) \tag{3}$$

In order to assign diffusion probability according to topics on an edge, we need to determine function f in Equation 3. In our approach, we employ *regression analysis* to determine the function f.

3.3 Calculating Factor Values

Factor 1. The degree of propagation for documents of topic T_k between U_A and U_B can be thought of as the influence ratio that U_A can affect U_B for topic T_k (i.e., $InfRatio_{A \to B, T_k}$). We expect that, if this influence ratio is high, then U_B is likely to propagate documents in D_{A, T_k}.

To calculate $InfRatio_{A \to B, T_k}$ in Equation 4, we calculate the sum of the normalized degree of effective diffusion for all documents in D_{A, T_k}. This sum is considered as the full extent to which U_A is willing to propagate T_k. Next, to calculate the actual ratio of documents of T_k that will be actually propagated to U_B, we determine the sum of the normalized degree of effective diffusion for all documents in $D_{A \to B, T_k}$. Here, $Norm\_score(D_i)$ is the normalized degree of *effective diffusion* for D_i.

Via extensive preliminary experiments, we decided to use 1 for the weight of each action and the top 5% for the normalization constant in our experimental analysis.

$$InfRatio_{A \to B, T_k} = \frac{\sum_{D_i \in D_{A \to B, T_k}} Norm\_score(D_i)}{\sum_{D_j \in D_{A, T_k}} Norm\_score(D_j)} \quad (4)$$

Factor 2. The degree of propagation for the documents that are not related to topic T_k between U_A and U_B refers to the overall influence that U_A has on U_B. This is represented as $InfRatio_{A \to B, T - \{T_k\}}$ and is computed by using Equation 5. Here, $T - \{T_k\}$ is a set of all the topics just excluding T_k. We eliminate topic T_k here to avoid duplicating the meaning of Factor 1. If U_A strongly affects U_B with respect to other topics, then the propagation of D_{A, T_k} may happen after the time of analysis even though U_A does not affect U_B for topic T_k at the time of analysis. For this calculation, we use the normalized degree of effective diffusion for each document.

$$InfRatio_{A \to B, T - \{T_k\}} = \frac{\sum_{D_i \in D_{A \to B, T - \{T_k\}}} Norm\_score(D_i)}{\sum_{D_j \in D_{A, T - \{T_k\}}} Norm\_score(D_j)} \quad (5)$$

Factor 3. The degree of influence that U_A has on topic T_k over a blogosphere is the influence ratio that U_A propagates D_{A, T_k} to all blog users excluding U_B and is represented by $InfRatio_{A \to (U - B), T_k}$ and is calculated by using Equation 6. The reason for excluding U_A is to avoid duplicating Factors 1 and 2. $InfRatio_{A \to (U - B), T_k}$ represents the overall influence of U_A for a topic over a blogosphere. We can expect that U_A is very likely to propagate documents of topic T_k if U_A has a high degree of influence in the blogosphere. For this calculation, we use the normalized degree of effective diffusion for each document.

$$InfRatio_{A \to (U-B),T_k} = \frac{\sum_{D_i \in D_{A \to (U-B),T_k}} Norm\_score(D_i)}{\sum_{D_j \in D_{A,T_k}} Norm\_score(D_j)} \qquad (6)$$

Factor 4. The degree of interest that U_B has in topic T_k is represented by Int_{B,T_k}. If U_B has a higher degree of interest in topic T_k, then it is very likely that the documents of topic T_k are propagated from U_A to U_B. For this measure, we use $|D_{B,T_k}|$, the number of U_B's documents related with topic T_k

3.4   Calculating Topic-Oriented Information Diffusion Probability

In this section, we discuss how to calculate diffusion probability for each edge in order to predict topic-oriented information diffusion. We use Equation 3 for regression analysis [10] to calculate the topic-oriented information diffusion probability. To obtain the regression term f in Equation 3, we use diffusion probability $P_{A \to B,T_k}$ as a dependent variable and $arg1$, $arg2$, $arg3$, and $arg4$ as dependent variables.

To calculate diffusion probability for predicting topic-oriented information diffusion, we collect blog data while dividing the time period of the collection into interval 1 and interval 2. Blog data collected from interval 2 represent the history of information diffusion occurring after interval 1. We predict information diffusion in interval 2 by using blog data collected in interval 1. We first establish a blog network for interval 1 as explained in Section 3.2. In order to compute the diffusion probability in interval 2 for each blog user, we need the regression equation $f_{1 \to 2}$ that utilizes the factor values in interval 1 as explained in Section 3.3. However, in order to draw regression equation $f_{1 \to 2}$, we should know diffusion probability values in interval 2. For this, we use diffusion probability values obtained from the probability values by using the Naïve approach for the blog data in interval 2. By combining diffusion probability values with the factor values obtained from interval 1, we draw $f_{1 \to 2}$ by regression analysis. As a result, we can determine the topic-oriented diffusion probability for each edge in interval 2 with equation $f_{1 \to 2}$.

Now, we propose a method for predicting the *future* information diffusion probability using the above prediction strategy. For simplicity, we refer to any future time interval as interval 3. To predict information diffusion for interval 3, we first construct a blog network for interval 2 and calculate the diffusion probability on edges using the equation $f_{1 \to 2}$ as stated above. We use the equation $f_{1 \to 2}$ instead of $f_{2 \to 3}$ because the diffusion probability for interval 3 cannot be calculated by the Naïve approach since future blog data for interval 3 is not available.

4   Performance Evaluation

4.1   Experimental Environment

In the performance experimentation, we use the real-world blog data used in [11]. This data is a large set of blog documents collected from a Korean blog portal site over seven months and classified into three representative topics, i.e., 'travel,' 'English,' and 'cooking.' It is assumed that each document can have only one topic and each user be interested in several topics. Table 2 shows the number of blog users and documents on each topic.

Table 2. Number of documents and users in a topic

	travel	English	cooking
# of users	20,394	22,734	72,408
# of documents	38,229	52,257	451,532

We select four other methods of information diffusion analysis for performance comparison with our topic-oriented information diffusion (TID) analysis model. The ES (Effective Score) method analyzes information diffusion without considering topics, the Naïve approach employs the ES for information diffusion for every topic, CST1 (Constant 1%) gives 1% probability value equally on each edge, and CST5 (Constant 5%) gives 5% probability value equally on each edge. We predict information diffusion using the *independent cascade model* over a blog network constructed by each method and compare the predicted values with real information diffusion history obtained from real-world blog data.

For comparison metrics, we use *recall*, *precision*, and *F-measure* [12]. In a blog network, if we define A as a set of users to whom specific information is really diffused and B as a set of users predicted by a method to be evaluated, then *recall* and *precision* mean that $(A \cap B)/A$ and $(A \cap B)/B$, respectively. They are combined into a single metric as *F-measure*, i.e., the harmonic average of *recall* and *precision* as in Equation 7 [13][14].

$$F = \frac{2}{\frac{1}{P} + \frac{1}{R}} = \frac{2PR}{P + R} \tag{7}$$

In a real blogosphere, there is a considerable amount of information diffusion history. In this paper, we measure *recall*, *precision*, and *F-measure* for all information history and consider their average values for evaluating the accuracy in our performance evaluation.

4.2   Results and Analyses

In Experiment 1, we examine the accuracy of the information diffusion prediction with each method. In Experiment 2, we examine the accuracy of information diffusion prediction with each method for those user pairs that do not have any actual diffusion history until the time of analysis.

4.2.1   Information Diffusion Prediction

To examine the accuracy of the information diffusion prediction for each method, we divide real-world blog data into three time intervals of equal length and analyze for these three intervals. For our TID, we calculate the accuracy of the information diffusion prediction for interval 3 by substituting four factor values for interval 2 into regression equation $f_{1\rightarrow 2}$ generated by using blog data from intervals 1 and 2. For the other methods, we calculate the accuracy of the information diffusion prediction for interval 3 using blog data from intervals 1 and 2.

Figure 1 shows the average accuracy of the information diffusion prediction for the three topics listed in Table 2. In Figure 1, each method is listed on the x axis and the prediction accuracy is in the y axis. The experimental results show that our proposed method (TID) is more accurate in terms of *recall* and *F-measure* than the other methods, and the Naïve approach is most accurate with respect to *precision*. The reason why the Naïve approach shows high precision is that it has the possibility of information diffusion only when any information diffusion exists during the time of analysis. Thus, it composes a small-sized blog network only with those edges having high probability values for information diffusion.

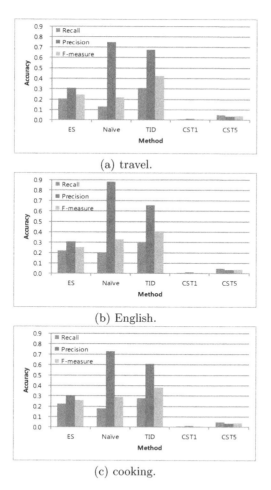

(a) travel.

(b) English.

(c) cooking.

Fig. 1. Accuracy of information diffusion prediction

4.2.2   Information Diffusion Prediction for User Pairs of No Previous Relationship

Unlike the Naïve approach, our TID can predict information diffusion for a specific topic through a current user relationship even when information diffusion has not occurred before the time of analysis. Figure 2 shows the average prediction accuracy of each approach for each topic. TID provides the highest *precision*, *recall* and *F-measure* among the five methods and the Naïve approach is unable to predict information diffusion as explained previously.

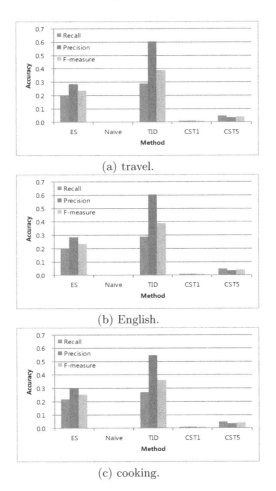

(a) travel.

(b) English.

(c) cooking.

Fig. 2. Accuracy of information diffusion prediction for user pairs of no previous relationship

5   Conclusions

The documents diffused over a blogosphere are classified into a number of topics. However, previous studies analyzing information diffusion in a blogosphere did not take their topics into account. Thus, the existing methods suffer from low accuracy in information diffusion prediction.

In this paper, we proposed the TID analysis model to predict information diffusion over a blogosphere while avoiding the problems described above. The proposed approach assigns edges not only when previous diffusion history exists between blog users, but also when information diffusion is expected even though previous diffusion history does not exist between blog users. We identified four factors that affect diffusion probability values (1) the degree of information diffusion for a given topic, (2) the degree of information diffusion for topics excluding a topic of interest, (3) the degree of influence that an information provider can have, and (4) the degree of interest that an information consumer can have. We also proposed a way to quantify diffusion probability on each edge using these four factors and regression analysis.

We used real-world blog data to evaluate the effectiveness of the proposed approach (TID). The experimental results indicate that the proposed approach significantly improves accuracy compared to existing approaches in predicting the future information diffusion.

Acknowledgement. This research was supported by MKE (The Ministry of Knowledge Economy), Korea and Microsoft Research, under the IT/SW Creative research program supervised by NIPA (National IT Industry Promotion Agency) (C1810-1102-0002) and also by MKE (Ministry of Knowledge Economy), Korea, under the Convergence-ITRC (Convergence Information Technology Research Center) support program (NIPA-2012-H0401-12-1001) supervised by NIPA (National IT Industry Promotion Agency).

References

1. Blogger.com Co., Ltd., http://blogger.com
2. MySpace Co., Ltd., http://www.myspace.com
3. NHN Co., Ltd., http://blog.naver.com
4. Daum Co., Ltd., http://blog.daum.com
5. SK Communications Co., Ltd., http://www.cyworld.com
6. Lim, S., Kim, S., Park, S., Lee, J.: Determining Content Power Users in a Blog Network: An Approach and Its Applications. IEEE Transactions on Systems, Man, and Cybernetics PART A 41(5), 853–862 (2011)
7. Kempe, D., Kleinberg, J., Tardos, É.: Maximizing the Spread of Influence Through a Social Network. In: Proc. ACM Int'l. Conf. on Knowledge Discovery and Data Mining, ACM SIGKDD, pp. 137–146 (2003)
8. Granovetter, M.: Threshold Models of Collective Behavior. American Journal of Sociology, AJS 86(6), 1420–1443 (1978)
9. Ellison, G.: Learning, Local Interaction, and Coordination. Econometrica: Journal of the Econometric Society 61(5), 1047–1071 (1993)
10. Chiang, C.L.: Statistical Methods of Analysis, p. 274. World Scientific (2003) ISBN 9812383107

11. Yoon, S., Shin, J., Park, S., Kim, S.: Extraction of a Latent Blog Community Based on Subject. In: Proc. ACM Int'l Conf. on Information and Knowledge Management, pp. 1529–1532 (2009)
12. Baeza-Yates, R., Ribeiro-Neto, B.: Modern Information Retrieval, pp. 75–84. ACM Press (1999)
13. van Rjisbergen, C.J.: Information Retrieval. Butterworth-Heinemann (1979)
14. Li, X., Wang, Y.-Y., Acero, A.: Learning query intent from regularized click graphs. In: Proc. ACM Int'l. Conf. on Research and Development in Information Retrieval, ACM SIGIR, pp. 344–345 (2008)

Trip Tweets Search by Considering Spatio-temporal Continuity of User Behavior

Keisuke Hasegawa, Qiang Ma, and Masatoshi Yoshikawa

Kyoto University, Kyoto, Japan 606-8501
{k.hasegawa@db.soc.,qiang@,yoshikawa@}i.kyoto-u.ac.jp

Abstract. A large amount of tweets about user experiences such as trips appear on Twitter. These tweets are fragmented information and not easy to share with other people as a whole experience. In this paper, we propose a novel method to find and organize such fragmented tweets at the level of user experiences. The notable feature of our method is that we find and organize tweets related to a certain trip experience by considering the spatio-temporal continuity of user-behavior of traveling. First, we construct a co-occurrence dictionary by considering the spatio-temporal continuity; i.e., the co-occurrence ratio of two terms is varying in time scopes and regions. Then, we use such dictionary to calculate the relatedness of a tweet to the trip experience from three aspects: content relatedness, temporal relatedness, and context relatedness. Tweets with high relatedness scores will be returned as search results. The experimental results showed our method performs better than conventional keyword-based methods.

Keywords: Microblog, Twitter Search, User experience, Spatio-temporal Continuity.

1 Introduction

Consumer Generated Media (CGM), such as microblogs and SNS, through which user publish information have recently be rapidly popularized on the Internet. Although a large amount of information about user experience is accumulated in CGM, developing the technology to organize and utilize the stored information has not been progressing sufficiently quickly.

Twitter[1] is a typical microblogging service that has two particular features. The first one is tweets on Twitter have length limitations, i.e., a tweet should be less than 140 characters. The second one is that users can post information on their experiences freely in real time. A typical example experience on which many users post tweets in real time is traveling, where they tweet their impressions of visiting tourist spots with photographs taken then and there.

However, the technology of sharing, organizing, and searching content posted on Twitter is in the process of being developed. Therefore, if users try to look

---

[1] http://twitter.com

S.W. Liddle et al. (Eds.): DEXA 2012, Part II, LNCS 7447, pp. 141–155, 2012.

back on past trips, for example, it is difficult for them to access old tweets because the only way of accessing these is to trace back tweets from the latest tweet. Moreover, sharing content on experience such as traveling is difficult because users post many tweets per day. In addition, information on the same experience is described in many tweets in most cases, and many tweets do not contain keywords that express the target experience due to the length limitations Twitter has. Therefore, even if one wants to find tweets about a traveling experience, the conventional method of keyword-based searches is not easy to find all target tweets without omissions.

Furthermore, there have been some cases where users who have postsed daily occurrences to blogs and SNS diaries up until now have come to post information mainly to Twitter. Such users often do not repost what they have tweeted to the blogs and SNS again. These circumstances make it difficult to share information on user experiences.

As one of the solutions, we propose a novel method of organizing tweets that represent traveling experiences to help users to share and organize these experiences. To a given trip experience (a series of spot names; as the query), we calculate the relatedness of a tweet to that trip experience from three aspects as follows and return the high scored tweets as the results.

– Content relatedness
  We use the degree of co-occurrence of location names and keywords in a tweet to consider content relatedness. It is based on the idea that the closer the word is related to a place, the higher the relatedness between the word and its place name. The degree of co-occurrence is calculated by using a co-occurrence dictionary constructed by taking into consideration the spatio-temporal continuity.
– Temporal relatedness
  We use the time interval of publish time between candidate tweets and the tweet containing given location names to calculate temporal relatedness by taking into consideration temporal continuity. It is based on the idea that the publish time of the tweets and the target trip experience are very close to each other.
– Context relatedness
  This is a criterion to take into consideration the influence of surrounding tweets. It is based on the idea that the tweets before and after a tweet related to a trip experience may have high probabilities of describing the same trip experience. Finally, the relatedness score is determined by integrating the influence of the surrounding tweets and the results of combining content relatedness and temporal relatedness.

2   Related Work

There has been much research on Twitter in recent years. Fujisaka et al. [1] proposed methods of estimating the regions of influence of social events found by

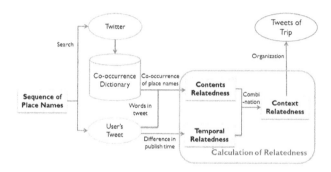

Fig. 1. Flow of organization tweets with our method

geo-tagging tweets. Wu et al. [2] investigated how information is propagated from celebrities and bloggers to ordinary users. Castillo et al. [3] proposed automatic methods of assessing credibility based on features extracted from users' posting behaviors of given sets of tweets.

Research on the mining of experiences has also been conducted. Kurashima et al. [4] proposed a method of extracting rules between five attributes of a person's experience, i.e., time, location, activity, opinion, and emotion from a large-scale set of blog entries. In addition, research on organizing a person's experience by using lifelogs has been carried out [5].

Arimitsu et al. [6] proposed a method of searching user experiences. They defined user experiences as several actions that are carried out in a particular sequence. However, we consider well both the temporal and regional continuity of user experiences.

3   Overview

We represent a trip experience as a sequence of place names. For example, when a user travels to Kyoto, s/he first visits Yasaka-jinja, next s/he visits Kodai-ji, and then s/he goes to Kiyomizu-dera. Tweets containing place names such as "Yasaka-jinja" and "Kiyomizu-dera" have high probability of being reporting that trip experience. This trip experience can be represented as a sequence of place name, i.e., "Yasaka-jinja, Kodai-ji, Kiyomizu-dera". Tweets representing trip experiences expressed in a sequence of place names are explained in this paper.

Fig.1 summarizes our process flow for organizing tweets. The input query is the sequence of place names that represent the target trip experience, and the output is the sequence of tweets describing the experience.

An input sequence of place names does not necessarily contain all the place names that represent the target trip experience. For example, even if users only input the place name "Kiyomizu-dera" as a sequence of place names, the system can find their tweets about "Yasaka-jinja" and "Kodai-ji" where they visited by taking into consideration spatio-temporal continuity. That is to say, one of the notable features of our method is that we can automatically complement the representation of trip experience and discovery information on it.

The first step is to search for candidate tweets. We generate a keyword based "OR" query by using the place names. Then, we use the search function provided by Twitter to collect the candidate tweets.

Then, the relatedness between the candidate tweets and the target trip experience is calculated. As we mentioned before, we calculate the degree of relatedness with these criteria: content, temporal and context relatedness. We organize tweets by regarding tweets whose relatedness score is higher than a threshold value as the tweet that represents the target trip experience.

4   Co-occurrence Dictionary

To compute the content relatedness, we construct a co-occurrence dictionary of region name and keywords appearing in tweets. It is based on the idea that the higher the relatedness between words in the tweet and location names representing the trip experience, the closer the tweet is related to the experience.

For example, if the word "Omi-kuji" (written oracle) appears in a tweet that represents the experience cast lots in Yasaka-jinja but the word "Yasaka-jinja" does not appear in the tweet, the tweet about the written oracle is not contained in the search results even if a search whose query is "Yasaka-jinja" with the conventional keyword-based search is carried out. However, by giving a high relatedness score to the tweet that contains "Omi-kuji" by taking into consideration the high relatedness between the word "Omi-kuji" and the place "Yasaka-jinja", the tweet about the experience cast lots in Yasaka-jinja is contained in the search results of Yasaka-jinja.

The co-occurrence dictionary is constructed with our method by tweets. We collected all tweets open to the public to collect many words highly related to the place names in costructing the co-occurrence dictionary.

There are two types of place names that are the targets for calculating the co-occurrence of words, i.e.,

1. A place name included in the query sequence of place names
2. The name of a location around the place in the query sequence of place names

For example, we calculate the co-occurrence of the names of locations around three places such as "Gion, Ninen-zaka, and Maruyama-kouen" in addition to the three place names in the sequence when calculating the degree of co-occurrence between words and the sequence of place names "Yasaka-jinja, Kodai-ji, and Kiyomizu-dera".

We search tweets that include at least one place name of these target place names with Twitter API[2] and we store up the tweets of all the place names gathered from the search results. The text of the tweets accumulated for all

---

[2] http://dev.twitter.com/

place names is conducted gathered, and words that appear in tweets and the degree of co-occurrence between place names and words are computed. Pairs of a word and the degree of co-occurrence are recorded for each place name in the co-occurrence dictionary.

Actually, the names of locations around places contained in the query sequence of place names should be retrieved automatically with Google Maps API[3]. However, currently, a few names of locations around places contained in the query sequence of place names were manually input into the system.

4.1 Calculation of Degree of Co-occurrence

$Co(p_k, w)$ is the degree of co-occurrence between place name p_k and target word w, and it is calculated by using the Jaccard index.

$$Co(p_k, w) = \frac{T_{p_k \cap w}}{T_{p_k \cup w}} = \frac{T_{p_k \cap w}}{T_{p_k} + T_w - T_{p_k \cap w}} \tag{1}$$

$T_{p_k \cap w}$ is the number of tweets that contain both p_k and w, $T_{p_k \cup w}$ is the number of tweets that contain either p_k or w, and T_{p_k} and T_w are the number of tweets that contain p_k and w.

4.2 Similarity of Dictionaries

Next, we will describe how to merge two co-occurrence dictionaries based on the spatio-temporal continuity.

Vector $\boldsymbol{d_{p_k}}$ expresses co-occurrence dictionary of place p_k, which records the degree of co-occurrence $Co(p_k, w_l)$ between words $w_1, w_2, ..., w_n$ and place name p_k.

$$\boldsymbol{d_{p_k}} = (Co(p_k, w_1), Co(p_k, w_2), \cdots, Co(p_k, w_n)) \tag{2}$$

The degree of similarity in two co-occurrence dictionaries d_i and d_j is defined based on cosine similarity.

$$S(\boldsymbol{d_i}, \boldsymbol{d_j}) = \frac{\boldsymbol{d_i} \cdot \boldsymbol{d_j}}{|\boldsymbol{d_i}||\boldsymbol{d_j}|} \tag{3}$$

The higher the degree of similarity $S(d_i, d_j)$, the more similar the content of the two co-occurrence dictionaries. If the content of the two co-occurrence dictionaries is temporally and spatially continuous, the content of the two co-occurrence dictionaries is considered to be similar. That is, if the degree of co-occurrence in two co-occurrence dictionaries very near or close in space or time is higher than a threshold value, these two co-occurrence dictionaries are temporally and spatially continuous, so we can organize tweets by using the new co-occurrence dictionary that is integrated from the two dictionaries.

---

[3] http://code.google.com/intl/ja/apis/maps/

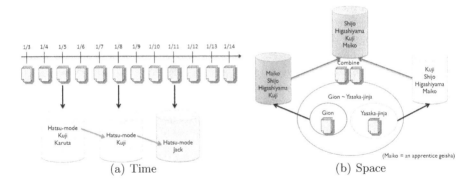

(a) Time   (b) Space

Fig. 2. Spatio-temporal continuity in co-occurrence dictionary

Merge Based on Temporal Continuity. Experiences at tourist spots vary with the seasons, such as when seasonal events occur and well-known products are offered unique to a season. Consequently, as we change the period of tweets used in constructing a co-occurrence dictionary to reflect the changes in the circumstances at tourist spots through time, more accurate organization of tweets is possible. Fig. 2(a) outlines temporal continuity in a co-occurrence dictionary.

For a certain place p_i, we constructed co-occurrence dictionaries in a certain time period, such as one day, one week, one month, and so on. Thus, we have a sequential dictionaries, $(d_{p_i}^{u_1}, \cdots, d_{p_i}^{u_m})$. $u_1, \cdots, u_m$ are certain time periods, such as day $1, \cdots$, day m, or week $1, \cdots$, week m. Then, we merge the adjacent dictionaries with high similarities incrementally. The similarity of two dictionaries is computed as follows.

$$St(d_{p_i}^{u_j}, d_{p_i}^{u_k}) = int(u_j, u_k) \times S(d_{p_i}^{u_j}, d_{p_i}^{u_k}) \qquad (4)$$

where $int(u_j, u_k) = (k - j) \times \delta$ is a function of time distance and δ is the unit of time period in which we construct the initial dictionaries.

If two dictionaries $d_{p_i}^{u_j}$ and $d_{p_i}^{u_j+1}$ have higher similarity than the threshold specified in advance, we merge them to construct a new one $d_{p_i}^{u_j'}$. The co-occurrence ratio of the new dictionary is defined as follows.

$$TCo(p_i, w) = \frac{1}{2} \left(Co^{u_j}(p_i, w) + Co^{u_k}(p_i, w) \right) \qquad (5)$$

Merge Based on Spatial Continuity. After the merge based on temporal continuity, we have a series of dictionaries, $(d_{p_1}, \cdots, d_{p_n})$. Then we merge them based on the spatial continuity. The basic idea is that if two regional dictionary have high similarity, these two place may be closed to each other and they should be merged to one bigger region. Fig. 2(b) outlines the combination of co-occurrence dictionaries while considering spatial continuity.

At first, we compute the similarity of two dictionaries of place p_i and p_j as follows.

$$Ss(\boldsymbol{d_{p_i}}, \boldsymbol{d_{p_j}}) = dist(p_i, p_j) \times S(\boldsymbol{d_{p_i}}, \boldsymbol{d_{p_j}}) \tag{6}$$

Where, $dist(p_i, p_j)$ is a distance function based on exponential degradation model, which has been employed by many researchers, such as Toda et al.[7] and Cui et al.[8].

$$dist(p_i, p_j) = e^{-\mu_d d(p_i, p_j)} \tag{7}$$

Where, $d(p_i, p_j)$ is the distance between p_i and p_j, and μ_d is a parameter for determining the rate of degradation in the similarity of content while increasing the distance between places.

If two dictionaries have higher similarity than the pre-specified threshold, we merge them to create a new dictionary $\boldsymbol{d_{p_{i,j}}}$. The co-occurrence ratio of the new dictionary is defined as follows.

$$Co(p_i, w) = \frac{1}{2} \left(Co(p_i, w) + dist(p_i, p_j) \times Co(p_j, w) \right) \tag{8}$$

$$Co(p_j, w) = \frac{1}{2} \left(Co(p_j, w) + dist(p_i, p_j) \times Co(p_i, w) \right)$$

5   Calculating Relatedness Score

To pick out the tweets related to a certain trip experience $e = (p_1, \cdots, p_n)$, which is represented by a sequence of place names $(p_1, \cdots, p_n)$, as mentioned before, we compute three relatedness scores: content relatedness, temporal relatedness, and context relatedness.

5.1   Content Relatedness

We use the co-occurrence dictionary to compute the content relatedness of the candidate tweet and a certain trip experience. It is based on the idea that the closer the word is related to a place, the higher the relatedness between the word and its place name.

The content relatedness of tweet t_i of trip experience e is computed as follows.

$$Rc(t_i) = \frac{1}{mn} \sum_{j=1}^{n} \sum_{k=1}^{m} Co(p_j, w_k) \tag{9}$$

where $w_1, \cdots, w_m$ are the keywords appearing in tweet t_i.

5.2   Temporal Relatedness

We use the time interval of a tweet that contains place names mainly to take into consideration temporal continuity. We calculate the interval of publish times between every candidate tweet that represents the target trip experience and tweets that contain the name of locations in the query sequence of place names and neighboring locations.

The publish time of tweets means a trip experience is close because one visits several places individually while moving on foot or by bus in a trip experience. For example, if there is a tweet that contains the place name "Kiyomizu-dera" in tweets that represent the trip experience, the following and preceding tweets whose publish time is close to the tweet probably describe Kiyomizu-dera.

We then calculate the relatedness score while considering temporal continuity by applying the exponential degradation model to the interval of publish times between the candidate tweet and the tweet that contains a place name. This is based on the idea that the shorter the interval of publish times between the candidate tweet and the tweet that contains a place name, the closer the tweet is related to that place.

If there are more than one tweets describing the same place, the basing publish-time used for computing the time interval is that of the tweet mostly related to that place. Currently, we choose such that tweet manually. The relatedness of a tweet and a place will be discussed in our future work.

Suppose that $q_1, q_2, ..., q_m$ are the names around locations of the places in the sequence of place names, $time(t_i)$ is the publish time of candidate tweet t_i, $time(t_j)$ is the publish time of basing tweet $t_j (j = 1, ..., N)$ that contains the name of place q_l near place p_k in the sequence, and there are N basing tweets that contain the place names preceding and following candidate tweet t_i. The temporal relatedness score, $Rt(t_i)$, of candidate tweet t_i computed by the time interval of the tweet with place names preceding and following tweet t_i is calculated as follows.

$$Rt(t_i) = \sum_{j=1}^{N} dist(p_k, q_l) \times e^{-\mu_t |time(t_i) - time(t_j)|} \qquad (10)$$

The temporal relatedness score in (10) is calculated with both spatial and temporal continuity by using the relatedness score computed with the time interval and weight $dist(k, l)$ by taking into consideration the spatial continuity in (7), where μ_t is a parameter for determining the rate of degradation of the similarity of content with increasing time intervals.

5.3   Context Relatedness

The simplest way to combine content and temporal relatedness is by multiplying them. Other combination methods will be discussed in our future work. $Rtc(t_i)$, which is the relatedness between tweet t_i and trip experience e without considering the influence of tweets preceding and following tweet t_i, is calculated as follows.

$$Rtc(t_i) = Rt(t_i)Rc(t_i) \qquad (11)$$

There are some tweets in those representing the target trip experience that contain a word with a very low degree of co-occurrence and tweets without a word co-occur with place names. As such tweets can be recognized as being related to the experience by the content of surrounding tweets, we need to consider the influence of the relatedness score of surrounding tweets when calculating the relatedness score. τ_i is the weighting representing how much we need to consider the influence of surrounding tweets, when calculating the relatedness score of tweet t_i by considering the influence of preceding and following X tweets, where $R(t_i)$ is the context relatedness between tweet t_i and the target trip experience, which is calculated by (12).

$$R(t_i) = \sum_{x=-X}^{X} \tau_{i+x} Rtc(t_{i+x}) \tag{12}$$

We have only considered the influence of tweets preceding and next to tweet t_i in this paper. $R(t_i)$, which is conclusive relatedness, is calculated with (13), which is $X = 1$ of (12).

$$R(t_i) = \tau_{i-1} Rtc(t_{i-1}) + \tau_i Rtc(t_i) + \tau_{i+1} Rtc(t_{i+1}) \tag{13}$$

We organize tweets representing the target experience by putting together tweets whose relatedness score is higher than a threshold value.

6   Experiments

We carried out experiments validating the spatio-temporal continuity of co-occurrence dictionaries and evaluated the efficiency of our method to search for tweets on a certain trip experience.

We used names of tourist spots in Kyoto to collect the test data.

6.1   Experiment on Spatio-temporal Continuity of Co-occurrence Dictionary

Temporal Continuity. We investigated the content of co-occurrence dictionaries constructed with tweets containing "Yasaka-jinja" tweeted from January 3 to 16 to validate how temporal continuity was reflected in co-occurrence dictionaries. We constructed 14 dictionaries from a set of tweets for each day by dividing tweets for the period by a day to investigate daily changes of content of co-occurrence dictionaries.

There were words whose degree of co-occurrence was always high in 14 co-occurrence dictionaries, and words whose degree of co-occurrence changed greatly on a daily basis. Examples of words whose degree of co-occurrence was always high were names around places such as "Higasiyama" and "Gion" and words concerning Yasaka-jinja such as "Omairi" (visit shrines) and "Sanpai" (worship). These words were always closely related to Yasaka-jinja, and this demonstrates that words related to Yasaka-jinja were contained in co-occurrence dictionaries constructed from tweets on Twitter.

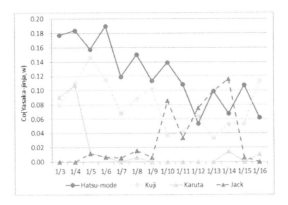

Fig. 3. Change in degree of co-occurrence Co(Yasaka-jinja, w)

However, we found that words whose degree of co-occurrence changed greatly day by day reflected events that occurred in the period. Fig. 3 expresses the transition in the degree of co-occurrence of words whose degree of co-occurrence changed greatly on a daily basis. As "hatsu-mode" (the practice of visiting a shrine or temple at the beginning of the new year) and "kuji" (lottery) were words closely related to hatsu-mode, the degree of co-occurrence from January 3 to 9 when many people visited Yasaka-jinja for hatsu-mode was high; after that, the degree of co-occurrence decreased.

Moreover, the degree of co-occurrence of "Karuta" (Japanese playing cards) from January 3 to 4 and "Jack" from January 10 to 14 suddenly rose. Karuta-hajime-shiki[4] (Ceremony for the first Karuta play of the new year), one of the annual events of Yasaka-jinja was held on January 3, and Kimono Jack[5], an event of occupying a certain place with many people wearing kimonos was also held. As many tweets on these events were tweeted, the degree of co-occurrence suddenly rose. It shows that words whose degree of co-occurrence changes suddenly reflect the occurrence of events for a short time.

Spatial Continuity. We investigated the correlation between the distance of several tourist spots and the similarity of co-occurrence dictionaries of several tourist spots to validate how spatial continuity was reflected in co-occurrence dictionaries.

We constructed co-occurrence dictionaries for 12 tourist spots in Kyoto, i.e., Yasaka-jinja, Kiyomizu-dera, Kodai-ji, Sannen-zaka, Chion-in, Ginkaku-ji, Sanju-sangendo, Jinushi-jinja, Otowa-no-taki, Nanzen-ji, Heian-jingu, Kitano-tenmangu, by using tweets from February 11 to 19, 2012, and we computed the similarity of co-occurrence dictionaries and the distance between two places at the 12 tourist spots based on Yasaka-jinja and Kiyomizu-dera. Distance was computed as the direct distance between two places, and similarity was computed with (3).

---

[4] http://web.kyoto-inet.or.jp/org/yasaka/english/event/index.html
[5] http://www.kimonojack.com/

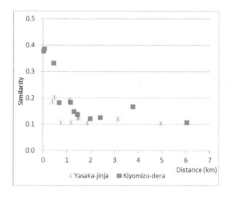

Fig. 4. Correlation between similarity and distance

Table 1. Correlation coefficient between similarity and distance

Place	Correlation Coefficient
Yasaka-jinja	-0.5384
Kiyomizu-dera	-0.6605

Fig.4 expresses the relationship of distance and similarity between tourist spots and Yasaka-jinja or Kiyomizu-dera. The similarity between the co-occurrence dictionary for Yasaka-jinja and that for Kodai-ji or Chion-in, which are spots near Yasaka-jinja is high, and the similarity between the co-occurrence dictionary for Kiyomizu-dera and that for Sannen-zaka or Jinushi-jinja, which are spots near Kiyomizu-dera is high. However, the similarity between the co-occurrence dictionary for Yasaka-jinja, that for Kiyomizu-dera, and that for Kitano-tenmangu, which are spots far from Yasaka-jinja and Kiyomizu-dera is low.

In addition, Table 1 summarizes the correlation between the distance and the similarity of co-occurrence dictionaries for Yasaka-jinja and Kiyomizu-dera. There is a negative correlation between similarity and distance for both Yasaka-jinja and Kiyomizu-dera. Therefore, the closer the distance between two places, the higher the similarity of co-occurrence dictionaries for the places. This means that the spatial continuity is reflected in the content of co-occurrence dictionaries.

6.2 Evaluation of Efficiency of Experience Search

We carried out experiments of evaluating the efficiency with which tweets were organized with our method.

Experimental Method. We organized the trip experience represented by a sequence of place names for "Yasaka-jinja, Kiyomizu-dera" in this experiment. We established Gion as a place around Yasaka-jinja, and Kodai-ji as a place around Kiyomizu-dera. We constructed co-occurrence dictionaries for Yasaka-jinja and Kiyomizu-dera by taking into consideration spatio-temporal continuity, and we calculated $Co(\text{Yasaka-jinja}, w)$ and $Co(\text{Kiyomizu-dera}, w)$, which are discussed in Section 4.2, by using the dictionaries.

We used MeCab as the Japanese morphological tool to extract keywords, place names from tweets. The $d(\text{Yasaka-jinja}, \text{Gion})$ in (7) was the linear distance between Yasaka-jinja and Gion computed by Google Maps API in this experiment.

We similarly combined the Kodai-ji and Kiyomizu-dera co-occurrence dictionaries. In addition, the period of tweets that we used to construct the co-occurrence dictionaries fit the day when the target trip experience occurred.

We organized tweets representing the one-day trip experiences A, B, and C of users U_a, U_b, and U_c that were manually chosen from users who posted both tweets containing Yasaka-jinja and those containing Kiyomizu-dera. We collected the tweets of users U_a, U_b, and U_c by using Twitter API, and we distinguished whether each tweet represented the target trip experience or not to specify the relevant results. The whole experience of visiting Kyoto on that day is called the target experience.

User U_a in experience A tweeted on tourist spots in Kyoto on the previous day, but these tweets were not the target tweets. User U_b in experience B, on the other hand, visited Osaka on his way home from Kyoto, and tweets on this were the target tweets because visiting Osaka was also contained in the one-day trip.

The closest tweet related to a place were chosen as the dominate ones for that place name. In this experiment, $\mu_d = \mu_t = 1$, μ_d and μ_t were the parameters for calculating the relatedness score in (7), (10), selecting the proper value will be discussed in our future work.

We used the recall ratio, the precision ratio, and the F-measure to evaluate our method. We computed scores of $Rc(t_i)$, $Rt(t_i)$, $Rtc(t_i)$, and $R(t_i)$ of tweets for each user experience (A, B, and C). We considered tweets whose scores were higher than the threshold value as the results representing the target experience. Based on this, we calculated the recall ratio, the precision ratio, and the F-measure of all results while changing the threshold value. We used the keyword-based OR search as the baseline.

Efficiency of Our Method. We compared the efficiency of trip experience search on the Twitter by using $R(t_i)$ with the baseline. We set the parameters to $\tau_i = 0.5$ and $\tau_{i-1} = \tau_{i+1} = 0.25$ in (13) to calculate $R(t_i)$ in this experiment. The discussion of variation of τ_i is another future work.

We compared the change in the F-measure of our method with the baseline by changing the threshold value of $R(t_i)$. Fig. 5 represents the results of comparison. In trip experiences A and C, the F-measure with our method is higher than that of the baseline, when the threshold value of $R(t_i)$ is lower than about 0.07. In trip experience B, the F-measure of our method is higher than that of the baseline when the threshold value of $R(t_i)$ is lower than about 0.09.

Table 2 summarizes the threshold value of $R(t_i)$, the precision ratio, the recall ratio, and the F-measure when the F-measure is maximum. Our method improved the F-measure by an average of five times that of the keyword-based method. Especially, our method could improve the recall ratio by an average of ten times than the keyword-based method while keeping degradation in the precision ratio to about 10% to 30%. The results revealed that our method greatly improved the efficiency of trip experience search.

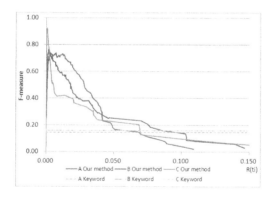

Fig. 5. Change in F-measurewith our method and baseline

Table 2. Precision and recall ratios with maximum F-measure of $R(t_i)$

Experience	$R(t_i)$	Precision	Recall	F-measure
A Our method	0.0039	0.6618	0.9091	0.7660
A Keyword	—	0.7273	0.0808	0.1455
B Our method	0.0031	0.7067	0.7681	0.7361
B Keyword	—	1.0000	0.0870	0.1600
C Our method	0.0080	0.9111	0.9318	0.9213
C Keyword	—	0.6667	0.0909	0.1600

Context Relatedness. We have proposed two combination ways of content and temporal relatedness: simply multiplying them, $Rtc(t_i)$, and considering context relatedness, $R(t_i)$. The results are shown in Fig. 6.

In the comparison considering whether or not the surrounding tweets had any influence, the maximum F-measure in all three experiences improved by an average of about 10%. This is because tweets with very low context relatedness, despite being related to the target experience. For example, there were tweets that only contained the URL of pictures taken at tourist spots, and there were tweets that contained collapsed informal expressions, from which the exact results of morphological analysis could not be collected; such tweets were not included in the results of organization by only using the relatedness of them because the content relatedness of such tweets was very low.

The result indicated that considering the influence of surrounding tweets by calculating context relatedness improved the efficiency with which tweets were searched.

Method of Constructing Co-occurrence Dictionaries. We compared the change in the efficiencies of tweets were searched by using co-occurrence dictionaries with and without considering spatio-temporal continuity.

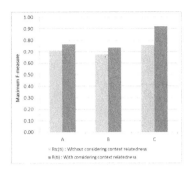
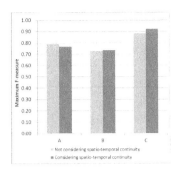

Fig. 6. Difference in the maximum F-measure by considering context relatedness

Fig. 7. Difference in maximum F-measure by using spatio-temporal continuity of dictionaries

We used co-occurrence dictionaries constructed by considering spatio-temporal continuity and combining those for Yasaka-jinja and Gion, and those for Kiyomizu-dera and Kodai-ji from tweets posted on the day when each target experience occurred, and we used co-occurrence dictionaries created constructed by Yasaka-jinja and Kiyomizu-dera without combining them with dictionaries for around places without considering spatio-temporal continuity from tweets posted from January 2 to 17, 2012. We constructed three co-occurrence dictionaries while considering the spatio-temporal continuity for the three experiences, and one co-occurrence dictionary without considering the spatio-temporal continuity for any of the three experiences.

Fig. 7 shows the change in the maximum F-measure. In comparison with trip experience search by using co-occurrence dictionaries without considering spatio-temporal continuity, the method using co-occurrence dictionaries while considering spatio-temporal continuity improved the maximum F-measure by about 4% for experience C, changed the maximum F-measure slightly for experience B, and decreased the maximum F-measure by about 3% for experience A.

The effectivity of spatio-temporal continuity based co-occurrence dictionary is not clear. It is necessary to carry out more experiments to investigate how considering it would affect the relatedness in reality, and this remains as future work.

We can organize tweets more accurately if the efficiency of organization is improved by improving the method of considering the spatio-temporal continuity in constructing co-occurrence dictionaries. One way of improving the method is by normalizing the relatedness score in combining co-occurrence dictionaries by taking into consideration the difference in the number of tweets containing place names. A few hundred tweets on famous spots, such as Kiyomizu-dera or Gion, were collected per day in this experiment, but only several dozens of tweets on not very famous spots, such as Kodai-ji, were collected per day. As such differences in the number of tweets at individual places were not considered in this research, we think that content relatedness can be calculated with higher accuracy by weighting by taking into consideration the number of tweets.

7   Conclusion

We proposed a novel method of organizing fragmented tweets related on a certain traveling experience. The most notable feature of our method is that we took into consideration the spatio-temporal continuity of trip experiences. We organized tweets by calculating the content relatedness, temporal relatedness, and context relatedness between the target trip experience and the tweets. In addition, we proposed a novel method of constructing co-occurrence dictionaries to calculate content relatedness by taking into consideration spatio-temporal continuity.

The results from experiments that we used to evaluate our method revealed that spatio-temporal continuity is reflected in co-occurrence dictionaries constructed with our approach, and the organization of tweets with our method is better than that with the keyword-based one. Moreover, taking into consideration the influence of surrounding tweets improved organization even further.

Future work is to more fully evaluate the influence of taking into consideration the spatio-temporal continuity in constructing co-occurrence dictionaries, general improvements to our method, and more detailed organization of trip experiences.

References

1. Fujisaka, T., Lee, R., Sumiya, K.: Discovery of user behavior patterns from geo-tagged micro-blogs. In: ICUIMC, pp. 246–255 (2010)
2. Wu, S., Hofman, J.M., Mason, W.A., Watts, D.J.: Who says what to whom on twitter. In: WWW, pp. 705–714 (2011)
3. Castillo, C., Mendoza, M., Poblete, B.: Information credibility on twitter. In: WWW, pp. 675–684 (2011)
4. Kurashima, T., Fujimura, K., Okuda, H.: Discovering Association Rules on Experiences from Large-Scale Blog Entries. In: Boughanem, M., Berrut, C., Mothe, J., Soule-Dupuy, C. (eds.) ECIR 2009. LNCS, vol. 5478, pp. 546–553. Springer, Heidelberg (2009)
5. Ushiama, T., Watanabe, T.: An Automatic Indexing Approach for Private Photo Searching Based on E-mail Archive. In: Gabrys, B., Howlett, R.J., Jain, L.C. (eds.) KES 2006, Part II. LNCS (LNAI), vol. 4252, pp. 1111–1118. Springer, Heidelberg (2006)
6. Arimitsu, J., Ma, Q., Masatoshi, Y.: A User Experience-oriented Microblog Retrieval Method. In: DEIM (2011) (in Japanese)
7. Toda, H., Kitagawa, H., Fujimura, K., Kataoka, R.: Topic structure mining using temporal co-occurrence. In: ICUIMC, pp. 236–241 (2008)
8. Cui, C., Kitagawa, H.: Topic activation analysis for document streams based on document arrival rate and relevance. In: SAC, pp. 1089–1095 (2005)

Incremental Cosine Computations for Search and Exploration of Tag Spaces

Raymond Vermaas, Damir Vandic, and Flavius Frasincar

Erasmus University Rotterdam
P.O. Box 1738, NL-3000 DR, Rotterdam, the Netherlands
info@raymondvermaas.nl, {vandic,frasincar}@ese.eur.nl

Abstract. Tags are often used to describe user-generated content on the Web. However, the available Web applications are not incrementally dealing with new tag information, which negatively influences their scalability. Since the cosine similarity between tags represented as co-occurrence vectors is an important aspect of these frameworks, we propose two approaches for an incremental computation of cosine similarities. The first approach recalculates the cosine similarity for new tag pairs and existing tag pairs of which the co-occurrences has changed. The second approach computes the cosine similarity between two tags by reusing, if available, the previous cosine similarity between these tags. Both approaches compute the same cosine values that would have been obtained when a complete recalculation of the cosine similarities is performed. The performed experiments show that our proposed approaches are between 1.2 and 23 times faster than a complete recalculation, depending on the number of co-occurrence changes and new tags.

1 Introduction

User-based content is becoming increasingly available on the Web. This content is often annotated using tags and then uploaded on social sites, like the photo sharing service Flickr. Because users can choose any tag they like, there is a large amount of unstructured tag data available on the Web. The unstructured nature of these tags makes it hard to find content using current search methods, which are based on lexical matching. For example, if a user searches for "Apple", (s)he could be looking for the fruit or for the company that makes the iPod.

There are several approaches available that aim to solve the previously identified problem [2, 4, 9–11]. In this paper, we focus on the Semantic Tag Clustering Search (STCS) framework [4, 9, 11]. The STCS framework utilizes two types of clustering techniques that allow for easier search and exploration of tag spaces. First, syntactic clustering is performed by using a graph clustering algorithm that employs the Levenstein distance measure in order to compute the dissimilarity between tags. As result of syntactic clustering, e.g., terms like "waterfal", "waterfall", and "waterfalls" are clustered. This means that when a user searches for one of these terms, all the terms that are syntactically associated will show up in the results. Second, semantic tag clustering is performed, where the aim is

S.W. Liddle et al. (Eds.): DEXA 2012, Part II, LNCS 7447, pp. 156–167, 2012.
© Springer-Verlag Berlin Heidelberg 2012

to cluster tags that are semantically related to each other, e.g., "tree", "plant", and "bush". The STCS semantic clustering algorithm makes use of the cosine similarity measure, applied on tag co-occurrence vectors. Due to the similarities between the STCS framework and the other clustering approaches that rely on the cosine similarity, the results presented in this paper can be easily applied on these approaches as well.

An important issue with the STCS framework is that it cannot be applied in an incremental way. This negatively influences the scalability of the approach. In other words, if new pictures are added, all steps in the framework have to be executed again, which can take a significant amount of time when handling large amounts of data. Another reason for updating the cosines incrementally is that when a small number of new tags are added, most of the new calculations will yield the same result as the previous computations. This is a natural consequence of the fact that the tag co-occurrence matrix remains the same for a large number of tags.

In this paper, we investigate how the cosine similarity computation can be done incrementally, for the purpose of scaling semantic clustering algorithms in tag spaces. We consider two approaches for incrementally computing the cosine similarity. The first approach only recalculates the cosine similarities of tag pairs that are affected by a change in the tag co-occurrence matrix. The second approach computes the cosine similarities by reusing the previously computed cosines. The second approach refines the first approach as only the cosine similarities of the tags affected by the changes in the tag co-occurrence matrix are computed. For the evaluation of the proposed approaches, we use a reference approach that computes all cosine similarities in a given data set. The evaluation is based on a simulated process in a photo sharing Web application (e.g., Flickr.com), where the system receives a set of new pictures that need to be processed. These new pictures can be annotated using existing (known) tags but also with new (unknown) tags. For the evaluation, we measure the execution time of each approach and compare that to the approach that performs a complete recalculation of all cosines. We use a data set from the photo sharing site Flickr.com to perform the evaluation. The data set has been collected by Li [8] and contains 3.5 million pictures and 580,000 tags, uploaded in the period 2005-2008. We use a subset of 50,000 pictures for the initial data set and between 2,500 and 75,000 pictures as the sets of new pictures that are pushed through the hypothetical photo sharing Web application. The reason for using a subset of the complete data set of 3.5 million pictures is that the computation of all cosines for the baseline takes too long for large data sets.

2   Related Work

There are a small number of approaches available in the literature that address the incremental (exact) computation of cosine similarities. One such approach is proposed by Jung and Kim [7]. The authors develop several methods for the incremental computation of similarity measures, including the cosine similarity, in the context of merging clusters. The aim is to merge two clusters without calculating all similarities between the new cluster and all the other clusters. To achieve this, the geometrical properties of the cosine similarity are used to calculate the similarity of the new merged cluster using the previously computed similarities. The results of this approach showed the usefulness of this solution with respect to achieving a significant speed-up and also a good accuracy. Unfortunately, we cannot use this approach for our purposes as we consider the incremental computation of cosines irrespective of the clustering method that is being used.

Another approach to incremental cosine computation is proposed by Friedman et al [5]. The authors develop three techniques for incremental cosines: crisp, fuzzy, and local fuzzy. In the crisp cosine computation, the cosines are calculated using only the dot-product. Every time a new vector is added, the cosine similarities with all cluster centroids are calculated. In the fuzzy cosine clustering technique, a degree of cluster membership is assigned to each new vector. This process takes the size of the new vector into account. The last technique that is proposed by the authors is the local fuzzy-based cosine, which is a modified version of the fuzzy cosine computation. This technique takes the difference between small vectors and a large centroid into account when assigning multiple degrees of membership. We do not consider this approach as it is focused on clustering scalability issues and not on the incremental computation of cosine similarities in general.

Literature in the field of incremental clustering shows more approaches that might be related to our work. The evolutionary clustering technique proposed in [3] considers a trade-off between low history cost (similarity with a previous clustering iteration) and high quality of each iteration. This is achieved by considering the similarity between the old clusters and new clusters (history cost), and the cosine similarity between the newly added elements into account. The authors only tried their technique on k-means clustering and agglomerative clustering, but suggest that their approach should also work for other clustering techniques. We do not consider this approach as it also does not distinguish between the cluster-based computations and the cosine similarity computations.

Locality sensitive hashing (LSH), as proposed by Giones et al [6], can also be used to perform incremental cosine computations. The idea behind LSH is that elements that are close to each other in a high dimensional space have a high probability of having a similar hash. The process of locality sensitive hashing consists two parts: a pre-processing part and a querying part. In the pre-processing part, the hashes are calculated for all the elements in the data set. Next, all the hashes are transformed to their bitwise representation and the similarity between the hashes is measured using the Hamming distance. Subsequently, all the elements with a similar hash are clustered in buckets using the k-nearest neighbour method. In the querying part, the hash for the query is calculated and a bucket containing similar elements is returned. LSH has proven to have sub-linear query time and a small error compared to other approaches that aim to solve similarity problems. We do not consider LSH in our work as it is an approximate technique and we aim to develop an exact technique for the incremental computation of cosines.

3   Incremental Cosine Computation Design

In this section, we describe the different approaches that we propose for the incremental computation of cosine similarities. First, we discuss the details of the cosine similarity and the operations that should be taken into account when calculating the cosine similarity incrementally. Then, we describe two approaches for incremental cosine computations, the recalculation approach and the delta cosine approach.

3.1   Cosine Similarity

The cosine similarity measure is used to measure the similarity between two vectors. This done by measuring the cosine of the angle between the two vectors. The cosine similarity ranges between 0 and 1 if the values in the vectors are positive. A cosine similarity of 1 represents complete similarity between the two vectors (i.e., the vectors point in the same direction) and a cosine of 0 represents complete dissimilarity (i.e., the vectors point in orthogonal directions). The cosine similarity is defined as:

$$\cos(\mathbf{a}, \mathbf{b}) = \frac{\mathbf{a} \cdot \mathbf{b}}{||\mathbf{a}|| \times ||\mathbf{b}||} \tag{1}$$

where $||\mathbf{a}||$ and $||\mathbf{b}||$ are the Euclidean norm of vectors $\mathbf{a}$ and $\mathbf{b}$, respectively, and $\mathbf{a} \cdot \mathbf{b}$ is the dot product between the vectors $\mathbf{a}$ and $\mathbf{b}$. The cosine similarity is calculated by dividing the dot product between the two vectors by the product of the Euclidean norm for both vectors, as shown in Equation 1. The dot product calculates the similarity between the vectors and the product of the Euclidean norms is used as a normalization factor.

In the STCS framework [4, 9, 11], the tag co-occurrence matrix is a central concept. This matrix contains for every tag pair the co-occurrence (i.e., how often two tags co-occur on pictures in the data set). A tag is represented as a vector of its co-occurrences with all other tags. This vector also includes a tag's co-occurrence with itself, which is by convention set to zero. The number of cosine computations that have to be computed between tag vectors grows quadratically with the number of tags in the data set. For n tags we have:

$$\text{Number of cosines to be calculated} = \frac{n^2 - n}{2} \tag{2}$$

This quadratic growth is a large bottleneck for the scalability of approaches that utilize the cosine similarity. Incremental cosine similarity computations are a possible solution to this problem, since there is no need to calculate the cosine similarity for every tag pair when new tags are added.

3.2   Approach 1: Incremental Recalculation Approach

The first approach is based on the incremental recalculation of cosines. By considering the changed tag co-occurrences, this method determines which cosines need to be recalculated. Incremental cosines are based on two types of update operations: an update of existing tag co-occurrences or the addition of new tags.

The update operation of already existing tags is quite simple. For every tag pair (i,j) that is updated (every tag has an index assigned), all the cosines for tag pairs of which tag i or tag j are part, need to recalculated. This is due to the fact that the cosine similarity uses the whole tag co-occurrence vector, rather than individual co-occurrences. As result of this, for every tag co-occurrence update of already existing tags, $2 \times n - 2$ cosines need to be updated (as 2 vectors have changed). Let us consider the co-occurrence matrix presented in Table 1. An example of an update on this matrix is shown in Table 2. In this example, the co-occurrence for the tag pair $(3,4)$ is changed (from 2 to 3). This change causes the cosine similarities for tag pairs of which tag 3 or 4 are part to be recalculated. The "-" in Table 1 and 2 are co-occurrences of a tag with itself, which we consider to be zero by default.

Table 1. The original co-occurrence matrix, which contains 5 tags

Tag	1	2	3	4	5	6
1	-	2	1	5	2	0
2	2	-	7	1	1	0
3	1	7	-	2	0	2
4	5	1	2	-	1	0
5	2	1	0	1	-	6
6	0	0	2	0	6	-

Table 2. The gray cells are the cosines of tag combinations that need to be recalculated after the co-occurrence between tag 3 and 4 changes

Tag	1	2	3	4	5	6
1	-	2	1	5	2	0
2	2	-	7	1	1	0
3	1	7	-	3	0	2
4	5	1	3	-	1	0
5	2	1	0	1	-	6
6	0	0	2	0	6	-

The addition of new tags is slightly more complicated, since it consists of two sub-operations. First, all cosine similarities between the new tag and all the other tags need to be calculated. An example is shown in Table 3, where tag 7 is being added as a new tag. Second, all combinations between existing tags that have a non-zero co-occurrence with the new tag need to recalculated. This has to be done in order to ensure that also the new tags are considered in the cosine similarities of vectors involving old tags. For example, it could happen that a certain existing tag pair had a similarity of 0 before the new tag was added, but now has a similarity larger than 0, because of the newly introduced co-occurrences (from the new tag). A visual example of this operation is shown in Table 4. Note that although many cells are gray, which indicates where re-calculation is needed, only half of the cosines are recalculated on account of the symmetrical properties of the tag co-occurrence matrix. After both operations are performed, we have a list of unique tag combinations for which the cosine similarity needs to be recalculated.

Table 3. Tag 7 is added to the tag co-occurrence matrix, new tag combinations are shown in gray.

Tag	1	2	3	4	5	6	7
1	-	2	1	5	2	0	1
2	2	-	7	1	1	0	0
3	1	7	-	3	0	2	6
4	5	1	3	-	1	0	1
5	2	1	0	1	-	6	0
6	0	0	2	0	6	-	0
7	1	0	6	1	0	0	-

Table 4. Gray indicates the existing combinations that need to be recalculated because tag 7 is added.

Tag	1	2	3	4	5	6	7
1	-	2	1	5	2	0	1
2	2	-	7	1	1	0	0
3	1	7	-	3	0	2	6
4	5	1	3	-	1	0	1
5	2	1	0	1	-	6	0
6	0	0	2	0	6	-	0
7	1	0	6	1	0	0	-

3.3   Approach 2: Delta Cosine Approach

The second approach first calculates the dot-product and the Euclidean distance for each incremental operation and then, when all the incremental operations are done, calculates the final cosine similarity. It uses both the changed and new co-occurrences in an iteration to determine for which tag pairs the cosine similarity needs to be recalculated, similar to the first proposed approach.

In contrast to the first approach, the two incremental operations are split in the second approach. First, the update operation for the existing tags is executed. During this operation the change to the dot-product for all affected cosines is considered for each update to the co-occurrences of the existing tags. In this step we consider the change to the $n \times n$ tag co-occurrence matrix A. The changes to the tag co-occurrences and the new co-occurrences are defined in a $m \times m$ matrix ΔA. The sizes of these matrices are such that $n \leq m$ and $m - n$ is the number of new tags. The set U_a contains indices of tag vector a for which the values are updated. Equation 3 is used to calculate the change to the dot-product of tag a and every other unchanged tag b ($a \neq b$).

$$(\Delta t_a + t_a) \cdot t_b = t_a \cdot t_b + \sum_{i \in U_a} \Delta t_{ai} \times t_{bi} \tag{3}$$

Tags t_a and t_b are both existing tag vectors of matrix A. This approach lets us update only effected elements of the dot-product of the cosine rather than the whole dot-product, which addresses the scalability of this method. In case both tags t_a and t_b change, the updated dot product formula becomes:

$$(\Delta t_a + t_a) \cdot (\Delta t_b + t_b) = t_a \cdot t_b + \left(\sum_{i \in U_a} \Delta t_{ai} \times t_{bi} \right) + \left(\sum_{i \in U_b} t_{ai} \times \Delta t_{bi} \right)$$

$$+ \left(\sum_{i \in U_{ab}} \Delta t_{ai} \times \Delta t_{bi} \right) \tag{4}$$

where U_{ab} contains indices of tag a and tag b for which the values are simultaneously updated, and U_a and U_b represent the indices of tag a or tag b for which the values are updated, but not for the other tag (i.e., b and a, respectively). The Euclidean norm is updated with the following equation for each changed co-occurrence:

$$||\Delta \mathbf{t}_a + \mathbf{t}_a|| = \sqrt{||\mathbf{t}_a||^2 + \sum_{i \in U_a} (\Delta \mathbf{t}_{ai} + \mathbf{t}_{ai})^2 - \sum_{i \in U_a} \mathbf{t}_{ai}^2} \qquad (5)$$

In case new tags are added, the dot-product of the new tag with the existing tags is calculated using the normal dot-product formula, shown in Equation 6, since there is no previous information available to calculate this using the difference with respect to previous iterations.

$$\mathbf{t}_a \cdot \mathbf{t}_b = \sum_{i=0}^{n} \mathbf{t}_{ai} \times \mathbf{t}_{bi} \qquad (6)$$

The changes of the dot-products between existing tags as result of addition of a new tag can be calculated using the difference (or delta) between the new and old vectors. For this purpose we use Equations 3 or 4 with vectors that have increased with one dimension due to the new tag. Also in this case only the new part is calculated and later added to the existing dot-product of the two tags. Equation 5 should be used to update the Euclidean norm for already existing tags in case a new tag is added, where vectors have increased with one dimension due to the new tag. The original equation for the Euclidean norm, given in Equation 7, should be used to calculate the Euclidean norm for new tags.

$$||\mathbf{t}_a|| = \sqrt{\sum_{i=0}^{n} \mathbf{t}_{ai}^2} \qquad (7)$$

After all the changes are processed, the cosines for the changed tag combinations are recalculated using the updated dot-product and the updated Euclidean norm.

4   Implementation

In this section we discuss the implementation of our proposed approaches. First, we describe the data cleaning process, after which we present the incremental recalculation approach and the delta cosine approach for computing cosines. The implementation of the incremental cosine approaches is done in Java and we make use of MySQL for data storage. The data processing is done in PHP.

4.1   Data Cleaning

The data we used was made available by Li [8]. It contains 3.5 million unique pictures and 570,000 tags, gathered from the photo sharing site Flickr.com. Since this data set contained some noise, we had to perform data cleaning. The following steps were performed:

- Remove tags longer than 32 characters. This operation is necessary, since we only want words and not complete sentences. Probability of sentences occurring in multiple pictures is quite low and therefore it does not have any effect on our result and therefore can be considered as noise.
- Remove tags that contain non-Latin characters. These tags are removed, since we only focus on English text in this paper.
- Pictures with no tags. Since our approach works with tags, we cannot process pictures without tags.
- Remove tags with low occurrence. We use the formula $\mu - 1.5 \times IQR$, where μ is the average tag occurrence and IQR is the inter-quartile range of the tag occurrences, to determine the minimum number of times a tag has to occur. In the original STCS approach [4], this threshold was set once for the complete data set. Here we calculate the minimum number of occurrences for a tag for every incremental data set and the initial data set. This cleaning rule is only applied to new tags. Already existing tags from previous iterations with a low occurrence in a certain iteration are not affected by this rule.

In order to calculate the cosines for the initial data set and to compute all cosines after updates for reference purposes, we implemented a separate cosine calculator. This program can calculate the cosine similarity for all tag combinations for a given co-occurrence matrix. The implementation was done in Java and for the matrix we used our own customized version of the JAMA [1] matrix library. This customized version uses a one-dimensional array rather than a two-dimensional array to store matrices. Compared to the original implementation of JAMA, our implementation offers faster matrix access with a lower memory footprint.

4.2   Incremental Recalculation Approach

The incremental recalculation approach implementation consists of three steps. The first step is the import of the new data of the current iteration, where the new data is cleaned using the cleaning rules described in Section 4.1 and then stored in the database. The new pictures are stored in a temporary table. This temporary table allows us to easily create a co-occurrence matrix with only the differences in co-occurrence of tags appearing in existing tags and the co-occurrences of new tags.

The next step in the process is the selection of cosines that need updating. This happens for each non-zero value in a matrix containing the co-occurrences of tags included in the new pictures. A unique list of tag pairs for which the cosine needs to be recalculated is the output of this step.

The last step is the recalculation of cosines from the list of tag pairs. Here we make use of the cosine calculator from the previous section, only instead of calculating every cosine, it uses the list of the previous step to calculate the correct cosines.

4.3 Delta Cosine Approach

In the delta cosine approach the import of new pictures is the same as in the previous approach. However, the initial import for the delta cosine approach differs from the recalculation approach. The cosine calculator, the dot-product calculator, and a script to store the Euclidean norm to database, have to be run for the initial import of pictures.

The next step is to compute the changes to the cosines. First, we calculate the changes to the dot-product and the changes to the Euclidean norm due to the tag co-occurrence updates of existing tags. Then, the dot-products and Euclidean norms for the new tags (if any) are calculated and the changes to dot-product and Euclidean norm of the existing tags caused by the addition of new tags are determined. After the update and new tag operations are finished, the affected cosines are recalculated using the updated dot-products and the Euclidean norms.

5 Evaluation and Results

In this section we evaluate the two proposed incremental cosine computation approaches. First, we discuss the evaluation set-up and the used data set. Then, the performance of each approach is discussed and compared to the baseline approach, which recalculates all cosines each time new pictures are added.

5.1 Evaluation Set-Up and Data Structure

The performance of the incremental approaches is measured using the execution time. By considering the execution time, we are able to investigate if there is any performance gain when using our proposed approaches for the incremental computation of cosine similarities. The test is done using a subset of the data set of Li [8]. The initial data set consists of the first 50,000 pictures (1,444 tags) in the original data set.

The incremental data set is composed of randomly selected pictures from the remaining data set. We chose to use 8 incremental data sets of different sizes, which are shown in Table 5. In this table, we give also the number of new pictures as percentage of the total number of pictures, including the corresponding co-occurrence update counts. The reason that we chose for relatively small data set sizes is that we needed to compute all cosine similarities for the baseline, which is a computationally-intensive process. Using these 8 data set sizes, we are able to determine at which point it becomes feasible to use an incremental cosine computation approach. The execution of the complete evaluation is done on a computer with an Intel Core i5 480M with 4 gigabyte of RAM.

5.2 Results of the Incremental Cosine Approaches

The execution time for each approach is measured for each set of newly added pictures. This allows us to compare it to the complete recalculation of all cosines.

Table 5. Properties of the different incremental data sets

New pictures	2,500	5,000	12,500	25,000	37,00	50,000	62,500	75,000
New pictures (%)	5%	10%	25%	50%	75%	100%	125%	150%
New tags	1	22	183	712	1,408	2,193	2,890	3,682
Total number of tags	1,445	1,466	1,627	2,156	2,852	3,637	4,334	5,126
Updated co-occurrences	10,319	33,730	52,351	79,482	98,643	112,905	137,723	140,145
Updated co-occurrences (%)	1.0%	3.2%	5.0%	7.6%	9.5%	10.8%	13.2%	13.5%

Table 6 shows the execution times for the incremental recalculation approach and complete recalculation approach of the cosine similarity, as well as the speed-up that is obtained by using the incremental calculation approach. This speed-up is used for the comparison with the other approach. The incremental recalculation is faster for all sets. If 2,500 new pictures are added it is 1.44 times faster to use an incremental recalculation than a complete recalculation. For 75,000 new pictures this is 1.23 times faster.

Table 6. Comparison in performance between incremental recalculation and complete recalculation

Number of new pictures	2,500	5,000	12,500	25,000	37,00	50,000	62,500	75,000
Time incremental calculation (s)	16	17	24	58	144	304	551	922
Time complete recalculation (s)	23	23	31	74	179	378	677	1 137
Speed-up	1.44	1.35	1.29	1.27	1.24	1.24	1.23	1.23

Table 7 shows the execution times for the delta cosine approach and complete recalculation approach of the cosine similarities. The delta cosine approach is faster for all data sets of newly introduced pictures. If 2,500 new pictures are added it is nearly 23 times faster to use incremental recalculation over complete recalculation. For 75,000 new pictures this is 1.24 times faster.

Table 7. Comparison in performance between the delta cosine approach and complete recalculation

Number of new pictures	2,500	5,000	12,500	25,000	37,00	50,000	62,500	75,000
Time delta cosine (s)	1	1	8	39	120	288	538	919
Time complete recalculation (s)	23	23	31	74	179	378	677	1 137
Speed-up	23	23	3.9	1.9	1.49	1.31	1.26	1.24

Figure 1 shows a plot of the execution times that are shown in Tables 6 and 7. As we can see, the delta cosine approach performs best on all data sets. However, the execution time of the incremental recalculation approach is approaching the execution time of the delta cosine approach for larger updates. This can be explained by the fact the delta cosines approach becomes similar to the incremental recalculation approach when the vectors are changed in many positions. As we

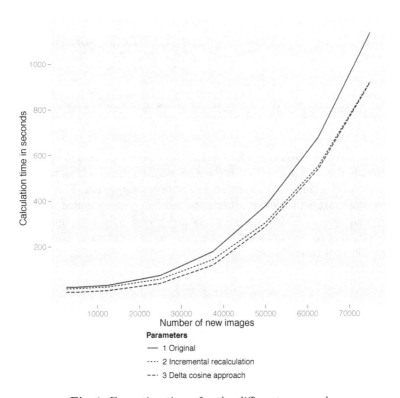

Fig. 1. Execution times for the different approaches

can observe in Figure 1, the delta cosine approach is more interesting if a small part of the co-occurrences change (for a data set size of 2,500, the delta cosine approach is faster). We can also see in Figure 1 that the delta cosine approach and the incremental recalculation approach are showing an exponential growth in execution time with respect to the number of pictures that are being added. However, this happens also for the complete recalculation approach and we can also notice that our proposed approaches grow with a smaller factor than the complete recalculation approach.

6   Conclusion

In this paper, we propose a method for the incremental calculation of the cosine similarity measure, originating from a scalability problem for tag spaces on the Web. We proposed two approaches for this purpose. The first approach is the incremental recalculation approach. In this approach, we consider updating only the cosine values that are affected by changes in the tag co-occurrences, and cosines that needed to be calculated for the first time, because new tags were added. The speed-up for the incremental recalculation was between 1.23 and 1.44 when comparing it to a complete recalculation of all cosine similarities.

The second approach is the delta cosine approach and improves on the first approach. In this solution, we calculated the change to the cosine similarity for each change in the co-occurrences matrix and each added tag. Compared to the complete recalculation, the speed-up was 23 for few changes in the co-occurrences and 1.23 for many changes to the tag co-occurrences matrix.

In future work, we would like to investigate how both approaches perform when sequentially adding new pictures. It would also be useful to research how one can perform the syntactic and semantic clustering techniques of the STCS framework in an incremental fashion. Furthermore, we would like to add our proposed approaches to the STCS framework and evaluate these techniques in a setting with a real-time flow of new pictures.

References

1. Java matrix package, http://math.nist.gov/javanumerics/jama/
2. Begelman, G.: Automated Tag Clustering: Improving Search and Exploration in the Tag Space. In: Collaborative Web Tagging Workshop at WWW 2006 (2006), http://www2006.org/workshops/#W06
3. Chakrabarti, D., Kumar, R., Tomkins, A.: Evolutionary Clustering. In: 12th ACM SIGKDD International Conference on Knowledge Discovery and Data Mining (KDD 2006), pp. 554–560. ACM (2006)
4. van Dam, J.W., Vandic, D., Hogenboom, F., Frasincar, F.: Searching and Browsing Tag Spaces Using the Semantic Tag Clustering Search Framework. In: Fourth IEEE International Conference on Semantic Computing (ICSC 2010), pp. 436–439. IEEE Computer Society (2010)
5. Friedman, M., Last, M., Makover, Y., Kandel, A.: Anomaly Detection in Web Documents Using Crisp and Fuzzy-based Cosine Clustering Methodology. Information Sciences 177(2), 467–475 (2007)
6. Gionis, A., Indyk, P., Motwani, R.: Similarity Search in High Dimensions via Hashing. In: 25th International Conference on Very Large Data Bases (VLDB 1999), pp. 518–529. Morgan Kaufmann Publishers Inc. (1999)
7. Jung, S.Y., Kim, T.S.: An Agglomerative Hierarchical Clustering Using Partial Maximum Array and Incremental Similarity Computation Method. In: IEEE International Conference on Data Mining (ICDM 2001), pp. 265–272. IEEE Computer Society (2001)
8. Li, X.: Flickr-3.5M Dataset (2009), http://staff.science.uva.nl/~xirong/index.php?n=DataSet.Flickr3m
9. Radelaar, J., Boor, A.-J., Vandic, D., van Dam, J.-W., Hogenboom, F., Frasincar, F.: Improving the Exploration of Tag Spaces Using Automated Tag Clustering. In: Auer, S., Díaz, O., Papadopoulos, G.A. (eds.) ICWE 2011. LNCS, vol. 6757, pp. 274–288. Springer, Heidelberg (2011)
10. Specia, L., Motta, E.: Integrating Folksonomies with the Semantic Web. In: Franconi, E., Kifer, M., May, W. (eds.) ESWC 2007. LNCS, vol. 4519, pp. 624–639. Springer, Heidelberg (2007)
11. Vandic, D., van Dam, J.W., Hogenboom, F., Frasincar, F.: A Semantic Clustering-Based Approach for Searching and Browsing Tag Spaces. In: 26th ACM Symposium on Applied Computing (SAC 2011), pp. 1693–1699. ACM (2011)

Impression-Aware Video Stream Retrieval System with Temporal Color-Sentiment Analysis and Visualization

Shuichi Kurabayashi and Yasushi Kiyoki

Faculty of Environment and Information Studies, Keio University
5322 Endoh, Fujisawa, Kanagawa, 252-0882, Japan
{kurabaya,kiyoki}@sfc.keio.ac.jp

Abstract. To retrieve Web video intuitively, the concept of "impression" is of great importance, because many users consider feelings and moods to be one of the most significant factors motivating them to watch videos. In this paper, we propose an impression-aware video stream retrieval system for querying the visual impression of video streams by analyzing the temporal change in sentiments. As a metric of visual impression, we construct a 180-dimensional vector space called as color-impression space; each dimension corresponds to a specific adjective representing humans' color perception. The main feature of this system is a context-dependent query processing mechanism to generate a ranking by considering the temporal transition of each video's visual impressions on viewers' emotion. We design an impression-aware noise reduction mechanism that dynamically reduces the number on non-zero features for each item mapped in the high-dimensional color-impression space by extracting the dominant salient impression features from a video stream. This system allows users to retrieve videos by submitting emotional queries such as "Find videos whose overall impression is happy and which have several sad and cool scenes". Through this query processing mechanism, users can effectively retrieve videos without requiring detailed information about them.

Keywords: video-search, impression, visualization, sentiment analysis.

1   Introduction

With the growing popularity of video-sharing websites, video data is becoming increasingly prevalent on the Web. Cisco reported that video data accounted for 40% of consumer Internet traffic in the world, and predicted that this share would increase to 54% by 2014 [1]. Video-sharing websites host a very large amount of content provided by amateur users as well as some content provided by media corporations and other organizations. For ease of use, such websites require powerful search engines because the results of a video search engine query can easily exceed the practical time limits a user has available to watch the retrieved videos.

To retrieve target video data intuitively, the concept of "impression" is of great importance because it has been shown that many users consider feelings and moods to be one of the most significant factors motivating them to watch videos [2]. In the field

S.W. Liddle et al. (Eds.): DEXA 2012, Part II, LNCS 7447, pp. 168–182, 2012.

of image database system, an impression-based image search system [3][4] is highly effective to retrieve image intuitively. Although many video search engines have already been proposed [5][6][7], extensive manual tweaking and heuristic trial-and-error is still required to find videos appropriate to users' impressions. There are two factors that contribute to this difficulty. The first is that user-generated videos have insufficient and/or unreliable metadata, which is generally provided in the form of titles and tags. It is thus desirable to develop an automatic annotation mechanism for video data from the viewpoint of "impression." The second factor is a "temporal" change in a video's impression, because the contexts and temporal transitions of a video deeply affect viewers' emotions. Traditional impression-based image retrieval mechanisms [8][9] cannot be directly applied to video retrieval. Conventional video retrieval approaches are unsuitable for an entire video stream because they focus on fragments of video data, such as the shot boundary and similarity of video shots [5][6][7]. Hence, a new analysis method for the temporal transition of sentiment in a video stream is required.

This study proposes an impression-aware video stream retrieval system for querying the overall impression of video streams by analyzing the temporal change of sentiments most often found in fictional television dramas, animation, and movies. Here, video "stream" retrieval means that the system aims to find an entire stream of video content, rather than a single scene or frame. This video stream retrieval mechanism is highly effective for finding new content that is unknown to users because it focuses on the "impression" in the query, and so does not require detailed information about the target videos. This system extracts spatiotemporal tendencies of colors from the entire video stream, and uses a psychological method of visual analysis, which defines relationships between colors and emotional adjective words, to automatically generate color-impression metadata for each frame in a video. As a metric of visual impression, the system builds a 180-dimensional vector space, called as the *color-impression space*, by leveraging traditional filmmaking principles and color psychology theory [10][11][12]. In the color-impression space, each dimension corresponds to a specific adjective representing humans' perception of colors. This vector space identifies 2^{180} contexts of color-impressions. The system generates a $180 \times n$ ($n = $ number of frames) matrix by computing the relevance between the color data of each video frame and the 180 impression adjectives. This matrix is called as the *time-series color-impression matrix*.

The main feature of this system is a context-dependent query processing mechanism to generate a ranking by considering the temporal transition of each video's visual impressions on viewers' emotions. Our system supports two types of impression contexts: the dominant context, which represents the overall impression of a video, and the salient context, which represents the local impression of several scenes. To achieve this context-dependent query processing, the system provides *impression-aware noise reduction operators* that eliminate the ambiguity of impressions caused by summarizing an entire video stream into a feature vector. The noise reduction operators dynamically reduce the number on non-zero features for each item mapped in the high-dimensional color-impression space according to the transition of sentiment in a video stream. The generated vectors have a higher capability for computing the semantic distance between the emotional adjectives than the original vectors.

The remainder of the paper is structured as follows. In section two, we briefly review related work on impression-based video retrieval. Section three discusses the query-by-impression model for video streams and how the color-impression-based image retrieval and ranking method is implemented. In section four, we present an experimental study, and in section five, we give our concluding remarks.

2   Related Work

The crux of an impression-based video search is to give impression annotations to the video. Videos can be annotated by two approaches: (1) direct approaches, in which humans manually annotate a video, and (2) indirect approaches, in which the system automatically generates impression annotations by using a defined correlation between video features, such as colors and impressions. Our approach is an indirect approach and extends this approach to provide the context-dependent noise reduction operators that remove the ambiguity of the automatically generated metadata. In terms of a direct approach, Nakamura and Tanaka [13] employed social annotation, in which many anonymous users annotate video, to extract impression annotations. Through this direct generation method, it is possible to generate highly accurate metadata that is compatible with the direct impressions experienced by humans while watching media. However, it is difficult to assign impression metadata that targets a large media group, as it is necessary for humans to actually watch the media. In addition, it becomes difficult to maintain consistency among the words assigned to impressions if there are a large number of people providing metadata. Due to the limited quality of the metadata annotated to video content on the Web, recent works on indirect approaches have typically considered the use of psychological or film-making methods to analyze the video content. As a method of analyzing fictional videos, Lehane et al. [14] proposed a video indexing method that generates the event-based structure of an entire movie by leveraging film-making techniques. Psychologists Russell and Mehrabian have proposed a three-dimensional emotion model, based on Pleasure, Arousal, and Dominance (P-A-D), for the analysis of human emotions [15]. Arifin and Cheung [16] applied the P-A-D model to a video indexing process. This introduces affect-based video segmentation that recognizes six types of emotional affections —amusement, violence, neutral, fear, sadness, and happiness—in a video stream by analyzing motions and colors. We proposed MediaMatrix [17][18] that is an impression-based video retrieval system that uses data mining techniques to extract a query context, which is expressed as a combination of multiple videos; it then uses methods developed in color psychology research to extract color-emotions hidden within video content.

3   Impression-Aware Video Analysis, Visualization, and Search

The framework of an impression-aware video stream retrieval system is shown in Fig. 1. It gathers video data from the Web and gives emotional annotations–sadness, happiness, warmth, dauntlessness, vividness, and so on–to each frame in a video stream by applying data mining techniques to the color information in the gathered video. As the generated annotation can only explain a frame, the system extracts the contextual impression of the entire video stream, such as the dominant impression and the salient impression, by analyzing the temporal change in the generated annotations. We design two noise reduction operators corresponding to the two types of impression contexts: (1) the *Dominant Impression Context*, which represents the overall impression of a video, and (2) the *Salient Impression Context*, which represents the local impression of several scenes. The noise reduction operators is applied to the color-impression space to select appropriate features according to the context-dependent weight of each impression adjective while considering the temporal behavior of the video stream. Our approach achieves impression-based context-dependent video stream retrieval by reduce the number on non-zero features to 20–50, on average, in the 180-dimensional color-impression space. Our concept of feature selection is that a user's impression query should be interpreted in the video's own context, because a different context gives an impression with a different meaning. The system does not define the dominant and salient feature sets statically. The temporal impression-aware noise reduction operators select a video's own feature axis set dynamically according to the content of the video in order to build a metric space for each video file. Finally, the system computes the relevance with respect to a query, which is given as emotive keywords or video files, by using the noise-reduced space according to the dominant/salient feature selection.

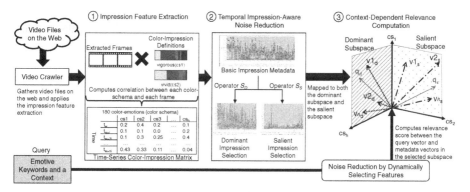

Fig. 1. Overview of Impression-Aware Video Stream Retrieval System that includes Impression Feature Selectors by Time-Series Impression Transition Analysis

3.1   Color-Sentiment Analysis Using Color-Impression Space

The proposed system uses two fundamental data structures for color-sentiment analysis: (1) a color-impression space and (2) a time-series color-impression matrix. The color-impression space is a 180-dimensional vector space that defines the correlation between colors and human emotions. The system computes the relevance between each frame in a video stream and each adjective in the color-impression space to generate a time-series color-impression matrix.

The color-impression space K is a $g \times h$ data matrix, where h is the number of color-impression words cs and g is the number of basic Munsell colors (c_1, c_2, ..., c_g). Each impression word cs_i ($i = 1$, ..., h) corresponds to a specific adjective that represents humans' color perception [11][12]. The impression word cs is a g-dimensional vector $cs = [f_1, f_2, ..., f_g]$ defining the relationship between the impression word and each basic color. Each feature value f is weighted from 0 to 1, where 0 means that the word cs and the color c are completely uncorrelated and 1 means that they are maximally correlated. Each feature vector cs is L2-normalized. Fig. 2 shows a partial example of the 180-dimensional color-impression space that is built using the color-impression data in [12]. This space defines the relationship between 180 impression adjectives and 130 basic Munsell colors [12][19]. In Fig. 2, the ratio of each color corresponds to the weight of the color on its impression adjective. For example, red has a large weight for the impression adjective "vigorous," and yellow has a large weight for "vivid."

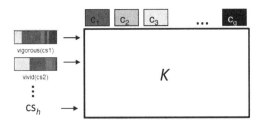

Fig. 2. Structure of the Color-Impression Space (as in [12] for Defining 180 Aesthetically Identified Sets of Color Schema Related to 130 Color Variations)

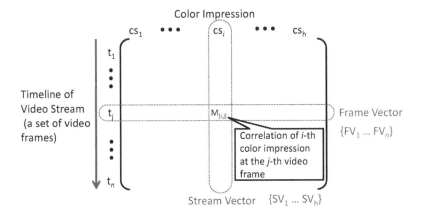

Fig. 3. Time-Series Color-Impression Matrix Structure Representing a Transition of Color-Impression along with a Video Stream Timeline

The time-series color-impression matrix is a $180 \times n$ (n = number of frames) matrix representing the relevance between the color data of each video frame and the 180 impression adjectives, as shown in Fig. 3. This matrix can be represented by two "views": (1) an h-dimensional frame vector FV that represents the color-impression of a specific frame, where each dimension is a color-impression cs defined in the color-impression space (Fig. 2), (2) an n-dimensional stream vector SV that represents the temporal transition of a specific impression's weight value, where each dimension is a specific color-impression's weight extracted from a frame. The system generates the time-series color-impression matrix by performing the following steps:

1. Decode an inputted video data stream into m frames. The system converts RGB color values to HSV color values per pixel of each image. HSV is a widely adopted space in image and video retrieval because it describes perceptual and scalable color relationships.
2. Create a color histogram that represents the color distribution of the frame by clustering every pixel into the closest of the predefined 130 Munsell basic colors c_j ($j = 1,..., 130$), which are defined in the color-impression space. We use Godlove's color difference formula [20] to calculate the color distance.
3. Generate a 180-dimensional color-impression vector for each frame by calculating the correlation between 180 adjectives cs_i ($i = 1, ..., 180$) and 130 colors. The correlation between color histogram r and the color-impression cs is defined as follows:

$$f_{map}(r, cs) := \left[\sum_{i=0}^{m} r_{[i]} \cdot cs_{1[i]}, \quad ..., \quad \sum_{i=0}^{m} r_{[i]} \cdot cs_{180[i]} \right] \tag{1}$$

where $r_{[i]}$ is the value of the i-th bin of the color histogram, cs_n is the n-th color-impression, and $cs_{n[i]}$ denotes the weight of the i-th color and n-th color-impression. The number of colors in the color-impression space is given by m.

Fig. 4. Visualization of the Temporal Transition of an Analyzed Color-Impression

Finally the system generates a time-series color-impression matrix, which is a chronologically ordered set of the generated color-impression vector. Fig 4 shows an example visualization of the generated time-series color-impression matrix.

The horizontal axis corresponds to the video stream timeline and the vertical axis corresponds to the correlation score of the 180 color-impression vectors. As the plain time-series color-impression matrix has the contextual ambiguity of containing both dominant impressions and salient impressions, our system provides the two feature selection operators in the purpose of noise reduction as follows: dominant impression selector S_D and salient impression selector S_S.

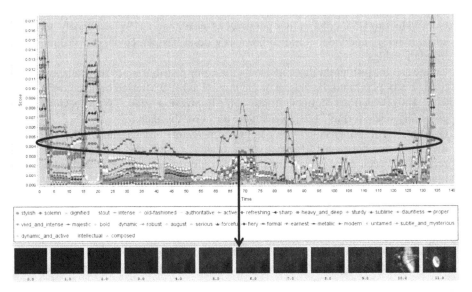

Fig. 5. Visualization M_D Generated by Applying the Dominant Impression Feature Selector S_D to the Video File Used in Fig.4

3.2   Dominant Impression Feature Selector: S_D

The temporal impression-aware feature selection operators select a set of color-impression features from a time-series color-impression matrix. Dominant impression selector S_D that inputs the time-series color-impression matrix M and returns the matrix M_D consisting of the selected features is defined as follows:

$$
S_D : M \rightarrow M_D =
\begin{bmatrix}
FV_{1[1]} \cdot \dfrac{\sum_{j=1}^{n} SV_{1[j]}}{n} & \cdots & FV_{1[180]} \cdot \dfrac{\sum_{j=1}^{n} SV_{180[j]}}{n} \\
\vdots & \ddots & \vdots \\
FV_{n[1]} \cdot \dfrac{\sum_{j=1}^{n} SV_{1[j]}}{n} & \cdots & FV_{n[180]} \cdot \dfrac{\sum_{j=1}^{n} SV_{180[j]}}{n}
\end{bmatrix}
\tag{2}
$$

where $FV_{j[i]}$ corresponds to the i-th color-impression extracted from the j-th frame in a video stream, and n is the number of frames. $SV_{i[j]}$ corresponds to the i-th color-impression's score in the j-th frame. Thus, this operator multiplies each frame's $FV_{j[i]}$ ($j = 1,...n$) by the average score of SV_i. As a result, the more frequently the color-impression appears in the video, the higher the color-impression scores. Fig.5 shows a visualization of M_D generated from the same video used in Fig.4. In this figure, the 33 features are selected from the 180 impression features. The horizontal axis corresponds to the video stream timeline and the vertical axis corresponds to the correlation score of the 33 selected color-impression vectors. We can see that the dominant color-impressions appear at the beginning and ending of the video, and this affects the background impression of this video.

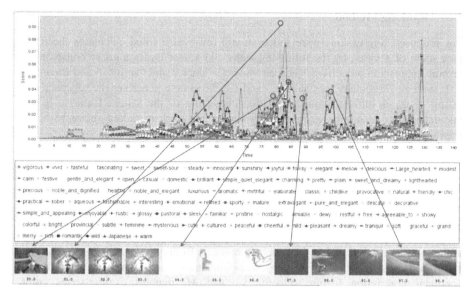

Fig. 6. Visualization of M_S Generated by Applying the Salient Impression Feature Selector S_S to the Video File Used in Fig.4

3.3 Salient Impression Feature Selector: S_S

S_S selects color-impression features by detecting emotionally salient points, in other words: rarity, of color-impressions in the entire video stream. Whereas S_D measures the color-impression popularity, S_S measures the color-impression specificity. S_S is defined as follows:

$$S_S : M \rightarrow M_S = \begin{bmatrix} FV_{1[1]} \cdot \log \dfrac{1}{\sum_{j=1}^{n} SV_{1[j]}} & \cdots & FV_{1[180]} \cdot \log \dfrac{1}{\sum_{j=1}^{n} SV_{180[j]}} \\ \vdots & \ddots & \vdots \\ FV_{n[1]} \cdot \log \dfrac{1}{\sum_{j=1}^{n} SV_{1[j]}} & \cdots & FV_{n[180]} \cdot \log \dfrac{1}{\sum_{j=1}^{n} SV_{180[j]}} \end{bmatrix} \quad (3)$$

Thus, this operator multiplies each frame's $FV_{j[i]}$ ($j = 1,...n$) by the logarithm of the reciprocal of the summed SV_i scores. This operator increases the weight in proportion to the appearance ratio of a color-impression in the frame, but is offset by the frequency of the color-impression in the entire video stream. Fig.6 shows a visualization of M_S generated from the same video used in Fig.4. In this figure, values below 0 are omitted as noise. The 78 features are selected as salient impressions from the 180 impression features. The horizontal axis corresponds to the video stream timeline and the vertical axis corresponds to the correlation score of the 78 selected color-impression vectors. Fig. 6, when compared with Fig. 4 (normal visualization) and Fig. 5 (visualization of dominant features), shows that the S_S operator successfully selects features that appear strongly in specific scenes. The system detects that the salient impressions appear in the middle of the video stream.

3.4   Ranking Method

Our proposed impression-aware video stream retrieval method calculates the relevance score of a video by the following steps: (1) a user issues a query consisting of several impression adjectives, such as "warm," "dark," and "stylish," and their context, such as "dominant" or "salient," and (2) the system generates the subspace consisting of the dominant features and that consisting of the salient features. It then calculates the distance between a query vector and each metadata vector to obtain relevance scores. The relevance score is defined as follows:

$$f_{relevance}(q_D, q_S, w_D, w_S, M_D, M_S) := w_D \left(\sum_{i=1}^{n} \sum_{j=1}^{180} q_{D[j]} \cdot M_{D[i,j]} \right) + w_S \left(\sum_{i=1}^{n} \sum_{j=1}^{180} q_{S[j]} \cdot M_{S[i,j]} \right) \quad (4)$$

where q_D denotes a query vector that is mapped into the dominant sub-space, q_S denotes a query vector that is mapped into the salient sub-space, where w_D denotes a weight of the relevance calculated in the dominant sub-space, and w_S denotes weight of the relevance calculated in the salient sub-space. Both query vectors q_D and q_S have 180-dimensions. M_D denotes a time-series color-impression matrix generated by the dominant impression feature selector S_D, and M_S denotes a time-series color-impression matrix generated by the salient impression feature selector S_S. The system thus acquires two relevance scores: one using dominant features and another by using salient features.

4   Evaluation Experiment

This section evaluates the effectiveness of our impression-aware video stream retrieval system when applied to commercial animations distributed by video sites on the Web (http://www.nicovideo.jp). The experimental data comprise 24-min episodes of animation. Animated shows are chosen because their story and concepts are deeply affected by colors. A total of 128 items of data, consisting of 4,423,680 frames in all, are collected. We perform two evaluation experiments: Experiment 1 is an evaluation of noise reduction based by the time series analysis of impression changes, and Experiment 2 is an evaluation of video search precision. To perform these experiments, we have implemented the proposed system open-source software, and it is available via http://web.sfc.keio.ac.jp/~kurabaya/mediamatrix.html.

4.1   Experiment 1: Evaluation of Noise Reduction by S_D and S_S

Experiment 1 applied a dominant impression feature selector, S_D, and salient impression feature selector, S_S, to the experimental data and verified whether these operators could select features appropriately. The experimental method was as follows: the 128 video data were input to our system, creating 128 corresponding time-series color-impression matrices. The system applied S_D and S_S to each of the 128 matrices. For each of the 128 video data, the number of impression features chosen was plotted on a graph which allowed us to confirm that an appropriate number of features were selected in response to changes in the video data content.

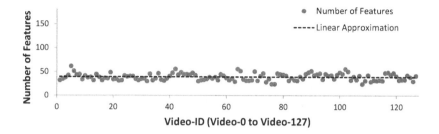

Fig. 7. Experimental Results 1-1: Number of Dominant Features Selected by Applying S_D

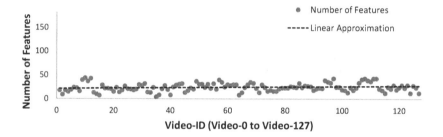

Fig. 8. Experimental Results 1-2: Number of Salient Features Selected by Applying S_S

Fig. 7 shows the graph of the number of features selected after the application of S_D. The system identified an average of 39.1 features (21.7% of the total 180 features) as dominant impression features. The highest number of features selected was 64; the lowest was 24, with 39 being the median value. Similarly, Fig. 8 shows the graph of the number of features selected after the application of S_S. The system identified an average of 24.3 features (13.5% of the total 180 features) as the salient impression features. The highest number of salient features selected was 44; the lowest was 4, with 24 being the median value. This graph shows that the operators S_D and S_S successfully reduced the number of features even if the video data contain different color-impression features. S_D takes all the stream vectors SV and calculates the average correlation for each feature selection; because this calculation method is weighted on the average correlation, the value can accumulate relatively easily. In contrast, feature selection based on S_S is weighted by the inverse log of the total number of features for a stream vector SV corresponding to a certain feature. This implies that the weighted value cannot increase easily. Even though there was a slight difference in the number of selections of dominant and salient features, the video search ranking was generated independently for each feature; hence, this difference did not pose a problem.

The number of times each impression word was chosen during Experiment 1 for S_D and S_S is shown in Fig. 9 and Fig. 10 respectively. This result shows that S_D and S_S successfully generated the noise-reduced impression metadata vector, by choosing the different features in a context-dependent manner. Fig. 9 shows the number of selections for each of the dominant impression features that were selected five or more times in Experiment 1-1. Fig. 10 shows the number of selections for each of the salient impression features that were selected five or more times. We found that only two features—the impression words "sporty" and "festive"—were shared between the two sets of results.

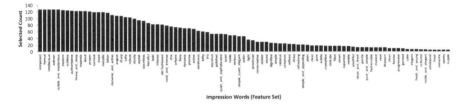

Fig. 9. Number of Times Each Impression Feature was Chosen during Experiment 1-1 for S_D

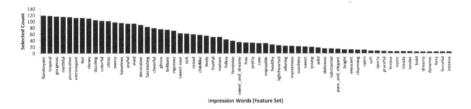

Fig. 10. Number of Times Each Impression Feature was Chosen during Experiment 1-2 for S_S

4.2   Experiment 2: Evaluation of Video Search Accuracy

Experiment 2 was carried out to evaluate the precision of the video search using the implemented system. The methodology for the evaluation was as follows: the degree of correlation between the query and each of the 128 pieces of video data was calculated, and a corresponding ranking of this correlation was generated. The ratio of the data that included correct answers was calculated for the top-k items of the search results. The correct answer data set for the experimental evaluation was created in the following manner: ten test subjects (six male and four female) watched the 128 test videos. Subjects were asked to judge whether the dominant impression queries D1–D3 and the salient impression queries S1–S3 matched their impressions. Querying scenarios are as follows: D1: the impressions about the school scenery in Japan (21 correct answers), D2: the impressions of the early 20th century Europe (24 correct answers), D3: the impressions of a futuristic city in the outer space (30 correct answers), S1: the sci-fi or fantasy impressions that included battle scenes, explosions, and magic (20 correct answers), S2: the shocking impressions such as those found in horror or suspense movies(16 correct answers), and S3: the gentle and pleasant impressions where soft or cuddly characters and objects appeared (17 correct answers). The judgment was formed by assigning a score as follows: 1 Does not match, 2 Cannot say either way, 3 Matches more than it does not, and 4 Matches. The scores assigned by each test subject were averaged, and when the overall average was greater than or equal to 3, the video was added to the set of correct answers for the relevant case. The correct answer ratio for the top-k items in the search result rankings, denoted as P_{qk}, can be defined as follows:

$$P_{qk} = \frac{\left| R_{qk} \cap A_q \right|}{k} \tag{5}$$

where, R_{qk} denotes the top-k items in the search result ranking for query q, and A_q denotes the correct answers corresponding to the query q. In this experiment, for the sake of comparison, we manually assigned 5–10 words describing impressions to the videos. The ratio of correct answers was calculated for a search result ranking by performing a pattern match search, with the set of words describing impressions as the target. For the data used in the pattern match search, we selected annotation classifications used by commercial video broadcasters as a reference, and had the same ten test subjects assign words to describe their impressions. For those words that were assigned by three or more test subjects, 5–10 such words were applied to each video as metadata. The precision of our system P_{qk} was tested by comparing the correct answer ratios from the impression search result ranking using our system (QBI: query-by-impression) and the search result ranking based on a pattern match search using impression words. The experimental results for the dominant impression queries are shown in Fig. 11, whereas those for the salient impression search queries are shown in Fig. 12. The vertical axes of these graphs show the correct answer ratio, and the horizontal axes show the ranking k. Further, the termination of the broken line is indicative of the point where the system detected all the correct answer items for the corresponding query.

Those results show that our system (QBI) achieves highly better precision than the conventional keyword search system. The impression search results obtained using our system had a higher correct answer ratio for the top search result rankings than searches carried out using human-derived impression metadata. This is because that our system could reduce the context-dependent noise of each metadata vector. Fig. 11 shows the high search accuracy for the ranking of the top-k items using the dominant impression search; the average correct answer ratio for searches using our system (QBI) for the top five items for cases D1–D3 was 95.7%; the average correct answer ratio for searches using our system (QBI) for the top ten items for cases D1–D3 was 81.3%. This confirms that the top ranked search results matched the dominant user-requested impressions. On the other hand, the average correct answer ratio for searches using the top five keywords for cases D1–D3 was 1.3%, and the average correct answer ratio for searches using the top ten keywords was 2.8%.

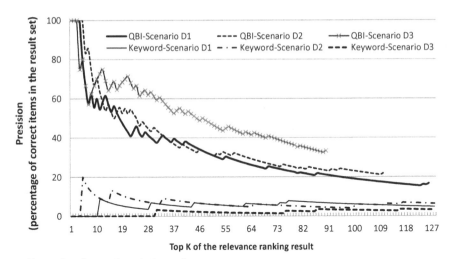

The number of correct items in the queries are:
Query-D1: 21 items (16%), Query-D2: 24 items (19%), Query-D3: 30 items (23%)

Fig. 11. Experimental Results 2-1: Evaluation of Precision in Top-*K* Ranking Results by the Dominant Impression Search

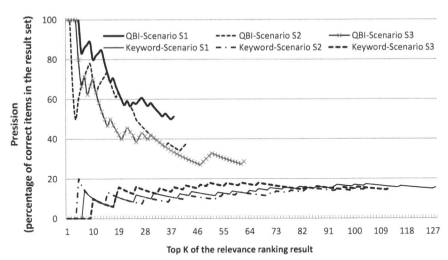

The number of correct items in the queries are:
Query-S1: 20 items (16%), Query-S2: 16 items (13%), Query-S3: 17 items (13%)

Fig. 12. Experimental Results 2-2: Evaluation of Precision in Top-*K* Ranking Result by the Salient Impression Search

Fig. 12 shows the high search precision for the ranking of the top-k items for the salient impression search; the average correct answer ratio for searches carried out using our system (QBI) for the top five items in cases S1–S3 was 90.4%; the average correct answer ratio for searches carried out using our system (QBI) for the top ten items in cases S1–S3 was 82.7%. This confirms that the top-ranked search results matched the salient user-requested impressions. On the other hand, the average correct answer ratio for searches using the top five ranking keywords in queries S1–S3 was 1.3%; the average correct answer ratio for searches using the top ten ranking keywords was 4.7%.

5   Conclusion

This paper proposed the impression-aware video stream retrieval and visualization system that supports a new, efficient search method for an entire video stream by analyzing sentiment features of colors in a video. The aim of this system is to find target video streams by the emotional impression they make along with the transition of visual appearance, such as warmth, coolness, happiness, and sadness. One unique feature of this system is a context-dependent noise reduction mechanism to extract dominant impressions and salient impressions from a video stream. Our system allows users to retrieve videos by submitting impression-based queries such as: "Find videos that are similar in an overall impression to the input video" and "Find videos whose impression is 'happy and funny' in local dramatic presentations." This query processing mechanism is effective at retrieving unfamiliar videos because users do not require detailed information about the target videos. In future work, we will develop an acceleration mechanism for analyzing video data, and the S_D and S_S operators by using GPGPU (General-Purpose Computing on Graphics Processing Units).

Acknowledgments. This research was supported by SCOPE: Strategic Information and Communications R&D Promotion Programme of the Ministry of Internal Affairs and Communications, Japan: "Kansei Time-series Media Hub Mechanism for Impression Analysis/Visualization Delivery Intended for Video/Audio Media."

References

1. Cisco: Cisco Visual Networking Index: Forecast and Methodology, 2009-2014 (2010), http://www.cisco.com/en/US/solutions/collateral/ns341/ns525/ns537/ns705/ns827/white_paper_c11-481360.pdf
2. Cunningham, S.J., Nichols, D.M.: How people find videos. In: Proceedings of the 8th ACM/IEEE-CS Joint Conference on Digital Libraries, pp. 201–210 (2008)
3. Kiyoki, Y., Kitagawa, T., Hayama, T.: A metadatabase system for semantic image search by a mathematical model of meaning. ACM SIGMOD Record 23(4), 34–41 (1994)
4. Kiyoki, Y., Kitagawa, T., Hitomi, Y.: A fundamental framework for realizing semantic interoperability in a multidatabase environment. Journal of Integrated Computer-Aided Engineering 2(1), 3–20 (1995)

5. Lew, M.S., Sebe, N., Djeraba, C., Jain, R.: Content-based multimedia information retrieval: State of the art and challenges. ACM TOMCCAP 2(1), 1–19 (2006)
6. Smeulders, A.W.M., Worring, M., Santini, S., Gupta, A., Jain, R.: Content-based image retrieval at the end of the early years. IEEE Transactions on Pattern Analysis and Machine Intelligence 22(12), 1349–1380 (2000)
7. Smeaton, A.F.: Techniques used and open challenges to the analysis, indexing and retrieval of digital video. Information Systems 32(4), 545–559 (2007)
8. Hou, X., Zhang, L.: Color conceptualization. In: Proceedings of the 15th International Conference on Multimedia, pp. 265–268. ACM (2007)
9. Corridoni, J.M., Del Bimbo, A., Pala, P.: Image retrieval by color semantics. Multimedia Systems 7(3), 175–183 (1999)
10. Valdez, P., Mehrabian, A.: Effects of color on emotions. Journal of Experimental Psychology: General 123(4), 394–409 (1994)
11. Kobayashi, S.: The aim and method of the color image scale. Color Research & Application 6(2), 93–107 (1981)
12. Kobayashi, S.: Color Image Scale. Oxford University Press (1992)
13. Nakamura, S., Tanaka, K.: Video Search by Impression Extracted from Social Annotation. In: Vossen, G., Long, D.D.E., Yu, J.X. (eds.) WISE 2009. LNCS, vol. 5802, pp. 401–414. Springer, Heidelberg (2009)
14. Lehane, B., O'Connor, N.E., Lee, H., Smeaton, A.F.: Indexing of Fictional Video Content for Event Detection and Summarisation. EURASIP Journal on Image and Video Processing, Article ID 14615, 15 pages (2007)
15. Russell, J.A., Mehrabian, A.: Evidence for a three-factor theory of emotions. Journal of Research in Personality 11, 273–294 (1977)
16. Arifin, S., Cheung, P.Y.K.: A computation method for video segmentation utilizing the pleasure-arousal-dominance emotional information. In: Proceedings of the 15th ACM International Conference on Multimedia, pp. 68–77 (2007)
17. Kurabayashi, S., Ueno, T., Kiyoki, Y.: A Context-Based Whole Video Retrieval System with Dynamic Video Stream Analysis Mechanisms. In: Proceedings of the 11th IEEE International Symposium on Multimedia (ISM 2009), pp. 505–510 (2009)
18. Kurabayashi, S., Kiyoki, Y.: MediaMatrix: A Video Stream Retrieval System with Mechanisms for Mining Contexts of Query Examples. In: Kitagawa, H., Ishikawa, Y., Li, Q., Watanabe, C. (eds.) DASFAA 2010, Part II. LNCS, vol. 5982, pp. 452–455. Springer, Heidelberg (2010)
19. Newhall, S.M., Nickerson, D., Judd, D.B.: Final Report of the O.S.A. Subcommittee on the Spacing of the Munsell Colors. Journal of the Optical Society of America 33(7), 385–411 (1943)
20. Godlove, I.H.: Improved Color-Difference Formula, with Applications to the Perceptibility and Acceptability of Fadings. Journal of the Optical Society of America 41(11), 760–770 (1951)

Dynamic Workload-Based Partitioning
for Large-Scale Databases

Miguel Liroz-Gistau[1], Reza Akbarinia[1], Esther Pacitti[2], Fabio Porto[3],
and Patrick Valduriez[1]

[1] INRIA & LIRMM, Montpellier, France
{Miguel.Liroz_Gistau,Reza.Akbarinia,Patrick.Valduriez}@inria.fr
[2] University Montpellier 2, INRIA & LIRMM, Montpellier, France
Esther.Pacitti@lirmm.fr
[3] LNCC, Petropolis, Brazil
fporto@lncc.br

Abstract. Applications with very large databases, where data items are
continuously appended, are becoming more and more common. Thus, the
development of efficient workload-based data partitioning is one of the
main requirements to offer good performance to most of those applications that have complex access patterns, e.g. scientific applications.
However, the existing workload-based approaches, which are executed in
a static way, cannot be applied to very large databases. In this paper,
we propose *DynPart*, a dynamic partitioning algorithm for continuously
growing databases. *DynPart* efficiently adapts the data partitioning to
the arrival of new data elements by taking into account the affinity of
new data with queries and fragments. In contrast to existing static approaches, our approach offers a constant execution time, no matter the
size of the database, while obtaining very good partitioning efficiency. We
validated our solution through experimentation over real-world data; the
results show its effectiveness.

1 Introduction

We are witnessing the proliferation of applications that have to deal with huge
amounts of data. The major software companies, such as Google, Amazon, Microsoft or Facebook have adapted their architectures in order to support the
enormous quantity of information that they have to manage. Scientific applications are also struggling with those kinds of scenarios and significant research
efforts are directed to deal with it [3]. An example of these applications is the
management of astronomical catalogs; for instance those generated by the Dark
Energy Survey (DES) [1] project on which we are collaborating. In this project,
huge tables with billions of tuples and hundreds of attributes (corresponding to
dimensions, mainly double precision real numbers) store the collected sky data.
Data are appended to the catalog database as new observations are performed
and the resulting database size is estimated to reach 100TB very soon. Scientists around the globe can query the database with queries that may contain a
considerable number of attributes.

S.W. Liddle et al. (Eds.): DEXA 2012, Part II, LNCS 7447, pp. 183–190, 2012.

The volume of data that such applications hold poses important challenges for data management. In particular, efficient solutions are needed to partition and distribute the data in several servers. An efficient partitioning scheme would try to minimize the number of fragments accessed in the execution of a query, thus reducing the overhead associated to handle the distributed execution. Vertical partitioning solutions, such as column-oriented databases [7], may be useful for physical design on each node, but fail to provide an efficient distributed partitioning, in particular for applications with high dimensional queries, where joins would require data transfers between nodes. Traditional horizontal approaches, such as hashing or range-based partitioning [4,5], are unable to capture the complex access patterns of scientific applications, especially since they usually make use of complex relations over big sets of columns, and are hard to be predefined.

One solution is to use partitioning techniques based on the workload. Graph-based partitioning is an effective approach for that purpose [6]. A graph (or hypergraph) that represents the relations between queries and data elements is built and the problem is reduced to that of minimum k-way cut problem, for which several libraries are available. However, this method always requires to explore the entire graph in order to obtain the partitioning. This strategy works well for static applications, but scenarios where new data are inserted to the database continuously, which is the most common case for scientific computing, introduce an important problem. Each time a new set of data is appended, the partitioning should be redone from scratch, and as the size of the database grows, the execution time of such operation may become prohibitive.

In this paper, we are interested in dynamic partitioning of large databases that grow continuously. After modeling the problem of data partitioning in dynamic datasets, we propose a dynamic workload-based algorithm, called *DynPart*, that efficiently adapts the partitioning to the arrival of new data elements. Our algorithm is designed based on a heuristic that we developed by taking into account the affinity of new data with queries and fragments. In contrast to the static workload-based algorithms, the execution time of our algorithm does not depend on the total size of the database, but only on that of the new data and this makes it appropriate for continuously growing databases. We validated our solution through experimentation over real-world data sets. The results show that it obtains high performance gains in terms of partitioning execution time compared to one of the most efficient static partitioning algorithms.

The remainder of this paper is organized as follows. In Section 2, we describe our assumptions and define formally the problem we address. In Section 3, we propose our solution for dynamic data partitioning. Section 4 reports on the results of our experimental validation and Section 5 concludes.

2   Problem Definition

In this section, we state the problem we are addressing and specify our assumptions. We start by defining the problem of static partitioning, and then extend it for a dynamic situation where the database can evolve over time.

2.1 Static Partitioning

The static partitioning is done over a set of *data items* and for a *workload*. Let $D = \{d_1, ..., d_n\}$ be the set of data items. The workload consists of a set of queries $W = \{q_1, ..., q_m\}$. We use $q(D) \subseteq D$ to denote the set of data items that a query q accesses when applied to the data set D. Given a data item $d \in D$, we say that it is *compatible* with a query q, denoted as $comp(q, d)$, if $d \in q(D)$. Queries are associated with a relative frequency $f : W \to [0, 1]$, such that $\sum_{q \in W} f(q) = 1$.
Partitioning of a data set is defined as follows.

Definition 1. *Partitioning of a data set D consists of dividing the data of D into a set of fragments, $\pi(D) = \{F_1, ..., F_p\}$, such that there is no intersection between the fragments and the union of all fragments is equal to D.*

Let $q(F)$ denote the set of data items in fragment F that are compatible with q. Given a partitioning $\pi(D)$, the set of *relevant fragments* of a query q, denoted as $rel(q, \pi(D))$, is the set of fragments that contain some data accessed by q, i.e. $rel(q, \pi(D)) = \{F \in \pi(D) : q(F) \neq \emptyset\}$.
To avoid a high imbalance on the size of the fragments, we use an *imbalance factor*, denoted by ϵ. The size of the fragments at each time should satisfy the following condition: $|F| \leq \frac{|D|}{|\pi(D)|}(1 + \epsilon)$.
In this paper, we are interested in minimizing the number of query accesses to fragments. Note that the minimum number of relevant fragments of a query q is $minfr(q, \pi(D)) = \left\lceil \frac{|q(D)|}{(|D|/|\pi(D)|)(1+\epsilon)} \right\rceil$. We define the *efficiency of a partitioning* for a workload based on its efficiency for queries. Let us first define the *efficiency of a partitioning for a query* as follows:

Definition 2. *Given a query q, then the efficiency of a partitioning $\pi(D)$ for q, denoted as $eff(q, \pi(D))$ is computed as:*

$$eff(q, \pi(D)) = \frac{minfr(q, \pi(D))}{|rel(q, \pi(D))|} \tag{1}$$

In the equation above, when the number of accessed fragments is equal to the minimum possible, i.e. $minfr(q, \pi(D))$, the efficiency is 1. Using $eff(q, \pi(D))$, we define the efficiency of a partitioning $\pi(D)$ for a workload W as follows.

Definition 3. *The efficiency of a partitioning $\pi(D)$ for a workload W, denoted as $eff(W, \pi(D))$, is*

$$eff(W, \pi(D)) = \sum_{q \in W} f(q) \times eff(q, \pi(D)) \tag{2}$$

Given a set of data items D and a workload W, the goal of static partitioning is to find a partitioning $\pi(D)$ such that $eff(W, \pi(D))$ is maximized.

2.2   Dynamic Partitioning

Let us assume now that the data set D grows over time. For a given time t, we denote the set of data items of D at t as $D(t)$.

During the application execution, there are some events, namely *data insertions*, by which new data items are inserted into D. Let $T_{ev} = (t_1, \ldots, t_m)$ be the sequence of time points corresponding to those events. In this paper, we assume that the workload is stable and neither the queries nor their frequencies change. However, the queries may access new data items as the data set grows.

Let us now define the problem of dynamic partitioning as follows. Let $T_{ev} = (t_1, \ldots, t_m)$ be the sequence of time points corresponding to data insertion events; $D(t_1), \ldots, D(t_m)$ be the set of data items at $t_1, \ldots, t_m$ respectively; and W be a given workload. Then, the goal is to find a set of partitionings $\pi(D(t_1)), \ldots, \pi(D(t_m))$ such that the sum of the efficiencies of the partitionings for W is maximized. In other words, our objective is as follows:

Objective: Maximize $\left(\sum_{t \in T_{ev}} \sum_{q \in W} (f(q) \times \mathit{eff}(q, \pi(D(t)))) \right)$

3   Affinity Based Dynamic Partitioning

In this section, we propose an algorithm, called *DynPart*, that deals with dynamic partitioning of data sets. It is based on a principle that we developed using the partitioning efficiency measure described in the previous section.

3.1   Principle

Let d be a new inserted data item. From the definition of partitioning efficiency, we infer that if we place d in a fragment F, then the total efficiency varies according to the following approximation[1]:

$$\mathit{eff}(W, \pi(D \cup \{d\})) \approx \mathit{eff}(W, \pi(D)) -$$
$$\sum_{q : q(F) = \emptyset \wedge comp(q,d)} f(q) \frac{minfr(q, \pi(D))}{|rel(q, \pi(D))| \, (|rel(q, \pi(D))| + 1)}, \quad (3)$$

Thus, the partitioning efficiency is reduced whenever there are queries that did not access F but after the insertion of d to F have to access it, thereby increasing the number of relevant fragments. The lower the number of those queries, the less the resulting loss of efficiency. Based on this idea, we define *the affinity between the data d and fragment F*, which we denote as

$$\mathit{aff}(d, F) = \sum_{q : q(F) \neq \emptyset \wedge comp(q,d)} f(q) \frac{minfr(q, \pi(D))}{|rel(q, \pi(D))| \, (|rel(q, \pi(D))| + 1)} \quad (4)$$

---

[1] Note that this approximation is an equality in all cases but when the increment in $|q(D)|$ makes $minfr(q, \pi(D))$ to be increased by 1, which happens very rarely.

Using (4), we can develop a heuristic algorithm that places the new data items in the fragments based on the maximization of the affinity between the data items and the fragments.

3.2 Algorithm

Our *DynPart* algorithm takes a set of new data items D' as input and selects the best fragments to place them. For each new data item $d \in D'$, it proceeds as follows (see the pseudo-code in Algorithm 1). First, it finds the set of queries that are compatible with the data item. This can be done by executing the queries of W on D' or by comparing their predicates with every new data item. Then, for each compatible query q, *DynPart* finds the relevant fragments of q, and increases the fragments affinity by using the expression in (4). Initially the affinity of fragments is set to zero.

Algorithm 1. *DynPart* algorithm

---

for each $d \in D'$ do
    for each $q : comp(q, d)$ do
        for each $F \in rel(q, \pi(D))$ do
            if $feasible(F)$ then
                $//aff(F)$ is initialized to 0
                $aff(F) \leftarrow aff(F) + f(q)\frac{minfr(q,\pi(D))}{|rel(q,\pi(D))|(|rel(q,\pi(D))|+1)}$
    if $\exists F \in \pi(D) : aff(F) > 0$ then $dests \leftarrow \arg\max_{F \in \pi(D)} aff(F)$
    else $dests \leftarrow \{F \in \pi(D) : feasible(F)\}$
    $F_{dest} \leftarrow$ select from $\arg\min_{F \in dests} |F|$
    move d to F_{dest} and update metadata

After computing the affinity of the relevant fragments, *DynPart* has to choose the best fragment for d. Not all of the fragments satisfy the imbalance constraints, thus we must only consider those that do meet the restrictions. We define the function $feasible(F)$ to determine whether a fragment can hold more data items or not. Accordingly, *DynPart* selects from the set of feasible fragments the one with the highest affinity. If there are multiple fragments that have the highest affinity, then the smallest fragment is selected, in order to keep the partitioning as balanced as possible.

DynPart works over a set of new data items D', instead of a single data item. This particularly reduces the amortized cost of finding the set of queries that are compatible with each of the inserted items. This is why, in practice, we wait until a given number of items have been inserted and then execute our algorithm for partitioning the new data.

Let $comp_{avg}$ be the average number of compatible queries per data item, and rel_{avg} be the average number of relevant fragments per query. Then, the average execution time of the algorithm is $O(comp_{avg} \times rel_{avg} \times |D'|)$, where $|D'|$ is the number of new data items to be appended to the fragments. The complexity can be $O(|W| \times |\pi(D)| \times |D'|)$ in the worst case. However, in practice, the averages are usually much smaller than the worst case values. The reason is that the queries usually access a small portion of the data, thus the average number of compatible queries is low. In any case, in order to reduce the number of queries, we may use a threshold on the frequency, so that only queries above that threshold are considered. In addition, the partitioning efficiency of our approach is good, so the average number of relevant fragments per query is low.

4   Experimental Evaluation

To validate our dynamic partitioning algorithm, we conducted a thorough experimental evaluation over real-world data. In Section 4.1, we describe our experimental setup. In Section 4.2, we report on the execution time of our algorithm and compare it with a well known static workload-based algorithm. In Section 4.3, we study the effect our heuristic on the partitioning efficiency.

4.1   Set-Up

For our experimental evaluation we used the data from the Sloan Digital Sky Survey catalog, Data Realease 8 (DR8) [2], as it is being used in LIneA in Brazil[2]. It consists of a relational database with several observations for both stars and galaxies. We obtained a workload sample from the SDSS SkyServer SQL query log data, which stores the information about the real accesses performed by users. In total, the database comprises almost 350 million tuples, that take 1.2 TB of space. The workload consists of a total of 27000 queries.

All queries were executed on the database and the tuple ids accessed by each of them were recorded. Only tuples accessed by at least one query were considered. We simulated the insertions on the database by selecting a subset of the tuples as the initial state and appending the rest of the tuples in groups. We varied the following two parameters: 1) the number of tuples inserted to the database before each execution of our algorithm, $|D'|$; and 2) the number of fragments in which the database is partitioned, $|\pi(D)|$. On each of the experiments, the specific numbers are detailed. Throughout the experiments we chose an imbalance factor of 0.15. All experiments were executed in a 3.0 GHz Intel Core 2 Duo E8400, running Ubuntu 11.10 64 bits with 4GB of memory.

4.2   Execution Time

In this section, we study the execution time of the *DynPart* algorithm (DP in the figure) and compare it with a static graph partitioning algorithm (SP). For the later, we use PaToH[3], an hyper-graph partitioner. Figure 1(a) shows the

---

[2] Data from the DES project is still unavailable, so we have used data from SDSS, which is a similar, previous project
[3] http://bmi.osu.edu/~umit/software.html

comparison of the execution time for 16 fragments. In the case of the dynamic algorithm, we executed the algorithm with two values for $|D'|$: 0.5 and 1 million tuples. Similar results are obtained for different values of $|\pi(D)|$. As the difference between execution times of the static and the dynamic algorithms is significant, we use a logarithmic scale for the y-axis in order to show the results. The results are only depicted until a database size of 20 million tuples, as the memory requirements for the static partitioning are bigger than the memory of our servers. The dynamic algorithm does not cause any problem as the memory footprint depends on $|D'|$, which is constant throughout the experiment.

As we can seen, execution time increases for the static algorithm as the size of the database increases, provided that the size of the graph increases accordingly. For the *DynPart* algorithm, on the other hand, the execution time stays at the same level, as it is always executed for the same number of data items.

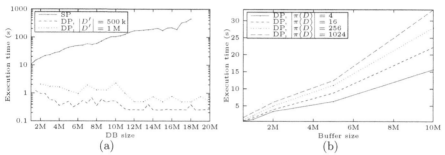

Fig. 1. a) Execution times of the static (SP) and dynamic (DP) approaches as DB size increases ($|\pi(D)| = 16$), b) *DynPart* execution time vs. $|D'|$

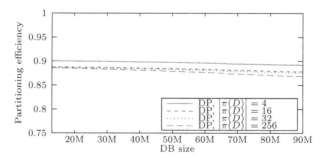

Fig. 2. Comparison of partitioning efficiency as the size of the DB grows ($|D'| = 1M$)

We executed our algorithm for different sizes of D'. Figure 1(b) shows the average execution time of the *DynPart* algorithm as $|D'|$ increases for different number of fragments. As expected, the execution time is linearly related to the number of tuples before execution. Also, the higher number of fragments, the higher the execution time. This increase is not linear since the number of relevant fragments does not increase at the same pace. In fact, the number of relevant fragments does not exceed 8 for $|\pi(D)| = 256$ and 16 for $|\pi(D)| = 1024$.

4.3   Partitioning Efficiency

One of the important issues to consider for the *DynPart* algorithm is how our heuristic algorithm affects the partitioning efficiency. We executed the *DynPart* algorithm as the database is fed with new data after an initial partitioning using the static approach. With $|D'| = 1$ M, Fig. 2 shows how the partitioning efficiency evolves as the database grows for different number of fragments. Similar results were obtained for other configurations of $|D'|$. The efficiency decreases as the database grows, as expected, but this reduction is very small. For example, in the worst case, $|\pi(D)| = 256$, the partitioning efficiency decreases 2.23×10^{-3} in average for each 10 million new tuples.

5   Conclusions

In this paper, we proposed *DynPart*, a dynamic algorithm for partitioning continuously growing large databases. We modeled the partitioning problem for dynamic datasets and proposed a new heuristic to efficiently distribute new arriving data, based on its affinity with the different fragments in the application.

We validated our approach through implementation, and compared its execution time with that of a static graph-based partitioning approach. The results show that as the size of the database grows, the execution time of the static algorithm increases significantly, but that of our algorithm remains stable. They also show that although the *DynPart* algorithm is designed based on a heuristic approach, it does not degrade partitioning efficiency considerably.

The results of our experiments show that our dynamic partitioning strategy is able to efficiently deal with the data of our application. But, we believe that its use is not limited to this application, and it can be used for data partitioning in many other applications in which the data items are appended continuously.

References

1. The dark energy survey, `http://www.darkenergysurvey.org/`
2. Sloan digital sky survey, `http://www.sdss3.org`
3. Ailamaki, A., Kantere, V., Dash, D.: Managing scientific data. Communications of the ACM 53(6), 68–78 (2009)
4. Chang, F., Dean, J., Ghemawat, S., Hsieh, W.C., Wallach, D.A., Burrows, M., Chandra, T., Fikes, A., Gruber, R.E.: Bigtable: a distributed storage system for structured data. ACM Transactions on Computer Systems 26(2), 1–26 (2008)
5. Cooper, B.F., Ramakrishnan, R., Srivastava, U., Silberstein, A., Bohannon, P., Jacobsen, H.A., Puz, N., Weaver, D., Yerneni, R.: PNUTS: Yahoo!'s hosted data serving platform. Proceedings of the VLDB Endowment 1(2), 1277–1288 (2008)
6. Curino, C., Jones, E., Zhang, Y., Madden, S.: Schism: a workload-driven approach to database replication and partitioning. Proceedings of the VLDB Endowment 3(1), 48–57 (2010)
7. Stonebraker, M., Abadi, D.J., Batkin, A., Chen, X., Cherniack, M., Ferreira, M., Lau, E., Lin, A., Madden, S., O'Neil, E., O'Neil, P., Rasin, A., Tran, N., Zdonik, S.: C-store: a column-oriented DBMS. In: Proceedings of the 31st International Conference on Very Large Data Bases, VLDB 2005, pp. 553–564 (2005)

Dynamic Vertical Partitioning of Multimedia Databases Using Active Rules

Lisbeth Rodríguez and Xiaoou Li

Department of Computer Science, CINVESTAV-IPN, Mexico D.F., Mexico
lisbethr@computacion.cs.cinvestav.mx, lixo@cs.cinvestav.mx

Abstract. Vertical partitioning is a design technique widely employed in relational databases to reduce the number of irrelevant attributes accessed by the queries. Currently, due to the popularity of multimedia applications on the Internet, the need of using partitioning techniques in multimedia databases has arisen in order to use their potential advantages with regard to query optimization. In multimedia databases, the attributes tend to be of very large multimedia objects. Therefore, the reduction in the number of accesses to irrelevant objects would imply a considerable cost saving in the query execution. Nevertheless, the use of vertical partitioning techniques in multimedia databases implies two problems: 1) most vertical partitioning algorithms only take into account alphanumeric data, and 2) the partitioning process is carried out in a static way. In order to address these problems, we propose an active system called DYMOND, which performs a dynamic vertical partitioning in multimedia databases to improve query performance. Experimental results on benchmark multimedia databases clarify the validness of our system.

Keywords: Multimedia databases, Dynamic vertical partitioning, Active rules.

1 Introduction

Vertical partitioning divides a table into a set of fragments, each fragment with a subset of attributes of the original table and defined by a vertical partitioning scheme (VPS). The fragments consist of smaller records, therefore, fewer pages from disk are accessed to process queries that require only some attributes from the table, instead of the entire record. This leads to better query performance [1,2]. For multimedia databases (MMDBs) many of the attributes tend to be of very large size objects (e.g. images, video, audio) and should not be retrieved if they are not accessed by the queries. Vertical partitioning techniques can be applied to these databases to provide faster access to multimedia objects [3–5].

Most vertical partitioning techniques only consider alphanumeric data. To the best of our knowledge vertical partitioning of MMDBs only has been addressed in [5,6]. Nevertheless, these techniques are static based on query patterns of the database which are available during the analysis stage. MMDBs are dynamic

S.W. Liddle et al. (Eds.): DEXA 2012, Part II, LNCS 7447, pp. 191–198, 2012.

because they are accessed by many users simultaneously [7], therefore their query patterns tend to change over time and a static vertical partitioning technique would be inefficient in these databases. In this context, it is more effective to periodically check the goodness of a VPS to determine whenever refragmentation is necessary [8].

In dynamic vertical partitioning, the database must be continuously monitored to detect the changes on query patterns, the relevance of these changes needs to be evaluated, and then a new partitioning process has to be initiated.

An effective way to monitor the database is using active rules because they enable continuous monitoring inside the database server and have the ability to automatically take actions based on monitoring, also due to their simplicity, active rules are amenable to implementation with low CPU and memory overheads [9]. Active rules are used for programming active systems, which have the ability to detect events and respond to them automatically in a timely manner [10]. They define the events to be monitored, the conditions to be evaluated and the actions to be taken.

We propose an active system for dynamic vertical partitioning of MMDBs, called DYMOND (DYnamic Multimedia ON line Distribution), to achieve efficient retrieval of multimedia objects at any time. DYMOND uses active rules to timely detect changes in query patterns, determine if a refragmentation is necessary, and execute the refragmentation process. DYMOND uses a vertical partitioning algorithm for MMDBs, called MAVP [6], which takes into account the size of the attributes to get an optimal VPS.

2   Related Work

A dynamic vertical partitioning approach to improve the performance of the queries in relational databases is presented in [12]. This approach is based on the feedback loop (it consists of observation, prediction and reaction) used in automatic performance tuning. First, it observes the change in the workload to detect the periods of time when the system is relatively idle, and then it predicts the future workload based on the features of the current workload and implements the new vertical fragments.

Reference [11] integrates both vertical and horizontal partitioning into automated physical database design. The main disadvantage of this work is that it only recommends which fragments must be created, but the database administrator (DBA) has to create the fragments. DYMOND has a partitioning generator which automatically creates the fragments on disk.

Autopart [13] is an automated tool that partitions the relations in the original database according to a representative workload. Autopart receives as input a representative workload and designs a new VPS, one drawback of this tool is that the DBA has to give the workload to Autopart. In contrast, DYMOND collects the SQL statements at the moment when they are executed.

Most dynamic vertical partitioning (attribute clustering) techniques consist of the following modules: a Statistic Collector (SC), that accumulates information about the queries run and data returned. The SC is in charge of collecting, filtering, and analyzing the statistics. It is responsible for triggering the Cluster Analyzer (CA).The CA determines the best possible clustering given the statistics collected. If the new clustering is better than the one in place, then the CA triggers the reorganizer that physically reorganizes the data on disk [8].

The problem of dynamic vertical partitioning in MMDBs has largely been ignored. Most dynamic vertical partitioning approaches do not consider multimedia data [8, 11–13]. The approaches for MMDBs are static [5, 6]. DYMOND takes into account both multimedia data and a dynamic vertical partitioning.

3   Dynamic Vertical Partitioning Using Active Rules

A vertical partitioning algorithm for MMDBs, called MAVP, was proposed in [6] to achieve efficient retrieval of multimedia objects in MMDBs. MAVP gets a VPS according to a set of queries, their frequencies, and the size of the attributes. MAVP only gets the VPS but the DBA has to develop the analysis phase to collect the information that MAVP needs as input, to execute MAVP in order to get the VPS, and to materialize the VPS, i.e., to create the fragments on disk.

MMDBs are accessed by many users simultaneously, therefore, queries and their frequencies tend to quickly change over time. In this context, the DBA would have to develop the analysis of the database, to execute MAVP and to materialize the VPS every time when the access patterns suffer enough changes to avoid degrading the database performance. So, a mechanism to know what changes are considered enough to trigger a new partitioning process is needed.

The main objective of DYMOND is to allow efficient retrieval of multimedia objects from the database at any time. DYMOND implements the MAVP algorithm to get a VPS. It adapts the technique of setting a performance threshold of the MMDB after each partitioning and a variable determined by the system, which measures the percentage of change of input data to MAVP [14].

DYMOND continuously monitors the changes in the database scheme and in the access patterns using active rules and evaluates the performance of the MMDB based on the current VPS. The performance of the MMDB is measured in terms of: 1) the cost of accessing irrelevant attributes (IAAC), and 2) the transportation cost (TC). DYMOND uses this measure to determine a performance threshold of the MMDB and when to trigger a new partitioning.

The process of dynamic vertical partitioning in DYMOND is presented in Figure 1. It is implemented using active rules. When an user sends a query to the MMDB or the DBA modifies the schema of the database, the statistic collector of DYMOND evaluates the information of the queries and the attributes, and compares the new information with the previous one in order to determine if the changes are enough to trigger the performance evaluator. When the performance evaluator is triggered, the cost of the VPS is calculated, if its cost is greater than the performance threshold, then MAVP is triggered, which obtains a new VPS. Finally, the partitioning generator materializes the new VPS.

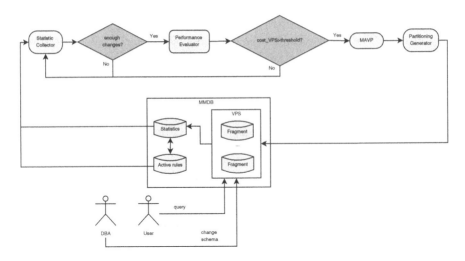

Fig. 1. Process of dynamic vertical partitioning in DYMOND

3.1 Statistic Collector

The statistic collector accumulates all the information that MAVP needs to get a VPS, such as the number of attributes (NA), the size of the attributes (S), the number of queries (NQ), and the access frequency of the queries (F), also registers all the changes in the input data over time and compares the new changes with the previous changes to determine if they are enough to trigger the performance evaluator.

A change in any input data NA, S, NQ, F is calculated as the percentage of the original data. This means that in a MMDB with 8 attributes, if an attribute is added or deleted, this is defined as 1/8*100=12.5% of change. The change in S is calculated as the maximum of the total of change divided by the size that the attribute had in the last partitioning multiplied by 100. The change in NQ is calculated as the sum of the number of new queries which access the MMDB or queries which leave to access it divided by the total number of queries which originally accessed the MMDB multiplied by 100. The change in F is calculated as the maximum change in the frequency of a query, divided by the total access frequency for this query, multiplied by 100.

For example, the rules for the detection of changes in S are defined in the following way:

update_att
ON update **attribute**.*size*
IF *prev_size*>0
THEN update **stat_attributes** set
$$currentchanges = \frac{attribute.size - attribute.prev\_size}{attribute.prev\_size} * 100$$

update_currentchanges
ON before update **stat_attributes**.*currentchanges*
IF *currentchanges*<0

THEN update **stat_attributes** set *currentchanges=currentchanges**-1
IF *currentchanges<old.currentchanges*
THEN update **stat_attributes** set *currentchanges=old.currentchanges*

after_update_currentchanges
ON after update **stat_attributes**.*currentchanges*
IF *currentchanges>previouschanges*
THEN update **stat_flags** set *flag_cs=t*

The performance evaluator is triggered when at least one of the flags of change stored in **stat_flags** is true. This is defined using the following active rule:

update_flag
ON after update **stat_flags**
IF (*flag_att=t* AND *flag_queries=t* AND *flag_frequency=t*)
       AND (*flag_cf=t* OR *flag_cq=t* OR *flag_ca=t* OR *flag_cs=t*)
THEN call Performance_Evaluator

3.2   Performance Evaluator

When the performance evaluator is triggered by the statistic collector, it calculates the execution cost of the queries in the current VPS. If the new cost is greater than the performance threshold then MAVP is triggered. This is implemented using the rule:

update_cost_vps
ON after update **stat_performance**.*cost_vps*
IF *cost_vps>performance_threshold*
THEN call MAVP

3.3   MAVP

MAVP gets a VPS, then it stores in the statistics the average changes in the input data NQ, F, S, NA and updates the value of the current changes to zero. After that, it calculates the performance threshold. To do this, it uses the cost of the new VPS (*best_cost_vps*) stored in the statistics and a variable of change of the MMDB (s). The threshold is calculated as *best_cost_vps* multiplied by s.

The variable of change s represents the frequency of change in the input data of MAVP that can affect the performance, s is calculated as the inverse of the percentage of average change in the input to MAVP, multiplied by 50 to allow a reasonable margin of performance degradation plus 1 (for example, in a MMDB with 100% of change in NQ, 80% of change in F, 60% of change in S, and 50% of changes in NA, the value of s is calculated as $s = 1 + \frac{50}{(100+80+60+50)/4} = 1.68$). The value of s always will be greater than 1 to get a threshold greater than the cost of the new VPS. A threshold lower than the cost of the new VPS always will trigger a new partitioning which is not desirable. Finally, MAVP triggers the partitioning generator.

3.4 Partitioning Generator

The new VPS is materialized by the partitioning generator. To do this, it deletes the previous fragments and creates the new fragments. It implements the fragments as indexes, in this way the indexes are automatically used by the DBMS when a query is executed and it is not necessary change the definition of the queries to use the fragments instead of the complete table. So the use of the fragments is transparent for the users.

4 Experiments

We use the cost model presented in [6] to compare DYMOND versus MAVP. For the first experiment we use the relation **Equipment**(*id, name, graphic, audio, video*). Using the AUM of the Table 1, DYMOND gets as optimal VPS *(id, audio, video) (name, image, graphic)*, with a cost of 361700. If after the partitioning, changes in the MMDB are produced, we have the AUM of the Table 2.

We also use the MMDB **Albums**(*name, birth, genre, location, picture, songs, clips*) presented in [15]. The AUM of **Albums** is presented in Table 3. After

Table 1. AUM of the relation Equipment

Q/A	id	name	image	graphic	audio	video	F
q_1	0	1	1	1	0	0	15
q_2	1	1	0	0	1	1	10
q_3	0	0	0	1	1	1	25
q_4	0	1	1	0	0	0	20
S	8	20	900	500	4100	39518	

Table 2. AUM of the relation Equipment after partitioning (Equipment2)

Q/A	id	name	image	graphic	audio	video	F
q_1	0	1	1	1	0	0	15
q_2	1	1	0	0	1	1	15
q_3	0	0	0	1	1	1	25
q_4	0	1	1	0	0	0	20
q_5	0	1	1	0	0	0	20
S	8	20	900	600	4100	39518	

Table 3. AUM of the relation Albums

Q/A	name	birth	genre	location	picture	songs	clips	F
q_1	0	0	1	0	0	1	0	25
q_2	0	0	0	0	1	0	0	30
q_3	0	0	0	1	1	0	1	28
S	20	10	15	20	600	5000	50000	

Table 4. AUM of the relation Albums after partitioning (Albums2)

Q/A	name	birth	genre	location	picture	songs	clips	F
q_1	0	0	1	0	0	1	0	50
q_2	0	0	0	0	1	0	0	20
q_3	0	0	0	1	1	0	1	60
q_4	1	1	0	0	1	0	0	15
S	20	10	15	20	600	5500	60000	

Table 5. VPSs obtained by MAVP and DYMOND

R	MAVP	DYMOND
E	(id, audio, video)(name, image, graphic)	(id, audio, video)(name, image, graphic)
E2	(id, audio, video)(name, image, graphic) (animation)	(id, audio, video, graphic) (name, image, animation)
A	(name, birth)(genre, songs) (location, clips) (picture)	(name, birth)(genre, songs) (location, clips)(picture)
A2	(name, birth)(genre, songs) (location, clips)(picture)	(name, birth, picture, location, clips) (genre, songs)
	R=Relation, E=Equipment, A=Albums	

Table 6. Cost comparison of the VPSs obtained by MAVP and DYMOND

Relation	MAVP			DYMOND		
	IAAC	TC	Cost	IAAC	TC	Cost
Equipment	47200	314500	361700	47200	314500	361700
Equipment2	69700	747500	817200	547960	64500	612460
Albums	0	470400	470400	0	470400	470400
Albums2	0	2166750	2166750	2103100	0	2103100

partitioning, some changes are made to **Albums** as we can see in Table 4. The VPSs obtained by MAVP and DYMOND are presented in Table 5. Finally the cost comparison between such schemes is shown in Table 6.

5 Conclusion

We presented an active system for dynamic vertical partitioning of MMDBs, called DYMOND, which uses active rules to continuously monitor the changes in the access patterns to the MMDB, to evaluate the changes and to trigger the partitioning process.

The main advantages of DYMOND are:

1. It improves the overall performance of the MMDB because it timely detects when the performance falls under a threshold to trigger the partitioning process.

2. It reduces the activities that the DBA must develop such as the analysis of the MMDB, the determination of the VPS and the creation of the fragments.
3. It provides an efficient retrieval of multimedia objects from the MMDB at all time.

References

1. Navathe, S., Ceri, S., Wiederhold, G., Dou, J.: Vertical Partitioning Algorithms for Database Design. ACM TODS 4, 680–710 (1984)
2. Guinepain, S., Gruenwald, L.: Automatic Database Clustering Using Data Mining. In: DEXA 2006 (2006)
3. Fung, C., Karlapalem, K., Li, Q.: An Evaluation of Vertical Class Partitioning for Query Processing in Object-Oriented Databases. IEEE Transactions on Knowledge and Data Engineering 14, 1095–1118 (2002)
4. Fung, C., Karlapalem, K., Li, Q.: Cost-driven Vertical Class Partitioning for Methods in Object Oriented Databases. The VLDB Journal 12(3), 187–210 (2003)
5. Fung, C., Leung, E.W., Li, Q.: Efficient Query Execution Techniques in a 4DIS Video Database System for eLearning. Multimedia Tools and Applications 20, 25–49 (2003)
6. Rodriguez, L., Li, X.: A Vertical Partitioning Algorithm for Distributed Multimedia Databases. In: Hameurlain, A., Liddle, S.W., Schewe, K.-D., Zhou, X. (eds.) DEXA 2011, Part II. LNCS, vol. 6861, pp. 544–558. Springer, Heidelberg (2011)
7. Van Doorn, M.G.L.M., De Vries, A.P.: The Psychology of Multimedia Databases. In: Proc. of the Fifth ACM Conf. on Digital Libraries, pp. 1–9. ACM (2000)
8. Guinepain, S., Gruenwald, L.: Research Issues in Automatic Database Clustering. SIGMOD Record 34, 33–38 (2005)
9. Chaudhuri, S., Konig, A.C., Narasayya, V.: SQLCM: a Continuous Monitoring Framework for Relational Database Engines. In: Proc. 20th Int. Data Engineering Conf., pp. 473–484 (2004)
10. Paton, N.W., Díaz, O.: Active Database Systems. ACM Comput. Surv. 31, 63–103 (1999)
11. Agrawal, S., Narasayya, V., Yang, B.: Integrating Vertical and Horizontal Partitioning into Automated Physical Database Design. In: Proc. of ACM SIGMOD, pp. 359–370 (2004)
12. Zhenjie, L.: Adaptive Reorganization of Database Structures Through Dynamic Vertical Partitioning of Relational Tables, MCompSc Thesis, University of Wollongong (2007)
13. Papadomanolakis, S., Ailamaki, A.: AutoPart: automating schema design for large scientific databases using data partitioning. In: Proc. 16th Int. Scientific and Statistical Database Management Conf., pp. 383–392 (2004)
14. Ezeife, C., Zheng, J.: Measuring the Performance of Database Object Horizontal Fragmentation Schemes. In: Int. Symposium Database Engineering and Applications, pp. 408–414 (1999)
15. Chbeir, R., Laurent, D.: Towards a Novel Approach to Multimedia Data Mixed Fragmentation. In: Proc. of the Int. Conf. on Manage. of Emergent Digital EcoSyst., MEDES (2009)

RTDW-bench: Benchmark
for Testing Refreshing Performance
of Real-Time Data Warehouse

Jacek Jedrzejczak[1], Tomasz Koszlajda[2], and Robert Wrembel[2]

Poznań University of Technology, Institute of Computing Science, Poznań, Poland
jacek@psi-poznan.com.pl,
{Tomasz.Koszlajda,Robert.Wrembel}@cs.put.poznan.pl

Abstract. In this paper we propose a benchmark, called *RTDW-bench*, for testing a performance of a real-time data warehouse. The benchmark is based on TPC-H. In particular, *RTDW-bench* permits to verify whether an already deployed RTDW is able to handle without any delays a transaction stream of a given arrival rate. The benchmark also includes an algorithm for finding the maximum stream arrival rate that can be handled by a RTDW without delays. The applicability of the proposed benchmark was verified in a RTDW implemented in Oracle11g.

1   Introduction

A data warehouse architecture has been developed in order to integrate and analyze data coming from multiple distributed, heterogeneous, and autonomous data sources deployed throughout an enterprise. New business domains of DW applications require more advanced DW functionalities, often combining the transactional and analytical features [11]. For example, monitoring unauthorized credit card usage, monitoring telecommunication networks and predicting their failures, monitoring car traffic, analyzing and predicting share rates, require accurate and up to date analytical reports. In order to fulfill this demand, one has to assure that the content of a DW is synchronized with the content of data sources with a minimum delay, e.g., seconds or minutes, rather than hours. To this end, the technology of a real-time data warehouse (RTDW) has been developed, e.g., [16].

1.1   Related Work

A high refreshing frequency of a RTDW imposes new research and technological challenges, including approximate query answering, efficient refreshing, and scheduling RTDW activities. In order to assure rapid (or on-line) answers to analytical queries, multiple research works proposed to apply various approximation techniques, like for example sampling [2,3] or wavelet-coefficient synopses [4]. In [6] the on-line aggregation of data was supported by means of the MapReduce technique in the Hadoop framework.

S.W. Liddle et al. (Eds.): DEXA 2012, Part II, LNCS 7447, pp. 199–206, 2012.
© Springer-Verlag Berlin Heidelberg 2012

In order to minimize a DW refreshing time, multiple incremental refreshing algorithms have been proposed, e.g., [5]. In [10], the authors developed an ETL optimization technique based on queuing systems' theory. In [13], the algorithm was proposed for optimizing joins between a data stream and a table, under in a limited memory.

Another important research issue encompasses scheduling refreshing processes and analytical queries. Multiple scheduling algorithms have been developed, aiming at guaranteeing the quality of service or quality of data, e.g., [16], or satisfactory performance, e.g., [14], or minimizing data staleness, e.g., [8].

Another issue is testing a performance of either traditional or a real-time data warehouse. Several benchmarks have been proposed for testing a performance of databases and data warehouses, e.g., [1,9,12] as well as ETL, e.g., [17,15,18]. Determining the maximum arrival rate of a data or transaction stream that a given RTDW is able to handle is also of a great practical importance. However, to the best of our knowledge, benchmarks for testing a performance of a RTDW have not been proposed yet.

1.2   Paper Contribution

In this paper we propose a *RTDW-bench* benchmark for testing a performance of a RTDW. By means of *RTDW-bench* we are able to determine two critical metrics: (1) *Critical Insert Frequency* that represents the maximum average stream arrival rate below which a RTDW is is capable of updating its tables without delays and (2) *Critical View Update Frequency* that represents the maximum average stream arrival rate below which a RTDW is capable of updating without delays its tables and refreshing all its materialized views.

2   Definitions and Problem Statement

We assume that a RTDW is supplied by a *stream* of transactions composed of inserts, updates, and deletes. Let transactions $T_1, T_2, \ldots, T_n$ be separated by inter-arrival time whose average value is denoted as t_g. The stream is characterized by its arrival rate, i.e., the number of operations delivered in a time unit. Executing T_i in a RTDW consumes time $t_{T_i}^e$, which is the average for the whole stream. Refreshing time of all materialized views impacted by T_i consumes time $t_{T_i}^r$, which is the average for the whole stream.

We say that a RTDW is *overloaded* when for a given transaction stream the RTDW is not able to execute T_i (modify its tables) before next transaction T_{i+1} arrives, i.e., $t_g < t_{T_i}^e$. If a RTDW includes also materialized views, then the following inequality is true for an overloaded RTDW: $t_g < t_{T_i}^e + t_{T_i}^r$. The consequence of a RTDW overloading is its inability to be fully synchronized with the arriving stream. Notice that a RTDW can receive a peek load when a current stream arrival rate changes from lower to higher. In such a case, arriving transactions can be buffered and executed when the current stream arrival rate decreases. For this reason, while testing the performance of a RTDW one have to consider average arrival rates, rather than current ones.

We consider the two following RTDW settings. In *setting 1* a RTDW does not contain any materialized view or the views are not being refreshed. In *setting 2* a RTDW contains materialized views that are refreshed. These two settings allow us to distinguish three zones of a RTDW operation, as shown in Figure 1, where the horizontal axis denotes the average arrival rate of a transaction stream.

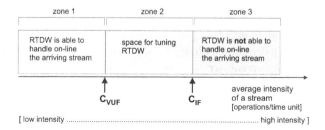

Fig. 1. Borderline values of the average arrival rate of a transaction stream

In zone 1, a RTDW is able to: (1) execute the arriving transactions that modify the RTDW tables and (2) refresh all the affected materialized views. The maximum average stream arrival rate below which the RTDW is capable of updating its tables without delays and refreshing all its materialized views is denoted as *Critical View Update Frequency* (C_{VUF}). With further increase of the stream arrival rate, the working characteristic of the RTDW enters zone 2. In this zone, in order to handle on-line the stream, a RTDW administrator has to apply various optimization techniques (from implementing advanced materialized view refreshing algorithms to dropping some of the materialized views and/or indexes). In zone 3, the RTDW is not able to maintain its content at all, even if all the materialized views were dropped. The maximum average stream arrival rate below which the RTDW is capable of updating on-line only its tables but it is not capable of refreshing its materialized views is denoted as *Critical Insert Frequency* (C_{IF}).

Problem Statement. Knowing the average arrival rate of transaction stream T_S, knowing the average number of DML operations in every transaction, and having an already deployed RTDW, figure out whether: (1) the RTDW can handle without delays stream T_S and (2) what the values of C_{VUF} and C_{IF} are.

3   RTDW-bench: Real-Time Data Warehouse Benchmark

The *RTDW-bench* that we propose permits to find the values of C_{IF} and C_{VUF} for a given, deployed RTDW. It must be stressed that in general the values of C_{IF} and C_{VUF} can be determined within a certain range of values.

The *RTDW-bench* environment is composed of: (1) a data warehouse refreshed in real time, (2) a parameterized transaction stream generator, and (3) an algorithm for finding the values of C_{IF} and C_{VUF}. A data warehouse schema uses the standard TPC-H benchmark schema. Based on the TPC-H tables, 28 materialized views were created. The views join multiple tables and compute aggregates. They are refreshed incrementally on commit. A transaction stream is characterized by a parameterized arrival rate and duration time. The stream arrival rate can be in turn parameterized by: (1) the number of transactions per a time unit and (2) the number of DML operations in a transaction.

The crucial component of *RTDW-bench* is the algorithm that finds the values of C_{IF} and C_{VUF}. The algorithm is implemented by means of two procedures, namely *Below Critical Point (BCP)* and *Critical Point Searching (CPS)*. *BCP* is responsible for: (1) defining the current value of a stream arrival rate and (2) deciding whether the current arrival rate is lower or greater than the borderline arrival rate.

The main idea of *BCP* operation is as follows. It executes 250 transactions, each of which inserts 1 row into table *Orders* and a variable number of rows into *LineItem*. The minimum and maximum number of rows inserted into *LineItem* are set up by means of input parameters. The number of rows inserted into *LineItem* is randomly generated with an even distribution. Transactions generated by *BCP* are executed in random intervals with an even distribution. The arrival rate of the transactions may vary from -0.3 to +0.3 of the arrival rate currently selected by *BCP*.

BCP maintains the sum of idle times and delay times. Determining whether the current stream arrival rate is greater or lower than the borderline value (C_{IF} or C_{VUF}) is based on comparing these times. If the sum of idle times is greater than the sum of delay times then the current stream arrival rate is lower than the borderline value.

The second procedure, i.e., *CPS*, searches for and returns the value (the range of values) of the borderline stream arrival rate (C_{IF} and C_{VUF}). *CPS* is invoked the first time for a RTDW without materialized views, in order to determine the value (range) of C_{IF}. *CPS* is invoked the second time for a RTDW that includes the set of materialized views, in order to determine the value (range) of C_{VUF}. *CPS* accepts two input parameters, namely an initial stream arrival rate, which is the starting point for searching, and the maximum estimation error of C_{IF} and C_{VUF}.

The main idea of *CPS* processing is as follows. *CPS* assumes that the value of the borderline arrival rate is within range $< 0, \infty >$. By narrowing the range, it finds the borderline stream arrival rate. First, a given arrival rate is selected. If it is lower than the borderline value, then the next arrival rate is selected by multiplying the initial arrival rate by 2^n (where n increases by 1 with every run). Notice that, for deciding whether the currently selected stream arrival rate is lower or greater than the borderline value, the *BCP* procedure is applied. When *CPS* founds the first arrival rate that is greater than the borderline value, then *CPS* uses the bisection method for determining the borderline value with the given estimation error.

4   Testing the Applicability of RTDW-bench

We tested the applicability of *RTDW-bench* to evaluating the performance of a RTDW, which was implemented in Oracle11g. To this end, we designed and executed four test scenarios that aimed at: (1) determining the values of C_{IF} and C_{VUF}, (2) finding out how the RTDW refreshing time depends on the number of materialized views that need to be refreshed, (3) finding out how the refreshing time depends on the complexity (the number of joins) of materialized views, (4) determining time overhead of system-defined and user-defined algorithms that are responsible for incrementally refreshing materialized views. The experiments were run on Oracle11g (11.2.0.1) installed on a HP DL585G1 server (4 2-core 2.4GHz AMD Opteron 850 processors, 16GB RAM, SGA size 1642MB). The TPC-H database was created with the scale factor equal to 1.

Due to a limited space, in this paper we discuss only the first test scenario, where we were interested in determining the values of C_{IF} and C_{VUF} of a stream whose duration time was fixed and equal to 5 minutes. We measured: (1) overall time (t_t) a transaction is executed, i.e., from its begin to its end and (2) time (t_c) measured from issuing the *commit* command until the end of the transaction. Notice that, in the presence of materialized views being refreshed on commit, the transaction that modified data was also responsible for refreshing all views affected by the modifications. Therefore, in the RTDW with materialized views, t_c represents the time spent on refreshing the materialized views as well.

4.1   Determining the Value of C_{IF}

In order to determine the value of C_{IF} we applied a RTDW without materialized views. To this end, *RTDW-bench* generated a stream of transactions. Every transaction encompassed 6 inserts (1 to *Orders* and 5 to *LineItem*).

The obtained results for the most interesting cases (for 7450 and 7475 transactions/minute) are shown in the two upper charts in Figure 2. They show transaction duration time t_c as well as an average value of t_c, which was computed for the whole test interval. The vertical axis represents the duration time of the test. The chart for 7450 trans/min presents the characteristic of the non-overloaded RTDW and the chart for 7475 trans/min presents the characteristic of the overloaded RTDW. For the discussed cases, the average value of t_c is equal to 44 msec (for 7450 trans/min) and 387 msec (for 7475 trans/min), respectively. In these experiments, the *CPS* procedure determined the value of C_{IF} in the range between 7460 and 7475 trans/min.

4.2   Determining the Value of C_{VUF}

In order to determine the value of C_{VUF} we deployed the RTDW with a set of 28 materialized views refreshed on commit. The views included a mixture of self-maintainable views based on one table and non-self-maintenable views based on joins from 2 to 7 tables. The queries defining self-maintainable views selected from either *LineItem* or *Orders*. The self-maintainable views grouped data by

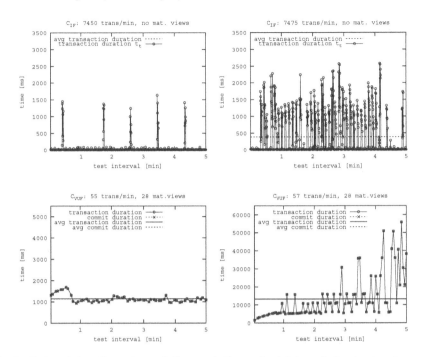

Fig. 2. Determining the values of C_{IF} and C_{VUF} [test interval: 5 minutes; transactions/minute for C_{IF}: 7450 (44700 inserts/min) and 7475 (44850 inserts/min); transactions/minute for C_{VUF}: 55 (330 inserts/min) and 57 (342 inserts/min); 28 materialized views for C_{VUF}]

the number of dimensions that varied from 1 to 4 and computed aggregates using sum, avg, and count. Similarly, the non-self-maintainable views grouped data by the number of dimensions that varied from 1 to 4 and computed aggregates using sum, avg, count, stddev, variance, and median. The aggregates were computed at different levels of dimension hierarchies, i.e., from the lowest to the highest level.

As an input to the RTDW, *RTDW-bench* generated transaction streams of different arrival rates. Every transaction in the stream included 6 inserts, as in the former test. For each of these streams we measured transaction duration time t_t and commit duration time t_c.

The obtained results for the most interesting cases (for 55 and 57 transactions/minute), are shown in the two lower charts in Figure 2. Each chart shows t_t and t_c as well as their average values, which were computed for the whole test interval. The vertical axis represents the duration of the test. The chart for 55 trans/min presents the characteristic of the non-overloaded RTDW and the chart for 57 trans/min presents the characteristic of the overloaded RTDW. In these experiments the CPS procedure determined the value of C_{VUF} in the range between 56 and 57 trans/min.

5   Conclusions

In this paper we proposed the *RTDW-bench* benchmark for evaluating a performance of a real-time data warehouse. As the performance measures we proposed: (1) the maximum transaction stream arrival rate for which a RTDW is capable of loading the transaction stream into its tables without delays (C_{IF}) and (2) the maximum transaction stream arrival rate for which a RTDW is capable of refreshing all its materialized views without delays (C_{VUF}). The core component of *RTDW-bench* is the algorithm for determining the values of C_{IF} and C_{VUF}.

We applied *RTDW-bench* to a RTDW implemented in Oracle11g. In this paper, we discussed tests for determining the values of C_{IF} and C_{VUF}. From the tests we drew the following conclusions.

- *RTDW-bench* is able to determine the values of C_{IF} and C_{VUF} with a certain precision. For example, the precision of C_{IF} was determined in the range between 7460 and 7475 trans/min, whereas the precision of C_{VUF} was determined in the range between 56 and 57 trans/min. The reason for that is some instability in the performance of the RTDW. From our experiments we observed that n consecutive executions of *RTDW-bench* resulted in n slightly different processing characteristics of the RTDW.
- The tests showed that even for a small RTDW composed of 28 materialized views and for a non intensive stream of 57 transactions (342 inserts) per minute, the RTDW gets overloaded. In real applications we may expect transaction streams of substantially higher arrival rates, e.g., Wal-Mart records 20 million of sales transactions daily (which yields 14000 operations/minute), Google receives 150 million searches daily (104000 operations/minute), AT&T records 275 million of calls daily (191000 operations/minute) [7], Internet auction service Allegro.pl handles over 2500000 auctions daily (1736 inserts/minute).

Acknowledgement. This work was supported from the Polish National Science Center (NCN), grant No. 2011/01/B/ST6/05169.

References

1. Transaction processing performance council, `http://www.tpc.org`
2. Acharya, S., Gibbons, P.B., Poosala, V.: Congressional samples for approximate answering of group-by queries. In: Proc. of ACM SIGMOD Int. Conf. on Management of Data, pp. 487–498 (2000)
3. Acharya, S., Gibbons, P.B., Poosala, V., Ramaswamy, S.: The Aqua approximate query answering system. SIGMOD Rec. 28, 574–576 (1999)
4. Chakrabarti, K., Garofalakis, M., Rastogi, R., Shim, K.: Approximate query processing using wavelets. The VLDB Journal 10, 199–223 (2001)
5. Colby, L.S., Kawaguchi, A., Lieuwen, D.F., Mumick, I.S., Ross, K.A.: Supporting multiple view maintenance policies. In: Proc. of ACM SIGMOD Int. Conf. on Management of Data, pp. 405–416 (1997)

6. Condie, T., Conway, N., Alvaro, P., Hellerstein, J.M., Gerth, J., Talbot, J., Elmele-egy, K., Sears, R.: Online aggregation and continuous query support in mapreduce. In: Proc. of ACM SIGMOD Int. Conf. on Management of Data, pp. 1115–1118. ACM (2010)

7. Domingos, P., Hulten, G.: Catching up with the data: Research issues in mining data streams. In: Proc. of ACM SIGMOD Workshop on Research Issues in Data Mining and Knowledge Discovery (2001)

8. Golab, L., Johnson, T., Shkapenyuk, V.: Scheduling updates in a real-time stream warehouse. In: Proc. of Int. Conf. on Data Engineering (ICDE), pp. 1207–1210. IEEE Computer Society (2009)

9. Graefe, G., König, A.C., Kuno, H.A., Markl, V., Sattler, K.-U.: Robust query processing. Dagstuhl Seminar Proceedings, vol. 10381. Schloss Dagstuhl - Leibniz-Zentrum für Informatik, Germany (2011)

10. Karakasidis, A., Vassiliadis, P., Pitoura, E.: Etl queues for active data warehousing. In: Proc. of Int. Workshop on Information Quality in Information Systems, pp. 28–39. ACM (2005)

11. Krueger, J., Tinnefeld, C., Grund, M., Zeier, A., Plattner, H.: A case for online mixed workload processing. In: Proc. of Int. Workshop on Testing Database Systems (DBTest). ACM (2010)

12. Poess, M., Nambiar, R.O., Walrath, D.: Why you should run tpc-ds: a workload analysis. In: Proc. of Int. Conf. on Very Large Data Bases (VLDB), pp. 1138–1149. VLDB Endowment (2007)

13. Polyzotis, N., Skiadopoulos, S., Vassiliadis, P., Simitsis, A., Frantzell, N.: Supporting streaming updates in an active data warehouse. In: Proc. of Int. Conf. on Data Engineering (ICDE), pp. 476–485. ACM (2007)

14. Sharaf, M.A., Chrysanthis, P.K., Labrinidis, A., Pruhs, K.: Algorithms and metrics for processing multiple heterogeneous continuous queries. ACM Trans. Database Syst. 33, 5:1–5:44 (2008)

15. Simitsis, A., Vassiliadis, P., Dayal, U., Karagiannis, A., Tziovara, V.: Benchmarking ETL Workflows. In: Nambiar, R., Poess, M. (eds.) TPCTC 2009. LNCS, vol. 5895, pp. 199–220. Springer, Heidelberg (2009)

16. Thiele, M., Fischer, U., Lehner, W.: Partition-based workload scheduling in living data warehouse environments. Information Systems 34(4-5), 382–399 (2009)

17. Tziovara, V., Vassiliadis, P., Simitsis, A.: Deciding the physical implementation of etl workflows. In: Proc. of ACM Int. Workshop on Data Warehousing and OLAP (DOLAP), pp. 49–56 (2007)

18. Wyatt, L., Caufield, B., Pol, D.: Principles for an ETL Benchmark. In: Nambiar, R., Poess, M. (eds.) TPCTC 2009. LNCS, vol. 5895, pp. 183–198. Springer, Heidelberg (2009)

Middleware and Language for Sensor Streams

Pedro Furtado

University of Coimbra,
Polo II, Coimbra
pnf@dei.uc.pt

Abstract. This paper describes a new stream-based sensor networks programming approach - language + middleware. Since many data processing aspects of sensor networks can be expressed intuitively as streams, commands and events, we apply such constructs to program individual nodes. Arbitrarily defined groups of nodes can be programmed with the same language and as a single program. We show that the language is indeed able to express different intentions – we are able to configure flexibly how the sensor network will process data and how data will flow; that it is able to generate efficient code with small memory footprints; and that it provides easy integration with platform code.

Keywords: distributed systems, wireless sensor networks.

1 Introduction

Wireless Sensor Network (WSN) embedded device nodes have limited amounts of memory, processing power and energy, therefore they typically run small OSs for embedded devices (e.g. tinyOS). The devices communicate wirelessly and one or more is connected to the internet through the USB adapter and a serial read/write driver as interface, with some form of gateway software above that. We argue that a stream processing style can represent the typical sensor network functionality, and that this is very advantageous for easy deployment in real-world applications, without the need of professional WSN programmers.

Current solutions for deploying the heterogeneous systems that include WSN nodes and the rest of the world involve either programming every detail of processing and communication by hand, both within the WSN and outside of it, or to use some middleware that works as a single black-box for the whole system (and must be ported to other systems as a whole). The state-of-the-art middleware approaches fail to consider the infrastructure as a heterogeneous distributed system that could be programmed by a simple stream model and with a uniform node component over heterogeneous physical nodes. We propose TinyStreams, a model to enable such vision. By considering a single node component that can be installed in any node, including nodes outside of the WSN, and a stream-like data management approach, we ensure that all nodes will have at least a uniform configuration interface (API or SQL-like) and important remote configuration and processing capabilities without any further programming or gluing together. As an

S.W. Liddle et al. (Eds.): DEXA 2012, Part II, LNCS 7447, pp. 207–214, 2012.

immediate advantage of our proposal, a control station and indeed any node outside of the WSN will have at least the same configuration and processing capabilities and the same interface as the remaining nodes without any custom programming, and it will be easy to command the system to do whatever data acquisition and operation we desire. More generically, in a heterogeneous deployment with different types of sensor devices, control stations and sub-networks, the approach offers easy and immediate homogeneity over heterogeneous infrastructures.

The paper is organized as follows: section 2 discusses related work. Section 3 and 4 discuss the architecture of the approach, then section 5 presents experimental results and section 6 concludes the paper.

2   Related Work

WSN programming approaches either use the platform language and the stream processing engine works on the workstation code to process data extracted from the WSN, or approaches such as TinyDB [6] go a bit further in providing a query engine for the WSN itself. Yet, those are not a language for programming the WSN nodes, rather a query engine commanding and using predefined and largely fixed protocols, optimizer and query processor. Since the ability to configure and program explicitly how the data is to be processed in nodes or groups of nodes, how data should flow between nodes or groups, what, when and how nodes should do something, is present in a typical WSN programming paradigm, TinyStreams offers a CQL-like stream processing language for programming the WSN.

[7] defines a taxonomy for high-level programming of sensor networks distinguishing, among others, communication and computation perspectives, programming idiom and distribution model. Distribution models can be classified as database-oriented, where SQL-like queries are used as in a relational database (e.g. TinyDB [6]), data sharing-oriented, where nodes can read or write data in the shared memory space (e.g. Kairos [4] and Abstract Regions [8]); or as message passing, based on exchanging messages between nodes (e.g. NesC [3], DSWare [5] or Contiki [2]). Database-oriented approaches such as TinyDB are very intuitive to use, but message passing paradigms are typically much more flexible, since they allow the programmer to specify exactly what is exchanged and how. TinyStreams has the flexibility of message passing, since it is possible to specify data exchanges between individual or groups of nodes, and has the advantage of database-oriented, since it uses an intuitive stream model.

3   Data Processing Model and Configuration

A stream processing model is ideal to represent the kinds of operations that exist typically in data collection sensor networks. A stream is modeled as a window of one or more values, and streams can receive their values from sensors or from other streams. Communication primitives are completely hidden to users, since the specification that a stream in a node reads form another stream in one or more nodes means that the data must be routed from the producer nodes to the consumer node. The stream model is quite expressive, since it contains the same clauses as the sql model, plus the explicit timing and data routing clauses. Commands and events do the rest of the job to represent remote configuration and actuation commands.

The following example shows how sensor signals are inserted into streams, from then on operations specify manipulations over streams. The creation of streams is through either a call to an API method, or a corresponding SQL-like command syntax. This is exemplified next for a list of sensor addresses represented as "Zone1SensorNodes". This commands acquisition every second:

```
API dialect:
CreateStream(Zone1SensorNodes,
"pressureStreamfromZone1SensorNodes",
1s, {(PRESSURE, VALUE)});


SQL-like dialect:
create stream pressureStreamfromZone1SensorNodes
in Zone1SensorNodes as
select pressure
from sensors
sample every 1 second;
```

Streams can be created with conditions, in which case they are called conditional streams or alarm. In the following example, a control station receives signals from the sensors and determines an alarm when pressure rises above a specified threshold:

```
API dialect:
CreateAlarm( controlSation1, "ServerPressureAlarm",
pressureStreamfromZone1SensorNodes,
CONDITION( PRESSURE.VALUE, >, 5));


SQL-like dialect:
create alarm ServerPressureAlarm in controlSation1 as
select NODEID, pressure
from pressureStreamfromZone1SensorNodes
where pressure>5;
```

The next example concerns a node called "Merger" that receives the streams from "Zone1SensorNodes", and computes the average and maximum of those values. Then a control station receives and displays those values.

```
API dialect:
    CreateStream( relay1, "Zone1MergerStream",
pressureStreamfromZone1SensorNodes,
NO CONDITION,
"avg(PRESSURE.VALUE),max(PRESSURE.VALUE)" );
```

```
    CreateStream( controlStation1, "ServerStream",
Zone1MergerStream,
NO CONDITION);

SQL-like dialect:
create stream Zone1MergerStream in relay1 as
select avg(pressure), max(pressure)
from pressureStreamfromZone1SensorNodes;

create ServerStream in controlStation1 as
select avg(pressure), max(pressure)
from Zone1MergerStream;
```

Actuation commands can be submitted by specifying the target node, an identifier of a DAC and a value:

```
    Actuate( nodeX, "ActuationNodeX", {(VALVE1, 100)});
```

Conditional actuation commands can be submitted by specifying the sensing nodes, de decision node, the actuator nodes, an identifier of a DAC and a value:

```
CreateActuation(sinkNode,
"Actuate1",pressureStreamfromZone1SensorNodes,
CONDITION( PRESSURE.VALUE, >, 5),
nodeX,(VALVE1, 100));
```

Internally, the stream-based data processor resident in the ConfigProcess component manages a circular window per stream, where incoming values are placed and computed upon.

4   Middleware Components

Fig. 1 shows the components of the middleware that supports the stream model. In the figure we can see middleware nodes (mw node, nodes outside wsns), gateway nodes – nodes that link a wsn with the cabled part of a system – and wsn nodes, which include sink nodes and other wsn nodes. A node middleware component called ConfOperate is developed only once for each specific platform, and it implements the configuration and operating functionality of the stream model. In the figure this is represented by ConfOperate-java, ConfOperate-tinyOS and ConfOperate-contiki.

A SysConfig application interacts and controls the nodes through the ConfOperate components, and offers a Config API for external applications to configure the system (e.g. this API can be exposed as webservices). The SysConfig UI in **Fig. 1** is an application interfacing with the SysConfig API and offers the programming interface where the commands that we described for the model are written.

Finally, the Data interface is an interface to get and put data into any node in the system. It can be implemented using a publish-subscribe pattern.

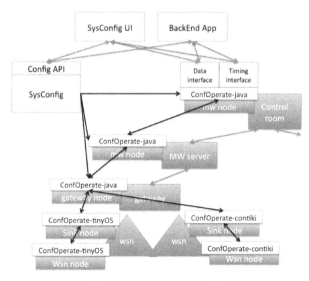

Fig. 1. Node and Config SW

4.1   ConfOperate

The ConfOperate component does two main things: responds to network message arrival events; operates periodically. It has a simple structure for keeping and organizing data. These are described next.

NetArrivalEvent - Fig. 2 shows a diagram of the NetArrival event callback. A message arrival from the network triggers this callback. It does the following:

1) If the arrived message is a data message, it will have a stream identifier associated with it, so the data will enter the stream window for the identified stream;

2) If the arrived message is an actuation command, the actuator DAC will be identified in the message and a value coming in the message will be written to the DAC;

3) If the message is a configuration command, the command will configure (add, remove, change, start, stop) some structure.

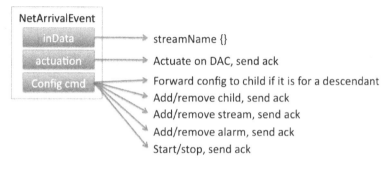

Fig. 2. Functionality of the NetArrivalEvent

Data Containers- Fig. 3 shows a diagram of the data container for each stream. It is simply a circular array of values (with a configurable array size). There may be any number of streams in a node. The main operations over the stream are "create and drop" for creation and deletion of the stream, and insert and get for adding or getting data to/from the stream.

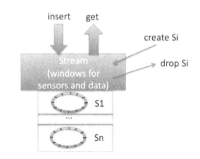

Fig. 3. Stream Windowed Data Container

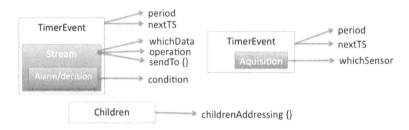

Fig. 4. ConfOperate Lowest Common Denominator Structures

Operation - Operation implements periodic processing of data values. It is controlled through timer events, and there may be several simultaneous timer events in a node. When the time comes for an event, the operation procedure processes whatever are the event specifics and reschedules the timer for the next period. The lowest common denominator for those events is the acquisition event (for periodic data acquisition from sensors), the stream event (for periodic sending of data to destinations, or to periodically operate on data and send the result), and the alarm or decision event (for testing some condition on data, raise an alarm or decision value, and send it to some destination). Fig. 4 shows the lowest common denominator structures of ConfOperate, with an indication of what information they must possess:

– TimerEvent must know the period and next triggering timestamp;

– Acquisition must be a TimerEvent and have information on which sensor to sample;

– Stream must be a TimerEvent and know which data is stored, who to send the data to when the time event is triggered and what operations should be done on the data from one or more streams. Examples of operations: window average, sum, maximum;

– Alarm/decision is a Stream plus a condition. The event will test the condition, compute a value and send it if the condition resolves to true;

– Children keeps information on which are the node children (there is also a parent structure identifying the parent(s).

5   Experimental Evaluation

The testbed is a network with 16 TelosB nodes organized hierarchically in a 3-2-1 tree and a control station receiving the sensor samples. The WSN nodes run the Contiki operating system with a tdma network protocol to provide precise schedule-based communication. An epoch size of 900 msecs is divided into 10 msecs slots. The protocol schedules slots for all nodes of the tree to send their data upstream in round-robin fashion (2 slots, one for the message, another one for retry), then adds a downstream slot for commands to be sent down. The following code start a sensor collection stream with a 3 seconds sampling rate, to change the rate to 2 seconds and to stop the stream.

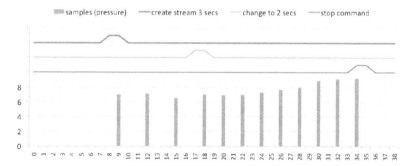

Fig. 5. Sensor Readings and Reconfigure

1. Create stream with readings every three seconds:
```
createStream(Node1, "pressureSensor1", 1s,
{(PRESSURE, VALUE)});
CreateStream( Node1, "pressureStreamfromNode1",
pressureSensor1, 3 seconds,
NO CONDITION,
"avg(PRESSURE.VALUE),max(PRESSURE.VALUE)" );

createStream(ControlStation,"pressureSensor",
pressureStreamfromNode1);
```
2. Modify the sampling rate to every two seconds:
```
setSamplingRate(pressureStreamfromNode1, 2s);
```
3. Stop sampling the sensor (then restart):
```
stop(pressureStreamfromNode1);
```

Fig. 5 shows the results of the commands. From the figure it is possible to see that pressure samples started to be collected at a rate of a sample every 3 seconds after command 1 configured the sensor; 9 seconds after that the sampling rate was changed to a sample every 2 seconds and the chart shows that the collected samples rate changed accordingly to one every 2 seconds; finally the stop command at second 34 was successful at stopping sensor sampling.

References

1. Abadi, D.J., et al.: Aurora: a new model and architecture for data stream management. The VLDB Journal 12(2), 120–139 (2003)
2. Dunkels, A., Gronvall, B., Voigt, T.: Contiki - A Lightweight and Flexible Operating System for Tiny Networked Sensors. In: Proceedings of the 29th Annual IEEE International Conference on Local Computer Networks, pp. 455–462. IEEE Computer Soc. (2004)
3. Gay, D., et al.: The nesC language: A holistic approach to networked embedded systems. In: Proceedings of the ACM SIGPLAN 2003 Conference on Programming Language Design and Implementation, pp. 1–11. ACM, San Diego (2003)
4. Gummadi, R., Gnawali, O., Govindan, R.: Macro-programming Wireless Sensor Networks Using *Kairos*. In: Prasanna, V.K., Iyengar, S.S., Spirakis, P.G., Welsh, M. (eds.) DCOSS 2005. LNCS, vol. 3560, pp. 126–140. Springer, Heidelberg (2005)
5. Li, S., Son, S., Stankovic, J.: Event Detection Services Using Data Service Middleware in Distributed Sensor Networks. Telecommunication Systems 26, 368, 351 (2004)
6. Madden, S.R., et al.: TinyDB: an acquisitional query processing system for sensor networks. ACM Trans. Database Syst. 30(1), 122–173 (2005)
7. Mottola, L.: Programming Wireless Sensor Networks: From Physical to Logical Neighborhoods. Politecnico di Milano (Italy) (2008), http://www.sics.se/~luca/
8. Welsh, M., Mainland, G.: Programming sensor networks using abstract regions. In: Proceedings of the 1st Conference on Symposium on Networked Systems Design and Implementation, vol. 1, p. 3. USENIX Association, San Francisco (2004)
9. Yao, Y., Gehrke, J.: The cougar approach to in-network query processing in sensor networks. SIGMOD Rec. 31(3), 9–18 (2002)

Statistical Analysis of the `owl:sameAs` Network for Aligning Concepts in the Linking Open Data Cloud

Gianluca Correndo, Antonio Penta, Nicholas Gibbins, and Nigel Shadbolt

Electronics and Computer Science, University of Southampton, UK
{gc3,ap7,nmg,nrs}@ecs.soton.ac.uk
http://www.ecs.soton.ac.uk/people/{gc3,ap7,nmg,nrs}

Abstract. The massively distributed publication of linked data has brought to the attention of scientific community the limitations of classic methods for achieving data integration and the opportunities of pushing the boundaries of the field by experimenting this collective enterprise that is the linking open data cloud. While reusing existing ontologies is the choice of preference, the exploitation of ontology alignments still is a required step for easing the burden of integrating heterogeneous data sets. Alignments, even between the most used vocabularies, is still poorly supported in systems nowadays whereas links between instances are the most widely used means for bridging the gap between different data sets. We provide in this paper an account of our statistical and qualitative analysis of the network of instance level equivalences in the Linking Open Data Cloud (i.e. the **sameAs** network) in order to automatically compute alignments at the conceptual level. Moreover, we explore the effect of ontological information when adopting classical Jaccard methods to the ontology alignment task. Automating such task will allow in fact to achieve a clearer conceptual description of the data at the cloud level, while improving the level of integration between datasets.

Keywords: Linked Data, ontology alignment, owl:sameAs.

1 Introduction

The increasing amount of structured information published on the Web in linked data is rapidly creating a voluminous collective information space formed of inter-connected data sets; the Linking Open Data cloud (LOD henceforth). The last version of the LOD diagram (2011/09/19) included 295 data sets, ranging from topics like encyclopaedic knowledge, to e-government, music, books, biology, and academic publications. These data sets are linked, most of the times, at the instance level where URIs representing entities are reused or aligned towards external URIs using `owl:sameAs` properties to link equivalent entities. According to OWL semantics [2], all entities within the closure set of the `owl:sameAs` relation are indistinguishable, thus every statement including one entity can be rewritten by replacing any of the equivalent element.

S.W. Liddle et al. (Eds.): DEXA 2012, Part II, LNCS 7447, pp. 215–230, 2012.

The problem of discovering "same" entities in different data sets, known as the record linkage problem, is quite well known in database community where a large body of literature can be found on the topic [20]. Semantic Web community has built upon the database research and proposed its set of solutions [7]. The discovery of equivalent entities in the Web of Data is therefore supported by automatic tools which exploit, similarly to ontology matching or record linkage tools, lexical and/or structural similarities between the entities of different data sets [10,18]. Semi-automated approaches has been also implemented in tools like Google Refine[1], where linkages found are subject to user approval. The collaborative effort of data publishers in inter-connecting their data sets has created a network of equivalences between instances which is a matter of study on its own, the **sameAs** network [4]. Studying the properties of this **sameAs** network in conjunction with the network of Class-Level Similarity, or CLS network as defined in [4] (i.e. the network of classes which overlaps because sharing same, or equivalent, instances), can lead us to a better understanding of how heterogeneous data sets can be integrated together.

Despite of the great amount of linkages between instances and the high availability of tools for aligning vocabularies, little effort has been devoted to provide authoritative alignments between the ontologies present in the LOD. As a representative example, in DBpedia the only alignments between ontologies, retrieved by querying the public endpoint, have been published by using `owl:sameAs` properties between concepts in `opencyc.org`[2], and `owl:equivalentClass` properties between `schema.org`[3] concepts.

The availability of ontology alignments in the LOD would allow the use of tools that exploit schema level mappings for achieving data integration [19], fuelling in this way a wider use of published linked data. The work described in this paper starts from the above consideration and attempts to exploit the available **sameAs** network in order to deduce statistically sound dependencies between concepts which have common instances taking into consideration the semantics attributed to the `owl:sameAs` property [2].

The work we presented in this paper is an account of our first attempts to adopt a well known instance-based technique (i.e. Jaccard coefficient) in discovering alignments between concepts in the LOD cloud. The vast amount of entity alignments present in the LOD cloud, under form of `owl:sameAs` statements, provides a good asset to experiment such an approach. Although, applying statistical techniques to a potentially very noisy data set for aligning heterogeneous ontologies could prove to be unreliable to some extent. This paper reports our attempts to study the behaviour of such basic technique on a real scenario. The rationales behind this approach are to be found in a previous work in the instance based ontology matching field [9], which did not addressed specifically the LOD, and an analysis on the deployment of `owl:sameAs` networks [4].

---

[1] http://code.google.com/p/google-refine/

[2] http://sw.opencyc.org/

[3] http://schema.org

The paper starts with Section 2 which provides some background information on instance based ontology alignments, how they are implemented in this work by exploiting `owl:sameAs` alignments, and finally describes the data used in this experiment. Section 3 provides some initial analysis, quantitative as well as qualitative, on the data used and on the alignments found in the CLS network. Section 4 provides an account of the behaviour of Jaccard based measures under different hypothesis by studying the usual indices from Information Retrieval (i.e. number of alignments, precision, and recall). Section 5 provides an account of similar works in the area of Linked Data and finally our conclusions are presented in Section 6.

2   Alignment Based on sameAs Network Analysis

The ontology alignment task has been widely studied in the last decade by the scientific community [7]. Ontology matching tools usually exploit a number of information sources such as lexical or structural similarity applied to the ontologies alone in order to produce a measure of the semantic distance between concepts. In recent years, methods based on statistical information (e.g. machine learning, bayes, etc.) have been also studied and proved to produce promising results [5,9].

The high level of inter-linking within the LOD cloud induces us to consider statistical techniques for ontology alignment as a promising approach to resolve semantic heterogeneity. The assumption we adopted in this work is that `owl:sameAs` equivalence bundles [8] can be treated as singleton instances whose interpretation is provided by following `owl:sameAs` semantics. Therefore all equivalent instances, hosted by different data sets, will be considered as a unique instance which is classified differently in different data sets (as seen in Rule 1 where *type* is `rdf:type` and *sameas* is `owl:sameAs`).

$$type(?x, ?xt) \wedge sameas(?x, ?y) \wedge type(?y, ?yt) \rightarrow type(?b, ?xt) \wedge type(?b, ?yt) \quad (1)$$

Leveraging the `owl:sameAs` inference we are then able to treat equivalence bundles as instances and compute the degree of overlapping between concepts by processing the typing statements (i.e. statements in the form $type(?b, class)$). In our approach we used the Jaccard coefficient [11] ($J(A, B)$ in Equation 2) in order to measure the similarity between two concepts when interpreted as sets of instances.

$$J(A, B) = \frac{|A \cap B|}{|A \cup B|} = \frac{|A \cap B|}{|A| + |B| - |A \cap B|} \quad (2)$$

Here the cardinality of the intersection set $|A \cap B|$ is computed in our triple store by counting the cardinality of the set $\{\langle x, y \rangle : type(x, A) \wedge same(x, y) \wedge type(y, B)\}$. The cardinality of the union set is then computed by summing the cardinality of the set of instances for the two concepts A and B (i.e. $\{x : type(x, A)\}$ and $\{x : type(x, B)\}$ respectively) and then subtracting the intersection as previously defined.

2.1 Definition of Ontology Alignment

In the work here described we reused and modified the framework proposed in
Isaac et al. [9] for representing instance-based alignments. In [9] an alignment
between a source ontology S and a target ontology T is a triple $t \in S \times T \times R$
where R is a relation taken from the set $\{\equiv, \sqsubseteq, \sqcap, \bot\}$ which expresses respectively equivalence, subsumption, overlap and disjointness. Such definition fits a
scenario where describing some informal degree of relatedness, measurable by
sets overlapping, is acceptable and even desirable. Given the target objective of
our work, data integration, we set for a less richer framework where it is possible
to distinguish only between $\{\equiv, \sqsubseteq, \bot\}$ since we could not make any use of information about overlapping concepts for integrating different data sets into an
homogeneous vocabulary. Moreover, when we state that two concepts are equivalent (i.e. $A \equiv B$), since we are not taking into account the concepts definitions
but merely the possibility of them covering the same set of instances, we will
intend that the two concepts are in `owl:equivalentClass` relationship, and not
`owl:sameAs`. Hence, the two concepts can still have different definitions without
causing any inconsistency.

In the subsequent evaluation of the alignments (see Section 4) we will consider
a successful alignment, a 'true positive, one which correctly correlate a couple of
concepts which are equivalent or in a relation of subsumption (i.e. one subsumes
non trivially the other). Any alignment provided which includes two disjoint
concepts is a false positive. This shrink in the power of discrimination is due
to the nature of the Jaccard measure itself which has bees devised to measure
concepts equivalence only.

2.2 Experimental Setup Based on LOD Entities

In order to experiment the usefulness of the Jaccard coefficient as a means for
measuring the semantic similarity between concepts in the LOD cloud a source
data set and a number of target data sets, aligned to the selected source by
`owl:sameAs` links, have been considered.

Because of its centrality in the LOD cloud, `DBpedia` is the natural candidate as a source data set while a number of target data sets has been selected
based on their abundance of instance alignments and diversity of size in terms
of concepts to align. The target data sets considered for our experiments are
described in Table 1 where for the source data set is reported the number of
info box concepts used to classify the DBpedia instances[4], and for each one of
the target data set is reported the number of equivalence links connecting it to
DBpedia and the number of concepts contained. It noteworthy that for for the
`nytimes` data set, although containing a rich hierarchy of terms, only two concepts are found. This is due to the fact that all the entities aligned are instances
of `skos:Concept` and some of `geonames:Feature`. Therefore, not knowing any
background information about the dataset, it is not possible to recognize a valid

---

[4] The dump used in this experiment is the DBpedia version 3.7.

OWL concept hierarchy since it is encoded in a vocabulary (i.e. SKOS [15]) which encodes concept hierarchies only between instances and not between concepts. This implies that every instance mapping from an instance in `DBpedia` to one in `nytimes` will support a correspondence between an OWL concept to the concept `skos:Concept` which is not very informative as an alignment between ontologies.

Table 1. Data sets considered for the experiments

source		number of concepts
DBpedia		9237320
target	number of `owl:sameAs`	number of concepts
opencyc	20362	314671
nytimes	9678	2
drugbank	4845	4
diseasome	2300	2
factbook	233	1
dailymed	43	1

Once identified the source and target data sets, we proceeded to download from the respective websites the triples belonging to: the **sameAs** network; the **type** network; and the **concepts hierarchy**. As we already mentioned, the **sameAs** network of a data set D is the set of triples contained in D which connect two entities by the property `owl:sameAs` (i.e. consistently with the notation already used: $sameas(D) = \{same(s, p) \in D\}$). The **type** network of a dataset D is the set of triples contained in D which connect every entity with one or more concept by the property `rdf:type` (i.e. $type(D) = \{type(a, b) \in D\}$). Finally, the **concepts hierarchy** is the set of triples contained in D wich connect two concepts by the property `rdfs:subClassOf` (i.e. $hierarchy(D) = \{subclassof(a, b) \in D\}$).

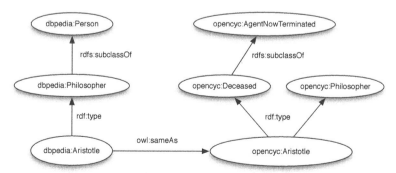

Fig. 1. Example of different networks extracted from DBpedia neighbours

An example of the networks taken into consideration in this paper are depicted in Figure 1. For the sake of the statistical analysis conducted in the experiments we did not take into consideration any other property which could describe the entities and the concepts (e.g. labels, abstracts or other properties) since only the `owl:sameAs` bundles are considered. In Figure 1 we can see how the bundle {`dbpedia:Aristotle`, `opencyc:Aristotle`} belongs to the intersection of the concepts `dbpedia:Philosopher` and `opencyc:Philosopher`, and `dbpedia:Philosopher` and `opencyc:Deceased`. By computing the number of co-occurrences of concepts connected by a common bundle in the way showed by Figure 1 we are able to compute the size of the intersection set and then to compute the Jaccard measure for each couple of concepts.

3   Experiment Scenario Analysis

In order to better understand the characteristics of such collected network, we decided to study a couple of aspects before processing the data in trying to discover concept alignments. The first thing we decided to look into is the size of the sameas bundles collected. Since the number of concepts reached from a single DBpedia entity can be reasonably related to the size of its equivalent class computed via `owl:sameAs` links, studying the distribution of such parameter can give us an insight about the variance we can expect in processing such bundles. The distribution of the frequency of the bundles' size is depicted in the graph in Figure 2, where the dimension of the y axes is reported in logarithmic scale. Considering the distribution in Figure 2 we can see that the size of bundles follows a logarithmic distribution where the more frequent size is 2, i.e. only one other entity except the source entity, and where bundles of size greater than 10 are very infrequent.

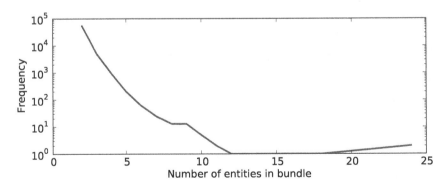

Fig. 2. Frequency of sameas bundles by size

The second aspect we studied, once we computed the Jaccard coefficient from the collected data, is the ratio of the cardinalities between aligned concepts. The hypothesis we formulated, given our past experience in handling linked data, is that the cardinality of overlapping concepts in the LOD cloud would be highly heterogeneous and therefore we would have a high level of asymmetry between the aligned data sets. In Figure 3 is reported the frequency of alignments plotted against the ratio (expressed in percentage) of the cardinality of the two concepts aligned[5]. Looking at Figure 3 we can say that concepts with similar cardinality would be nearer the right end of the graph, while concepts dissimilar in cardinality would be nearer to the left end of the graph; the graph reported makes clear that the vast majority of alignments produced are between concepts dissimilar in cardinality. Although this result is particular to the scenario under scrutiny, it is also true that DBpedia is a typical example of a general domain hub data set whose behaviour in terms of inter linkages is likely to be seen in other hub data sets.

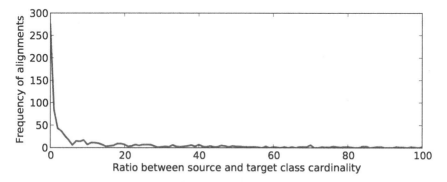

Fig. 3. Ratio of cardinalities between aligned concepts

3.1 Qualitative Analysis of Jaccard Alignments

Before discussing into detail the quality of the Jaccard coefficient as a means for producing ontology alignments in the LOD cloud (topic covered in Section 4), we would like to provide a qualitative analysis of the first batch of results. Among the alignments we have manually checked for judging their quality, in order to compute the precision and recall of the procedure, we noticed that some of the alignments produced, which were supported by statistical evidence, were quite interesting in nature.

Many of the alignments produced, even with a high value of Jaccard coefficient, had the `owl:Thing` as a source concept. This is due to the fact that many DBpedia instances have multiple types associated, and the top of the hierarchy is directly, and quite frequently, mentioned in the type network, whatever the level of abstraction of the entity is. This fact hinders us in identifying a canonical classification of entities and introduces some noise in discovering concept alignments. The root of the OWL hierarchy is in fact non trivially equivalent to any other concept in any other ontology, at the same time is the superclass of all OWL based hierarchies, therefore any mapping would provide very little information gain.

---

[5] The ratio has been normalised between [0,1].

Table 2. Concept alignment patterns

source[dbpedia]	target[opencyc]
Model	Woman
Writer	Male
Philosopher	Dead organism
Monument	Bell

Similarly, as mentioned earlier for the `nytimes` data set, encoding all entities as `skos:Concept` instances, implies that the only mapping one can find within the data set is to that concept, rendering useless any alignment effort. We may expect to find the same results every time we try to exploit entities alignments between domain concept instances and knowledge organization systems as thesauri and classification schemes.

The last consideration we did on the alignments found is on some related patterns that seem to be quite common and which could be justified by a cultural and contextual interpretation of the data, and alignments. An account of some of the unusual patterns discovered by processing the concepts co-occurrences is provided in Table 2. As we can see, the first two alignments are indicative of a statistical preference of representing female models and male writers[6]. The alignment between concept *Philosopher* and *Dead organism* is proposed as less likely than the correct alignment (i.e. opencyc *Philosopher* concept), and it is probably due to the fact that the vast majority of the philosophers described in DBpedia are actually deceased. The last alignment is due to the fact that in DBpedia, listed as entities of type *Monument* are just historical bells (e.g. the Liberty Bell in Philadelphia). Therefore, although odd, the wrong alignment reflects the extensional definition of the concept which clearly conflicts with the semantics we would expect from the *Monument* concept.

4   Evaluation

In order to study the behaviour of Jaccard alignments we collected the usual measures from Information Retrieval under different conditions. We proposed in fact two scenarios that affect either the way the alignments are produced or the way the alignments are used, and we measured the performances of Jaccard for each scenario. The measures under scrutiny are: the **Number of alignments** computed (either correct or incorrect), the **Precision** of the alignments computed, **Recall** of the alignments computed, the **F-measure**[7] of the results, and finally the **Precision at** n^{th} of the alignments[8].

The first scenario explored the gain we have when we take into account (or not) the concept hierarchy when we compute the cardinality of the two sets, A

---

[6] Note, this is not due only to the particular source data considered but also to the instance alignment performed on the target data set.

[7] The harmonic mean of precision and recall

[8] The precision computed for the first n^{th} alignments

and B respectively, as defined in Section 2. For doing this we used 4sr, a reasoner that efficiently implements the reasoning over rdfs:subClassOf axioms (sc_0 and sc_1 rules in [17]).

The second scenario studied the different performances we gain when relaxing the acceptance criteria from **equivalence** only (i.e. an alignment is considered correct if the two concepts are equivalent) to **equivalence** or **subsumption** (i.e. an alignment is considered correct even when two concepts are not trivially subsuming one another). Around a thousands of the generated concept alignments have been manually checked and classified as: erroneous, subclass/ uperclass, or correct. The precisions have been computed by considering as successful either subclass/uperclass and correct or correct only. The legends of Figure 4, 5, 6, 7, and 8 reports **Jaccard** when no hierarchy information is used and only equivalent entities are considered as correct, **h Jaccard** when hierarchy information is used and equivalent concepts are considered, **s Jaccard** when no hierarchy information is used and with subclasses considered as correct, and finally **hs Jaccard** when hierarchy information is used and with sub concepts considered as correct. Finally, the recall of the respective measures have been computed by taking the maximum number of correct alignments found as the reference limit. That is why in Figure 6 the legend reports **Relative** recall in the label.

4.1 Number of Alignments

The comparison here is made by using or not the concepts hierarchy in the computation of the Jaccard coefficients, since the acceptance of the produced alignments does not influence their generation. A first superficial analysis of the distribution of the number of alignments found per different values of thresholds (see Figure 4), the number of alignments produced increases exponentially when lowering the threshold value and it is noteworthy the fact that the most of the alignments are produced with very little values of threshold. This implies that it is important to maintain a good quality of the alignments even at low values of thresholds since the amount of false positives could hinder the usability of the produced alignments to a point where human intervention could not be feasible any more.

Moreover, comparing the distribution of generated alignments we can notice that, even if both distributions are inversely exponential, including hierarchical information increases drastically the number of produced alignments. A superficial analysis showed that the rate between the two distributions increases from 1, for higher values of thresholds, to 15, for lower thresholds.

4.2 Precision

In Figure 5 it is shown the graph of the distribution of the precision of the alignments, under different conditions as described earlier, by varying the value of threshold. The comparison of Jaccard performances (acceptance for equivalence) with and without hierarchical information shows that when decreasing

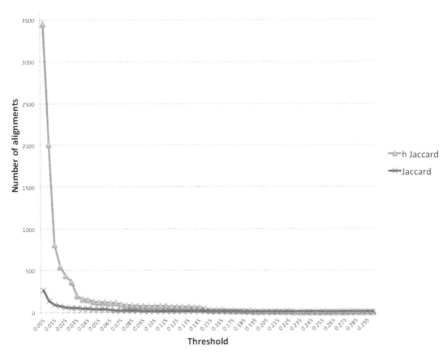

Fig. 4. Number of alignments found per threshold value

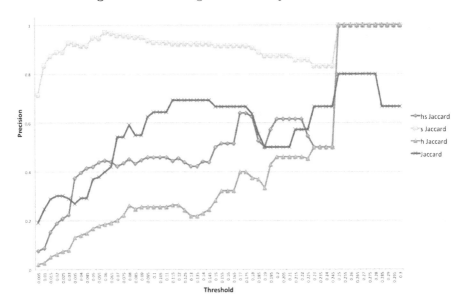

Fig. 5. Precision per threshold value

the threshold level, the more informative measure (the one with the hierarchical information) drops its precision level drastically and from that point on its precision is always worse than the less informative measure (i.e. the one without hierarchy information).

Comparing instead the two acceptance criteria, the one for equivalence and the one with equivalence or subsumption (see Figure 5 **Jaccard** and **s Jaccard**), we can notice that Jaccard provides increasingly imprecise equivalence alignments when lowering the threshold, while the precision of the method is steadily high if we are satisfied with alignments that we can refine later on. Even then though, by using hierarchical information (see Figure 5 **hs Jaccard**) the precision of the method drops quickly to unacceptable levels.

One striking fact from all the precision distributions depicted in Figure 5 is that such distributions are not monotone non-decreasing as one would expect. In fact for all distributions there are frequent local maxima and only the general trend is, for all plots, increasing. This strange behaviour could be caused by the high level of noise within the **sameAs** network, although further experiments are needed to confirm that.

4.3   Relative Recall

In Figure 6 we can see the graph for the relative recall of each measure. Not surprisingly we can see that all distributions are monotonic decreasing and that, roughly for all levels of thresholds, **hs Jaccard** provides the highest recall, followed by **s Jaccard**, **h Jaccard**, and finally **Jaccard** which is the less prolific method. For lowest levels of thresholds we can see that measures that share the same acceptance criteria provides more similar recall values while the use of

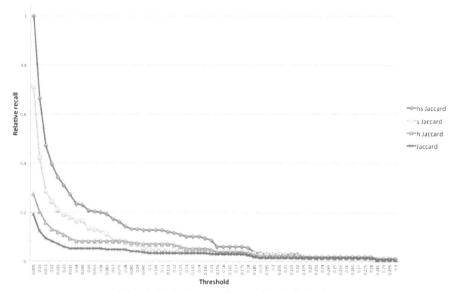

Fig. 6. Relative recall per threshold value

hierarchical information, although it increases the recall of a method, it affects less heavily the overall behaviour of a method.

4.4   F-Measure

The value of F-measure computed as the harmonic mean of precision and (relative) recall depicted respectively in Figure 5 and Figure 6 are reported in Figure 7. The most remarkable thing when considering the plots in Figure 7 is that the two less informed measures (i.e. **s Jaccard** and **Jaccard**) shows the same monotonic non-increasing trend while the two most informed measures (i.e. **hs Jaccard** and **h Jaccard**) have a local max before decreasing.

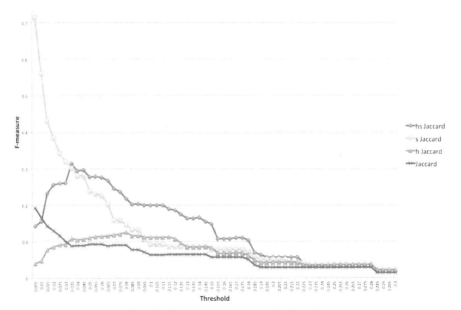

Fig. 7. F-measure per threshold value

Although F-measure is only an indication of the overall performances of an information retrieval method, the results of the experiments conducted by the authors seem to suggest that, when the more alignments are retrieved by lowering the threshold value it is best not to use hierarchical information. In this way in fact the overall performances, precision and recall wise, seem improving steadily making still rewarding the consideration of alignments even for low levels of thresholds (i.e. when more noise is expected).

4.5   Precision at n

The actual usability of Jaccard alignment in a user engaging scenario can also be judged by looking at the average precision of the measures for the first n

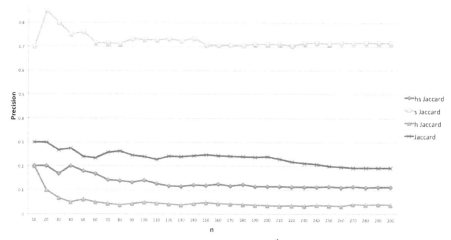

Fig. 8. Average precision at n^{th} alignment

alignments. The precision at n for all the considered scenarios is plotted in Figure 8 where we can clearly see that for all scenarios the average precision quickly stabilize after the first 50 alignments. **s Jaccard** provides the best precision with 7 good alignments out of 10. All the other scenarios perform quite poorly: **Jaccard** with 2 good alignments out of 10, **hs Jaccard** with 1 out of 10, and finally **h Jaccard** with 1 out of 20 correct alignments.

5   Related Work

The alignment problem, which is finding correspondences among concepts in ontologies, is well studied research problem in the Semantic Web community [7], as well as in the Database community to support the integration of heterogeneous sources [13] or to solve record linkage problems [6]. In particular, a lot of work is done by researchers in the last years in the context of Ontology Matching [7], where the alignments are the output of these systems. Different techniques have been proposed to address this problem in order to improve the performance related to the Ontology Merging, Integration and Translation systems. An evaluation competition [9] is also proposed to compare those matching systems using common testbeds. These techniques can be described in four main categories : i) *lexical*, which means that are based on detecting similarities between the concept descriptions such as labels; ii) *structural*, which means that are based on the knowledge descriptions; iii) *instance mapping*, which means that are based on the knowledge expressed in the ABox. Most of the proposed techniques comes from the Machine Learning research area [7,16,5,14]. In literature [3,1] also different measures are proposed to evaluate the semantic similarities among concepts that takes into account :i) the expressive power of the description logic used by the knowledge bases, ii) the information content assigned to the observed classes. Despite the different studies in the theoretical background, we observe

---

[9] http://oaei.ontologymatching.org/

a marginal effort in evaluate these approaches in the Linked Data Cloud, which is the most concrete realization of the Semantic Web vision nowadays and they are a valuable resources for different application domain. In literature, the most recent works that use Linked Data in the simalar context of our paper but with different purposes are [4,12]. In particular the first evaluates the implications of `owl:sameAs` assertions in Linked Data data sets and the second uses the LOD to evaluate an ontology matching system. In our paper we boost some of these previous studies in order to give a real evaluation in the context of Linked Data Cloud. From the instance-based matching techniques, the closest paper is [9], our contribution differs from the previous one in the richness of measures adopted and in evaluation proposed in the Linked Data environment.

6   Conclusion

In this paper we conducted some experiment with Jaccard-based concept similarity measures based on the analysis of the instance alignments provided by the `sameAs` network that connects **DBpedia** with some of the neighbourhood data sets. Being the chosen domain very broad (i.e. **DBpedia** concepts) and the alignments not focused on any specific application, we assumed to have very noisy results which suggested the use of statistical methods a natural choice.

The first analysis on the experimental data showed the typical signs of a power-based network, where a small number of `sameAs` bundles contained many entities and the vast majority contained no more than five instances (see Figure 2). We devised four different scenarios under which analyse the behaviour of classical Jaccard similarity measure, studying the influence of hierarchical information in producing the alignments and the difference when choosing a broader acceptance criteria.

The experimental results showed that Jaccard, for this particular DBpedia experiment, provided very low values which makes it difficult to choose a good threshold value which produced a fair amount of good alignments. The results outlined that the use of hierarchical information in computing concepts similarity measures increased drastically the number of alignments found but unfortunately dropped the precision of the results as well making increasingly inconvenient to consider further alignments below a given threshold. Conversely, by considering the concepts detached by a subclass hierarchy, Jaccard measures improve steadily.

The relaxation of the acceptance criteria on the other hand, did not influence the overall performance of the measures while giving better performances of the respective more restrictive measures. This is not surprising since the alignments found and the coefficients computed are the same when hierarchy is counted in or not, and it changes only the criteria for the acceptance, and one criteria includes the other. Ultimately, a less restrictive acceptance criteria, without hierarchy information, gives us a better overall performance and stably produces a fair amount of sensible alignments.

This scenario suits best an approach where alignments can be proposed to users for classification and where more elaborate alignments (i.e. not only concept equivalence) can be exploited for integrating data.

Future work will include a better study of the sources of noise in Jaccard-based methods when applied to the in order to provide a robust methodology for aligning ontologies at Web scale.

Acknowledgments. This work was supported by the EnAKTing project funded by the Engineering and Physical Sciences Research Council under contract EP/G008493/1.

References

1. Al-Mubaid, H., Nguyen, H.A.: Measuring semantic similarity between biomedical concepts within multiple ontologies. IEEE Transactions on Systems, Man, and Cybernetics, Part C: Applications and Reviews 39(4), 389–398 (2009)
2. Bechhofer, S., van Harmelen, F., Hendler, J., Horrocks, I., McGuinness, D.L., Patel-Schneider, P.F., Stein, L.A.: OWL Web Ontology Language Reference. W3C Recommendation (February 2004)
3. d'Amato, C., Fanizzi, N., Esposito, F.: A semantic similarity measure for expressive description logics. Computing Research Repository-arxiv.org (2009)
4. Ding, L., Shinavier, J., Shangguan, Z., McGuinness, D.L.: SameAs Networks and Beyond: Analyzing Deployment Status and Implications of owl:sameAs in Linked Data. In: Patel-Schneider, P.F., Pan, Y., Hitzler, P., Mika, P., Zhang, L., Pan, J.Z., Horrocks, I., Glimm, B. (eds.) ISWC 2010, Part I. LNCS, vol. 6496, pp. 145–160. Springer, Heidelberg (2010)
5. Doan, A., Madhavan, J., Domingos, P., Halevy, A.: Ontology matching: A machine learning approach. In: Handbook on Ontologies in Information Systems, pp. 397–416. Springer (2003)
6. Elmagarmid, A.K., Ipeirotis, P.G., Verykios, V.S.: Duplicate record detection: A survey. IEEE Transactions on Knowledge and Data Engineering 19(1), 1–16 (2007)
7. Euzenat, J., Shvaiko, P.: Ontology Matching. Springer, Heidelberg (2007)
8. Glaser, H., Jaffri, A., Millard, I.: Managing co-reference on the semantic web. In: WWW 2009 Workshop: Linked Data on the Web, LDOW 2009 (April 2009)
9. Isaac, A., van der Meij, L., Schlobach, S., Wang, S.: An Empirical Study of Instance-Based Ontology Matching. In: Aberer, K., Choi, K.-S., Noy, N., Allemang, D., Lee, K.-I., Nixon, L.J.B., Golbeck, J., Mika, P., Maynard, D., Mizoguchi, R., Schreiber, G., Cudré-Mauroux, P. (eds.) ASWC 2007 and ISWC 2007. LNCS, vol. 4825, pp. 253–266. Springer, Heidelberg (2007)
10. Isele, R., Jentzsch, A., Bizer, C.: Silk Server - Adding missing Links while consuming Linked Data. In: 1st International Workshop on Consuming Linked Data (COLD 2010), Shanghai, China (November 2010)
11. Jaccard, P.: Étude comparative de la distribution florale dans une portion des Alpes et des Jura. Bulletin del la Société Vaudoise des Sciences Naturelles 37, 547–579 (1901)
12. Jain, P., Hitzler, P., Sheth, A.P., Verma, K., Yeh, P.Z.: Ontology Alignment for Linked Open Data. In: Patel-Schneider, P.F., Pan, Y., Hitzler, P., Mika, P., Zhang, L., Pan, J.Z., Horrocks, I., Glimm, B. (eds.) ISWC 2010, Part I. LNCS, vol. 6496, pp. 402–417. Springer, Heidelberg (2010)

13. Lenzerini, M.: Data integration: a theoretical perspective. In: Proceedings of the Twenty-First ACM SIGMOD-SIGACT-SIGART Symposium on Principles of Database Systems, pp. 233–246 (2002)
14. Mao, M., Peng, Y., Spring, M.: Ontology mapping: as a binary classification problem. Concurrency and Computation: Practice and Experience 23(9), 1010–1025 (2011)
15. Miles, A., Pérez-Agüera, J.R.: SKOS: Simple Knowledge Organisation for the Web. Cataloging & Classification Quarterly 43(3), 69–83 (2007)
16. Niepert, M., Meilicke, C., Stuckenschmidt, H.: A probabilistic-logical framework for ontology matching. In: Proceedings of the Twenty-Fourth AAAI Conference on Artificial Intelligence (2010)
17. Salvadores, M., Correndo, G., Harris, S., Gibbins, N., Shadbolt, N.: The Design and Implementation of Minimal RDFS Backward Reasoning in 4store. In: Antoniou, G., Grobelnik, M., Simperl, E., Parsia, B., Plexousakis, D., De Leenheer, P., Pan, J. (eds.) ESWC 2011, Part II. LNCS, vol. 6644, pp. 139–153. Springer, Heidelberg (2011)
18. Salvadores, M., Correndo, G., Rodriguez-Castro, B., Gibbins, N., Darlington, J., Shadbolt, N.R.: LinksB2N: Automatic Data Integration for the Semantic Web. In: Meersman, R., Dillon, T., Herrero, P. (eds.) OTM 2009, Part II. LNCS, vol. 5871, pp. 1121–1138. Springer, Heidelberg (2009)
19. Schultz, A., Matteini, A., Isele, R., Bizer, C., Becker, C.: LDIF - Linked Data Integration Framework. In: 2nd International Workshop on Consuming Linked Data (COLD 2011), Bonn, Germany (October 2011)
20. Winkler, W.E.: Overview of record linkage and current research directions. Technical report, Bureau of the Census (2006)

Paragraph Tables: A Storage Scheme Based on RDF Document Structure

Akiyoshi Matono and Isao Kojima

National Institute of Advanced Industrial Science and Technology
1-1-1 Umezono, Tsukuba, Japan
a.matono@aist.go.jp, kojima@ni.aist.go.jp

Abstract. Efficient query processing for RDF graphs is essential, because RDF is one of the most important frameworks supporting the semantic web and linked data. The performance of query processing is based on the storage layout. So far, a number of storage schemes for RDF graphs have already been proposed. However most approaches must frequently perform costly join operations, because they decompose an RDF graph into a set of triples, store them separately, and need to connect them to reconstruct a graph that matchs the query graph, and this process requires join operations. In this paper, we propose a storage scheme that stores RDF graphs as they are connected, without decomposition. We focus on RDF documents, where adjacent triples have a high relationship and may be described for the same resource. So we define a set of adjacent triples that refer to the same resource as an RDF paragraph. Our approach constructs the table layout based on the RDF paragraphs. We evaluate the performance of our approach through experiments and demonstrate that our approach outperforms other approaches in query performance in most cases.

1 Introduction

The Resource Description Framework (RDF) [7,12] is a flexible and concise model for representing the metadata of resources on the web. RDF is the most essential format used with linked data [1] and for achieving the Semantic Web vision [5]. Using RDF, a particular resource can be described by RDF statements, each of which forms a triple (subject, predicate, and object). The *subject* is the resource being described, the *predicate* is the property being described with respect to the resource, and the *object* is the value for the property. RDF data consists of a set of RDF triples and thus the structure of RDF data is a labeled, directed graph, in which subjects and objects correspond to the nodes, and predicates correspond to the labeled, directed edges.

The amount of RDF data has become extremely large, for example, the linked data has grown over the years. Thus the query processing of RDF data is an essential issue. The performance of the query processing is based on the storage layout. Various storage schemes for RDF data have already been proposed.

---

[1] Linked Data: http://linkeddata.org/

S.W. Liddle et al. (Eds.): DEXA 2012, Part II, LNCS 7447, pp. 231–247, 2012.

Triple store

Subject	Predicate	Object
id:codd	type	FullProfessor
id:codd	name	Codd
id:codd	teach	id:course1
id:course1	name	AAA
id:course1	room	0123
id:codd	teach	id:course2
id:course2	name	BBB
id:course2	room	0124
id:codd	email	codd@...
id:jim	type	AssistantProfessor
id:xxx	type	Paper
id:xxx	name	XXX
id:xxx	author	id:codd
id:yyy	type	Book
id:yyy	name	YYY
id:yyy	author	id:codd
id:yyy	author	id:jim

Vertical partitioning

resource	type
id:codd	FullProfessor
id:jim	AssistantProfessor
id:xxx	Paper
id:yyy	Book

resource	teach
id:codd	id:course1
id:codd	id:course2

resource	room
id:course1	0123
id:course2	0124

resource	name
id:codd	Codd
id:course1	AAA
id:course2	BBB
id:xxx	XXX
id:yyy	YYY

resource	email
id:codd	codd@...

resource	author
id:xxx	id:codd
id:yyy	id:codd
id:yyy	id:jim

Property table

resource	type	name	teach	email
id:codd	FullProfessor	Codd	{id:course1, id:course2}	codd@...
id:jim	AssistantProfessor	...	...	...

resource	name	room
id:course1	AAA	0123
id:course2	BBB	0124

resource	type	name	author
id:xxx	Paper	XXX	{id:codd}
id:yyy	Book	YYY	{id:codd, id:jim}

Fig. 1. RDF storage schemes; (a) Triple store (b) Vertical partitioning (c) Property tables

Existing storage schemes are classified into three categories [16]: *triple store* [2,6,8,9,18,13], *vertical partitioning* [2,1], and *property tables* [9,19,11,17]. These are illustrated in Fig. 1, which depicts an example of storing RDF data using each of these schemes.

The triple store is the simplest storage scheme and uses a big, flat table that consists of three columns to store all RDF triples in it as shown in Fig. 1(a). This scheme requires many self-join operations to construct answers from the separated triples to match the given query graph. However, join ordering, which is one of the most efficient query optimization methods, cannot be applied to this scheme, because it stores properties as well as resources and values, and thus it cannot estimate the statistics of each property. Moreover, the table is so huge that the performance of selection operations also declines.

The vertical partitioning approach has been proposed to address this problem. It consists of one or more two-column tables, in each of which the first column contains the subjects and the second column contains the objects as shown in Fig. 1(b). In this approach, one table is created for each property and the number of tables is equal to the number of types existing for the predicates. This approach can maintain statistics about property values more easily and hence its query performance is better than that of the triple store. Moreover, only those properties required by the query need to be read. However, this approach also decomposes RDF data into RDF triples and stores them in the tables corresponding to their predicates. Therefore, the number of join operations preformed during query processing is the same as that of the triple store.

In order to improve the performance of query processing significantly, we have to reduce the number of join operations. The property table is a solution that has been proposed to address this issue. The property table shown in Fig. 1(c) is composed of a set of multi-column tables, whose attributes are related to a resource, that is, a tuple in a table consists of a set of triples that have the same resources. One typical version of the property table approach stores all triples that have the same subject as a tuple, where the first column contains the same subjects and the other columns contain the objects of those subjects for each property. In RDF, a resource can have one or more values for a property, and therefore all the columns, except the first one, use an array type to store the objects. For example, in the typical property table approach, given a book as a resource, the ID of the book is stored in the first column, and it's bibliographic information, such as the title, authors, publishing date, isbn, and so on, are stored in the second or later columns. In order to store multiple authors, the type of the second or later columns is defined as an array.

The property table approach resembles materialized views used in relational databases because the layout of the property table is similar to one that has been merged by performing join operations among some two-column tables of the vertical partitioning. In fact, Oracle [9] uses materialized views to construct property tables. Oracle assumes that user demand and query workloads are given and the structure of property tables is determined based on them. However, user demand and query workload can not be obtained easily before the database runs as a service. Furthermore, Levandoski and Mokbel [11], and FlexTable [17] use RDF instance data to determine the structure of property tables. However, the instance data has already been decomposed, and all connections are treated equivalently. Hence the structure determined is based on only the instance data, and does not take human intention into account at all. Therefore, we propose an idea in which the structure of RDF documents is utilized to determine the structure of property tables.

1.1   Overview

An RDF data forms a graph structure, but the graph structure cannot be written as it is in RDF documents. Generally, an RDF document which is represented in structured languages, such as RDF/XML, turtle, notation3, etc., consists of a sequence of triples. We claim that the order of the triples in an RDF document has a meaning, but the triples in a document almost never occur randomly. Thus the adjacent triples may be described for the same resource, so that they can connect and construct a graph that consists of the two triples. We call such a graph that consists of a set of adjacent and connectable triples as an *RDF paragraph* in this paper.

We believe that an RDF document is written to be easily understood by humans, and query structures must also be structured in human-readable form. Therefore, using our approach, we hypothesize that the query performance improves because the structures of RDF paragraphs and queries are similar and the number of join operations can be reduced. In this paper, we propose the use of one of the property table approaches, called the *paragraph table*, that stores RDF paragraphs taken from RDF documents as they are without modification.

Fig. 2 depicts an example of an RDF document and the paragraph table approach. In the document, there are four paragraphs. The first two paragraphs are stored in paragraph table t1, and the last two are stored in t2. In addition to tables t1 and t2, there is also a *schema table* and a *connection table*. The schema table is used to map the structure of an RDF paragraph to table columns, because the structure of the paragraph is represented as a graph but the paragraph must be stored into a table flatly. The connection table is used to store the information about the pairs of two columns that may be joined because a column of a table may be joined to any column of any table. In this example, the first tuple of the connection table means that there may be common resources between c1 of t1 and c4 of t2.

In this paper, we do not consider the case where the paragraph tables are merged or decomposed, but our approach can be applied to the existing approaches proposed in [17] and [11]. We evaluate the performance of our approach through a series of experiments comparing the paragraph table with other storage schemes using relational databases as the back-end. Based on the results, our approach outperforms other approaches in query performance, in most cases.

The rest of this paper is organized as follows. We first define some terms for our approach in Section 2. In Section 2.1, we explain the extraction algorithms of a schema and instance tuples from an RDF paragraph. In Section 3, we explain the usage of our proposed tables using an example. We evaluate the performance

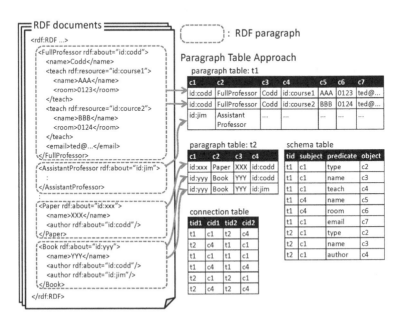

Fig. 2. an RDF document and its storage, based on the paragraph table approach

of Paragraph Tables though a series of experiments in Section 4. Finally, we conclude in Section 5.

2   RDF Paragraphs

We present the definition of RDF paragraphs in this section. RDF paragraphs are extracted from RDF documents. In general, adjacent triples in a document have a high relationship and may be described for the same resource. So we briefly define a set of adjacent triples that have the same resources in a document as an RDF paragraph.

An *RDF document* is a plain text file used to describe and/or exchange RDF data. There are some syntaxes, such as RDF/XML [20], Notation3 [4], Turtle [3], and N-Triples [10]. All languages can describe a graph by a set of triples. When parsing an RDF document using any language, triples can be extracted one by one in the order of reading the document. We can thus use the information of the order of triples in order to construct RDF paragraphs. We generate an RDF paragraph from a set of adjacent triples that have the same resources.

Alg. 1. Generate RDF paragraphs from a document

   Input: RDF document D
   Output: a set of paragraph $\mathbb{P}$

1   a set of paragraph $\mathbb{P} \leftarrow \emptyset$
2   current paragraph $P \leftarrow \emptyset$
3   a list of triples $T \leftarrow$ parse the document D
4   **for** $t \in T$ **do**
5   **if** $t.s_t$ = subject or object of a triple $\in P$ **then**
6        $P \leftarrow P \cup t$ /* t is added into P */
7   **else**
8        $\mathbb{P} \leftarrow \mathbb{P} \cup P$ /* P is added into $\mathbb{P}$ */
9        $P \leftarrow \{t\}$ /* create a new paragraph */
10   $\mathbb{P} \leftarrow \mathbb{P} \cup P$ /* The last process P is added into $\mathbb{P}$ */
11   **return** $\mathbb{P}$

In order to map a paragraph to a table, we extract a schema graph and some instances from a paragraph, then determine a table schema based on the schema graph extracted, and store all the instances into the table as tuples. An RDF paragraph forms a labeled directed graph with a root, $G = (r, f, V, E, l_V, l_E)$, where V is a set of nodes, E is a set of edges, f is a function that maps edges to a pair of two nodes, l_V and l_E are label functions that map nodes and edges to their labels, and r is a root node. We define a *path* the list of all edges that connect from the root to a node in a paragraph. A set of nodes with the same path contracts to a node, which we call a *schema node*, and also a set of edges with the same start node and the same labels contracts to an edge, which we call a *schema edge*. A graph that is composed of a set of schema nodes and schema edges is called a *schema graph* or *schema* $\hat{S} = (\hat{r}, f, \hat{V}, \hat{E}, l_{\hat{V}}, l_E)$, where $\hat{r}$ is a root schema node $(\hat{r} \in \hat{V} \vee \hat{r} \mapsto \forall \hat{v} \in \hat{V})$, $\hat{V}$ is a set of schema nodes, $\hat{E}$ is a set of schema edges and $l_{\hat{V}}$ is a label function that maps schema nodes to labels.

2.1   Extraction of Schema and Tuples from a Paragraph

In this section, we discuss how to extract a schema graph and tuples from a paragraph. Alg. 2 shows the pseudo code explaining how to extract a schema from a given paragraph G. Moreover, Alg. 3 shows how to extract an instance tuple from a paragraph based on its schema. The extraction of a schema has to be performed before that of tuples. Both functions are called recursively, that is, these processes traverse in a depth-first manner. The normalization of the order of edges for each node of a given paragraph have to be performed before the extraction of a schema because there are many schemata in a denormalized paragraph.

Alg. 2. Extraction of the schema graph `getSchemaRecursive`

Input: schema $\hat{S}$, node v, schema node $\hat{v}$, int p, int i
Output: the number of instances

1 $\hat{S}.\hat{V} \leftarrow \hat{S}.\hat{V} \cap \hat{v}$
2 $s \leftarrow p$
3 **for** *Same label edges* $E \in v.edgeSet$ **do**
4 $\quad$ $\hat{e} \leftarrow$ Create a schema edge from E
5 $\quad$ $\hat{S}.\hat{E} \leftarrow \hat{S}.\hat{E} \cap \hat{e}$
6 $\quad$ **for** $e \in E$ **do**
7 $\quad\quad$ $f \leftarrow p - i$
8 $\quad\quad$ Set f into e as the first access number
9 $\quad\quad$ $v_c \leftarrow$ Get the node that is indicated from e
10 $\quad\quad$ $\hat{v}_c \leftarrow$ Create a schema node from v_c
11 $\quad\quad$ Connect from $\hat{v}$ to $\hat{v}_c$ by $\hat{e}$
12 $\quad\quad$ $p \leftarrow$ `getSchemaRecursive`$(\hat{S}, v_c, \hat{v}_c, p, i)$
13 $\quad\quad$ $l \leftarrow p$
14 $\quad\quad$ Set l into e as the last access number
15 $\quad\quad$ **if** e *is not the last of* E **then**
16 $\quad\quad\quad$ $p \leftarrow p + i$
17 $\quad$ $i \leftarrow p - s + 1$
18 **return** p

Alg. 3. Extraction of the instance tuple `getInstanceRecursive`

Input: instance T, node v, schema node $\hat{v}$, int c

1   $T \leftarrow T \cap (\hat{v} \prec v)$
2   **for** $\hat{e} \in \hat{v}.edgeSet$ **do**
3      $E \leftarrow$ Get edges with the same label $\hat{e}$ from v
4      $e[0] \leftarrow$ Get the first edge of E
5      $f \leftarrow$ Get the first access number from $e[0]$
6      $l \leftarrow$ Get the last access number from $e[0]$
7      $|E| \leftarrow$ the number of edges E
8      $m \leftarrow \lfloor c/(l - f) \rfloor \% |E|$
9      $e[m] \leftarrow$ Get the mth edge of E
10     $\hat{v}_c \leftarrow$ Get the schema node from $\hat{e}$
11     $v_c \leftarrow$ Get the node that is indicated from $e[m]$
12   `getInstanceRecursive`$(T, v_c, \hat{v}_c, c)$
13 **return** T

Alg. 2 has two purposes: extraction of the schema, and labeling as preparation to extract instances. For the arguments of Alg. 2, $\hat{S}$ is an empty schema graph, v is the root node of G, $\hat{v}$ is new schema root node generated based on v, and p and i are set to 1. For the process of schema extraction, in lines 1, 4-5, and 10-11, a schema graph is generated. The labeling process is done in lines 2, 7-8, and 13-17. The extracted schema graph is stored in the schema table by decomposing it to triples (see the schema table in Fig. 2). The labeling process is to assign two integers f (*first access number*) and l (*last access number*) to all edges which are used to extract instances as in Fig. 3. f and l are calculated from integers p and i.

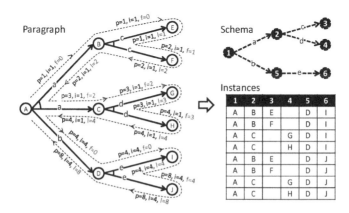

Fig. 3. Labeling as preparation for extracting instances from a paragraph

Alg. 3 takes as input an empty instance T, a node v, a schema node $\hat{v}$ and an integer c. This algorithm makes it possible to get the cth instance tuple. So this algorithm yields the number of instances with a different input for c. This process uses the integers f and l labeled by Alg. 2.

3   Query Example

In this section, we explain the usage of our proposed tables through an example of a query. Assuming we want to "find papers that are written by someone who teaches a course named AAA.", the query can be described at List 1 in SPARQL [15].

List 1. SPARQL query

```
1  SELECT ?x ?y ?z
2  WHERE { ?x type Paper.  ?x author ?y.
3          ?y teach ?z.    ?z name "AAA". }
```

We first generate the preparatory SQL query in List 2 to get information on which columns may be joined from the schema table and the connection table. The query looks complicated but the cost is not large, because the sizes of both tables are small.

The result of List 2 is shown in List 3. The first line indicates the attributes and the second or lower lines indicate the instance tuples. The 1st tuple contains four tables t1, t2, t3, and t4, the 2nd contains t5 and t6, and the 3rd contains t7. In other words, in the 3rd, t7 can return a part of the answers without any join.

Based on the results List 3, we then generate the final SQL query List 4. The SQL contains three subqueries merged by UNION in FROM clause. Each subquery corresponds to a tuple in List 3.

4   Experimental Evaluation

We conduct two experiments; one is to investigate whether useful paragraphs can be extracted from actual RDF data, the other is to evaluate the performance of the paragraph table approach compared to other approaches.

4.1   Statistics of Linked Data

We use the linked data as actual RDF data. We first download about 400 RDF documents from the linked data project, and then we measure the statistics of the paragraphs for the RDF documents in linked data. Fig. 4 depcts the scatter plot between the average number of triples and the average depth of paragraphs

List 2. Preparatory SQL query used to get joinable columns information

```
1  SELECT
2    t1.tid T1, t1.subject s1, t1.object o1,
3    t2.tid T2, t2.subject s2, t2.object o2,
4    t3.tid T3, t3.subject s3, t3.object o3,
5    t4.tid T4, t4.subject s4, t4.object o4
6  FROM
7    schema AS t1, schema AS t2,
8    schema AS t3, schema AS t4,
9    connection AS x, connection AS y,
10   connection AS z
11 WHERE t1.predicate = 'type'
12   AND t2.predicate = 'author'
13   AND t3.predicate = 'teach'
14   AND t4.predicate = 'name'
15   AND x.tid1 = t1.tid AND x.cid1 = t1.subject
16   AND x.tid2 = t2.tid AND x.cid2 = t2.subject
17   AND y.tid1 = t2.tid AND y.cid1 = t2.object
18   AND y.tid2 = t3.tid AND y.cid2 = t3.subject
19   AND z.tid1 = t3.tid AND z.cid1 = t3.object
20   AND z.tid2 = t4.tid AND z.cid2 = t4.subject
```

List 3. Result of List 2

T1	s1	o1	T2	s2	o2	T3	s3	o3	T4	s4	o4
t1	c1	c2	t2	c1	c2	t3	c1	c2	t4	c1	c2
t5	c1	c2	t5	c1	c4	t6	c1	c4	t6	c4	c5
t7	c1	c2	t7	c1	c3	t7	c3	c4	t7	c4	c5

for every RDF documents in linked data. The y-axis shows the average number of triples in log scale. The x-axis shows the average depth of paragraphs. Each cross mark means an RDF document which has many paragraphs. The red squere represents the scatter plot between the average number of triples and the average depth of all paragraphs and the values are 23.15 and 1.82 respectively.

From this figure, we know that the size of paragraph correlates with the depth of paragraph. Moreover paragraphs tends to be widespread graphs because the average number of triples par paragraph is 23.15 though the average depth par paragraph is 1.82. Additionally we measured that the percentage of cyclic graphs is only 2%. Hence, the results show that most paragraphs that are extracted from actual RDF documents shaped widespread low trees. Therefore we can say that the paragraph as a unit can be applied to actual RDF data, and is useful.

List 4. SQL query generated based on List 3

```
1  SELECT c.x, c.y, c.z
2  FROM   (
3      SELECT t1.c1 x, t3.c1 y, t4.c1 z FROM t1,t2,t3,t4
4      WHERE t1.c2='Paper' AND   t4.c2='AAA'
5      AND   t1.c1=t2.c1 AND   t2.c2=t3.c1 AND   t3.c2=t4.c1
6  UNION
7      SELECT t5.c1 x, t6.c1 y, t6.c4 z FROM t5,t6,
8      WHERE t5.c2='Paper' AND   t6.c5='AAA'
9      AND   t5.c4=t6.c1
10 UNION
11     SELECT t7.c1 x, t7.c3 y, t7.c4 z FROM t7,
12     WHERE t7.c2='Paper' AND   t7.c5='AAA'
13 ) AS c
```

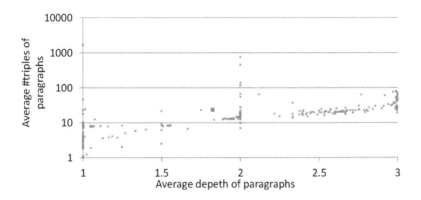

Fig. 4. The scatter plot between the average number of triples and the average depths of paragraphs for each document in linked data

4.2   Performance Evaluation

We evaluate the performance of the paragraph table approach comparing some existing approaches. We utilize PostgreSQL 9.1 as a back-end database. We use a machine where the CPU is an Intel Core2 Quad Q9450, and the main memory size is 4 Gbytes in this experiment.

We compare the following strategies:

TripleStore is the simplest storage scheme and where we create two tables: the id-value mapping table and the triple storing table. The former is used to map string values to integer ids and contains two b-tree indices for the two columns. The latter stores all triples flatly and contains 6 b-tree indices of subject, predicate, object, subject-predicate, predicate-object, and subject-object.

VerticalPartitioning has also an id-value mapping table and has some two-column property tables, each of which is created for each predicate and stores tuples, each of which consists of a subject and an object. We create two b-tree indices of both columns for each two-column property table.

SameSubject is one of the property table approaches, where we create an id-value mapping table and some multi-column property tables, each of which stores a set of triples that have the same subjects as a tuple. The first column stores the id of a resource, and the second or later columns are of the int array type, because RDF supports multi-valued properties. We create some indices for each multi-column property table: one is a b-tree index for the first column, the others are GIN indices for the second or later columns using the intarray module provided as a contribution of PostgreSQL.

FlexTable is another one of the property table approaches, which was implemented by us based partly on [17]. The schema layout is similar to that of SameSubject. FlexTable measures similarities between all combinations of two tables using a method based on TF-IDF, and then two tables whose similarity exceeds a threshold are merged to one table. The original FlexTable uses the GiST index, but we use the GIN index for int array typed columns because GiST index causes an error when storing large values in our environment.

ParagraphTable is our proposed paragraph table scheme, which is the another of property table approaches. As with the other approaches, we create an id-value mapping table. We create a schema table that has two b-tree indices, where the one is for the table id column and the other is for predicate, subject and object columns. The connection table has two b-tree indices for the first two columns and the last two columns. A paragraph table creates indices for each column.

We use LUBM [14] as the experimental dataset. LUBM uses metadata about universities, and you can give the number of universities as a parameter when generating data. In particular, metadata about students, professors, lectures, courses and publications for each university is generated in RDF documents. In our experiments, we generated RDF documents for many universities, computed the statistics of each file, and then we made 6 datasets, each of which consists of 500k, 1M, 2M, 4M, 8M or 16M triples. The statistics are shown at Table 1.

Table 1. Experimental datasets generated from LUBM

	#triples	#docs	size[bytes]	#paragraphs	#universities
500 k	500,973	72	40 M	83,490	4
1 M	998,230	144	80 M	166,180	7
2 M	2,002,594	290	162 M	333,137	14
4 M	3,999,967	579	324 M	665,591	29
8 M	8,003,576	1,159	649 M	1,332,517	57
16 M	16,002,910	2,316	1,299 M	2,663,974	115

As experimental queries, we use the test queries that are defined by LUBM. However, some queries are defined to evaluate the performance of inference functions. For example, query 6 "(?X rdf:type Student)" queries about only one class, but it assumes both an explicit subClassOf relationship between UndergraduateStudent and Student and the implicit one between GraduateStudent and Student. So the result values bounded by ?X contain instances not only of Student but also of UndergraduateStudent and GraduateStudent. The storing schemes compared, including our approach, do not support such inference queries, so we did not evaluate the inference queries. In particular, we use queries 1, 2, 4, 7, 8, 9, and 14 in our experiment. In the appendix, these queries are shown in SPARQL format.

Fig. 5 shows the results of the query processing times for the LUBM benchmark. The vertical axis of all figures depicts the processing times in log scale. The horizontal axis of all figures only in Fig. 5(a). We omitted the legends of represents the number of triples of the dataset. The legend is included the other figures.

In queries 1, 7, and 8, the results are similar, because only our approach and VerticalPartitioning are stable without relation to the number of triples of the RDF data. In contrast, in the other approaches, the times increase in proportion to the growth of the RDF data. In query 4, in a similar way, the times of SameSubject and FlexTable are increasing according to the growth of the data, but the other schemes including our approach are stable. In other words, VerticalPartitioning and our approach have scalability in queries 1, 4, 7, and 8.

The reason for this is that SameSubject and FlexTable use the integer array type to store multi-value properties so that the cost of a join operation between an integer array in the property tables and an integer in the id-value table is very large. In addition, TripleStore has to do some self-joins, each of which is high cost because of the large number of records, and can not use join ordering optimization.

Queries 7 and 8 are similar, so the results are similar too. The difference between them is that query 8 is more complicated than query 7, and our approach is worse than the vertical partitioning method in query 7, but our approach is better than that in query 8. From this, we can say that our approach is suitable for complicated queries.

In query 4, there are three stable schemes, our approach, VerticalPartitioning, and TripleStore. In the three approaches, our approach is stable at the lowest level. Thus, with a larger number of triples, our approach surpasses the other approaches. The reason for this is that, in query 4, our approach needs no join operations at all to construct the answer structure, but the other approaches need as many as four join operations.

In the other queries, 2, 9, and 14, the times of all schemes increase proportionately. Only in query 14 is our approach inferior, because query 14 is the simplest query, so it does not use any join operations, and our approach needs some union operations. However our approach is comparable to the other approaches in even the simplest query. Both queries 2 and 9 are structured as a triangle by join operations. So the results of them are similar. The difference is

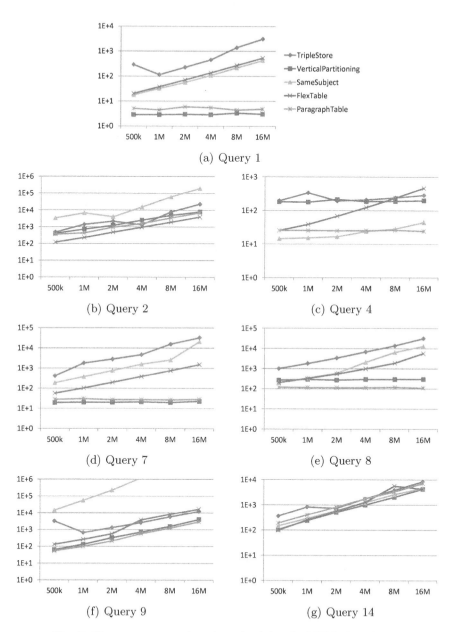

(a) Query 1

(b) Query 2

(c) Query 4

(d) Query 7

(e) Query 8

(f) Query 9

(g) Query 14

Fig. 5. The query processing times (ms) for the LUBM benchmark

that the selectivity of query 9 is lower than that of query 2. From this, we know that our approach is better than the other three approaches when the selectivity is lower, since our approach shows the best performance in query 9.

In all queries, except for query 14, our approach demonstrates the best or the second best performance. In particular, when the structure of the query matches to that of the paragraph in query 4 and 8, and when a query does not include low-selectivity joins as in query 9, our approach outperforms the other approaches.

5   Conclusion

In this paper, we present a novel idea holding that the structure of RDF documents includes human intention and this should be used to determine the structure of the storage layout. To successfully implement our idea, we proposed various algorithms and special tables. In the results of a performance evaluation, our approach achieves the best or the second best performance in all queries except for one.

As future work, we must measure and evaluate the performance of data loading. We have to consider the merging of similar tables and/or the division of large sparse tables. Moreover, in our approach, the table layout is based on the order of triples in documents, so we have to discuss the effects of the order of triples.

Acknowledgments. This work was supported by Grants-in-Aid for Scientific Research (24680010 and 24240015).

References

1. Abadi, D.J., Marcus, A., Madden, S., Hollenbach, K.J.: Scalable semantic web data management using vertical partitioning. In: Koch, C., Gehrke, J., Garofalakis, M.N., Srivastava, D., Aberer, K., Deshpande, A., Florescu, D., Chan, C.Y., Ganti, V., Kanne, C.C., Klas, W., Neuhold, E.J. (eds.) VLDB, pp. 411–422. ACM (2007)
2. Alexaki, S., Christophides, V., Karvounarakis, G., Plexousakis, D., Tolle, K.: The ICS-FORTH RDFSuite: Managing voluminous RDF description bases. In: SemWeb (2001)
3. Beckett, D., Berners-Lee, T.: Turtle - terse rdf triple language (January 2008), http://www.w3.org/TeamSubmission/turtle/ (W3C Team Submission January 14, 2008)
4. Berners-Lee, T., Connolly, D.: Notation3 (n3): A readable rdf syntax (January 2008), http://www.w3.org/TeamSubmission/n3/ (W3C Team Submission January 14, 2008)
5. Berners-Lee, T., Hendler, J., Lassila, O.: The semantic web. Scientific American 284(5), 34–43 (2001)
6. Broekstra, J., Kampman, A., van Harmelen, F.: Sesame: A Generic Architecture for Storing and Querying RDF and RDF Schema. In: Horrocks, I., Hendler, J. (eds.) ISWC 2002. LNCS, vol. 2342, pp. 54–68. Springer, Heidelberg (2002)

7. Candan, K.S., Liu, H., Suvarna, R.: Resource description framework: Metadata and its applications. SIGKDD Explorations 3(1), 6–19 (2001)
8. Carroll, J.J., Reynolds, D., Dickinson, I., Seaborne, A., Dollin, C., Wilkinson, K.: Jena: Implementing the semantic web recommendations. Technical Report HPL-2003-146, HP Labs (2003)
9. Chong, E.I., Das, S., Eadon, G., Srinivasan, J.: An efficient SQL-based RDF querying scheme. In: Böhm, K., Jensen, C.S., Haas, L.M., Kersten, M.L., Larson, P.Å., Ooi, B.C. (eds.) VLDB, pp. 1216–1227. ACM (2005)
10. Grant, J., Beckett, D.: Rdf test cases: N-triples (February 2004),
   `http://www.w3.org/TR/rdf-testcases/#ntriples`
   (W3C Recommendation February 10, 2004)
11. Levandoski, J.J., Mokbel, M.F.: RDF data-centric storage. In: ICWS, pp. 911–918. IEEE (2009)
12. Manola, F., Miller, E.: RDF primer (February 2004),
   `http://www.w3.org/TR/2004/REC-rdf-primer-20040210/` (W3C Recommendation February 10, 2004)
13. Neumann, T., Weikum, G.: RDF-3X: a RISC-style engine for RDF. PVLDB 1(1), 647–659 (2008)
14. Projects, S.: The lehigh university benchmark (LUBM),
   `http://swat.cse.lehigh.edu/projects/lubm/`
15. Prud'hommeaux, E., Seaborne, A.: SPARQL query language for RDF (January 2008), `http://www.w3.org/TR/rdf-sparql-query/` (W3C Recommendation January 15, 2008)
16. Sakr, S., Al-Naymat, G.: Relational processing of RDF queries: a survey. SIGMOD Rec. 38(4), 23–28 (2009)
17. Wang, Y., Du, X., Lu, J., Wang, X.: FlexTable: Using a Dynamic Relation Model to Store RDF Data. In: Kitagawa, H., Ishikawa, Y., Li, Q., Watanabe, C. (eds.) DASFAA 2010, Part I. LNCS, vol. 5981, pp. 580–594. Springer, Heidelberg (2010)
18. Weiss, C., Karras, P., Bernstein, A.: Hexastore: sextuple indexing for semantic web data management. PVLDB 1(1), 1008–1019 (2008)
19. Wilkinson, K.: Jena property table implementation. Technical Report HPL-2006-140, HP Labs (2006)
20. World Wide Web Consortium: Rdf/xml syntax specification (revised) (2004), `http://www.w3.org/TR/rdf-syntax-grammar/` (W3C Recommendation February 10, 2004)

Appendix

The following SPARQL queries were used in our experiments and are based on the LUBM query set. We chose queries that are not used for performance evaluation of inferences and modified them in order not to use inference processing.

Query 1 bears large input and high selectivity. It queries for just one class and one property and does not assume any hierarchy information or inference.

```
SELECT ?X WHERE {
  ?X rdf:type ub:GraduateStudent .
  ?X ub:takesCourse
  "http://www.Department0.University0.edu/GraduateCourse0".}
```

Query 2 increases in complexity: 3 classes and 3 properties are involved. And there is a triangular pattern of relationships between the objects involved.

```
SELECT ?X ?Y ?Z WHERE {
  ?X rdf:type ub:GraduateStudent .
  ?Y rdf:type ub:University .
  ?Z rdf:type ub:Department .
  ?X ub:memberOf ?Z .
  ?Z ub:subOrganizationOf ?Y .
  ?X ub:undergraduateDegreeFrom ?Y.}
```

Query 4 has small input and high selectivity. Another feature is that it queries for multiple properties of a single class. We changed the query from Professor to FullProfeseor because we do not evaluate the performance of inference functions.

```
SELECT ?X ?Y1 ?Y2 ?Y3 WHERE {
  ?X rdf:type ub:FullProfessor .
  ?X ub:worksFor  "http://www.Department0.University0.edu".
  ?X ub:name ?Y1 .
  ?X ub:emailAddress ?Y2 .
  ?X ub:telephone ?Y3.}
```

Query 7 is similar to Query 6, but it increases in the number of classes and properties and its selectivity is high. We changed it from Student to UndergraduateStudent.

```
SELECT ?X ?Y WHERE {
  ?X rdf:type ub:UndergraduateStudent .
  ?Y rdf:type ub:Course .
  ?X ub:takesCourse ?Y .
  "http://www.Department0.University0.edu/AssociateProfessor0"
    ub:teacherOf ?Y.}
```

Query 8 is additionally more complex than Query 7 by including one more property. We changed it from Student to UndergraduateStudent.

```
SELECT ?X ?Y ?Z WHERE {
  ?X rdf:type ub:UndergraduateStudent.
  ?Y rdf:type ub:Department.
  ?Y ub:subOrganizationOf "http://www.University0.edu".
  ?X ub:memberOf ?Y.
  ?X ub:emailAddress ?Z.}
```

Query 9 has a triangular pattern of relationships like query 2. In the original query, the types of X, Y and Z are defined as Student, Faculty, and Course, respectively. We deleted them since they are for inference functions. As a result, the selectivity is lower than that of query 2.

```
SELECT ?X ?Y ?Z WHERE {
  ?X ub:advisor ?Y.
  ?Y ub:teacherOf ?Z.
  ?X ub:takesCourse ?Z.}
```

Query 14 is the simplest in the test set. This query represents those with large input and low selectivity and does not assume any hierarchy information or inference.

```
SELECT ?X WHERE {?X rdf:type ub:UndergraduateStudent.}
```

Continuously Mining Sliding Window Trend Clusters in a Sensor Network

Annalisa Appice, Donato Malerba, and Anna Ciampi

Dipartimento di Informatica, Università degli Studi di Bari Aldo Moro
via Orabona, 4 - 70126 Bari, Italy
{appice,malerba,ciampi}@di.uniba.it

Abstract. The trend cluster discovery retrieves areas of spatially close
sensors which measure a numeric random field having a prominent data
trend along a time horizon. We propose a computation preserving algorithm which employees an incremental learning strategy to continuously maintain sliding window trend clusters across a sensor network.
Our proposal reduces the amount of data to be processed and saves the
computation time as a consequence. An empirical study proves the effectiveness of the proposed algorithm to take under control computation
cost of detecting sliding window trend clusters.

1 Introduction

The trend cluster discovery, as a spatio-temporal aggregate operator, may play
a crucial role in the decision making process of several sensing applications.
Initially formulated for data warehousing, trend cluster discovery gathers spatially clustered sensors whose readings for a numeric random field show a similar
trend (represented by a time series) along a time horizon. In our previous studies, we resorted to a segmentation of the time in consecutive windows such that
trend clusters can be discovered window per window [1]; the representation of
a trend can be compressed by applying some signal processing techniques [2];
trend clusters can be used to feed a trend based cube storage of the sensor data
[3]. Anyway, trend clusters are always discovered along the time horizons of non-
overlapping widows; in this way trend clusters discovered in a window do not
share, at least explicitly, any knowledge with trend clusters discovered in any
other window. While in a data stream system, sliding window computation [4]
is often considered. Thus, in this paper we explore sliding window trend cluster
discovery which seeks for trend clusters over the latest data constrained by a
sliding window.

The main challenge of maintaining sliding window knowledge is how to minimize the computation cost (memory and time usage) during the discovery process. We face this challenge for the trend cluster discovery by a novel technique,
called Sliding WIndow Trend cluster maintaining algorithm (SWIT), that efficiently maintains accurate sliding window trend clusters which arise in a sensor
network. Each time new data are collected from the network, they are temporally buffered in a graph data synopsis. For each trend cluster which is presently

S.W. Liddle et al. (Eds.): DEXA 2012, Part II, LNCS 7447, pp. 248–255, 2012.

maintained, the oldest time point is discarded. Clusters which are spatially close and share a similar trend along the time horizon under consideration are merged. Finally, trend clusters are spatially split in case they do not fit data trends until the present.

The paper is organized as follows. Section 2 report basic definitions, while Section 3 presents the algorithm SWIT. The empirical evaluation of SWIT on real streams is illustrated in Section 4 and conclusions are drawn.

2 Basics and Problem Definition

Sensor readings are arranged in data snapshots which feed a sensor data stream.

Definition 1 (Snapshot). *Let Z be a numeric field, the snapshot of Z at the time t is modeled by a field function $z_t : K_t \mapsto Z$ where K_t ($K_t \subseteq \mathbb{R}^2$) is the finite set of 2D points which geo-reference sensors that produced a datum in t. z_t assigns a sensor position $(x, y) \in K_t$ to a reading of Z at the time t.*

Definition 2 (Sensor data stream). *Let K be a sensor network which measures Z; T be the time line discretized in transmission points; $K: T \mapsto P(K)$ be a function which assigns a point $t \in T$ to the sensor set K_t ($K_t \subseteq K$) which are switched-on in t. The stream $z(T, K)$ is the time unbounded sequence of snapshots $z_t(\cdot)$ which are produced from K at the consecutive time points of T.*

The sliding window model of a sensor network stream can be defined.

Definition 3 (Sliding window). *Let $z(T, K)$ be a sensor data stream, w be a window size ($w > 1$). The w-sized sliding window model of $z(T, K)$ is the sequence of windows defined as follows:*

$$slide(z(T, K), w) = \underset{t_1 \to t_w}{\omega} z(T, K), \underset{t_2 \to t_{w+1}}{\omega} z(T, K), \dots, \underset{t_{i-w+1} \to t_i}{\omega} z(T, K), \dots \tag{1}$$

with each window $\underset{rt_{i-w+1} \to t_i}{\omega} z(T, K) = \langle z_{t_{i-w+1}}(\cdot), z_{t_{i-w+2}}(\cdot), \dots, z_{t_i}(\cdot) \rangle$.

A trend cluster is defined in [1] as a spatio-temporal summarizer of a window of snapshots of a numeric random field Z produced through a sensor network.

Definition 4 (Trend Cluster). *Let $z(T, K)$ be a sensor data stream, a trend cluster in $z(T, K)$ is a triple $(t_i \to t_j, c, z)$, where $t_i \to t_j$ is the time horizon; c enumerates the "spatially close" sensors whose readings exhibit a "similar temporal variation" from t_i to t_j; z is the time series of Z readings aggregated on c. Aggregates are computed at each time point from t_i to t_j.*

As in [1], a distance-based definition of the spatial closeness relation between sensors is adopted.

Definition 5 (Sensor Spatial Closeness Relation). *Let d be a user-defined distance bandwidth, a sensor A is close to a sensor B if A is far at worst d from B.*

Similarly, the trend similarity can be bounded by a similarity threshold δ.

Definition 6 (Trend Similarity Measure). *Let δ be a domain similarity threshold, z_A and z_B be two time series of Z having time horizon H; Then*

$$trendSim(z_A, z_B) = 1 - \frac{1}{|H|} \sum_{t_i \in H} I(z_{t_i}(A), z_{t_i}(B)), \qquad (2)$$

where $I(z_{t_i}(A), z_{t_i}(B)) = 0$ if $\|z_{t_i}(A), z_{t_i}(B)\| \leq \delta$; otherwise 1. $trendSim(z_A, z_B) = 1$ means that A and B share a similar trend over H.

3   Sliding Window Trend Cluster Maintaining

A buffer consumes a present snapshot and pours it into SWIT to slide the previous discovered set of trend clusters to the present time. Let w be the size of a sliding window, δ be the domain similarity threshold, d be the distance bandwidth for the spatial closeness relation between sensors, and t_i be the present time. The input data to the discovery process are the present snapshot and the set of sliding window trend clusters as it was detected at the previous time t_{i-1}. The output is the set of sliding window trend clusters as it is updated to the time t_i. The learning strategy is three-stepped. First, the snapshot $z_{t_i}(\cdot)$ is buffered into a data synopsis and trend values now maintained for the oldest time point t_{i-w} are discarded. Then, spatially close trend clusters which share a similar trend along $t_{i-w+1} \to t_{i-1}$ are merged. Finally, trend clusters are split according to data in the present snapshot. It is noteworthy that d, δ and w can be set-out according to the general knowledge we have on the network design (e.g. average distance between sensors) and the random field in analysis.

3.1   Snapshot Acquisition Phase

The acquisition phase loads the snapshot into the graph data synopsis γ^K and updates the set of trend clusters which is maintained in the graph data synopsis γ^{TC} by discarding the trend information associated to the oldest time point.

Storing the Snapshot in γ^K. The data synopsis γ^K loads snapshotted data and logs transmission status (switched on/off) of network sensors during the sliding window (see Figures 1(a)-1(b)). A graph structure models the spatial arrangement of sensors which transmitted at least once along the sliding window: graph nodes represent sensors and graph edges virtually link sensors which are spatially close (see Definition 5) across the network. Each graph node is one-to-one associated to an auxiliary data structure which comprises a real valued data slot (*snapshot value slot*) and a w-sized binary bucket (*window transmission bucket*). The snapshot value slot loads the sensor reading, if any, which is comprised in $z_{t_i}(\cdot)$. The window transmission bucket loads 0/1 for each time point in the sliding window: 0 for a missing reading; 1 otherwise. A datum is maintained in its slot until a new snapshot arrives; the binary information stored

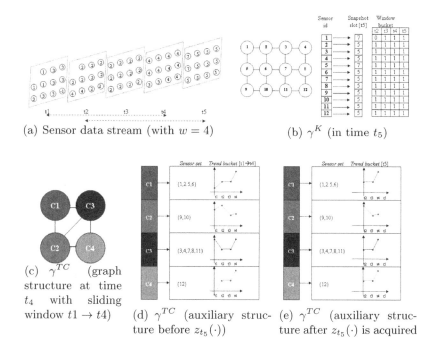

(a) Sensor data stream (with $w = 4$)

(b) γ^K (in time t_5)

(c) γ^{TC} (graph structure at time t_4 with sliding window $t1 \to t4$)

(d) γ^{TC} (auxiliary structure before $z_{t_5}(\cdot)$)

(e) γ^{TC} (auxiliary structure after $z_{t_5}(\cdot)$ is acquired

Fig. 1. A sensor data stream. The numeric reading of a random field Z is plotted in the circle representing the sensor (Figure 1(a)). A watch on γ^K (Figure 1(b)) once the snapshot $z_{t_5}(\cdot)$ is poured in SWIT. A watch on γ^{TC} between t_4 and t_5 (Fig. 1(c)- 1(d)).

in the bucket is subject to the sliding window mechanism. The storage of the snapshot $z_{t_i}(\cdot)$ in γ^K is three-stepped. In the first step, sensor readings are stored in γ^K. Each window transmission bucket is shifted-on-left such that the oldest time point is discarded and 0/1 is added at the present time. In particular, for each sensor $A \in K_{t_i}$, SWIT distinguishes two cases.

(1) A is already modeled by a graph node. $z_{t_i}(A)$ is stored in the snapshot value slot of its auxiliary data structure. 1 is stored at the present time of the window transmission bucket.

(2) A is not yet modeled by a graph node. A new graph node to represent A and new edges to connect A to the neighbor sensors are created. The node is associated to its auxiliary data structure with a zero-valued window transmission bucket. $z_{t_i}(A)$ is loaded in the snapshot value slot and 1 is is stored at the present time of the window transmission bucket.

In the second step, SWIT purges sensors which were continually inactive along the window. In particular, sensors which index a zero-valued window transmission bucket are purged from γ^K. To purge a sensor from γ^K, the associated node, edges and auxiliary data structure are disposed. Finally, in the third step, missing data in $z_{t_i}(\cdot)$ are interpolated. A datum is expected to be known at t_i if it should come from a sensor that was active in the window (whose graph node survived to the purge). The interpolate is the inverse distance weighted (IDW[1]) sum of spatially close readings in the snapshot. Interpolated data are loaded in γ^K like the real readings of the transmitting sensors. Proposition 1 analyzes the size of γ^K .

Proposition 1. *Let $z(K, T)$ be the sensor stream, w be the sliding window size. For each time point $t_i \in T$, then $size(\gamma^K) \leq size(window)$ with $w \geq 2$.*

Proof. Let n be the number of sensors transmitting in the window. We consider that (1) $size(window) = size(g) + n \times w \times size(float)$ where $size(g)$ is the size of the graph structure (edges and nodes) and (2) $size(\gamma^K) = size(g) + n \times size(float) + n \times w \times size(bit)$. Based on (1) and (2), $size(\gamma^K)$ can be equivalently written as $w \geq \frac{size(float)}{size(float)-1}$. As it is reasonable that $\frac{size(float)}{size(float)-1} \leq 2$, then Proposition is always satisfied with $w \geq 2$.

Sliding Trend Clusters in γ^{TC}. The data synopsis γ^{TC} uses a graph to maintain the set of trend clusters (see Figures 1(d)-1(e)). Each graph node of γ^{TC} represents a trend cluster as it was discovered at the previous time point. The node indexes an auxiliary data structure which maintains the set of sensors (*sensor set*) grouped in the cluster and the w-sized time series (*trend bucket*) which describes the data trend along the window. Trend is described by a series of triples. Each triple is timestamped at a transmission point of the window and collects the minimum, the maximum and the mean of the clustered readings for the sensor set in the time point. Finally, graph edges link spatially close trend clusters (i.e. trend clusters grouping at least a pair of spatially close sensors). Once $z_{t_i}(\cdot)$ arrives, trend clusters obtained at time t_{i-1} and now maintained in γ^{TC} are slid. For each graph node of γ^{TC}, the oldest trend triple (i.e. that timestamped with t_{i-w}) in the associated trend bucket is discarded. The sensor set is updated by removing sensors which are no more maintained in γ^K. If the sensor set of a trend cluster runs out of sensors, the trend cluster is purged from γ^{TC}, that is, its graph node, edges and auxiliary data structure are disposed.

3.2 Merging Phase

The merging phase inputs γ^{TC}; identifies spatially close trend clusters which share a similar trend over $t_{i-w+1} \rightarrow t_{i-1}$; and applies the merge operator $\mu(\cdot, \cdot)$ to obtain a trend cluster which replaces them in γ^{TC}. For each pair of spatially close trend clusters (i.e. edged nodes in γ^{TC}), the similarity condition is checked.

Definition 7 (Trend Cluster Similarity Condition). *Let $tc_a = (H, c_a, z_a)$ and $tc_b = (H, c_b, z_b)$ be two trend clusters along the time horizon H with z_a and z_b be time series of triples (mean, min, max) . $trendClusterSim(tc_a, tc_b) = true$ iff $\forall t_i \in H, \max(max_a[t_i], max_b[t_i]) - \min(min_a[t_i], min_b[t_i]) \leq \delta$; otherwise $trendClusterSim(tc_a, tc_b) = false$.*

The merge operator is defined in the followings.

---

[1] IDW[5] is a simple interpolate which is frequently used in spatial statistics.

Definition 8 (Trend Cluster Merge Operator μ). *Let* $tc_a = (H, c_a, z_a)$ *and* $tc_b = (H, c_b, z_b)$ *be two trend clusters. The merge operator* μ *inputs* tc_a *and* tc_b *and outputs a new trend cluster* $tc = (H, c, z)$ *such that* $c = c_a \cup c_b$ *and for each* $t_i \in H$, $mean_z[t_i] = \frac{mean_{z_a}[t_i] \times |c_a| + mean_{z_b}[t_i] \times |c_b|}{|c_a| + |c_b|}$, $min_z[t_i] = \min(min_{z_a}[t_i], min_{z_b}[t_i])$ *and* $max_z[t_i] = \max(max_{z_a}[t_i], max_{z_b}[t_i])$.

We observe that the operator μ, when applied to spatially close trend clusters tc_a and tc_b, which satisfy the trend similarity condition, outputs a trend cluster (according to Definition 4). The proof is a consequence of the existence of a spatial closeness relation and trend similarity relation between $tc.a$ and $tc.b$.

The merging process repeatedly applies the merge operator $\mu(\cdot, \cdot)$ to trend clusters maintained in γ^{TC} each time the merge conditions are satisfied. For each graph node tc_a of γ^{TC}, its neighborhood, η_{tc_a}, in γ^{TC} is constructed. η_{tc_a} collects graph nodes (trend clusters) directly linked (spatially close) to tc_a in γ^{TC}. Then, for each node $tc_b \in \eta_{tc_a}$, the trend similarity condition between tc_a and tc_b is evaluated. This similarity judgment accounts for trends along the time horizon $t_{i-w+1} \rightarrow t_{i-1}$; data at time t_i will be considered in the splitting phase only. In case the trend cluster similarity condition is satisfied, tc_a and tc_b are merged in a single trend cluster. The computed trend cluster $\mu(t_a, t_b)$ replaces tc_a in γ^{TC}, while tc_b is definitely removed from γ^{TC}. The sensor set and the trend bucket indexed by tc_a are updated in order to maintain the cluster and the trend as they are computed by the merge operator. Before removing tc_b, the edges connecting tc_b to the remaining nodes in the graph are moved on tc_a. Finally, the graph node tc_b (and the associated auxiliary data structure) is disposed. The neighborhood construction and the merging of its members to the current trend cluster are repeated until there is no further trend cluster which was merged to in last iteration.

3.3 Splitting Phase

The splitting phase inputs γ^K and γ^{TC}, splits trend clusters now stored in γ^{TC} according to the data in the snapshot, extends trend information until present time, creates a new trend cluster for each sensor which is switched-on just now. For each trend cluster tc now maintained in γ^{TC}, the sensors which are grouped in tc are considered and a spatial clustering of their readings in $z(t_i)$ is computed. Spatial clustering is done by a graph partitioning technique which splits sensors (nodes) in groups of spatially close (graph edged) sensors whose readings in t_i are similar through the clustering, that is, c is a cluster iff $|max(c) - min(c)| \leq \delta$. Detected spatial clusters drive the splitting of tc in distinct trend cluster whose trends are now fitted to the present snapshot. In particular, for each spatial cluster c, a new trend cluster $splitTc$ having time horizon $t_{i-w+1} \rightarrow t_i$ is created. $splitTc$ has the spatial cluster c and the trend of tc from t_{i-w+1} to t_{i-1} which is extended with the triple (mean, min, max) timestamped at t_i and computed on the clustered readings in t_i. Each new trend cluster is modeled as a new graph node in γ^{TC} , while tc will be purged from γ^{TC}. To complete the process, SWIT considers sensors modeled in γ^K which are not yet collected in any trend cluster of γ^{TC} (i.e. sensors which are switched-on just in t_i). For each one of these sensors, a new trend cluster is created; this trend cluster is associated to an unknown trend for past times in the window.

4 Experimental Study

SWIT is written in Java and evaluated on two real sensor data streams. The Intel Berkeley Lab [2] (ILB) stream collects in-door temperature (in Celsius Degrees) transmitted every 31 seconds from 54 sensors deployed in the Intel Berkeley Research lab between February 28th and April 5th 2004. A sensor is close to the nearest sensors with $d = 6mt$. Readings are noised by outliers. Several sensors are often switched-off during the transmission time. By a box plot, we deduce that temperature ranges in [9.75, 34.6]. The South American Air Climate [3] (SAAC) stream collects monthly-mean out-door air temperature (in Celsius Degrees) recorded between 1960 and 1990 for a total of 6477 sensors. A sensor is close to the nearest sensors with $d = 0.5^o$. Temperature ranges in [-7.6, 32.9].

Experiments are run on Intel Pentium M Processor@2.20GHz (2.0 GiB RAM Memory) running Windows 7 Professional 32 bit with the parameter settings in Table 1. SWIT is compared to W-by-W which performs trend cluster discovery from scratch in each window by running the algorithm in [1] in a sliding window framework. The learning time (in millisecs) quantifies the time spent to maintain sliding window trend clusters once a new snapshot is acquired in the stream. The accuracy error quantifies how accurate sliding window trend clusters are in summarizing recently windowed data. It is computed as the root mean square error of each sensor reading in the sliding window once the reading is interpolated by the trend of the cluster which groups in the sensor. The error rate is a misclassification error to measure how many times a sensor is grouped in an unexpected cluster. The expected clusters are the trend clusters discovered by W-by-W. Then for each trend cluster tc discovered by SWIT, we determine the baseline cluster which groups the majority of its sensors, and label tc with this majority baseline. This error is computed in percentage.

Experimental results, collected window by window, are averaged on the sliding windows and reported in Table 2. They suggest several considerations. The sliding window trend clusters provide an accurate insight in trends and clusters in the recent past. The incremental learning strategy speeds-up sliding window trend cluster discovery. This result gains relevance by enlarging window/network size. Trend clusters maintained by SWIT do not move too much from those discovered by W-by-W. In particular, the error rate is always low (with picks of 8% in IBL and 26% in SAAC). The error tends to decrease by enlarging w.

---

[2] http://db.csail.mit.edu/labdata/ labdata.html
[3] http://climate.geog.udel.edu/~climate/html_pages/archive.html

Table 1. SWIT vs W-byW: parameter settings

	d	δ	w		d	δ	w
IBL	$6mt$	2.5	32, 64, 128	SAAC	$0.25°$	4	6, 12, 24

Table 2. SWIT vs W-by-W: statistics averaged on sliding windows

			SWIT			W-by-W			
stream	w	δ	time (ms)	rmse	#trend clusters	time (ms)	rmse	#trend clusters	error rate%
IBL	32	2.5	2.38	0.75	3.53	24.31	0.74	3.57	9.67
IBL	64	2.5	4.01	0.73	4.68	49.07	0.73	4.41	8.24
IBL	128	2.5	11.48	0.70	7.22	104.89	0.70	5.46	4.46
SAAC	6	4	9595.95	1.43	56.08	44017.49	1.53	36.31	16.00
SAAC	12	4	9024.30	1.34	70.64	60952.52	1.45	46.72	17.50
SAAC	24	4	8456.80	1.35	96.60	95688.27	1.38	49.97	19.91

5   Conclusions

We have presented an incremental strategy to maintain sliding window trend clusters in a sensor network. The proposed strategy reduces the amount of processed data by saving computation time. We plan to integrate sliding window trend cluster discovery in a system to detect concept drift in a sensor network.

Acknowledgments. This work fulfills the research objectives of both the project: "EMP3: Efficiency Monitoring of Photovoltaic Power Plants" funded by "Fondazione Cassa di Risparmio di Puglia," and the PRIN 2009 Project "Learning Techniques in Relational Domains and their Applications" funded by the Italian Ministry of University and Research (MIUR).

References

1. Ciampi, A., Appice, A., Malerba, D.: Summarization for Geographically Distributed Data Streams. In: Setchi, R., Jordanov, I., Howlett, R.J., Jain, L.C. (eds.) KES 2010, Part III. LNCS, vol. 6278, pp. 339–348. Springer, Heidelberg (2010)
2. Ciampi, A., Appice, A., Malerba, D., Guccione, P.: Trend cluster based compression of geographically distributed data streams. In: CIDM 2011, pp. 168–175. IEEE (2011)
3. Ciampi, A., Appice, A., Malerba, D., Muolo, A.: Space-Time Roll-up and Drill-down into Geo-Trend Stream Cubes. In: Kryszkiewicz, M., Rybinski, H., Skowron, A., Raś, Z.W. (eds.) ISMIS 2011. LNCS, vol. 6804, pp. 365–375. Springer, Heidelberg (2011)
4. Gaber, M.M., Zaslavsky, A., Krishnaswamy, S.: Mining data streams: a review. SIGMOD Rec. 34(2), 18–26 (2005)
5. Tomczak, M.: Spatial interpolation and its uncertainty using automated anisotropic inverse distance weighting (IDW) - cross-validation/jackknife approach. Journal of Geographic Information and Decision Analysis 2(2), 18–30 (1998)

Generic Subsequence Matching Framework: Modularity, Flexibility, Efficiency

David Novak, Petr Volny, and Pavel Zezula

Masaryk University, Brno, Czech Republic
{david.novak,xvolny1,zezula}@fi.muni.cz

Abstract. Subsequence matching has appeared to be an ideal approach for solving many problems related to the fields of data mining and similarity retrieval. It has been shown that almost any data class (audio, image, biometrics, signals) is or can be represented by some kind of time series or string of symbols, which can be seen as an input for various subsequence matching approaches. The variety of data types, specific tasks and their solutions is so wide that their proper comparison and combination suitable for a particular task might be very complicated and time-consuming. In this work, we present a new generic Subsequence Matching Framework (SMF) that tries to overcome the aforementioned problem by a uniform frame that simplifies and speeds up the design, development and evaluation of subsequence matching related systems. We identify several relatively separate subtasks solved differently over the literature and SMF enables to combine them in a straightforward manner achieving new quality and efficiency. The strictly modular architecture and openness of SMF enables also involvement of efficient solutions from different fields, for instance advanced metric-based indexes.

1 Introduction

A large fraction of the data being produced in current digital era is in the form of *time series* or can be transformed into sequences of numbers. This concept is very natural and ubiquitous: audio signals, various biometric data, image features, economic data, etc. are often viewed as time series and need to be also organized and searched in this way.

One of the key research issues drawing a lot of attention during the last two decades is the *subsequence matching problem*, which can be basically formulated as follows: Given a query sequence, find the best-matching subsequence from the sequences in the database. Depending on the specific data and application, this general problem has many variants – query sequences of fixed or variable size, data-specific definition of sequence matching, requirement of dynamic time warping, etc. Therefore, the effort in this research area resulted in many approaches and techniques – both, very general and those focusing on a specific fragment of this complex problem.

The leading authors in this field identified two main problems that limit the comparability and cooperation potential of various approaches: the *data bias* (algorithms are often evaluated on heterogeneous datasets) and the *implementation*

S.W. Liddle et al. (Eds.): DEXA 2012, Part II, LNCS 7447, pp. 256–265, 2012.

bias (the implementation of the specific technique can strongly influence experiment results) [1]. The effort to overcome the data bias is expressed by founding a common set of data collections [2] which is publicly available and that should be used by any consequent research in this area. However, the implementation bias lingers, which also obstructs a straightforward combination of compatible approaches whose interconnection could be fruitful.

Analysis of this situation brought us to conclusion that there is a need for a unified environment for developing, prototyping, testing, and combination of subsequence matching approaches. After a brief overview and analysis of current state of the field (Section 2), we propose a generic subsequence matching framework (SMF) in Section 3. Section 4 contains a detailed example of design and realization of a subsequence matching algorithm with the aid of SMF. The paper concludes in Section 5 by future research directions that cover possible performance boost enabled by a straightforward cooperation of SMF with advanced distance-based indexing and searching techniques [3,4]. Due to space limitations, an extended version of this work is available as a technical report [5].

2   Subsequence Matching Approaches

The field opening paper by Faloutsos et al. [6] introduced a subsequence matching application model that has been used ever since only with smaller modifications. The model can be summarized in the following four steps that should be adopted by a subsequence matching application:

slicing of the time series sequences (both data and query) into shorter subsequences (of a fixed length),
transforming each subsequence into lower dimension,
indexing the subsequences in a multi-dimensional index structure,
searching in the index with a distance measure that obeys the lower bounding lemma on the transformed data.

Originally [6], this approach was demonstrated on a subsequence matching algorithm that used the *sliding window* approach to slice the indexed data and *disjoint window* for the query. The Discrete Fourier Transformation (DFT) was used for dimensionality reduction and the data was indexed using the minimum bounding rectangles in R-Tree [7]. The Euclidean distance was used for searching since it satisfies the lower bounding lemma on data transformed by DFT.

The *data representation* and the choice of *distance function* are fundamental questions for each specific application. Current approaches regarding these questions were thoroughly overviewed and analyzed [5] with a conclusion that the questions of data representation and distance function can be practically separate from the specific subsequence matching algorithm; it is important to pair the data representation and the distance function wisely in order to satisfy the lower bounding lemma [6].

The work by Faloutsos et al. [6] encouraged many following works. Moon et al. [8] suggested a dual approach for slicing and indexing sequences. This

DualMatch uses the sliding windows for queries and disjoint windows for data sequences to reduce the number of windows that are indexed. DualMatch was followed by the generalization of windows creation method called GeneralMatch [9]. Another significant leap forward was made by the effort of Keogh et al. in their work about exact indexing of Dynamic Time Warping [10]. They introduced a similarity measure that is relatively easy to compute and it lower-bounds the expensive DTW function. This approach was further enhanced by improving I/O part of the subsequence matching process using Deferred Group Subsequence Retrieval introduced in [11].

If we focus on the performance side of the system, we have to employ enhancements like indexing, lower bounding, window size optimization, reducing I/O operations or approximate queries. Lots of approaches for building subsequence matching applications often use the very same techniques for solving common sub-tasks included in the whole retrieval process and changes only some parts with some novel approach. As a result, the same parts of the process (like DFT or DWT) are implemented repeatedly which leads to the phenomenon of the *implementation bias* [1]. The modern subsequence matching approaches [11,10] employ many smaller tasks in the retrieval process that solve sub-problems like optimizing I/O operations. Implementations of routines that solve such sub-problems should be reusable and employable in similar approaches. This led us to think about the whole subsequence matching process as a chain of sub-tasks, each solving a small part of the problem. We have observed that many of the published approaches fit into this model and their novelty is often only in reordering, changing or adding new subtask implementation into the chain.

3   Subsequence Matching Framework

In this section, we describe the general Subsequence Matching Framework (SMF) that is currently available under GPL license at http://mufin.fi.muni.cz/smf/. The framework can be perceived on the following two levels:

- on the *conceptual level*, the framework is composed of mutually cooperating modules, each of which solves a specific sub-task, and these modules are cooperating within specific subsequence matching algorithms;
- on the *implementation level*, SMF defines the functionality of individual module types and their communication interfaces; a subsequence matching algorithm is implemented as a *skeleton* that combines module types in a specific way and these can be instantiated by actual module implementations.

In Section 3.1, we describe the common sub-problems (sub-tasks) that we identified in the field and we define corresponding module types (conceptual level). Further, in Section 3.2, we justify our approach by describing fundamental subsequence algorithms in terms of SMF modules and we present a straightforward implementation of these algorithms within SMF. Section 3.3 is devoted to details about implementation of the framework.

Table 1. Notation used throughout this paper

Symbol	Definition
$S[k]$	the k-th value of the sequence S
$S[i:j]$	subsequence of S from $S[i]$ to $S[j]$, inclusive
$S.len$	the length of sequence S
$S.id$	the unique identifier of sequence S
$S'.pid$	if S' is subsequence of S then $S'.pid = S.id$
$S'.offset$	if $S' = S[i:j]$ then $S'.offset = i$ and $S'.len = j - i + 1$
$D(Q, S)$	distance between two sequences Q and S

The key term in the whole framework is, naturally, a *sequence*. As we want to keep the framework as general as possible, we do not lay practically any restrictions on the components of the sequence – it can be integers, real numbers, vectors of numbers, or any more sophisticated structures. The sequence similarity functions are defined relatively independently of specific sequence type (see Section 3.3). In the following, we will use the notation summarized in Table 1.

3.1   Common Sub-problems: Modules in SMF

Studying the field of subsequence matching, we identified several common subproblems addressed by a number of approaches in some sense. Specifically, we can see the following sub-tasks that correspond to individual modules in SMF.

Data Representation (Module: Data Transformer). The raw data sequences entering an application are often transformed into other representation which can be motivated either by simple dimensionality reduction (DFT, DWT, SVD, PAA) [6,12,13,14] or also by extracting some important characteristics that should improve the effectiveness of the retrieval [15]. In either case, the general task can be defined simply as follows: *Transform given sequence S into another sequence S'*. We will use the symbol in Figure 1 (a) for this *data transformer* module. The following table summarizes information about this module and gives a few examples of specific approaches implementing this functionality.

data transformer	transform sequence S into sequence S'
DFT	apply the DFT on sequence of real numbers S [6]
PAA	apply the PAA on sequence of real numbers S [14]
Landmarks	extract *landmarks* from sequence S [15]

Windows and Subsequences (Module: Slicer). Majority of the subsequence matching approaches partitions the data and/or query sequences into subsequences of, typically, fixed length (windows) [6,8,9,11]. Again, this task can be isolated, well defined, and the implementation can be reused in many variants of subsequence matching algorithms. Partitioning a sequence S, each resulting subsequence $S' = S[i:j]$ has $S'.pid = S.id$, $S'.offset = i$, and $S'.len = j - i + 1$.

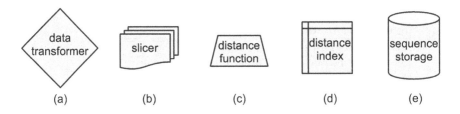

Fig. 1. Types of SMF modules and their notation

The module will be denoted as in Figure 1 (b) and its description and specific examples are as follows:

sequence slicer	partition S into list of subsequences $S'_1, \ldots, S'_n$
disjoint slicer	partition S disjointly into subsequences of length w [6]
sliding slicer	use sliding window of size w to partition S [6]

Sequence Distances (Module: Distance Function). There is a high number of specific distance functions D that can be evaluated between two sequences S and T. The intention of SMF is to partially separate the distance functions from the data and to use the specific distance function as a parameter of the algorithm (see Section 3.3 for details on realization of this independence). Of course, it is the matter of configuration to use appropriate function for respective data type, e.g. to preserve the lower bounding property. The distance functions symbol is in Figure 1 (c) and it can be summarized as follows:

distance function	evaluate dissimilarity of sequences S, T
L_p metrics	evaluate distance L_p on equally long number sequences
DTW	use DTW on any pair of number sequences S, T [16]
ERP	calculate Edit distance with Real Penalty on S, T [17]
LB_PAA, LB_Keogh	measures which lower-bound the DTW [10]

Efficient Indexing (Module: Distance Index). An efficient subsequence-matching algorithm typically employs an index to efficiently evaluate distance-based queries on the stored (sub-)sequences using the query-by-example paradigm (QBE). Again, we see the choice of the specific index as a relatively separate component of the whole algorithm and thus as an exchangeable module. Also, we see a space for improvement in boosting the efficiency of this component in future. We denote this module as in Figure 1 (d):

distance index	evaluate efficiently distance-based QBE queries
R-Tree family	index sequences as n-dimensional spatial data
iSAX tree	use a symbolic representation of the sequences [18,19]
metric indexes	index and search the data according to mutual distances [3]

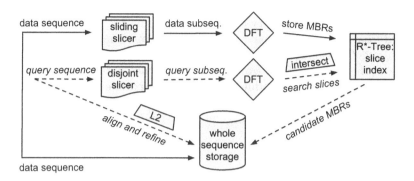

Fig. 2. Schema of the fundamental subsequence matching algorithm [6]

Efficient Aligning (Module: Sequence Storage). The approaches that use sequence slicing typically also need to store the original whole sequences. The slice index (for window size w) returns a set of candidate subsequences S', $S'.len = w$ each matching some query subsequence Q' such that $Q'.len = w$. If the query sequence Q is actually longer than w, the subsequent task is to align Q to corresponding subsequence $S[i : (i+Q.len-1)]$ where $i = S'.offset - Q'.offset$ and $S.id = S'.pid$. To do this aligning for each S' in the candidate set may be very demanding. For smaller datasets, this can be done in memory with no special treatment, but more advanced approaches are profitable on disk [11]. We will call this module *sequence storage* (Figure 1 (e)) and it is specified as follows:

sequence storage	store sequences S and return $S[i : j]$ for given $S.id$
hash map	basic hash map evaluating queries one by one
deferred retrieval	deferred group sequence retrieval (I/O efficient) [11]

3.2 Subsequence Matching Strategies in SMF

Staying at the conceptual level, let us have a look at the whole subsequence matching algorithms and their composition from individual modules introduced above. As an example, we take again the fundamental algorithm [6] for general subsequence matching of queries Q, $Q.len \geq w$ for an established window size w. The schema of a slight modification of this algorithm is in Figure 2. The solid lines correspond to data insertion and the dash lines (with italic labels) correspond to the query processing.

A data sequence S is first partitioned by the sliding window approach (*sliding slicer* module) into slices $S' = S[i : (i + w - 1)]$, these are transformed by Discrete Fourier Transformation (*data transformer* module DFT), and the Minimum Bounding Rectangles (MBR) of these transformed slices are stored in an R*-tree storage (*distance index* module); the original sequences S is also stored (*whole sequence storage* module). Processing a subsequence query, the query sequence Q is partitioned using the *disjoint slicer* module, each slice Q'

($Q'.len = w$) is transformed by DFT and it is searched within the slice index (using L_2 distance or a simple binary function *intersect*). For each of the returned candidate subsequences S', a query-corresponding alignment $S[i : (i+Q.len-1)]$ is retrieved from the *whole storage* (see above for details) and the candidate set is refined using L_2 *distance* $D(Q, S[i : (i + Q.len - 1)])$).

Preserving the skeleton of an algorithm (module types and their cooperation), one can substitute individual modules with other compatible modules obtaining a different processing efficiency or even a fundamentally different algorithm. For instance, swapping the sliding and disjoint slicer modules practically results in the DualMatch approach [8].

3.3   Implementation

The SMF was not implemented from scratch but with the aid of framework MESSIF [20]. The MESSIF is a collection of Java packages supporting mainly development of metric-based search approaches. SMF uses especially the following MESSIF functionality:

- encapsulation of the concept of data objects and distances,
- implementation of the queries and query evaluation process,
- distance based indexes (building, querying),
- configuration and management of the algorithm via text *config files*.

The *sequence* is in SMF handled very generally; it is defined as an interface which requires that each specific sequence type (e.g. a float sequence) must, among other, specify the distance between two sequence components $d(S[i], S'[j])$. For number sequences, this distances could be, naturally, absolute value of differences $d(S[i], S'[j]) = |S[i] - S'[j]|$, but one can imagine complex sequence components, for instance vectors where d could be an L_p metric. Implementation of a sequence distance $D(S, S')$ (for instance, DTW) then treats S and S' only as general sequences that use the component distance d and, thus, this implementation can be independent of specific sequence type.

4   Example of Subsequence Algorithm with SMF

An algorithm is within SMF implemented as a *skeleton* – module types, their connections via specified interfaces, and all algorithm-specific operations. The algorithm is then configured and instantiated by a text configuration file – module types required by the algorithm skeleton are filled by specific modules. Let us describe this principle on an example of a simple algorithm for general subsequence matching with variable query length – see Figure 3 for this skeleton schema. It uses two *slicer* modules, one *distance index* with a *distance function*, and a *sequence storage* for the whole sequences (again, with a *distance function*).

In order to run this algorithm, we have to instantiate these module types with specific modules. Figure 4 shows the key part of a SMF configuration file that starts such algorithm. On the first two lines, the *sliding slicer* module is defined

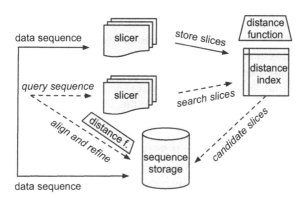

Fig. 3. Skeleton of a simple `VariableQueryAlgorithm` algorithm

```
slidingSlicer = namedInstanceAdd
slidingSlicer.param.1 = smf.modules.slicer.SlidingSlicer(<w>)

disjointSlicer = namedInstanceAdd
disjointSlicer.param.1 = smf.modules.slicer.DisjointSlicer(<w>)

index = namedInstanceAdd
index.param.1 = smf.modules.index.ApproxAlgorithmDistanceIndex(mIndex)

seqStorage = namedInstanceAdd
seqStorage.param.1 = smf.modules.seqstorage.MemorySequenceStorage()

startSearchAlg = algorithmStart
startSearchAlg.param.1 = smf.algorithms.VariableQueryAlgorithm
startSearchAlg.param.2 = smf.sequence.impl.SequenceFloatL2
startSearchAlg.param.3 = seqStorage
startSearchAlg.param.4 = index
startSearchAlg.param.5 = slidingSlicer
startSearchAlg.param.6 = disjointSlicer
startSearchAlg.param.7 = <w>
```

Fig. 4. SMF configuration file for `VariableQueryAlgorithm` algorithm

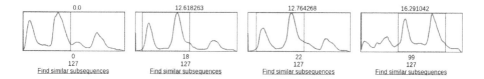

Fig. 5. Demonstration of `VariableQueryAlgorithm`: http://mufin.fi.muni.cz/subseq/

by an action called `namedInstanceAdd` which creates an instance `slidingSlicer` of class `smf.modules.slicer.SlidingSlicer` (with parameter w); the *disjoint slicer* is created accordingly. Then, the instance of *distance index* is created; it is a self-standing *algoritm*, namely a metric index M-Index [4] (we assume that the instance `mIndex` has been already created). In this example, the *sequence storage* is instantiated as a simple memory storage (`seqStorage`).

Finally, the actual `VariableQueryAlgorithm` is started passing the created module instances as parameters to the skeleton. The `param.2` of this action specifies that this particular algorithm instance requires sequences of floating point numbers and will compare them by Euclidean distance. Such SMF configuration files are processed directly by the MESSIF framework that enables creation of such algorithm and its efficient management.

This algorithm is demonstrated by a publicly available demo on a set of real number sequences compared by L_2 distance (http://mufin.fi.muni.cz/subseq/). Figure 5 shows screenshot of the GUI of this demo: The user can specify a subsequence (offset and width) of the query sequence and the most similar subsequences are located within the indexes set.

5   Conclusions and Future Work

The sequence data is all around us in various forms and extensive volumes. The research in the area of subsequence matching has been very intensive, resulting in many full or partial solutions in various sub-areas. In this work, we have identified several sub-tasks that circulate over the field and are tackled within various subsequence matching approaches and algorithms.

We present a generic subsequence matching framework (SMF) that brings the option of choosing freely among the existing partial solutions and combining them in order to achieve ideal solutions for heterogeneous requirements of different applications. Also, this framework overcomes the often mentioned implementation bias present in the field and it enables a straightforward utilization of techniques from different areas, for instance advanced metric indexes. We describe SMF on conceptual and implementation levels and present an example of a design and realization of subsequence algorithm with the aid of SMF. The SMF is available under GPL license at `http://mufin.fi.muni.cz/smf/`.

The architecture of the framework is strictly modular and thus one of natural directions of future development is implementation of other modules. Also, we will develop SMF according to requirements emerging from continuous research streams that utilize SMF. Finally and most importantly, we would like to contribute to the efficiency of the subsequence matching systems by involvement of advanced metric indexes. We believe in a positive impact of such cooperation of these two research fields that were so far evolving relatively separately.

Acknowledgments. This work was supported by national research projects GACR 103/10/0886, and GACR P202/10/P220.

References

1. Keogh, E., Kasetty, S.: On the Need for Time Series Data Mining Benchmarks: A Survey and Empirical Demonstration. In: Proceedings of ACM SIGKDD 2002, pp. 102–111. ACM Press (2002)
2. Keogh, E., Zhu, Q., Hu, B., Hay, Y., Xi, X., Wei, L., Ratanamahatana, C.A.: The UCR Time Series Classification/Clustering Homepage (2011)
3. Zezula, P., Amato, G., Dohnal, V., Batko, M.: Similarity Search: The Metric Space Approach. Springer (2006)
4. Novak, D., Batko, M., Zezula, P.: Metric Index: An Efficient and Scalable Solution for Precise and Approximate Similarity Search. Information Systems 36(4), 721–733 (2011)
5. Novak, D., Volny, P., Zezula, P.: Generic Subsequence Matching Framework: Modularity, Flexibility, Efficiency. Technical report, arXiv:1206.2510v1 (2012)
6. Faloutsos, C., Ranganathan, M., Manolopoulos, Y.: Fast Subsequence Matching in Time-Series Databases. ACM SIGMOD Record 23(2), 419–429 (1994)
7. Guttman, A.: R-Trees: A Dynamic Index Structure for Spacial Searching. ACM SIGMOD Record 14(2), 47–57 (1984)
8. Moon, Y.S., Whang, K.Y., Loh, W.K.: Duality-Based Subsequence Matching in Time-Series Databases. In: Proceedings of the 17th International Conference on Data Engineering, p. 263 (2001)
9. Moon, Y.S., Whang, K.Y., Han, W.S.: General Match: A Subsequence Matching Method in Time-series Databases Based on Generalized Windows. In: International Conference on Management of Data, p. 382 (2002)
10. Keogh, E., Ratanamahatana, C.A.: Exact indexing of dynamic time warping. Knowledge and Information Systems 7(3), 358–386 (2004)
11. Han, W.S., Lee, J., Moon, Y.S., Jiang, H.: Ranked Subsequence Matching in Time-series Databases. In: Proceedings VLDB 2007, pp. 423–434. ACM (2007)
12. Chan, K.P., Fu, A.W.C.: Efficient Time Series Matching by Wavelets. In: Proceedings ICDE 1999, pp. 126–133 (1999)
13. Korn, F., Jagadish, H.V., Faloutsos, C.: Efficiently supporting ad hoc queries in large datasets of time sequences. ACM SIGMOD Record 26(2), 289–300 (1997)
14. Keogh, E., Chakrabarti, K., Pazzani, M., Mehrotra, S.: Dimensionality Reduction for Fast Similarity Search in Large Time Series Databases. Knowledge and Information Systems 3(3), 263–286 (2001)
15. Perng, C.S., Wang, H., Zhang, S.R., Parker, D.S.: Landmarks: A New Model for Similarity-based Pattern Querying in Time Series Databases. In: Proceedings of ICDE 2000, pp. 33–42. IEEE Computer Society, Washington, DC (2000)
16. Sakoe, H., Chiba, S.: Dynamic programming algorithm optimization for spoken word recognition. IEEE Transactions on Acoustics Speech and Signal Processing 26(1), 43–49 (1978)
17. Chen, L., Ng, R.: On the Marriage of L_p-norms and Edit Distance. In: Proceedings of VLDB 2004, pp. 792–803 (2004)
18. Shieh, J., Keogh, E.: i SAX. In: Proceeding of the 14th ACM SIGKDD International Conference on Knowledge Discovery and Data Mining, KDD 2008, p. 623. ACM Press, New York (2008)
19. Camerra, A., Palpanas, T., Shieh, J., Keogh, E.: iSAX 2.0: Indexing and Mining One Billion Time Series. In: 2010 IEEE International Conference on Data Mining, pp. 58–67. IEEE (2010)
20. Batko, M., Novak, D., Zezula, P.: MESSIF: Metric Similarity Search Implementation Framework. In: Thanos, C., Borri, F., Candela, L. (eds.) Digital Libraries: R&D. LNCS, vol. 4877, pp. 1–10. Springer, Heidelberg (2007)

R-Proxy Framework for In-DB Data-Parallel Analytics

Qiming Chen, Meichun Hsu, Ren Wu, and Jerry Shan

HP Labs
Palo Alto, California, USA
Hewlett Packard Co.
{qiming.chen,meichun.hsu,ren.wu,jerry.shan}@hp.com

Abstract. R is a powerful programming environment for data analysis. However, when dealing with big data in R, a kind of main-memory based functional programming environment, the data movement and memory swapping become the major performance bottleneck. Therefore, executing a big-data-intensive R program could be many orders of magnitude less efficient than processing the SQL query directly inside the database for dealing with the same analytic task. Although there exists a number of "parallel-R" solutions, pushing R operations down to the parallel database layer, while retaining the natural R interface and the virtual R analytics flow, remains a very competitive alternative.

This has motivated us to develop the R-Vertica framework to scale-out R applications through in-DB, data-parallel analytics. In order to extend the R programming environment to the space of parallel query processing transparently to the R users, we introduce the notion of **R Proxy** - the R object with instance maintained in the parallel database as partitioned data sets, and schema (header) retained in the memory-based R environment. A function (such as aggregation) applied to a proxy is pushed down to the parallel database layer as SQL queries or procedures, with the query results automatically returned and converted to R objects. By providing the transparent 2-way mappings between several major types of R objects and database tables or query results, the R environment and the underlying parallel database are seamlessly integrated. The R object proxies may be created from database table schemas, in-DB operations, or the operations for persisting R objects to the database. The instances of the R proxies can be retrieved into regular R objects using SQL queries. With this framework, an R application is expressed as the analytics flow with the R objects bearing small data and the R proxies representing, but not bearing, big data. The big data are manipulated, or flow, underneath the in-memory R environment in terms of In-DB and data-parallel operations.

We have implemented the proposed approach and used it to integrate several large-scale R applications with the multi-node Vertica parallel database system. Our experience illustrates the unique feature and efficiency of this R-Vertica framework.

1   Introduction

R is an open source language for statistical computing and graphics. R provides a wide variety of statistical (linear and nonlinear modeling, classical statistical tests,

S.W. Liddle et al. (Eds.): DEXA 2012, Part II, LNCS 7447, pp. 266–280, 2012.

time-series analysis, classification, clustering) and graphical techniques, and is highly extensible. However, based on the in-memory computation, R programs are hard to scale-up or scale-out with big data.

We tackle this issue by the following steps: first, we support In-DB R analytics to offer the benefits of fast data access, reduced data transfer, minimized RAM requirement and SQL's rich expressive power [1-3]. Next, we rely on the Vertica system[12], HP's parallel database engine, to data-parallelize the sequential building blocks of R programs at the database layer instead of in the R environment. Further, targeting on the existing R users, we keep the natural R interface as the essential prerequisite. To integrate and abstract the above mechanisms, we propose the notion of **R proxy** object and developed the R-Vertica framework, an extension of the R environment to the underlying parallel database system.

1.1   The Problem

R programming is a kind of main-memory based functional programming, which makes it difficult to scale-up and scale-out with big data. Since the performance of functional programming is in general not comparable with the system programming used to build the DBMS, using an R program to execute the operations involving large data sets and heavy iterations, such as OLAP operations, could be many orders of magnitude less efficient than using the query processing. These scalability problems, due to their reasons, cannot be fully solved by increased CPU power and cores; but instead, can be solved by pushing the data-intensive analytics down to the database layer from the R layer. Our experience shows that for executing the typical multi-level, multi-dimensional OLAP operations with sizable input data, using a Vertica, a high-performance, column-based parallel database engine that reads data from DB, is 10000X faster than using the corresponding R program that reads data into the R environment and rollup the data over there. In that situation, even if the R program can be split into, say 100 parallel threads, the execution is still 100X slower than the query processing.

In addition to pushing R analytics down to the database layer, particularly to the parallel database layer, it is essential to retain the top-level R programming style and environment, which has given rise to the need for separating the "virtual" flow of the big data objects at the R layer, and the actual flow of them at the database layer.

1.2   The Prior Art

The scalability and efficiency of R programs in big-data analytics have been intensively studied in several dimensions with each having certain strengths and limitations. These approaches can be reviewed from the capability of scaling-out wrt both memory and computation, as well as the capability of retaining the natural R interface.

- Providing some R operators through procedural call from UDFs, such as Postgres-R (PL/R)[4,9]. While such an approach can please SQL programmers, the loss of the native R interface for R programmers is a problem.

- Parallelizing R computation on multi-cores and multi-servers, with memory sharing protocol (e.g. OpenMP) or distributed memory protocol (e.g. MPI, PVM)[5,6]. However, these efforts focus on enhanced computation power, but not on data partitioning, movement and buffering.

- Loading data and push certain computation down to the database layer[9,10,11]. The problem with the previous approaches in this regard is that the regular relational algebra and SQL are not the right abstraction for many of the statistical and numerical operations, and the conventional ways to store arrays, vectors, etc, in tables are not efficient for accessing.

- Develop array stores for analytics from scratch[6-8]. The problem of such approach is not leveraging the existing DBMS functionalities, particularly, the parallel data management and efficient data processing capabilities provided by a column-based parallel database system like Veritica.

1.3   Our Solution: R Proxy and R-Vertica Framework

Motivated by converging data analytics and data management platforms to support efficient and scalable analytics on big data, we provide the R-Vertica framework for scaling-out R applications through in-DB, data-parallel analytics.

The Vertica Analytic Database[12] is an innovative relational database management system optimized for read-intensive workloads. It provides extremely fast ad hoc SQL query performance for supporting data warehousing and Business Intelligence (BI).

Our solution is characterized by pushing the data-intensive R operations down to the Vertica database layer, and relying on the parallel query engine to data-parallelize the sequential building blocks of the analytics process at the database layer instead of at the R programming layer.

Further, targeting on the existing R users, we focus on the seamless integration of the two systems. In order to extend R analytics to the space of parallel query processing while keeping the natural R interface, we introduce the notion of **R Proxy** - the R object with instance maintained in the parallel database, possibly as partitioned data sets, with schema (header) retained in the memory-based R environment. A function (such as aggregation) applied to a proxy is pushed down to the parallel database layer as SQL queries or procedures to be executed efficiently by the parallel database engine, with the query results automatically converted to and returned as R objects. The R environment and the underlying parallel database are tightly integrated with the transparently provided, 2-way mappings between several major types of R objects, such as data frames, matrix, arrays, and database tables or query results. The R object proxies may be created from database table schemas, in-DB operations, or the operations for persisting R objects to the database. The instances of the R proxies can be retrieved into regular R objects using SQL queries.

With the R-Vertica framework, an R application is expressed as the analytics flow with the R objects bearing small data and the R proxies representing, but not bearing, big data. The big data are manipulated, or flow, underneath the in-memory R environment in terms of In-DB and data-parallel operations.

We have implemented the proposed R-Vertica package and used it to integrate several large-scale R applications with multi-node Vertica parallel database system. Our experience shows that for executing the typical multi-level, multi-dimensional OLAP operations with sizable input data, using Vertica, a high-performance, column-based parallel database engine that reads data from DB, is many orders magnitude faster than using the corresponding R program that reads and manipulates data in the R environment.

The rest of this paper is organized as follows: Section 2 describes how to push some data-intensive R operations down to the parallel database engine; Section 3 discusses the R-Vertica framework and introduces the notion of R-Proxy; Section 4 shows our experimental results; Section 5 concludes the paper.

2   In-DB and Data-Parallel Execution of R Operations

R is an integrated suite of software facilities for data manipulation, calculation and graphical display. It provides a collection of analytics tools as well as a well-developed extensibility mechanism, allowing new tools to be introduced as packages. Thus intuitively, for data-intensive tasks, new functions may be added which are implemented with SQL and actually executed by database engines. However, if the mapping between R objects and database objects are not automated, it would become the extra burden for R programmers; if big data still reside in the R programming space, and the data move between the R and database platforms is not reduced, the performance gain would be diminished.

We developed R packages for integrating R application with the Vertica Analytic Database, a multi-nodes, clustered parallel database management system optimized for read-intensive workloads and characterized by column storage, compression, data partition and parallel query processing. We chose Vertica as the executor of the R operations pushed down to the data management layer, to take the above advantages.

Let us consider a real application where a company gets revenues from selling hardware devices and providing support services for some of the delivered hardware devices. The hardware sale has certain characteristics such as customer, vertical domain, market segment, country, region, account class, channel, etc. The support service sale also has certain characteristics such as duration, service type, category, etc. One analytics problem is to know which customer characteristics, geography, hardware characteristics and service characteristics are most influential, either positively or negatively, in the sale of support services. For this purpose, one task is to match the supports with the corresponding hardware sales and to find out the total units and revenues of the supported hardware sales group by each support characteristics as well as their combinations. This is a kind of multilevel, multidimensional OLAP problem.

The original R implementation for this application has the major steps shown in Fig. 1.

- Load two big data sets (actually not big enough but already sufficient to show our argument) from files or databases into the R program as R's data-frames *hardware* and *support*, with each having 1M to 10M records (to avoid memory leak we used the sample data sets with 0.5M *hardware* records and 0.25M *support* records).

- Filter, extract and transform the input data to the required formats.
- Correlate the *support* data-frame and the *hardware* data-frame to identify the hardware sales involved in the support services.
- Aggegate the sales units and revenue of the hardware involved in the support services group by each, as well as by the combinations of the support characteristics.
- Other analytics tasks.

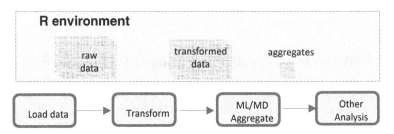

Fig. 1. An R application with big data loaded in memory, transformed and aggregated

In a conventional R program, the big data reside in the memory-based R environment, and operated by R programs. There exist several time consuming tasks: loading data; transforming data which requires scanning the whole data set; correlating *support* with *hardware* data-frames which involves Cartesian-product oriented, nested loop based R data manipulations; multi-level multi-dimensional aggregation of the hardware sales measures against each support characteristics; … etc. Our experiments show that on the above moderate-sized sample data, the correlation and aggregation with respect to each support characteristics consumes 17 minutes; if 16 support characteristics are considered, it takes 4-5 hours to complete the correlation-aggregation operations group by each individual support characteristics, excluding grouping by the combination of these characteristics. Such computation time, only on the sample data set, is way too long for providing near-real-time analytics service. When the full data are considered, the system simply fails to handle them at once.

There exist two general performance bottlenecks in implementing the above application by a conventional R program:

- Insufficient memory capacity for big data which may cause frequent page swap under the virtual memory management.
- Big data manipulation such as date transformation, correlation and aggregation which are not the strength of R programming.

However, the above weakness of R programming is just the strength of the database system. Using a database system to support R application essentially means

- Keep big data in the database thus avoid loading them to the memory based R programming environment.

- Manipulate big data on the database layer to take advantage of the query engine's efficient data processing capability for date transformation, correlation and aggregation, with only the required results, such as aggregates, returned to the R programming environment; in this way the data transfer is also greatly reduced.

- Further, using a parallel database system such as Vertica, the In-DB analytics query can be parallelized and scaled-out over multiple computer nodes. Viewed from R application, push operations down to the parallel database engine allows the data-parallel execution of sequential building blocks, in the highly efficient and scalable way.

These are illustrated in Fig. 2.

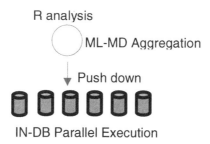

R analysis

ML-MD Aggregation

Push down

IN-DB Parallel Execution

Fig. 2. Push data-intensive R tasks down to parallel database for In-DB data-parallel analytics

More exactly, the above R application is implemented by pushing operations down to the parallel database engine, with the following scenario (Fig 3).

- It is unnecessary to load the instances of the two big data sets into R; instead, they reside in the database as tables "*hardware*" and "*support*". The table schemas, however, are loaded to R as the corresponding R data-frames, as the ***object proxies*** to be explained later. After that, there exist two headers-only (i.e. schema-only) data-frame proxies in R without big data instances.
- Filter, extract and transform the input data are performed by SQL queries; the simple transformation, for instance, can be performed by SELECT INTO or UPDATE queries. The data "flow" inside the database layer without transferring to R.
- Correlate the *support* data with the *hardware* data to identify the hardware sales involved in the support service, is performed by a more complicated SQL join query.
- Aggegate the sales units and revenue of the hardware involved in the support services group by each, as well as the combinations, of the support characteristics, is performed by a list of AGGREGATE-GROUP BY queries.

As shown in Fig 3, up to the aggregation step, there is no data instances of *hardware* and *support* tables loaded into R, these data sets are kept in the database, filtered, updated, transformed and aggregated in the database with SQL queries which may involve User Defined Functions (UDFs); only the aggregation results are returned to R which are much smaller than the input data.

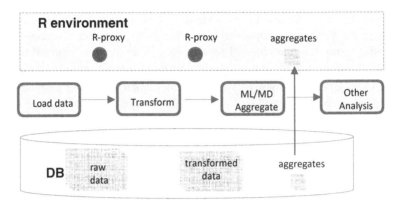

Fig. 3. The *hardware* and *support* tables with big data are stored, transformed, updated and aggregated in database; only aggregation results are returned to R. This way, the dataflow is actually represented by the flow of objects and proxies.

However, the R program at the function call level remains unchanged except applying a function to a ***proxy*** results in a query execution at the database layer. This way, the dataflow is actually represented by the flow of objects and proxies.

In the Vertica parallel database, the "*support*" table is *hash partitioned* to multiple nodes, and the "*ardware*" able is *replicated* to those nodes, which allows the join of two tables as well as the aggregations to be carried out in parallel. In this way, *we support "parallel R" indirectly at the database layer.*

Pushing R operations down to the database layer overcomes the difficulty of R programs in dealing with big data. It eliminates the need to load big data instances to R; it relies on the query engine to transform, update and derive data efficiently inside the database environment; it uses parallel query processing technique to speed up join and aggregate operations. Our experiments show that on the above sample data, the correlation and aggregation with respect to each support characteristics consumes only 0.2 second; compared with 17 minutes with R, this represents 5000X – 10000X performance gain (depends on the number of database nodes), and makes the provisioning of near-real-time analytics service possible.

3   R-Vertica Framework and R-Proxy

In order to extend R analytics to the space of parallel query processing while keeping the natural R interface for R users, we developed the R-Vertica framework for the seamless integration of the two systems.

We provide the automatic mappings between several major types of R objects and database relations, and introduce the notion of **R Proxy** as the R layer representation of the data instance stored in the database. With the regular R objects and the R proxies, an R analysis flow can be expressed naturally in the R programming, but the data instance related operations defined on R proxies are actually executed by the Vertica parallel query engine.

3.1   R-Vertica Connector

The R-Vertica connector is a package or library that provides two-way connections between R and the Vertica parallel database system (Fig 4). It allows an R program to send queries to Vertica and receive the query results as R data-frames. Both database DML operations and DDL operations are supported, such as create a table, truncate a table, update a table and retrieve a table. It also allows an R data-frame to be persisted in Vertica as a table.

The R database connectors with the similar functionality have been provided for other databases and reported in literatures. We plan to extend the connector functionality for accommodating parallel database engines. For example, when transferring an R data frame to an automatically generated Vertica table, we allow the options of partitioning or replicating data over multiple nodes. We also focus on abstracting the object mappings, making the database as the virtual extension of the R environment.

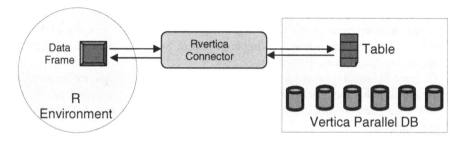

Fig. 4. R-Vertica Connector allowing database operations to be issued from R programs and query results returned as R data-frames

3.2   R-Vertica Object Mapping

On top of the R-Vertica connector, we provide an abstract object mapping layer to allow a compatible relation table or query result set to be converted to one of the major R object, such as a matrix, an array, in addition to a data-frame. This concept is illustrated in Fig 5. With the proposed R-Vertica mapping, some major types of R objects can be directly, i.e. with transparent data conversion, stored in the corresponding Vertica tables. For example, an array in R can be persisted to the corresponding array table in Vertica.

3.3   R-Proxy

In R, an object can have header (schema) and value (instance). In order to extend R analytics to the space of parallel query processing while keeping the natural R interface for R users, we introduce the notion of **R Proxy** - the R object with instance maintained in the parallel database, possibly as partitioned data sets, and schema (header) retained in the memory-based R environment.

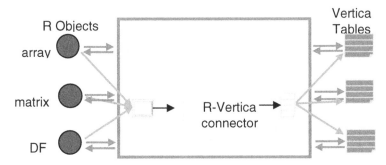

Fig. 5. R-Vertica Object Mapping virtualizes the R programming environment APIs

The purpose of introducing R proxy can be viewed from the following points:

- To avoid caching and manipulating big data in R programs but keep and manipulate them in the database; however, those data must be represented and referable in R programs.
- To provide the reference of the corresponding database object, typically a table or a SQL view, to which a function, implemented in SQL, for launching a database operation on it from the R program.
- To carry out meta data manipulation, since an R proxy, although not bearing data instances, holds the meta-data (R object header or relation schema) to which a meta-data manipulation function, such as *getNumberOfColumns(data-frame)*, may apply.

The notion of R proxy is illustrated in Fig 6 which can be explained in more detail as below.

- An R proxy represents the corresponding database object, typically a table or a SQL view. For example, an R data-frame proxy has its data stored in a table.
- An R proxy bears the meta-data, i.e. R object header or database table schema, but not necessarily the object value (data instance) unless the R proxy is explicitly instantiated in the R program. For example, a data-frame proxy in R contains the header of the data-frame only; that header is consistent to the relation schema of the table holding the instances of that data-frame.
- There is one-to-one mapping between the header of the R proxy and the schema of the table schema. For simplicity, currently we require the same column in the R proxy header and in the corresponding table schema to have the same name. We provide a simple version of *sync()* operation, i.e. the change of the R object header causes the corresponding change of the table schema, but not vice versa. For example, the sync() operation checks the column name consistency between an R proxy header and the corresponding table schema, and if the name of a column is altered in the R proxy header, an ALTER TABLE SQL statement is to be sent to the query engine to sync the table schema with the proxy header.
- A function (such as aggregation) applied to a proxy is pushed down to the parallel database layer as ad-hoc SQL queries or procedures.

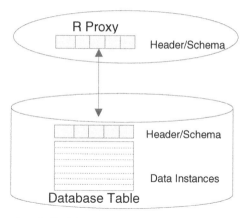

Fig. 6. R-Proxy representing the R object persisted in database and holding object header only

How R Proxy Is Created. In R, an object can have header (schema) and value (instance); conceptually an R proxy maps to a database table but remains its schema in R; practically an R proxy may be created in the following ways.

- An R proxy can be created from DB table schema. For example, an array is stored in table *matrix1* from that an R proxy, also referred to as *matrix1*, can be created by

    matrix1 ← **dbNewMatrixProxy** (.., "matrix1")

  where the header of the proxy is generated from the schema of the table.

- An R proxy can be derived from in-DB operations. For example, new tables and the corresponding R proxies may be generated such as

    matrix2 ← **dbMatrixTranspose** (.., matrix1)

    matrix3 ← **dbMatrixSub**(.., matrix1, dim/index parameters)

- Store the value of an R object to DB. When the value of an R object is persisted in the database, the R object degraded to a proxy with only header left in R. For example,

    array1 ← **dbCreateArrayTable**(.., array1)

- Create a corresponding proxy in R for holding a query result. For example

    df ← **dbNewDfProxyByQuery**(.., SQL stmt)

How R Proxy Instance Retrieved. The general way to retrieve the data instances associated with an R proxy includes the following.

- Instantiated by a SELECT * query using R-Vertica connector or object mapping, e.g. s

    df ← **dbQuery**(.., SQL stmt)

- Returned a subset of data from query and assign the result to another R object, e.g.

  df ← **dbQuery**(.., SQL stmt)

- Returned from a function that invokes a query. For example

  arrayObject← dbGetArrayInst(.., arrayProxy, dim/index…)

  where arrayProxy is a R proxy but arrayObject is a regular R object.

3.4   "Airflow" and "Seaflow"

While extending the R programming environment to the database space, we retain its integrity and natural R user interface, where the basic data citizens are the regular R data objects and the R proxies serving as the references of R objects persisted in the database. As a regular R object may be derived from an existing one, it can also be derived from the data content associated with a proxy. When a regular R data object is stored into the database, the corresponding proxy remains in the R environment.

A function may be defined on either a regular R data object or an R proxy. A function (such as aggregation) applied to a proxy is pushed down to the parallel database layer as ad-hoc SQL queries or procedures to be executed at the database layer, with the query results automatically converted to and returned as R objects.

The analysis flow logic for an application can be specified in the R environment with functions applied to the regular and proxy R objects. Since a function applied to a proxy is actually pushed down to the database layer, the corresponding dataflow at the R layer is *virtual*, namely, it expresses the logic but not necessarily the instances of dataflow. The state transitions of the big data actually manipulated in the database reflects the physical dataflow corresponding to a subset of the virtual dataflow in the R environment.

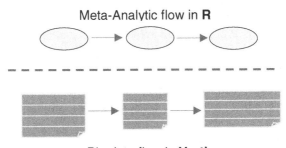

Fig. 7. R Objects and Proxies flow in R programming layer; big data flow in Vertica database layer

Intuitively, as shown in Fig 7, we refer to the virtual dataflow at the R layer that involves light data volumes, as "airflow", and the dataflow at the database layer that involves heavy data volumes, as "seaflow". Since the big data are manipulated, or flow, underneath the in-memory R environment in terms of In-DB and data-parallel operations, the separation of these two layers actually leverages the strengths of these two layers for supporting enhanced scalability.

4 Prototype and Experiments

This novel platform is designed and prototyped by integrating and extending several technologies we developed at HP Labs in in-DB analytics, UDF and dataflow management. The R programming environment is based on R2.1.3. Several HP Vertica 5.0 parallel database engines, with 1 node, 4 nodes and 8 nodes respectively, are utilized in the testing, they are built on the Linux servers with gcc version 4.1.2 20080704 (Red Hat 4.1.2-50), 8G RAM, 400G disk and 8 Quad-Core AMD Opteron Processor 2354 (2200.082 MHz, 512 KB cache).

The testing data are contained in two tables: a "support" table and a "hardware" table. The volume of the "support" table is around 200M; it contains 0.5 M tuples with each having 50 attributes. The volume of of the "hardware" table is around 50M; it contains 0.25M tuples with each having 20 attributes. By size, these data do not deserve "big-data" but already too big for the R programs to handle with reasonable efficiency (such that we can see the testing results in an acceptable waiting-time).

The R programs have two kinds of applications: continuous analytics and interactive query answering.

- In the continuous analytics, the loading, transformation, join and aggregation of big data are carried out in the database layer; the aggregated data are periodically retrieved to the R programming layer for mathematical analysis, with results either flowing in the R programming or storing back to the database.
- In the interactive query answering, a request with parameters is input from the user GUI; the request and its parameters are used to generate SQL queries, or instantiate *prepared queries*. The resulting query is sent to the Vertica database through the Rvertica connector, with query results returned as R objects (typically dataframes).

We choose to illustrate the comparison results of aggregation and OLAP operations. With R programming, these operations are very inefficient; however, pushing them down to the database layer, the query performance got improved tremendously.

The R function for connecting R and parallel DB is typically coded as below.

```
con.vertica <- function(node, db)
{
        library(DBI);
        library(rJava);
        library(RJDBC);
        library(Rvertica);
        con <- dbConnect(vertica(), user = "vertica", password = "", dbname = db,
                host = paste(node, ".hpl.hp.com", sep=""), port = 5433);
        con;
}
```

Then to connect to a database named "test" with leading node installed on machine "synapse-1", invoke

```
con8 <- con.vertica("synapse-1", "testdb");
```

As an example of formulating and executing a query, the following R function is executed where we assume that a user request information is captured in the R object mList from that a SQL query is generated. The query evaluation results in the R data-frame tb.

```
tb <- hardware.mquery(con8, mList);
```

The above utilities underlie the R proxy manipulation.

Below we compare the performance of running an operation by the R program and pushing the corresponding query down to the database layer, we run the query by the Vertica parallel databases with 1, 4 and 8 nodes respectively.

The first query is to get aggregates from the support table. The comparison is shown in Fig 8 where the performance gain of using in-DB analytics reaches 3 orders of magnitudes.

[Query (a)]
```
        SELECT s.Duration, SUM(s.support_dollars) AS total_support_dollars,
        SUM(s.support_units) AS total_support_units
        FROM support s GROUP BY s.Duration;
```

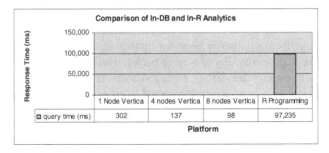

Fig. 8. Performance comparison of In-DB and In-R aggregation on the "support" data (Query a)

The second query is to correlate the hardware information in the support table while any duplicate is removed, which is much more expensive than the last query. The comparison is shown in Fig 9 where the performance gain of using in-DB analytics is almost 4 orders of magnitudes.

[Query (b)]
```
        SELECT Duration, SUM(hw_dollars), SUM(hw_units)
        FROM (SELECT DISTINCT Duration, s.customer_id, s.l4_customer_id, s.vertical,
                    s.segment, s.region, s.country, s.f_month, s.min_channel, s.max_channel,
                    s.GBU, s.platform, hw_dollars, hw_units
              FROM support s) r
        GROUP BY Duration
```

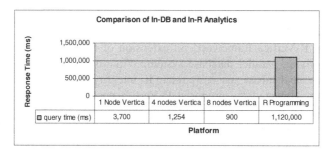

Fig. 9. Performance comparison of In-DB and In-R of query (b)

The effects of parallelism at the database layer is illustrated by Fig 10 where the query (b) is running on 1 node, 4 nodes and 8 nodes parallel database respectively.

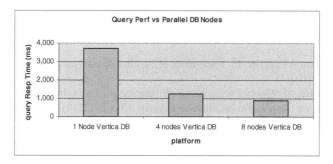

Fig. 10. Effects of parallelism at the database layer

The corresponding R program involves nested loops and rather inefficient. Since these R programs are quite tedious we do not list them here.

A query generated from user request is often filtered to certain factors the user is interested in, below is an example.

```
SELECT  coverage, duration, SUM(support_dollars) AS total_support_dollars,
SUM(support_units) AS total_support_units FROM support WHERE coverage IN (
'SBD', 'CTR', 'Other' ) AND duration IN ( '4y', '3y', '1y' ) GROUP BY  coverage, duration
```

Pushing such a query down to the database layer offers the similar ration of performance gain.

5   Conclusions

R is an open source language for statistical computing and graphics that provides a wide variety of statistical and graphical techniques. However, based on the in-memory computation, R programs are hard to scale-up or scale-out with big data. This has motivated a large number of research efforts in parallelizing R programs and pushing R functions down to the database layer. While this work is also characterized by in-DB analytics, we focus on retaining the natural R interface and supporting the "virtual analytics flow" at the R layer. We propose the notion of **R proxy** and developed the R-Vertica framework, an extension of the R environment integrated with the underlying parallel database system. While extending the R programming environment to the database space, we retain its integrity and natural R user interface.

We have implemented the proposed R-Vertica package and used it to integrate several large-scale R applications with multi-node Vertica parallel database system. Our experience shows that for executing the typical multi-level, multi-dimensional OLAP operations with sizable input data, using Vertica, a high-performance, column-based parallel database engine that reads data from DB, is many orders magnitude faster than using the corresponding R program that reads and manipulates data in the R environment.

References

1. Bryant, R.E.: Data-Intensive Supercomputing: The case for DISC. CMU-CS-07-128 (2007)
2. Chen, Q., Hsu, M., Zeller, H.: Experience in Continuous analytics as a Service (CaaaS). In: EDBT 2011 (2011)
3. Chen, Q., Hsu, M.: Query Engine Net for Streaming Analytics. In: Proc. 19th International Conference on Cooperative Information Systems, CoopIS (2011)
4. PL/R - R Procedural Language for PostgreSQL, http://www.joeconway.com/plr/
5. Schmidberger, M., Morgan, M., Eddelbuettel, D., Yu, H., Tierney, L., Mansmann, U.: State of the Art in Parallel Computing with R. Journal of Statistical Software 21(1) (2009)
6. Soroush, E., Balazinska, M., Wang, D.: ArrayStore: A Storage Manager for Complex Parallel Array Processing. In: ACM-SIGMOD 2011 (2011)
7. Stonebraker, M.: SciDB - A DBMS for Analytic Applications. In: ACM-SIGMOD 2011 (2011)
8. Stonebraker, M., SciDB Development Team: Overview of SciDB. In: ACM-SIGMOD 2010 (2010)
9. Teradata, In-database analytics with TeradataR (October 2010), http://developer.teradata.com/applications/articles/in-database-analytics-with-teradata-r
10. Zhang, Y., Zhang, W., Yang, J.: I/O-efficient statistical computing with RIOT. In: ICDE 2010 (2010)
11. Zhang, Y., Kersten, M., Ivanova, M., Nes, N.: SciQL: bridging the gap between science and relational DBMS. In: IDEAS 2011 (2011)
12. Vertica System, http://www.vertica.com

View Selection under Multiple Resource Constraints in a Distributed Context

Imene Mami[1], Zohra Bellahsene[1], and Remi Coletta[2]

[1] University Montpellier 2 - INRIA, LIRMM, France
[2] University Montpellier 2 - LIRMM, France
{mami,bella,coletta}@lirmm.fr

Abstract. The use of materialized views in commercial database systems and data warehousing systems is a common technique to improve the query performance. In past research, the view selection issue has essentially been investigated in the centralized context. In this paper, we address the view selection problem in a distributed scenario. We first extend the AND-OR view graph to capture the distributed features. Then, we propose a solution using constraint programming for modeling and solving the view selection problem under multiple resource constraints in a distributed context. Finally, we experimentally show that our approach provides better performance resulting from evaluating the quality of the solutions in terms of cost saving.

1 Introduction

View materialization is a widely used strategy in commercial database systems and data warehousing systems to improve the query performance. Indeed, answering queries using materialized views can significantly speed up the query processing since the access to materialized views is much faster than recomputing views on demand. However, whenever a base relation is changed the materialized views built on it have to be updated in order to compute up-to-date query results. The process of updating materialized views is known as view maintenance. Besides, materialized views need storage space.

The problem of choosing which views to materialize by taking into account three important features: query cost, view maintenance cost and storage space is known as the view selection problem. This is one of the most challenging problems in data warehousing [16]. For this reason the view selection problem has received significant attention in past research but most of these studies presented solutions in the centralized context [9].

In a distributed environment the view selection problem becomes more challenging. Indeed, it includes another issue which is to decide on which computer nodes the selected views should be materialized. Furthermore, resource constraints such as CPU, IO, network bandwidth have to be taken into consideration. The view selection problem in a distributed context may also be constrained by storage space capacities per computer node and maximum view maintenance cost.

S.W. Liddle et al. (Eds.): DEXA 2012, Part II, LNCS 7447, pp. 281–296, 2012.

To the best of our knowledge, no past work has addressed this problem under all these resource constraints. Our constraint programming based approach fills this gap. Indeed, all these resource constraints will easily be modeled with the rich constraint programming language. Furthermore, the heuristic algorithms which have been designed to solve the view selection problem in a distributed scenario are deterministic algorithms. For example greedy algorithm [3] and genetic algorithm [8], a type of randomized algorithms. These heuristic algorithms may provide near optimal solutions but there is no guarantee to find the global optimum because of their greedy nature or their probabilistic behavior. We have demonstrated in our recent work [10] the benefit of using constraint programming techniques for solving the view selection problem with reference to the centralized context in terms of the solution quality. Indeed, our approach is able to provide a near optimal solution to the view selection problem during a given time interval. The quality of this solution may be improved over time until reaching the optimal solution. Specifically, our main contributions are:

1. We propose an extension of the concept of the AND-OR view graph [15] in order to reflect the relation between views and communication network within the distributed scenario. We make use of the concept of the AND-OR view graph to exhibit common sub-expressions between queries of workload which can be exploited for sharing updates and storage space.
2. We describe how to model the view selection problem in a distributed context as a Constraint Satisfaction Problem (CSP). Its resolution is supported automatically by the constraint solver embedded in the constraint programming language such as the powerful version of CHOCO [1]. The view selection problem has been addressed under multiple resource constraints. The limited resources are the total view maintenance cost and the storage space capacity for each computer node. Furthermore, we consider the IO and CPU costs for each computer node as well as the network bandwidth.
3. We have implemented our approach and compared it with a randomized method i.e., genetic algorithm [8] which has been designed for a distributed setting. We experimentally show that our approach provides better performance resulting from evaluating the quality of the solutions in terms of cost saving.

The rest of this paper is organized as follows. Section 2 defines the view selection problem in a distributed scenario and discusses the settings for the problem. In section 3, we present the framework that we have designed specifically to a distributed setting. Section 4 describes how to model the view selection problem under multiple resource constraints in a distributed environment as a constraint satisfaction problem (CSP). In section 5, it is provided our experimental evaluation. Section 6 presents a brief survey of related work. Finally, section 7 contains concluding remarks and future work.

2 Preliminaries

2.1 View Selection Problem and Cost Model in a Distributed Context

View Selection Problem. The general problem of view selection in a centralized context is to select a set of views to be materialized that minimizes the cost of evaluating the query workload. In a distributed scenario, multiple computer nodes with different resource constraints (i.e., CPU, IO, storage space capacity, network bandwidth, etc.) are connected to each other. Moreover, each computer node may share data and issue numerous queries against other computer nodes. In this paper, we have examined the problem of choosing a set of views and a set of computer nodes at which these views should be materialized so that the full query workload is answered with the lowest cost. In our approach, the view selection is decided under multiple resource constraints. Resources may be storage space capacity per computer node and maximum view maintenance cost. Furthermore, we consider the IO and CPU costs for each computer node as well as the network bandwidth.

Cost Model. The cost model assigns an estimated cost e.g., query cost or view maintenance cost to any view (or query) in the search space. In a distributed system, a cost model should reflect CPU, IO and communication costs.

$$Estimated\ cost = IO\ cost + CPU\ cost + Communication\ cost$$

The two first components IO and CPU costs measure the local processing cost. This cost is computed as the sum of all execution costs incurred by the required relational operations. The CPU cost is estimated as the time needed to process each tuple of the relation e.g., checking selection conditions. The IO cost estimate is the time necessary for fetching each tuple of the relation. The third cost component is the communication cost which is the time needed to transfer data e.g., transmitting views on the communication network. In our cost model these costs are estimated according to the size of the involved relations and in terms of time.

2.2 Constraint Programming

Constraint Programming is known to be a powerful approach for modeling and solving combinatorial search problems such as scheduling and timetabling. More recently, constraint programming has been considered as beneficial in data mining setting [13]. By constraint programming, we mean the computer implementation of an algorithm for solving Constraint Satisfaction Problems (CSPs).

A CSP model is composed of a set of variables $VAR = \{var_1, var_2, ..., var_n\}$, each variable var_i has a set of values which is called the domain of values $DOM = \{d_{var_1}, d_{var_2}, ..., d_{var_n}\}$ and a set of constraints $CST = \{c_1, c_2, ..., c_n\}$ that describes the relationship between subsets of variables. Formally, a constraint C_{ijk} between the variables var_i, var_j, var_k is any subset of the possible combinations of values of var_i, var_j, var_k, i.e., $C_{ijk} \subset d_{var_i} \times d_{var_j} \times d_{var_k}$.

The subset specifies the combinations of values that the constraint allows. A feasible solution to a CSP is an assignment of a value from its domain to every variable, so that the constraints on these variables are satisfied. For optimization purpose some cost expression on these variables takes a maximal or minimal value.

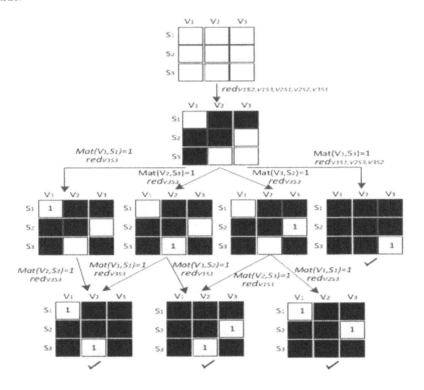

Fig. 1. Search tree using constraint propagation

Let us illustrate how the constraint programming can be applied to select and place materialized views. Figure 1 shows the domain reduction of nine variables $Mat(v_1, s_1)$, $Mat(v_1, s_2)$, $Mat(v_1, s_3)$, $Mat(v_2, s_1)$, $Mat(v_2, s_2)$, $Mat(v_2, s_3)$, $Mat(v_3, s_1)$, $Mat(v_3, s_2)$ and $Mat(v_3, s_3)$ where $Mat(v_i, s_j)$ denotes for each view v_i if it is materialized or not materialized on site s_j. It is a binary variable, $d_{Mat_{v_i, s_j}} = 0,1$ (0: v_i is not materialized on site s_j, 1: v_i is materialized on s_j). The problem is to select a set of views and a set of sites at which these views should be materialized under a maintenance cost constraint which guarantees that the total maintenance cost of the set of materialized views is less than 12 (knowing that $Mc(v_1, s_1)=8$, $Mc(v_1, s_2)=14$, $Mc(v_1, s_3)=18$, $Mc(v_2, s_1)=16$, $Mc(v_2, s_2)=15$, $Mc(v_2, s_3)=3$, $Mc(v_3, s_1)=12$, $Mc(v_3, s_2)=3$ and $Mc(v_3, s_3)=9$; where $Mc(v_i, s_j)$ denotes the cost of maintaining the view v_i on site s_j). At the beginning, the initial variable domains are represented by three columns of white squares meaning that every view can be materialized on any site. Considering the maintenance cost constraint, it appears that $Mat(v_1, s_2)$, $Mat(v_1, s_3)$,

$Mat(v_2, s_1)$, $Mat(v_2, s_2)$ and $Mat(v_3, s_1)$ cannot take the value 1 because otherwise the total maintenance cost will be greater than 12. Let $red_{v_i s_j}$ denotes the reduction of the domain of the variable $Mat(v_i, s_j)$. For instance in figure 1, $red_{v_1 s_2, v_1 s_3, v_2 s_1, v_2 s_2, v_3 s_1}$ filters the value 1 (the inconsistent value) from the domain of $Mat(v_1, s_2)$, $Mat(v_1, s_3)$, $Mat(v_2, s_1)$, $Mat(v_2, s_2)$ and $Mat(v_3, s_1)$. The deleted values are marked with a black square. After this stage some variable domains are not reduced to singletons, the solver takes one of these variables and tries to assign it each of the possible values in turn (i.e., $Mat(v_1, s_1)=1$). This enumeration stage triggers more reductions (i.e., $red_{v_3 s_3}$ where $Mat(v_1, s_1)=1$) which leads in our example to four solutions. These solutions are of various quality or cost. In addition to providing a rich constraint language to model a problem as a CSP and techniques such as constraint propagation to reduce the search space by excluding solutions where the constraints become inconsistent, constraint programming offers facilities to control the search behavior. This means that search strategies can be defined to decide in which order to explore the created child nodes in an enumeration tree which can significantly reduce the execution time. Furthermore, constraint programming provides ways to limit the tree search regarding different criteria. For instance performing the search until reaching a feasible solution in which all constraints are satisfied, or until reaching a search time limit or until reaching the optimal solution.

3   Distributed AND-OR View Graph

In order to exhibit common sub-expressions between queries of workload, the view selection is represented by using a AND-OR view graph [15,12]. Common sub-expressions can be exploited for sharing updates and storage space. The AND-OR view graph is a Directed Acyclic Graph (DAG) which is composed of two types of nodes: Operation nodes (Op-nodes) and Equivalence nodes (Eq-nodes). Each Op-node represents an algebraic expression (Select-Project-Join) with possible aggregate function. An Eq-node represents a set of logical expressions that are equivalent (i.e., that yield the same result). The Op-nodes have only Eq-nodes as children and Eq-nodes have only Op-nodes as children. The root nodes are equivalence nodes representing the queries and the leaf nodes represent the base relations. Equivalence nodes correspond to the views that are candidates to materialization.

The AND-OR view graph is the union of all possible execution plans of each query. Our motivation to consider all execution strategies is that it has been argued that a good selection of materialized views can only be found by considering the optimization of both global processing plans and materialized view selection [17]. The AND-OR view graph of the queries $q_1=$ P join PS join S and $q_2=$ PS join S join N where P, PS, S and N are the base relations [1] is shown in figure 2. Circles represent operation nodes and boxes represent equivalence nodes. For simplicity, we represent only two execution plans for the query q_1

---

[1] The subscripts P, PS, S and N denote respectively the base relations of TPC-H benchmark: Part, PartSupp, Supplier and Nation

and one execution plan for the query q_2. The remaining execution plans are just indicated by dashed lines. For example, view P-PS-S, corresponding to query q_1, can be computed from P-PS and S or P and PS-S. This dependence is indicated in figure 2 by AND and OR arcs.

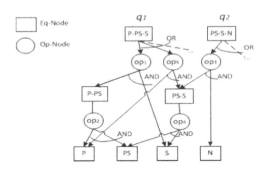

Fig. 2. AND-OR view graph of two queries q_1 and q_2

In this paper, we extend the concept of the AND-OR view graph to deal with distributed settings. Therefore, we propose the distributed AND-OR view graph to reflect the relation between views and communication network in the distributed scenario. We consider a distributed setting involving a set of sites (computer nodes) with different resource constraints (CPU, IO, storage space capacity, network bandwidth), a set of queries, a set of updates and their respective frequencies. For each query q, we consider all possible execution plans which represent its execution strategies. In this paper we consider selection-projection-join (SPJ) queries that may involve aggregation and a group by clause as well. Let us consider the query q defined over a simplified version of the TPC-H benchmark [2]. Query q finds the minimal supply cost for each country and each product having the brand name 'Renault'. The associated query is as follows:

Select	**P.partkey, N.nationkey, N.name, Min(PS.supplycost)**
From	**Part P, Supplier S, Nation N, PartSupp PS**
Where	**P.brand = 'Renault'**
and	**P.partkey = PS.partkey**
and	**PS.suppkey = S. suppkey**
and	**S.nationkey = N.nationkey**
Group by	**P.partkey, N.nationkey, N.name;**

A sample distributed AND-OR view graph is shown in figure 3. For simplicity, we consider a network of only three sites s_1, s_2, s_3 and we illustrate a part of the query q by considering only join operations and one execution strategy. Indeed, in figure 3 we consider only the join between Part (P) and PartSupp (PS) and the join between PartSupp (PS) and Supplier (S). The execution strategy that we have presented in figure 3 is ((P join PS) join S). We suppose that the base relations are stored on different sites.

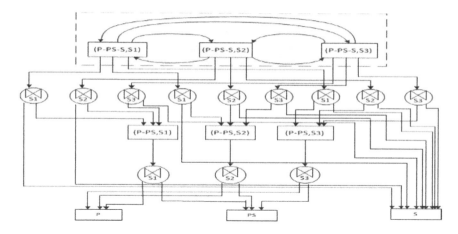

Fig. 3. Distributed AND-OR view graph

In order to represent the communication channels, every node is split into three sub-nodes, each of which denotes the view or the execution operation at one site. The communication edges between equivalence nodes of the same level (i.e., $(P - PS - S, S_1)$, $(P - PS - S, S_2)$ and $(P - PS - S, S_3)$), as shown in the dashed rectangle in figure 3, denote that a view can be answered from any other site if it is less expensive than computing this view from any children nodes. However, these edges are bidirectional creating cycles which no longer conforms to the characteristics of a DAG. In order to eliminate cycles, each sub-node (v_i, S_j), as illustrated in figure 4, has been artificially split into two nodes $(v_i, S_j)'$ and $(v_i, S_j)''$.

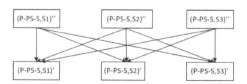

Fig. 4. Modified Distributed AND-OR view graph

4   Modeling View Selection Problem in a Distributed Context as a Constraint Satisfaction Problem (CSP)

In this subsection, we describe how to model the view selection problem in a distributed scenario as a Constraint Satisfaction Problem (CSP). Then, its resolution is supported automatically by the constraint solver embedded in the constraint programming language. All the symbols as well as the variables that we have used in our CSP model are defined in Table 1. The view selection in a distributed scenario can be formulated by the following constraint satisfaction model.

$$minimize \quad \sum_{(v_i,s_j) \in Q(G)} \left(f_q(v_i) * Qc(v_i, s_j) \right) \tag{1}$$

$$subject\ to \quad \forall s_j \in S \sum_{(v_i,s_j) \in V(G)} \left(Mat(v_i, s_j) * |v_i| * IO_j \right) \leq Sp_{max_j} \tag{2}$$

$$\sum_{(v_i,s_j) \in V(G)} \left(Mat(v_i, s_j) * f_u(v_i) * Mc(v_i, s_j) \right) \leq U_{max} \tag{3}$$

In our approach, the main objective is the minimization of the total query cost. The total query cost is computed by summing over the cost of processing each input query rewritten over the materialized views. Constraints (2) and (3) state that the views are selected to be materialized on a set of sites under a limited amount of resources. Constraint (2) ensures that for each site the total space occupied by the materialized views on it is less than its storage space capacity. Constraint (3) guarantees that the total maintenance cost of the set of materialized views is less than the maximum view maintenance cost.

Table 1. Symbols and CSP variables

Symbols of CSP model	
G	The distributed AND-OR view graph.
$Q(G)$	The query workload.
$V(G)$	The set of candidate views
U	The set of updates.
$\delta(v_i, s_j, u)$	denotes the differential result of view v_i on s_j, with respect to update u.
f_q	The frequency of a query.
f_u	The update frequency of a query (or view).
S	The set of sites which represent the computer nodes.
Sp_{max_i}	The storage space capacity of the site s_i.
U_{max}	The maximum view maintenance cost.
$|v_i|$	The size of v_i in terms of number of bytes.
$Bw(s_k, s_j)$	The bandwidth between s_j and s_k.

CSP variables and their domains	
$Mat(v_i, s_j)$	The materialization of the view v_i on site s_j. It is a binary variable $(d_{Mat(v_i,s_j)} = 0,1;\ 0: v_i$ is not materialized on s_j, 1: v_i is materialized on s_j).
$Qc(v_i, s_j)$	The query cost corresponding to the view v_i if it is computed or materialized on site s_j.
$Mc(v_i, s_j)$	The maintenance cost corresponding to the view v_i if it is updated on site s_j.
The costs are defined in terms of time (see subsection 2.1).	
Their domain is a finite subset of $\mathbb{R}_+^*$ ($d_{Qc(v_i,s_j)} \subset \mathbb{R}_+^*$ and $d_{Mc(v_i,s_j)} \subset \mathbb{R}_+^*$).	

The query and maintenance costs may be formulated as follows.

$$Qc(v_i, s_j) = \min_{s_k \in S} \left(Qc_{local}(v_i, s_k) + \frac{|v_i|}{Bw(s_k, s_j)} \right) \tag{4}$$

$$Qc_{local}(v_i, s_j) = \begin{cases} ComputingCost(v_i, s_j) & if \ \ Mat(v_i, s_j) = 0 \\ |v_i| * IO_j \ otherwise \end{cases} \tag{5}$$

$$ComputingCost(v_i, s_j) = \min_{op_l \in child(v_i, s_j)} \left(cost(op_l, s_j) + \right.$$
$$\left. \sum_{(v_m, s_n) \in child(op_l)} \left(Qc(v_m, s_n) + \frac{|v_m|}{Bw(s_n, s_j)} \right) \right) \tag{6}$$

Query Cost. The query cost includes the local processing cost and the communication cost. The local processing cost reflects CPU and IO costs (see subsection 2.1). Constraint (4) guarantees that a view is answered from the site that can provide the answer with the lowest cost. Constraint (5) and (6) ensure that the minimum cost path is selected for computing a given view on a given site. Each minimum cost path is composed of all the cost of executing the operation nodes on the path and the query cost corresponding to the related views or bases relations. The reading cost is considered if the view has been materialized.

$$Mc(v_i, s_j) = \begin{cases} 0 \ if \ \ Mat(v_i, s_j) = 0 \\ \sum_{u \in U(v_i, s_j)} \left(\min_{s_k \in S} \left(Mcost(v_i, s_k, u) + \frac{|v_i|}{Bw(s_k, s_j)} \right) \right) otherwise \end{cases} \tag{7}$$

$$Mcost(v_i, s_j, u) = \min_{op_l \in child(v_i, s_j)} \left(cost(op_l, s_j, u) + \right.$$
$$\left. \sum_{(v_m, s_n) \in child(op_l)} \left(UpdatingCost(v_m, s_n, u) + \frac{|v_m|}{Bw(s_n, s_j)} \right) \right) \tag{8}$$

$$UpdatingCost(v_m, s_n, u) = \begin{cases} Mcost(v_m, s_n, u) + \frac{|v_l|}{Bw(s_n, s_m)} & if \ \ Mat(v_m, s_n) = 0 \\ \delta(v_m, s_n, u) \ otherwise \end{cases} \tag{9}$$

View Maintenance Cost. The view maintenance cost is computed by summing the number of changes in the base relations from which the view is updated. We assume incremental maintenance to estimate the view maintenance cost. Therefore, the maintenance cost is the differential results of materialized views given the differential (updates) of the bases relations. Constraint (7) guarantees that a view with respect to the updates of the underlying base relations is updated from the site that can provide the differential results with the lowest cost. Constraints

(8) and (9) insure that the best plan with the minimum cost is selected to maintain a view. The view maintenance cost is computed similarly to the query cost, but the cost of each minimum path is composed of all the cost of executing the operation nodes with respect to update on the path and the maintenance cost corresponding to the related views.

5   Experimental Evaluation

In this section, we demonstrate the performance of our approach and a randomized method i.e., genetic algorithm which has been designed for a distributed setting [8]. The performance of view selection methods was evaluated by measuring the solution quality which results from evaluating the quality of the obtained set of materialized views in terms of cost saving.

5.1   Experiment Settings

For our experiments, we implemented a simulated distributed environment including a network of a set of sites (computer nodes). We assume that the different sites are divided into clusters so that there is a high probability that the sites which belong to the same cluster have similar query workloads. In our approach, for each cluster all the queries of the different workloads are merged into the same graph (see section 3) in order to detect the overlapping and capture the dependencies among them. Then, our method decides which views have to be selected and determine where these views should be materialized so that the full query workload is answered with the lowest cost under multiple resource constraints. The query workload are defined over the database schema of the TPC-H benchmark [2]. We then randomly assigned values to the frequencies for access and update based on a uniform distribution. In order to solve the view selection problem in a distributed context as a constraint satisfaction problem, we have used the latest powerful version of CHOCO [1]. For the randomized method, we have implemented the genetic algorithm presented in [8] by incorporating space and maintenance cost constraints into the algorithm. In order to let the genetic algorithm converge quickly, we generated an initial population which represents a favorable view configuration rather than a random sampling. Favorable view configuration such as the views which satisfy space and maintenance cost constraints are most likely selected for materialization. In the experimental results, the solution quality denoted by Q_s is computed as follows.

$$Q_s = 1 - \frac{\sum_{(v_i, s_j) \in Q(G)} \left(f_q(v_i) * Qc(v_i, s_j) \right)}{WM} \tag{10}$$

Where WM is the total query cost obtained using the "WithoutMat" approach which does not materialize views and always recomputes queries. The "WithoutMat" approach is used as a benchmark for our normalized results. Recall that $Qc(v_i, s_j)$ is the query cost corresponding to the view v_i on site s_j and $f_q(v_i)$ is the frequency of the view v_i corresponding to a single query.

In our approach, the view selection problem in a distributed environment is constrained by storage capacities $Sp_{max} = \{Sp_{max_i}, Sp_{max_j}, .., Sp_{max_n}\}$ where each site s_i has an associated storage space capacity Sp_{max_i} and maximum view maintenance cost U_{max}. Similar to [6] the storage space and maintenance cost limits are computed respectively as a function of the size (see equation 11) and total maintenance cost (see equation 12) of the query workload.

$$Sp_{max_i} = \alpha * Sp_i(AllM) \qquad (11)$$

$$U_{max} = \beta * Mc(AllM) \qquad (12)$$

Where $AllM$ is the "AllMat" approach which materializes the result of each query of the workload; α and β are constant. In our experiments, the storage space limit is per site and computed as a function of the size of the associated query workload. The view maintenance cost limit is calculated as a function of the total maintenance cost when all the queries are materialized.

Our approach to solve the view selection problem in a distributed setting is able to provide optimal solutions. However, computing optimal solutions may be very expensive because of the great number of comparisons between all possible subsets of views which are candidate to materialization. In this case, we use *timeout* condition to limit the search by considering that some solutions should not be explored. As mentioned in section 2.2, the constraint solver can find a set of feasible solutions in which all the constraints are satisfied before reaching the optimal solution. In the next experiments, the constraint solver performed a search until reaching the *timeout* condition. Indeed, our approach is able to provide a feasible solution at any time. The *timeout* condition was set to the time required by the genetic algorithm to solve the problem. This means that the constraint solver was left to run until the convergence of the genetic algorithm in the following experiments.

5.2 Experiment Results

We examined the effectiveness of our approach within three experiments. The first one compares the performance of our approach and the genetic algorithm for various values of storage space and maintenance cost limits. The second experiment evaluates the view selection methods with respect to different sizes of the distributed AND-OR view graph in terms of number of views (equivalence nodes). Finally, the last experiment evaluates our approach and the genetic algorithm with different network sizes in terms of the number of sites per cluster.

Performances under Resource Constraints. In this experiment, we examine the impact of space and maintenance cost constraints on solution quality. For this evaluation, each cluster includes 8 sites with different constraints of CPU, IO and network bandwidth and each site has an associated query workload. The values of α and β which define respectively the storage space capacities and the

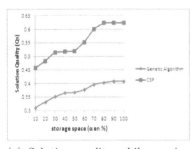
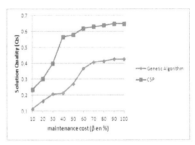

(a) Solution quality while varying the space constraint

(b) Solution quality while varying the maintenance cost constraint

Fig. 5. Evaluating the performance under resource constraints

view maintenance cost limit are varied from 10% to 100%. All the results are shown in figure 5.

Figure 5 (a) investigates the influence of space constraint on solution quality for each value of α where β was set to 60%. We note that the quality of the solutions produced by our approach and genetic algorithm improves when α increases, since there is storage space available for more views to be materialized. However, when $\alpha >= 80\%$ there is no improvement in the solution quality because the maintenance cost constraint becomes the significant factor.

Figure 5 (b) examines the impact of maintenance cost constraint on solution quality for each value of β where α was set to 80%. We can observe similarly to figure 5 (a) that we have better solutions when β increases since there is time to update the materialized views. The performance stabilizes when $\beta >= 90\%$ because the space constraint becomes the significant factor. We note from these experiments that our approach outperforms the genetic algorithm in the case where the resource constraints become very tight as well as in the case where we relax them. Indeed, for different values of α and β we can see that our approach generates solutions with cost saving more than 2 times more than the genetic algorithm.

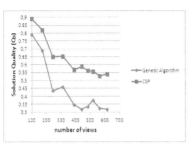
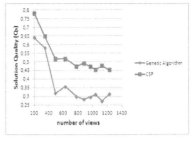

(a) Number of sites=4

(b) Number of sites=8

Fig. 6. Evaluating the performance over different number of views

Performance According to the Number of Views. Let us now evaluate the performance of our approach and the one of genetic algorithm while varying

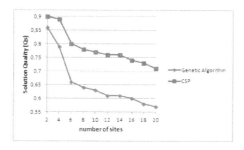

Fig. 7. Evaluating the performance over different number of sites

the size of the search space. Recall that the size of the search space is estimated according to the number of views (equivalence nodes) in the distributed AND-OR view graph described in section 3. Figure 6 illustrates the quality of the solutions produced by the two methods in a distributed environment. The number of sites per cluster is 4 sites in figure 6 (a) and 8 sites in figure 6 (b). The queries of the workload are randomly distributed over the network so that each site has an associated query workload. For instance, in figure 6 (b), the number of views in the distributed AND-OR view graph ranges from 200 to 1232 views. For each site, α was set to 40%. For the maintenance cost constraint, β was set to 60%. The experiment results depicted in figure 6 (a) and 6 (b) show that our approach provides the lowest query cost while varying the number of views. In fact, the cost saving is up to 27% more than the genetic algorithm. Therefore, our approach provides better performances compared with the genetic algorithm in terms of the solution quality.

Performance According to the Number of Sites. In order to evaluate the performance of view selection methods according to the number of sites, we conducted experiments with clusters of different sizes. For each cluster, we considered different number of sites with different constraints of CPU, IO and network bandwidth. The number of sites per cluster varies from 2 to 20. For each site, α was set to 40% and for the maintenance cost constraint, β was set to 60%. The experiment results are shown in figure 7. As in the previous experiments, we observe that our approach provides an improvement in the quality of the obtained set of materialized views in terms of cost saving compared with the genetic algorithm. Indeed, the cost saving is up to 15% more than the genetic algorithm.

6   Related Work

Several view selection methods have been proposed in the literature to select which views to materialize in a centralized context. They can be classified into four major groups.

Deterministic methods: Methods in this class take a deterministic approach by exhaustive search [14] or by some heuristics such as greedy [5,15]. However, greedy search is subjected to the known caveats, i.e., sub-optimal solutions may be retained instead of the globally optimal one since initial solutions influence the solution greatly.

Randomized methods: Typical algorithms in the context of view selection are genetic [8,7] or use simulated annealing [4,6]. Randomized algorithms can be applied to complex problems dealing with large or even unlimited search spaces. However, the quality of the solution depends on the set-up of the algorithm as well as the extremely difficult fine-tuning of algorithm that must be performed during many test runs. Furthermore, randomized algorithms do not guarantee to find the global optimum because of their probabilistic behavior.

Hybrid methods: Hybrid methods combine the strategies of deterministic and randomized algorithms in their search. A hybrid approach has been applied in [17] to the view selection problem which combine heuristic algorithms i.e., greedy algorithms and genetic algorithms. They prove that hybrid algorithms provide better solution quality. However, they are more time consuming and may be impractical due to their excessive computation time.

Constraint Programming methods: A constraint programming based approach has been presented in our previous work [10] to address the view selection problem in a centralized context. We have proved experimentally that our approach provides better performance compared with a randomized method i.e., genetic algorithm in term of cost savings. The success of using constraint programming for combinatorial optimization is due to its combination of high level modeling, constraint propagation and facilities to control the search behavior.

Analysis of view selection methods has shown that there is little work on view selection in a distributed scenario. The view selection problem is addressed in a distributed data warehouse environment in [3]. An extension of the concept of a data cube lattice to capture the distributed semantics has been proposed. Moreover, they extend a greedy based selection algorithm to the distributed case. However, the cost model that they have used does not include the view maintenance cost. Furthermore, the network transmission costs are not considered which is very important in a distributed context. The study presented in [8] deals with the view selection problem in distributed databases. This approach consists in applying a genetic algorithm to select a set of materialized views and the nodes of the network on which they will be materialized. However, this approach does not take into account neither the space nor the maintenance cost constraint. Besides, our approach provides better results compared with genetic algorithm in terms of the solution quality. A survey of view selection methods can be found in our previous work [11].

7   Conclusion

In this paper we have designed a constraint programming based approach to address the view selection problem under multiple resource constraints in a distributed environment. Furthermore, we have introduced the distributed AND-OR view graph to reflect the relation between views and communication network. We have performed several experiments over TPC-H queries and comparison with a genetic algorithm. The experiment results have shown that our approach provides better performance where the space and maintenance cost constraints become very tight as well as in the case where we relax them or when the number of views is high. Besides, our approach provides better solution quality in terms of cost saving when we consider diverse number of sites.

As a future work, we plan to design a set of pruning heuristics in order to reduce the search space of candidate views to materialization. This means that the size of the distributed AND-OR view graph will be small enough to allow its use for solving the view selection problem in a large scale distributed environments within reasonable execution time. The design of these heuristics will also guarantee the optimality of the solution where no time limit is imposed.

References

1. Choco, open-source software for csp, http://www.emn.fr/z-info/choco-solver
2. Tpc-h, http://www.tpc.org/tpch/spec/tpch2.14.3.pdf
3. Bauer, A., Lehner, W.: On solving the view selection problem in distributed data warehouse architectures. In: SSDBM, pp. 43–51 (2003)
4. Derakhshan, R., Stantic, B., Korn, O., Dehne, F.: Parallel Simulated Annealing for Materialized View Selection in Data Warehousing Environments. In: Bourgeois, A.G., Zheng, S.Q. (eds.) ICA3PP 2008. LNCS, vol. 5022, pp. 121–132. Springer, Heidelberg (2008)
5. Gupta, H., Mumick, I.S.: Selection of Views to Materialize under a Maintenance Cost Constraint. In: Beeri, C., Bruneman, P. (eds.) ICDT 1999. LNCS, vol. 1540, pp. 453–470. Springer, Heidelberg (1998)
6. Kalnis, P., Mamoulis, N., Papadias, D.: View selection using randomized search. Data Knowl. Eng. 42(1), 89–111 (2002)
7. Lee, M., Hammer, J.: Speeding up materialized view selection in data warehouses using a randomized algorithm. Int. J. Cooperative Inf. Syst. 10(3), 327–353 (2001)
8. Hueske, F., Böhm, K., Chaves, L.W.F., Buchmann, E.: Towards materialized view selection for distributed databases. In: EDBT, pp. 1088–1099. ACM, New York (2009)
9. Mami, I., Bellahsene, Z.: A survey of view selection methods. To appear in Sigmod Record (2012)
10. Mami, I., Coletta, R., Bellahsene, Z.: Modeling View Selection as a Constraint Satisfaction Problem. In: Hameurlain, A., Liddle, S.W., Schewe, K.-D., Zhou, X. (eds.) DEXA 2011, Part II. LNCS, vol. 6861, pp. 396–410. Springer, Heidelberg (2011)
11. Mami, I., Bellahsene, Z.: A survey of view selection methods. SIGMOD Record 41(1), 20–29 (2012)
12. Mistry, H., Roy, P., Sudarshan, S., Ramamritham, K.: Materialized view selection and maintenance using multi-query optimization. In: SIGMOD Conference, pp. 307–318 (2001)
13. De Raedt, L., Guns, T., Nijssen, S.: Constraint programming for itemset mining. In: KDD, pp. 204–212 (2008)

14. Ross, K.A., Srivastava, D., Sudarshan, S.: Materialized view maintenance and integrity constraint checking: Trading space for time. In: SIGMOD Conference, pp. 447–458 (1996)
15. Roy, P., Seshadri, S., Sudarshan, S., Bhobe, S.: Efficient and extensible algorithms for multi query optimization. In: SIGMOD Conference, pp. 249–260 (2000)
16. Widom, J.: Research problems in data warehousing. In: CIKM, pp. 25–30 (1995)
17. Zhang, C., Yao, X., Yang, J.: An evolutionary approach to materialized views selection in a data warehouse environment. IEEE Transactions on Systems, Man, and Cybernetics, Part C 31(3), 282–294 (2001)

The Impact of Modes of Mediation
on the Web Retrieval Process

Mandeep Pannu, Rachid Anane, and Anne James

Faculty of Engineering and Computing,
Coventry University, UK
{m.pannu,r.anane,a.james}@coventry.ac.uk

Abstract. This paper is concerned with the investigation of mediation between users and Web search engines and the impact of different modes of mediation on the Web search effectiveness. This involves the integration of explicit, implicit and hybrid modes of mediation within a content-based framework, facilitated by the adoption of the Vector Space Model. The work is supported by an experimental evaluation of the impact of different mediation modes on documents retrieval process in terms of recall and precision. The results of the experiments indicate that the mediation framework improves the quality of the retrieval process, and that the difference in the quality of the results is statistically significant.

Keywords: User profiling, Personalisation, Implicit profile, Explicit profile, content-based.

1   Introduction

Most current Web search engines are designed to serve a generic user irrespective of individual needs and interests. This raises the fundamental issue of how to identify and select the information that is relevant to a specific user. The retrieval process can be improved through personalisation of the search according to the specific needs and interests of the users. Implicit and explicit approaches can be used for user profiling. In the implicit approach, the behaviour of the users and their activities are observed and information is collected without the direct involvement of the user. On the other hand, explicit profile generation requires the users to directly provide specific information in order to create an individual user profile.

This research is an integral part of the effort aimed at overcoming the limitations of classic search engines. A mediation framework which is proposed allows the filtering of the results generated by classical search engines with the use of information contained in a user profile. The framework incorporates content-based information retrieval techniques, and it is facilitated by the adoption of the Vector Space Model (VSM). The proposed framework has been used to create a critical evaluation of variants of explicit, implicit and hybrid profiling techniques.

This paper is structured as follows. Section 2 is concerned with the background of this research. Section 3 presents the architecture of the proposed framework. Section 4 deals with the experimental evaluation of the framework in relation to the classical search engines, and offers pointers for further work. Section 5 concludes the paper.

S.W. Liddle et al. (Eds.): DEXA 2012, Part II, LNCS 7447, pp. 297–304, 2012.

2   Research Scope

The relation between a user query and Web pages is problematical and is driving the research in the field of information retrieval. Users have a variety of needs and the retrieval systems are often unable to satisfy adequately the requirements fulfil of an individual user.

2.1   Related Work

Personalised systems are designed to help users overcome the limitations of Web search by extracting keywords based on individual preferences. Personalisation can be implicitly or explicitly generated.

Explicit profile creation involves asking users for specific information in order to create an individual user profile. Salton et al. [1] considered user involvement as a powerful way of improving the relevance of the search results, and systems based on the information explicitly provided by users are constantly being developed [2-3-4]. In the explicit user profile generation users can build their own profile according to their specific interest and needs. Methods for generating explicit profile include asking the user to approve/disapprove a document [4], to give rating from a scale of values [5] or to engage in a dialogue in a natural language [3].

One of the first personalisation systems designed by Lieberman [1] was implicitly assuming an interest in a document if it was bookmarked, and a lack of interest if the document was left without saving or following hyperlinks inside it. Other approaches use techniques such as capturing mouse clicking [2] or tracking user mouse movements, as the mouse pointer can be used for reading [6]. A more recent approach used by Hussein and Elsayed [7] involves capturing the users' facial expression to estimate their interest in a document. Implicit feedback can be as effective as explicit feedback [8].

Although implicit methods are the focus of many research programmes, their reliability is still an issue [2]. Moreover, Paulson and Tzanavari [9] have pointed out that implicitly generated profiles are often not useful once users change their area of interest.

3   Proposed Framework

The primary goal of this research is to introduce a mediation framework, which act as an interface between a user and a classical search engine to provide personalised search results, without violating the privacy of the user. The framework can also be used as a vehicle for the investigation of different modes of user profiling. The major issue in evaluating an information retrieval approach is the amount of documents available and the quality of the results. Instead of developing a search engine, techniques can be evaluated by filtering only a subset of Web documents, where this subset would be retrieved from a base Web search engine API. The proposed framework is part of this endeavour.

3.1   Design Requirements

For the framework to be useful, it has to meet several objectives. First, it has to allow the implementation of custom methods for building user profiles. The framework should provide a programming interface that supports the tracking of actions detected in a Web browser, like navigating or clicking. A programmer modifying the framework in order to evaluate different filtering techniques should be able to do so by only modifying the filtering method, and by handling events from the browser to gather implicit or explicit information. Gathering explicit information may require some modification to the graphical user interface (GUI). The framework should support transparency for all other operations like retrieving search results from a base Web search engine or maintaining a database.

3.2   Mediation Framework Architecture

The framework combines a content-based approach and Vector Space Model for the filtering of results. The content-based approach was adopted because of its focus on the interaction between the profile of a single user and the content of a document. The VSM was used for the determination of the similarity between personalised profiles and documents. It offers formal clarity, efficiency in documents representation and consistent use of weights for the terms in documents, profiles, and query representations.

The overall architecture of the mediation framework is presented on the Figure 1. The user can use the provided interface to access documents on the Web. Every action in the browser is handled by the framework, and the information about it stored in the database, together with a VSM term vector containing keywords for visited document, the URL of the document and the time of the event. Similarly, the explicit feedback provided by the user is stored in the database.

When the user starts the search process by selecting one of the currently implemented profiling methods (explicit, implicit or hybrid), the application will generate a VSM representation of a selected profile, based on the information stored previously in the database. If the hybrid profile is selected, then both implicit and explicit vector representations are created separately and are then merged together.

Queries are used to retrieve a number of documents from a base Web search API. All the returned documents are indexed and represented in a vector form. The vector contains a list of terms extracted from the document, together with a weight for each of the term.

The system attempts to sort the documents by calculating the cosine similarity between the vector representing the user profile and the vectors representing each document. The cosine similarity is calculated by multiplying values for corresponding terms, and dividing the sum by the length of both vectors. The system then returns documents with the highest value of similarity. Depending on the configuration, the system can either return a constant number of documents, or only documents for which some similarity threshold is satisfied.

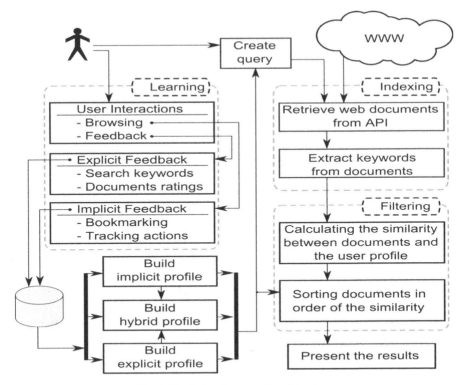

Fig. 1. Mediation framework architecture

As one of the objectives of the framework is to support the evaluation of different methods for user profile generation, the methods responsible for creating the user profile vector and for composing the profile vector with the entered query can be selected through the implementing of a single interface.

3.3   User Profile Generation

In order to evaluate the impact the framework, three different methods of profile generation have been investigated and implemented. In each case a user profile is represented in the VSM by a list of keywords with weights stored as a term vector.

An explicit user profile generation used for the evaluation is achieved by the submission of specific keywords by the users. The implicit profile is based on the observation of user behaviour and browsing history. The time spent on each page is assumed to be a good indicator of the user interest in a Web page. The system creates the profile vector by calculating the average time the user spent viewing each document, and adding together the vector representations for documents that were opened for longer than that time. The resulting vector is normalised in the final step.

In the hybrid profile the explicit and implicit profiles are generated independently and combined into a single term vector. For the purpose of this experiment the keywords from both vectors are simply added with equal weight.

The last editable part of the framework is the generation of a query. The associated method takes a list of keywords (as a query) entered by the user, and the generated user profile as parameters. Every keyword entered by the user is assigned a value equal to the highest value from the profile vector. Both vectors are then added and the result vector is truncated to 10 keywords. This vector is used by the framework to retrieve a list of documents from one of the base Web search API. Each document is in turn compared for similarity with the user profile.

4   Evaluation and Discussion

A comprehensive quantitative evaluation of the framework is presented. In the evaluation the performance of the three mediation systems is measured in terms of two metrics: precision and recall. The experiment was performed with 30 users with their own choice of keywords and areas of interests. To measure the system effectiveness the evaluation was conducted with Yahoo! and Google search APIs, and the mediation system with the three different types of profiling.

In the experiment, the system retrieves for each query 80 documents returned by each of the API; the framework was set to order these documents according to the similarity to the user profile and to return 20 documents with the highest similarity. For each base search APIs the first 20 returned documents were considered without any filtering.

4.1   Document Rating

To ensure that a consistent scale of scores is adhered to users were presented with an indication on how to assess a page depending on whether it was relevant or not. They were instructed to give 2 points to fully relevant documents, 1 point to documents containing relevant information as part of its contents, 0.5 point for documents that contained links to relevant information and 0 point if there was no relevant information [10]. The documents which could not be opened were ignored. The search results were presented in random order, and users were not aware of which search method generated the results. This process makes the scoring fairer, consistent and easier for the users.

4.2   Experimental Analysis

The experiment was conducted in two phases. In the first phase only a short time was given to build implicit profiles, while in the second phase this time was extended.

Experiment Phase 1. In the first phase the users were instructed to use the provided Web browser for 15 minutes so that the browsing behaviour could be recorded in the database. After the browsing session each user proceeded to enter the keywords for the search. The documents returned by each of the implemented systems and by the base Web Search APIs were combined into one list, sorted randomly and pre-opened in a Web browser. This approach was designed to avoid the situation where the ratings given by the users would be affected by their opinions of the retrieval systems. The documents were then rated by users.

Experiment Phase 2. The second phase of the experiment was performed to check how the retrieval effectiveness changes when a system had additional time to learn from the user behaviour. Each user was allocated the same user name through all the experiments so that new information could be added to an already stored browsing history. The additional learning time was set to 15 minutes per user, so that the total time allowed for system learning was doubled. The users rated search results in the same way as in the first phase of the experiment.

First Phase Results. Figure 2 presents the average precision and recall calculated for each of the systems after the first phase of the experiment.

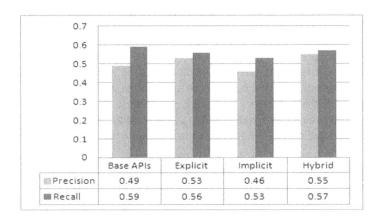

Fig. 2. Average precision and recall after the first phase

Both explicit and hybrid systems have improved the search performance in terms of precision, while the implicit system performance is worse than the performance of the base API. The recall values are almost unaffected by mediation, with the exception of the implicit system. Even through the implicit system alone has a negative effect on the performance, the hybrid system which uses the implicit information as part of the user profiling performs better than the explicit system which does not use it. This leads to a conclusion that while the implicit profile alone suffers from 'the cold start' problem, it can still be beneficial (in terms of precision) to use it in the filtering process, even after very limited learning.

Second Phase Results. The second phase of the experiment was designed to assess how precision and recall are affected when more time is given to the system to learn. As the explicit system being evaluated or the base APIs would not benefit from the additional time, the precision and recall in this phase were only measured for the implicit and hybrid systems.

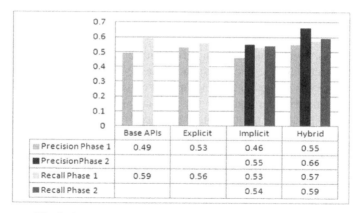

	Base APIs	Explicit	Implicit	Hybrid
▓ Precision Phase 1	0.49	0.53	0.46	0.55
■ PrecisionPhase 2			0.55	0.66
░ Recall Phase 1	0.59	0.56	0.53	0.57
■ Recall Phase 2			0.54	0.59

Fig. 3. Average precision and recall after the second phase

As shown in Figure 3, in the second phase the precision improved for both implicit and hybrid systems. It can be concluded that the results generated by implicit systems are more precise once a browsing history is generated. The values for relative recall calculated after the second phase appeared to have improved.

The Student's t-test was conducted to determine whether these results are significant. The significance of the results was calculated by comparing the performance of the hybrid system with the other systems after the second phase of the experiment. The results show that the change in precision is significant with 99% confidence, while the change in recall is not significant. This confirms that hybrid personalisation generates more relevant documents, but that the pool of relevant documents is however similar to the other systems.

4.3   Discussion

This section puts the work in context and identifies issues for further work. The experimental results indicate that the system using the hybrid profiling has better and more accurate results than the base APIs without the profiling. The hybrid profiling combines the explicitly stated interests with the observation of user behaviour and it can offer an effective way of dealing with information overload.

Although the aims and objectives of this research were met, a number of limitations have been identified. Useful documents can be ignored by their linguistic constraint as the calculation of the similarity is performed by exact match only. Secondly, efficiency issues which are also important were not addressed. The overheads caused by the need to download multiple documents before the framework can presents the search results were not investigated.

The proposed framework appears to be a viable mediator between users and the Web, but there is still scope for enhancing its effectiveness. Further work will seek to generate more accurate profiles, e.g. by widening the criteria of implicit observation. Currently, the framework is based on VSM with exact match for keywords. Adding support for synonyms or ontological context could help to identify the terms that are related to those stored in a user profile.

5   Conclusion

This research is an integral part of the effort aimed at overcoming the limitations of the classic search engines. The investigation has led to the proposal of a mediation approach which was applied in the development of three evaluated systems.

Conducted experiments indicate that mediation frameworks can improve the quality of the Web search results, with the choice of the mode of mediation being an important factor in enhancing search precision. With the high improvement in terms of precision and insignificant change in the recall, the personalised systems can definitely provide better users` experience and reduce the time they need to spend on searching.

References

1. Salton, G., Singhal, A., Mitra, M., et al.: Automatic Text Structuring and Summarization. Information Processing and Management 33, 193–207 (1997)
2. Rastegari, H., Shamsuddin, S.M.: Web Search Personalization Based on Browsing History by Artificial Immune System. Journal of Advances in Soft Computing and Its Applications 3(2), 282–301 (2010)
3. Stegmann, R.: Improving Explicit Profile Acquisition by Means of Adaptive Natural Language Dialog. In: Ardissono, L., Brna, P., Mitrović, A. (eds.) UM 2005. LNCS (LNAI), vol. 3538, pp. 518–520. Springer, Heidelberg (2005)
4. Swapna, P., Ravindran, R.B.: Personalized Web-page Rendering System. In: Das, G., Sarda, N.L., Reddy, K.P. (eds.) COMAD, pp. 30–39. Computer Society of India, India (2008)
5. Claypool, M., Le, P., Waseda, P., and Brown, D.: Implicit Interest Indicators. In: Proceeding of the 6th International Conference on Intelligent User Interface held at Santa Fe, New Mexico, United States, pp. 33–40 (2001)
6. Aoidh, E.M., Bertolotto, M., Wilson, D.C.: Implicit Profiling for Contextual Reasoning about Users Spatial Preferences, vol. 271 (2007)
7. Hussein, M., Elsayed, T.: Studying Facial Expressions as an Implicit Feedback in Information Retrieval Systems (2008)
8. Hopfgartner, F., Hannah, D., Gildea, N., and Jose, J.M.: Capturing Multiple Interests in News Video Retrieval by Incorporating the Ostensive Model. In: Proceeding of the Second International Workshop on Personalized Access, Profile Management, and Context Awareness in Databases held at Auckland, New Zealand, pp. 48–55 (2008)
9. Paulson, P., Tzanavari, A.: Combining Collaborative and Content-Based Filtering Using Conceptual Graphs. In: Lawry, J., Shanahan, J., Ralescu, A. (eds.) Modelling with Words. LNCS (LNAI), vol. 2873, pp. 168–185. Springer, Heidelberg (2003)
10. Kumar, S.B.T., Prakash, J.N.: Precision and Relative Recall of Search Engines: A Comparative Study of Google and Yahoo. Singapore Journal of Library & Information Management 38, 124–137 (2009)

Querying a Semi-automated Data Integration System

Cheikh Niang[1,2], Béatrice Bouchou[1], Moussa Lo[2], and Yacine Sam[1]

[1] Université François Rabelais Tours, Laboratoire d'Informatique
first.last@univ-tours.fr
[2] Université Gaston Berger de Saint-Louis, LANI
first.last@ugb.edu.sn

Abstract. A data integration system enables users to query a unified view of data sources through a global schema. We consider a semi-automatically built data integration system in the semantic web context, where data sources are annotated with ontologies. The global schema is also an ontology, expressed in *DL-Lite$_{\mathcal{A}}$*. After the semi-automated building of the global schema in a previous work, we focus here on the second part of this system, dedicated to the query answering process. We show how it can rely on either *GAV or LAV mappings, that are automatically computed*. We present algorithms for both cases, and we discuss the properties of this semi-automatically built data integration system.

Keywords: Data Integration, Description Logics, LAV/GAV Mappings.

1 Introduction

Data integration involves combining data residing in different sources and providing users with a unified view of these data. There are two main conceptual architectures [5]: the mediator approach, where data remain in their sources and are obtained when the system is queried, and the warehousing approach, where data are extracted from their sources, transformed and loaded in a warehouse before being queried. In this work we deal with a mediator-based data integration architecture, where the data integration system is seen as a triple $\langle \mathcal{G}, \mathcal{S}, \mathcal{M} \rangle$, such that [5]:

(*i*) $\mathcal{G}$ is the global schema, providing the conceptual representation of the application domain and an integrated view of the underlying sources.

(*ii*) $\mathcal{S}$ is the source schema, i.e., schemas of the sources where data are stored.

(*iii*) $\mathcal{M}$ is the mapping between $\mathcal{G}$ and $\mathcal{S}$, *i.e.* a set of assertions establishing the connection between the elements of the global schema and those of the source schema.

In general, one starts by designing the schemas $\mathcal{G}$ and $\mathcal{S}$ before choosing either GAV mappings, that associate to each element of the global schema a view over the sources, or LAV mappings, that associate to each source a view over the global schema.

Several integration systems relying on ontologies have been developed and some recent ones use some expressive description logics languages as formalisms to specify ontologies [4]. In this article, we focus on the *DL-Lite$_{\mathcal{A}}$* description logics [7], which has already been successfully used for building efficient ontology-based data access systems [2]. It is specifically tailored for both allowing a rich ontology description and preserving a reasonably low complexity of reasoning [3]. We have presented in [6] an

S.W. Liddle et al. (Eds.): DEXA 2012, Part II, LNCS 7447, pp. 305–313, 2012.

approach to build an ontology-based data integration system in a semi-automated way, based on this formalism.

Our main contribution is to show that the semi-automated semantic data integration task is no longer as elusive as it was, by presenting the query processing abilities of our semi-automatically built ontology-based integration system, thus completing our proposal in [6]. Before describing the querying process, we recall this proposal:

(i) We start with computing, for each source, an intermediate ontology called *Agreement*, that represents data to be shared. Agreements are alignments of local ontologies with a given *domain-reference* ontology[1]. Known ontology mapping methods are applied and a (local) expert validates this semi-automated step.

(ii) All agreements are then automatically conciliated in a global ontology via a relevant subsumption hierarchy of concepts, which is computed from the domain-reference ontology. We keep track of the origin of concepts in the global ontology.

Fig. 1 shows two agreements built from two agricultural sources, and the corresponding constructed global ontology. The first source treats information about agricultural-crop varieties and their production quantities, while the second one deals with the average cost prices of some agricultural products. Concepts represented by ovals with dash lines come from the domain-reference ontology. Considering such a global ontology, we present in the rest of this article *query rewriting* processes that reformulate a global query q_u into a query q_{rew} composed only of views of sources involved by q_u.

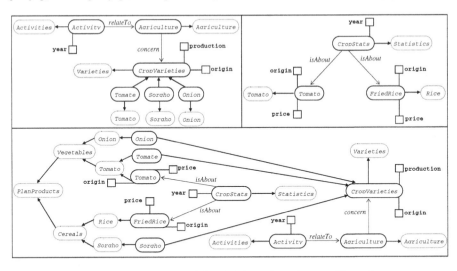

Fig. 1. Two agreements and the global ontology that conciliates them

2 Preliminaries

We express ontologies using *DL-Lite$_A$* [7], which belongs to the family of Description Logics (DLs) that allows representing the domain of interest in terms of *concepts*,

---

[1] The domain-reference ontology considered in our example is AGROVOC: http://www.fao.org/agrovoc.

denoting sets of objects, and *roles*, denoting binary relations between (instances of) concepts [7]. DLs differ in the constructs they offer to specify concepts and roles. *DL-Lite$_\mathcal{A}$* is known as one of the most expressive DL in the *DL-Lite* family that still allows efficient query answering [3]. It makes a distinction between objects and values, introducing attributes for describing properties of concepts represented by values (data properties) rather than objects.

A *DL-Lite$_\mathcal{A}$* Knowledge Base (*KB*) is a pair $\mathcal{K} = \langle \mathcal{T}, \mathcal{A} \rangle$, where $\mathcal{T}$ is the *TBox* (Terminological Box) and $\mathcal{A}$ is the *ABox* (Assertional Box): the former specifies concepts and roles and the latter represents their instances. The *DL-Lite$_\mathcal{A}$* **TBox** consists of a finite set $\mathcal{T}$ of inclusion assertions specified according to the following syntax: $B \sqsubseteq C \mid E \sqsubseteq F \mid Q \sqsubseteq R \mid U_C \sqsubseteq V_C$. A concept (respectively, value domain, role, and attribute) inclusion expresses that a basic concept B (respectively, basic value domain E, basic role Q, and atomic attribute U_C) is subsumed by a general concept C (respectively, value domain expression F, role R, attribute V_C). Inclusion assertions of the form $B_1 \sqsubseteq B_2$ or $Q_1 \sqsubseteq Q_2$ are called *positive inclusions* (PI), while inclusion assertions of the form $B_1 \sqsubseteq \neg B_2$ or $Q_1 \sqsubseteq \neg Q_2$ are called *negative inclusions* (NI). As an example, below is a part of the *DL-Lite$_\mathcal{A}$* TBox $\mathcal{T}$ of the ontology shown in bottom side of Fig. 1, that says that tomatoes and onions are two kinds of vegetables, and that they are disjoint. Crop statistics are a kind of statistics and is about[2] tomatoes. The two last assertions specify that tomatoes have a price which is a string.

$Tomato \sqsubseteq Vegetables$	$CropStats \sqsubseteq Statistics$	$Tomato \sqsubseteq \delta(price)$
$Onion \sqsubseteq Vegetables$	$\exists\, isAbout \sqsubseteq CropStats$	$\rho(price) \sqsubseteq xsd:string$
$Tomato \sqsubseteq \neg Onion$	$\exists\, isAbout^- \sqsubseteq Tomato$	

A *DL-Lite$_\mathcal{A}$* **ABox** consists of a finite set $\mathcal{A}$ of concept and role membership assertions of the form: $A(a)$, $P(a,b)$ and $U_C(a,b)$, where a and b are constants. For instance, the following assertions form a part of a *DL-Lite$_\mathcal{A}$* ABox $\mathcal{A}$, saying that the object identified by the constant *st7* denotes a crop statistics which is about a tomato identified by the constant *tm0*, *st7* is in 2007 and *tm0* has a price equals to *1€/Kg*:

$CropStats(st7)$	$year(st7,2007)$	$price(tm0,1 €/Kg)$
$isAbout(st7,tm0)$	$Tomato(tm0)$	

We consider conjunctive queries. Let x be a tuple of distinct variables, the so-called *distinguished variables*, and y be a tuple of distinct existentially quantified variables (not occurring in x), called the *non-distinguished* variables. $conj(x,y)$ is a conjunction of atoms with variables in x and y, whose predicates are atomic concepts ($A(x)$), atomic roles ($P(x,y)$), or atomic attributes ($U_C(x,y)$) of the TBox $\mathcal{T}$. A *conjunctive query* (CQ) over a *DL-Lite$_\mathcal{A}$* KB $\mathcal{K} = \langle \mathcal{T}, \mathcal{A} \rangle$ is an expression of the form: $q(x) \leftarrow conj(x,y)$. A *union of conjunctive queries* (UCQ) is a set of CQs with the same head.

Our notion of views for query processing is the same as what is known in data integration context, i.e they describe the content of a set of autonomous sources. In this article we consider conjunctive views, and therefore we define a view v_i as a CQ of the form: $v_i(x) \leftarrow conj(x_i,y_i)$.

Each view v has an extension with respect to a knowledge base, denoted $\mathcal{E}(v)$, which consists of a set of tuples of constants $v(t)$, of the same arity as v. When data are formalized using extension of views, the knowledge base can be expressed in the form

---

[2] The role *isAbout* has as *domain CropStats* and as *range Tomato*.

$\mathcal{K} = \langle \mathcal{T}, \mathcal{V}, \mathcal{E} \rangle$, where $\mathcal{V}$ is a set of views v, whose extensions are $\mathcal{E}(v) \in \mathcal{E}$. Since a *KB* $\mathcal{K}$ is characterized by a *set of models*, query answering over $\mathcal{K}$ can be seen as evaluating a query q over a set of interpretations I with respect to $\mathcal{K}$. However, one distinguishes the case where the extensions of predicates appearing in q are from an ABox from the case where data are inferred from extensions of views. In the latter case, the answers are called *certain answers* to q over $\mathcal{K}$.

In our system, data that a source shares with others are accessed using views, that are extracted from the agreement (cf. end of Section 1). We build (i) views for describing binary relations chains between local concepts, that we call *path views* and (ii) views for describing the properties of local concepts, called *star views* . Our purpose is to have more flexibility in the query rewriting process, by having more possibilities to combine different views when the properties of queried data are distributed across different sources. For example if we consider the agreement $\mathcal{T}_l$ shown graphically at the top-left part of Fig 1, a path view can be the following one:

$$v_{p_1}(x,y,z) \leftarrow Activity(x), relateTo(x,y), Agriculture(y), concern(y,z), CropsVarieties(z).$$

Similarly, one star view in $\mathcal{T}_l$ is:

$$v_{s_1}(x,a,b) \leftarrow CropsVarieties(x), production(x,a), origin(x,b).$$

Furthermore, we take into account the concepts that have specializations in $\mathcal{T}_l$. So, when a view v_{p_1} contains a concept A_1 that subsumes another concept A_2, we build another view $v_{p_{1.1}}$ from v_{p_1} by substituting A_2 to A_1 in v_{p_1}. We do the same for star views. For example, the concept *CropsVarieties* in the view v_{p_1} above subsumes the concept *Tomate* in $\mathcal{T}_l$, then we build also the two following views:

$$v_{p_{1.1}}(x,y,z) \leftarrow Activity(x), relateTo(x,y), Agriculture(y), concern(y,z), Tomate(z)$$
$$v_{s_{1.1}}(x,a,b) \leftarrow Tomate(x), production(x,a), origin(x,b)$$

On the one hand, views generated for a source S are used for extracting data that S can give as certain answers to a query expressed in terms of the global ontology. The values that correspond to views are computed according to how data are stored in the source (relational, XML, RDF triple store, ...). Due to space limitations, we can not present these parts of the system. On the other hand, computed path and star views may be stored in the semantic-mapping part of the Mediator, for being used for query processing following the LAV approach presented in Section 3. Indeed, as shown in [6], the automated global schema building process implies that all the concepts and roles that define views exist also in the global ontology. Thus, the views that we **automatically** generate for representing the content of a source S are also views over the global ontology, *i.e. they are LAV mappings*.

3   Semantic Query Rewriting

We consider the following assertions, in the TBox $\mathcal{T}_g$ of G in Fig 1, that are computed automatically during the global ontology building process, and say that tomato, tomate and onion products are also vegetables, rice products are also cereals, tomato products are disjoint to onion products and vegetable products are disjoint to cereal products:

Tomato ⊑ Vegetables	Tomate ⊑ Tomato	Tomato ⊑ ¬Onion
Onion ⊑ Vegetables	Rice ⊑ Cereals	Vegetables ⊑ ¬Cereals

We also consider the following user query q_u expressed in terms of $\mathcal{G}$:

$q_u(x,a,b) \leftarrow Vegetables(x), price(x,a), production(x,b)$

Our **LAV** query rewriting process is sketched in Algorithm 3.1. It takes as inputs the query on the global ontology q_u, the TBox of the global ontology $\mathcal{T}_g$, and $\mathcal{V}_s$, the set of LAV mappings that we just described (end of Section 2). For example, let's consider that we have computed the following views $v_{11}, v_{12} \in \mathcal{V}(S_1)$ and $v_{21}, v_{22} \in \mathcal{V}(S_2)$:

$v_{11}(x,a) \leftarrow Onion(x), production(x,a)$ $v_{21}(y,b) \leftarrow Tomato(y), price(y,b)$
$v_{12}(x,a) \leftarrow Tomate(x), production(x,a)$ $v_{22}(y,b) \leftarrow FriedRice(y), price(y,b)$

Algorithm 3.1 - LAV Query Rewriting

Input:
   (i) q_u: user query.
   (ii) $\mathcal{V}_s$: LAV mappings.
   (iii) $\mathcal{T}_g$: TBox of the global ontology.
Output:
   q_{rew}: rewritten query, to be distributed through sources.
Local Variables:
   (i) Q_u: set of reformulated queries.
begin
   1. $Q_u := PerfectRef(q_u, \mathcal{T}_g)$
   2. $q_{rew} := ConsistentMiniCon(Q_u, \mathcal{T}_g)$
end

Based on the views computed in the sources, that are the LAV mappings in $\mathcal{V}_s$, and the semantic assertions declared in $\mathcal{T}_g$, the algorithm finds from q_u a query expression q_{rew} that uses only the views in $\mathcal{V}_s$, such that (i) q_v is equivalent to q_u (or is the maximal query contained in q_u) and (ii) q_v is expressed in terms of data provided by sources. In the context of our example, the following query is the best rewriting of q_u:

$q_{rew}(x,a,b) \leftarrow v_{21}(x,a), v_{12}(x,b)$

Indeed, the view $v_{21} \in \mathcal{V}(S_2)$ provides the price of tomato that are vegetables, while the view $v_{12} \in \mathcal{V}(S_1)$ provides the production of tomate which are also vegetables. Notice however that the view $v_{11} \in \mathcal{V}(S_1)$ also provides the production of onion, which are vegetables, but in the TBox assertions $\mathcal{T}_g$ it is argued that tomato product is disjoint from onion product and so, tomato instances can not be joint to onion instances.

For computing the correct rewritings, we propose to *adapt the MiniCon algorithm* [8], which produces the *maximally-contained* conjunctive rewritings of a conjunctive query q using a set $\mathcal{V}$ of conjunctive views. Firstly, the principles of MiniCon must be extended to our semantic context: indeed, consider again our example with the query q_u and the views in $\mathcal{V}_s$, the subgoals of q_u are $SG = \{Vegetables, price, production\}$. No bucket will be formed from the subgoal *Vegetables*, because there is no view that contains the predicate *Vegetables* among the views in $\mathcal{V}_s$. However, if we consider the logical implication of (concepts/role) subsumption assertions in $\mathcal{T}_g$, we can reformulate the query q_u into a union of conjunctive queries (UCQ) Q_u such that $q_i \sqsubseteq_{\mathcal{T}_g} q_u$, for each $q_i \in Q_u$, and $Subgoals(Q_u) = \cup_{q_i \in Q_u} Subgoals(q_i)$. In our example, the UCQ Q_u that reformulates the query q_u w.r.t $\mathcal{T}_g$ contains the following queries:

$q_1(x,a,b) \leftarrow Vegetables(x), price(x,a), production(x,b)$
$q_2(x,a,b) \leftarrow Tomate(x), price(x,a), production(x,b)$

$q_3(x, a, b) \leftarrow Tomato(x), price(x, a), production(x, b)$
$q_4(x, a, b) \leftarrow Onion(x), price(x, a), production(x, b)$

With this reformulation, the subgoals of Q_u are $SG = \{Vegetables, price, production,$ $Tomate, Tomato, Onion\}$. It is clear that a bucket can be formed for the new subgoals in SG by comparing them to the predicates of the views in $\mathcal{V}_s$.

We propose to use the *PerfectRef* algorithm of [3] to achieve the query reformulation process. *PerfectRef* takes as input a UCQ Q and a TBox $\mathcal{T}$ and compiles the assertions of $\mathcal{T}$ into Q, returning a new UCQ Q' over $\mathcal{T}$ which is the perfect first-order logic (FOL) reformulation of Q w.r.t $\mathcal{T}$. Each conjunctive query q in Q is reformulated by using the positive inclusion assertions (PIs) in $\mathcal{T}$ as rewriting rules. Moreover, we propose to enhance MiniCon by extending its best underlying idea to the semantic level: the aim is to avoid putting in a bucket an atom that will only generate invalid rewritings. Due to the fact that assertions declared in the TBox may express disjunctness between instances of different atoms, even if their bindings of variables match, a test of consistency w.r.t the TBox assertions must be performed before adding a view in a bucket: we propose to apply the *Consistent* algorithm of [3]. This algorithm uses the negative inclusion constraints (NIs) declared in the TBox to check data consistency, by translating each negative inclusion (NI) assertion into a Boolean conjunctive query q_{unsat} that must be evaluated over the data. Function *ConsistentMiniCon* is the MiniCon modified to integrate the consistency checking, with respect to assertions in $\mathcal{T}_g$.

We consider now the following **GAV** mappings, assertions of the global ontology $\mathcal{G}$, **automatically** computed during the conciliation phase of our global ontology building process (cf. [6]):

$Tomato(x) \supseteq S_1.Tomate(x)$ $production(x, y) \supseteq S_1.Tomate(x), S_1.production(x, y)$
$Tomato(x) \supseteq S_2.Tomato(x)$ $production(x, y) \supseteq S_1.Onion(x), S_1.production(x, y)$
$Onion(x) \supseteq S_1.Onion(x)$ $price(x, y) \supseteq S_2.Tomato(x), S_2.price(x, y)$
$Rice(x) \supseteq S_2.FriedRice(x)$ $price(x, y) \supseteq S_2.FriedRice(x), S_2.price(x, y)$

Algorithm 3.2 below allows to compute answers to the user query q_u by using these GAV mappings. It rewrites q_u into source queries, by applying the so-known *query unfolding* method [5], but here again, it must be extended to deal with semantic aspects. As for LAV rewriting solution, we have first to consider the TBox assertions in $\mathcal{T}_g$ to obtain a complete unfolding of q_u. Here again, we propose to apply the *PerfectRef* algorithm to produce Q_u. Then, the consistency w.r.t. the TBox assertions must be checked for each unfolding. For this purpose, we propose to follow the approach presented in [1], first by expanding each rewriting, then by applying the q_{unsat} boolean queries computed from the NIs of $\mathcal{T}_g$.

Algorithm 3.2 - GAV Query Rewriting
Input:
    (i) q_u: user query.
    (ii) $\mathcal{T}_g$: TBox of the global ontology, containing GAV mappings.
Output:
    Q_{rew}: set of rewritten queries.
Local Variables:
    (i) Q_u: set of reformulated queries.
    (ii) $\mathcal{U}$: set of unfoldings.
begin
  1. $Q_u := PerfectRef(q_u, \mathcal{T}_g)$
  2. $\mathcal{U} := Unfold(Q_u, \mathcal{T}_g)$

3. $Q_{rew} := \emptyset$
4. **for each** q in $\mathcal{U}$ **do**
5. $Q_{rew} := Q_{rew} \cup Consistent(Expand(q, \mathcal{T}_g), \mathcal{T}_g))$
6. **end for**
7. $Reduce(Q_{rew})$
end

Consider for example the query $q_3 \in Q_u$, one of the unfoldings we obtain is:

$u_8(x, a, b) \leftarrow S_2.Tomato(x), S_2.FriedRice(x), S_2.price(x, a), S_1.Onion(x), S_1.production(x, b)$

The expansion of $u_8(x, a, b)$ gives the following result:

$Exp\_u_8(x, a, b) \leftarrow Tomato(x), Rice(x), price(x, a), Onion(x), production(x, b)$. Applying the *Consistency* algorithm leads to evaluate, among others, the boolean query $q_{unsat} \leftarrow Tomato(x), Rice(x)$ over the body of $Exp\_u_8(x, a, b)$: an inconsistency is detected, so the unfolding $u_8(x, a, b)$ is rejected. This process is repeated for all generated unfoldings. In our example, the only valid rewriting of the query $q_3 \in Q_u$ is:

$q_{rew3}(x, a, b) \leftarrow S_1.Tomate(x), S_1.production(x, b), S_2.Tomato(x), S_2.price(x, a)$.

4   Discussion and Related Work

A complete discussion on LAV and GAV approaches, with complexity results, can be found in [5]. The main novelty that we introduce to deal with our semantic Web context is the use of *DL-Lite*$_\mathcal{A}$ description logics, following [7]. The properties of the two query rewriting methods presented in Section 3 follows from the properties of the used algorithms, despite our adaptations to the semantic level. For these adaptations, we have built upon clear explanations of [1]. Several other works use *DL-Lite*$_\mathcal{A}$ in the context of data integration systems but, to the best of our knowledge, our proposal is the first that uses it for *building (semi-)automatically a data integration system* which avoids some drawbacks related to either GAV or LAV approaches.

Adding (or removing) data sources is harder with **classical GAV** integration systems because local and global relations that define the global schema are strongly coupled, which involves to consider how a new arrived source may be combined with all the existing data sources, in order to produce tuples of some global relation. This problem does not arise in our solution, because the sources keep their independence from each other in the global ontology, although connected by semantic links. Indeed, when observing the GAV mappings $\mathcal{M}$ shown in the previous section, it can be noticed that concepts and roles of the global ontology are never related to concepts and roles belonging to several different data sources, on contrary they are always related to concepts and roles of a single source. Relations between the global concepts involved in the mappings are only semantically established in the global TBox $\mathcal{T}_g$ through hierarchical links that come from the reference-domain ontology. For an illustration, if a new source S_3 must be added to S_1 and S_2, assuming that S_3 provides information about patatoe products and their production, by applying our solution the following new semantic information are added to $\mathcal{T}_g$:

$Patato \sqsubseteq Vegetables$	$Patato \sqsubseteq \neg Tomato$	$Patate \sqsubseteq \delta(production)$
$Patate \sqsubseteq Patato$	$Patato \sqsubseteq \neg Onion$	

and the followings GAV mappings are automatically generated:

$$Patato(x) \supseteq S_3.Patate(x) \qquad production(x,y) \supseteq S_3.Patate(x), S_3.production(x,y)$$

As illustrated in this example we do not need to combine any local relations of different sources to provide tuples of global relations when adding a new source. Therefore, our system allows GAV query processing whereas it is also flexible with respect to source adding or removing.

The price paid in general for this flexibility in **classical LAV** approaches is that the rewritings are more complicated to find and to deal with. Our solution allows to enhance this situation thanks to the views that we automatically generate on sources (cf. end of Section2). For example, to describe the fact that a source S has information about activities related to agricultural products, precisely tomatoes with their production and origin, instead of considering only the view $v:v(x,y,z,a,b) \leftarrow Activity(x),$
$relateTo(x,y), Agriculture(y), concern(y,z), Tomate(z), production(z,a), origin(z,b)$
we compute the set of views describing the same information as in v, w.r.t. our source $S1$ (Fig. 1):

$v_{p_{11}}(x) \leftarrow Activity(x)$	$v_{p_{31}}(x,y,z) \leftarrow Activity(x), relateTo(x,y), Agriculture(y),$
$v_{p_{12}}(x) \leftarrow Agriculture(x)$	$\qquad concern(y,z), Tomate(z)$
$v_{p_{13}}(x) \leftarrow Tomate(x)$	$v_{s_{11}}(x,a) \leftarrow Tomate(x), production(x,a)$
$v_{p_{21}}(x,y) \leftarrow Activity(x), relateTo(x,y), Agriculture(y)$	$v_{s_{12}}(x,b) \leftarrow Tomate(x), origin(x,b)$
$v_{p_{22}}(x,y) \leftarrow Agriculture(x), concern(x,y), Tomate(y)$	$v_{s_{21}}(x,a,b) \leftarrow Tomate(x), production(x,a), origin(x,b)$

If we considere the following user's query: $q(x,a,b) \leftarrow Tomate(x), production(x,a)$, the selection of the view v for rewriting the query q implies irrelevant parts in the answer. However, using the set of views resulting from our decomposition method in path and star views, we are able to select the most exact view in order to have precise answer: in our example $v_{s_{11}}$ describes only tomatoes and their production. Thus, we avoid a post-processing task consisting of pruning all irrelevant parts in the sources' answers.

5   Conclusion

We have presented the query processing part of our data integration system, that is semi-automatically built. We have shown how it combines the advantages of LAV and GAV approaches. The global ontology building process is now implemented in a prototype, where ontologies are expressed in RDF/OWL, and we are working on completing this prototype with the query processing part.

References

1. Abiteboul, S., Manolescu, I., Rigaux, P., Rousset, M.-C., Senellart, P.: Web Data Management. Cambridge University Press (2012)
2. Calvanese, D., De Giacomo, G., Lembo, D., Lenzerini, M., Poggi, A., Rodriguez-Muro, M., Rosati, R., Ruzzi, M., Savo, D.F.: The mastro system for ontology-based data access. Semantic Web 2(1), 43–53 (2011)

3. Calvanese, D., De Giacomo, G., Lembo, D., Lenzerini, M., Rosati, R.: Tractable Reasoning and Efficient Query Answering in Description Logics: The *L-Lite* Family. J. Autom. Reasoning 39(3), 385–429 (2007)
4. Horrocks, I.: Ontologies and the semantic web. Commun. ACM 51(12), 58–67 (2008)
5. Lenzerini, M.: Data integration: A theoretical perspective. In: PODS, pp. 233–246 (2002)
6. Niang, C., Bouchou, B., Lo, M., Sam, Y.: Automatic Building of an Appropriate Global Ontology. In: Eder, J., Bielikova, M., Tjoa, A.M. (eds.) ADBIS 2011. LNCS, vol. 6909, pp. 429–443. Springer, Heidelberg (2011)
7. Poggi, A., Lembo, D., Calvanese, D., De Giacomo, G., Lenzerini, M., Rosati, R.: Linking Data to Ontologies. In: Spaccapietra, S. (ed.) Journal on Data Semantics X. LNCS, vol. 4900, pp. 133–173. Springer, Heidelberg (2008)
8. Pottinger, R., Halevy, A.Y.: Minicon: A scalable algorithm for answering queries using views. VLDB J. 10(2-3), 182–198 (2001)

A New Approach for Date Sharing and Recommendation in Social Web

Dawen Jia[1], Cheng Zeng[1,*], Wenhui Nie[2], Zhihao Li[2], Zhiyong Peng[2]

[1] State Key Laboratory of Software Engineering, Wuhan University, China
[2] Computer School, Wuhan University, China
{brilliant,zengc,momo_nwh,z.h.li,peng}@whu.edu.cn

Abstract. The *social Web* is a set of social relations that link people through the World Wide Web. Typical social Web applications which include social media and social network services etc. have already become the mainstream of web application. User-oriented and content generated by users are pivotal characteristics of the social Web. In the circumstance of massive user generated unstructured data, data sharing and recommendation approaches take a more important role than information retrieval approaches for data diffusion. In this paper, we analyze the disadvantages of current data sharing and recommendation methods and propose an automatic group mining approach based on user preferences, which lead to sufficient data diffusion and improve the sociability between users. Intuitively, the essential idea of our approach is that users who have the same preferences towards a set of interested topics could be gathered together as a *Common Preferences Group (CPG)*. To evaluate the efficiency of the *CPG* mining algorithm and the accuracy of data recommendation based on our approach, the experiments use dataset collected from the most popular image sharing site *Flickr*. The experimental results prove the superiority of our new approach for data sharing and recommendation in social Web.

Keywords: Recommender System, Common Preference Group, Social Web.

1   Introduction

With the advent of Web 2.0, users are allowed to produce content in the Web. The social Web is a set of relationships that link together people over the World Wide Web. Typical Social Web applications which include social networking services, social media and online communities, etc. have already become the mainstream of web application. User interaction is a pivotal characteristic of the social Web. Starting with the social media sites, such as *Flickr* and *Youtube*, user-generated contents have taken over social web. Users and contents in these sites are still rapidly increasing every day. Traditional information retrieval technologies have limits when dealing with the problem of *information overload* in the social Web, mainly in that large amounts of unstructured data produced by users is more difficult to classify and retrieve. In the

---

* Corresponding author.

S.W. Liddle et al. (Eds.): DEXA 2012, Part II, LNCS 7447, pp. 314–328, 2012.

circumstance of massive user-generated unstructured data, data sharing and recommendation approaches take a more important role than information retrieval approaches for data diffusion in the Social Web. *Facebook* revealed that online sharing is growing at an exponential rate and users were sharing 4 billion 'things' on *Facebook* every day last year. 'Things' here normally refers to multimedia objects, such as textual web pages, videos and images. This kind of sharing is the *friend-to-friend* style. Only the friends of the distributor could see them. As we know, people who are friends do not mean they have the common interests. We might have hundreds of friends and get thousands of sharing things each day, but we might find out only few of them get our attention and the rest are all junk information for us. What we need is a personalized data sharing and recommendation based on our own interests. As the matter of fact, many social Web sites provide the *group mechanism* for personalized data sharing, where the user can manually create a group with a certain topic and the other users who are interested in the topic can join this group as a member and upload the relevant objects into the group pool. Literature [1] pointed out that more than half of users of *Flickr* participated in at least one group with their snapshot, which indicated that a large number of users engaged in group activities. Users create and join groups for social purposes, and the formation of groups has gained great popularity and attracted enormous number of users [2]. Basically, each group represents a specific topic, and users who are interested in this topic can join the group as a member and upload the relevant media data into the group pool. The study shows that adding photos into groups is one of the main reasons for photo diffusion [9, 10]. It is no doubt that group is a useful tool for media sharing and recommendation. However, the existing group mechanisms in current social web sites have many disadvantages. Some wildly mentioned problems are listed below [3, 4, 5, 7]:

- The most mentioned problem is that group is self-organized; one topic may have a huge number of corresponding groups. For example, there are about 31494 groups related to "dog" in *Flickr*[1]. The number of groups is still increasing.
- Furthermore, a data object can match more than one topic, and users have to match the subject of each image with various topics. Manually assigning each media object to the appropriate groups is tedious task.
- For each user, it is also a difficult task to choose the right groups to join. Sometimes users won't join the group on their own initiative, since they were not aware their interests yet.
- Since the data objects are added to certain group pool manually, no one can guarantee all the objects in a group pool comply with the group topics. As the matter of fact, the phenomenon that data objects are inconsistence with their group topics is ubiquitous.

These disadvantages have become the severely obstacle of data sharing and recommendation. Since a topic might match thousands of groups, at first, it is not practical for the user join all the relevant groups, the consequence is the objects in a certain group or the new objects adding to this group cannot be shared with the users who are not the member of this group; secondly, to share and recommend a object, users have

---

[1] http://www.flickr.com/groups/. Retrieved July 11, 2011.

to manually upload the media object to relevant groups, and it is also not practical for the user upload a media object to thousands of groups which are relevant to a same topic. Furthermore, users may not be aware of their preferences, which means users' behaviors indicate they are interested in some groups but they have not join those groups which they should join into, then objects in those groups could not be shared to those users.

To tackle these problems, we proposed a new approach to discover groups automatically based on user's preference. To distinguish two kinds of group in this paper, we call the group which is automatically generating by our approach the *Common Preference Group (CPG)*. This research is conducting under a hypothesis that a user like a data object because the user is interested in some semantic topics implied in the object. With this assumption, we switch interests of users from the objects to the semantic topics, and we group users who share common interests together as a *CPG*.

The brief process is described as follow: in the social Web, users have their own preferences about data objects. Currently, the systems provide some functions which allow users express their preferences, such as marking "*favorite*" or "*like*" to an object or give an object a rating score. Several semantic topics could be extracted from each object. The users' behaviors could reveal their preferences on each semantic topic. Intuitively, the essential of our grouping idea is that users who have the same preference on a set of semantic topics should be gathered together as a *CPG*.

The automatic grouping approach has the following features: *i)* a *CPG* is corresponding to a set of semantic topics, which indicate the users' preference about each semantic topic; *ii)* the *CPG* is automatically generated, and a user whose preference matches a *CPG* will be added to this *CPG* automatically; *iii)* The data object can be automatically added to the corresponding *CPG* pools as well. *CPG* can be used as social purposes and data recommendation. Compared with current group mechanism, the new features of *CPG* bring these advantages: *i)* the users could discover their own preferences and the other people who have the same preference with them; *ii)* different *CPG* can be recommended to users based on user's preferences; *iii)* objects are distributed to different *CPG* without human involved. In order to evaluate the accuracy and efficiency of our approach, we use dataset collected from the most popular image sharing site *Flickr*. The detail elaborates in the experimental section.

2   Related Work

To the best of our knowledge, little literatures have been published on automatic group generating based on users' preferences. We compare our work with related domains from two different viewpoints.

- Social Group Recommendation

The research on social group recommendation which called social group suggestion as well started from recent year. The most mentioned problem in social group recommendation domain is that groups are self-organized, which causes one topic may have a huge number of similar groups. It is a difficult task for users to either choose the right groups to join in by themselves or distribute a media object into appropriate group pools. Many researches focus on recommending groups for a given object according to its content [4, 5, 7], and there are some other research that focus on recommending groups to each user [3, 17]. Though those approaches could recommend the best groups for users and media objects, it still can't satisfy the data sharing and recommendation task. Users with common interests may still separate in different groups, so that lots of shared data objects cannot be seen by other group members. On the other side, some people obviously have common interests, but they just aren't aware of the existence of the corresponding groups or the corresponding groups haven't been manually created yet. As a result, they won't be connected together.

• Recommendation Techniques

There exist different kinds of recommendation techniques for various user tasks. Those techniques have a number of possible classifications, such as content-based [8], collaborative filtering (CF) [12, 14], hybrid [13] etc. Collaborative filtering is the most successful techniques in recommender systems. Cui et al. [6] propose a content/similarity-based technique to facilitate the recommendation application in social media environment. When applying these recommendation techniques in Social Web, we find a common fault of existing recommendation techniques. The existing techniques recommend each object to the individual user. These techniques focus on mainly the accuracy of recommendation algorithms, however they overlook the time efficiency and computational complexity. As we mentioned before, there exists immense amount of users and objects, and the numbers are rapidly increasing every day in Social Web. Though, the requirement of time efficiency in recommender system is not as crucial as in retrieval system, in the circumstance of mass data and users, it is necessary to consider the factor of time efficiency for recommendation approaches. Obviously, the social group mechanism which could realize batch recommendation has higher efficiency since the number of groups could be much less than the number of users.

3   Preliminaries

We argue that users' preferences on data objects are caused by their preferences on various semantic topics contained in the object. With this assumption, we can switch users' preferences from the objects to the semantic topics. At first, we establish the relation between the users and the semantic topics. Then we can cluster the users who share common interests together by *CPG* mining algorithm.

3.1   Semantic Topic Extraction

Extracting semantic topics which imply users' preferences from data objects is the foundation of our approach, and two things should be considered in this module: first, no matter where semantic topics come from, predefined or extracted from data object, these topics should be consistent with user's interests; second, the process of extracting semantic topics from data objects should be as accurate as possible. Different types of data objects, such as text, video, image and audio, have different semantic extraction methods. Semantic topics extracting from social object is a classic problem in multimedia domain [11, 15, 16]. In the circumstance of social Web, semantic topics could be extract either from data objects themselves or the rich useful clues surrounding data objects, such as tags, descriptions and comments. Moreover, the personal information of creator, uploader, viewer and even the geographical location reveals the semantics of data objects.

3.2   Establish the User Preference Model

In the previous subsection, we discussed the principles of extracting semantic topics from data objects. Generally, more than one semantic topic could be extracted from a data object. *TopicSpace* denotes all possible topics that could be predefined or extracted from all objects. o_{topic} is a set of semantic topics $\{t_i\}$ implied by the object o.

$$o_{topic} \subseteq TopicSpace \tag{1}$$

A user u may interact with a set of objects. *ObjectSpace* denotes all objects. $u_{interaction}$ is a set of objects $\{o_i\}$ which user u have interacted with.

$$u_{interaction} \subseteq ObjectSpace \tag{2}$$

Based on (1) and (2), the users' preference on semantic topics can be derived. $u_{preference}$ is used to describe users' preference on a set of semantic topics. d_i is the user's interestingness on each semantic topic t_i.

$$u_{preference} = \{< t_i, d_i >\} \tag{3}$$

Definition 1: User Preference Model (UPM)
The User Preference Model (UPM) is a model which records the users' interestingness on each semantic topic. The UPM can be represented as a 4-tuple < U, T, D, Λ >, where U is a set of users, T is a set of semantic topics, D is a set of possible levels of interestingness, and the Λ ⊆ U × T × D indicates the user preference relation (UPR) which is each user's interestingness on each semantic topic.

4   Automatic CPG Mining Approach

In this section, we first give the formal definition of *CPG* and *CPG* mining. Then we elaborate the process of *CPG* mining algorithm step by step with an example.

4.1   CPG Definition

Intuitively, the essential of our grouping idea is that users who have the same interestingness towards a set of semantic topics should be gathered together as a CPG. Since users' interestingness on each semantic topic has been record in *UPM*, we can design algorithms to mining *CPG* from *UPM*.

Definition 2: Common Preference Group (CPG)

A Common Preference Group (CPG) is a subset of UPM, which satisfies the condition that all users in CPG sharing the same $u_{preference}$. A CPG g_i can be represented as a pair $<U_i, cp(U_i)>$, where U_i is a set of users and $cp(U_i) = \cap_{u \in U_i} u_{preference}$ refers to the common preferences shared by the users in U_i.

In the paper, we also absorb the advantages of *Collaborative Filter (CF)* technology and combine it into *CPG* mining algorithm, called *CF-CPG*, which can tolerate the slight difference among user preferences in the same *CPG* and thus will gain better recommendation effect. We use *Ori-CPG* to represent the original *CPG* without *CF*. For more clearly describing our approach, we don't distinguish *CF-CPG* from *Ori-CPG* in mostly elaboration and uniformly use the concept *CPG*.

Definition 3: CPG Mining

The process of CPG Mining is discovering all CPG from a given UPM. With CPG, the Group Preference Relation (GPR) and User Group Relation (UGR) are easily inferred. UGR indicate the relationship between users and CPG, namely group members. GPR indicate the relationship between user's preference and CPG, namely the common preference sharing by all members in a CPG.

	t_1	t_2	t_3	t_4
u_1	d1	d2	d1	0
u_2	d1	0	0	d3
u_3	d1	d2	d1	d3
u_4	d1	d3	d3	d3
u_5	d2	d3	d3	d1
u_6	d2	0	0	0

(a) The User Preference Model (*UPM*)

	t_1	t_2	t_3	t_4
u_1	d1	d2	d1	0
u_2	d1	0	0	d3
u_3	d1	d2	d1	d3
u_4	d1	d3	d3	d3
u_5	d2	d3	d3	d1
u_6	d2	0	0	0

(b) Common Preference Group(*CPG*) Mining

	t_1	t_2	t_3	t_4
g_1	d1	d2	d1	0
g_2	d1	0	0	d3
g_3	0	d3	d3	0

(c) The Group Preference Relation(*GPR*)

	u_1	u_2	u_3	u_4	u_5	u_6
g_1	1	0	1	0	0	0
g_2	0	1	1	1	0	0
g_3	0	0	0	1	1	0

(d) The User-Group Relation(*UGR*)

Fig. 1. An example of *CPG* mining

We define two parameters *minUsers* and *minTopics* to denote the minimal requirements of the number of users and semantic topics to form a CPG.

4.2   The CPG Mining Algorithm

We elaborate the process of *CPG* mining algorithm step by step with the example given in Fig.1 in this section.

Step 1: Clustering users based on their interestingness on each semantic topic

With *UPM* in Fig.1 (a), we can easily cluster the users who have the common preference together based on their interestingness for each semantic topic, shown in

Table 1. Each element p in this table is a candidate *CPG* having one common seman-tic topic, which means the users in this element share the same interestingness on a certain semantic topic. We delete the elements in which the number of user is less than *minUsers* and add the rest to array *initialList* List.

Table 1. User Clustering Table

	t_1	t_2	t_3	t_4
d_1	u_1,u_2,u_3,u_4		u_1,u_3	$u_5(delete)$
d_2	u_5,u_6	u_1,u_3		
d_3		u_4,u_5	u_4,u_5	u_2,u_3,u_4

Table 2. Intersection Operation

	t_1	t_2	t_3	t_4
d_1	u_1,u_2,u_3,u_4		u_1,u_3	
d_2		u_1,u_3		
d_3		*(Delete)* u_4,u_5	*(Delete)* u_4,u_5	u_2,u_3,u_4

Algorithm 1: UserClustering	**Algorithm 2: CreateCandidateCPGList**
Input: *The user preference relation Λ*	**Input:** *p, initialList*
Output: *initialList*	**Output:** *cCList*
Description: *Clustering users who have the same interestingness for a semantic topic into element p and add p into initialList.*	**Description:** *Compare each element p with the other elements in initialList, we could generate a set of candidate CPG.*
1. extract users set U, topics set T and interestingness D in Λ;	1. $cCList = \{\}$;
2. **for** each interestingness $d_i \in D$ **do**	2. **for** each $p_j \in initialList$ & $p_j.t$ after $p.t$ **do**
3. **for** each topic $t_j \in T$ **do**	3. $users = p_j.users \cap p.users$;
4. **if** $(u(t_j) = d_i)$	4. **if** $(users.length >= minUsers)$
5. **add** u to $p(d_it_j)$;	5. **add** $<\{p_j.dt\},\{users\}>$ to $intersList(p)$;
6. **end if**	6. **endif**
7. **end for**	7. **for** each *element* $e_i \in intersList(p)$
8. **if** $(p(d_it_j).length >= minUsers)$	8. *MergeWithcCList(e_i, cCList);*
9. **add** $p(d_it_j)$ to *initialList*;	9. **add** $e_i \cap p$ to *cCList*;
10. **end if**	10. *RemoveOverlapped(cCList);*
11. **end for**	11. **end for**
	12. **Return** *cCList;*
	13. **end for**

Step 2: Mining the CPG which has multiple common semantic topics

We merge the elements in Table 1. to form *CPG* which has multiple common seman-tic topics. For each element $p(d_it_j)$ in *initialList*, we implement intersection operations between $p(d_it_j)$ and those elements with different columns t_j. We take the first element $p(d_1t_1)$ as an example. Table 2. shows the results of intersection operations between $p(d_1t_1)$ and other elements. We delete the results in which the number of user is less than *minUsers*, and put the effective results into intersection list (*intersList*). For each element in *interList*, we merge it with elements in the candidate *CPG* List (*cCList*) (depicted in algorithm *MergeWithcCList*) and put itself into *cCList* as well.

Algorithm 3: MergeWithcCList	Algorithm 4: RemoveOverlapped
Input: *e, cCList*	**Input:** *cCList (With redundant elements)*
Output: *cCList*	**Output:** *cCList (Without redundant elements)*
Description: *For an element e in intersList, compare it with those elements in cCList to see whether e could bring more semantic topics into current candidate CPG.*	**Description:** *An element will be deleted if it is fully covered by another element in the same cCList.*
1.　**for** each element $cCPG_i$ in *cCList* **do**	1.　**for** two elements $<set\{t\}, set\{u\}>$,
2.　　*users = e.users ∩ cCPG_i.users;*	$<set'\{t\}, set'\{u\}>$ in *cCList*
3.　　**if** *(users.length >= minUsers)*	2.　　**if** *($set'\{t\} ⊆ set\{t\}$ & $set'\{u\} ⊆ set\{u\}$)*
4.　　　**add** element $<\{cCPG_i.dt ∪ e.dt\}$, $\{users\}$ > **to** *cCList;*	3.　　　**remove** element $<set'\{t\}, set'\{u\}>$ **from** *cCList;*
5.　　**endif**	4.　　**endif**
6.　**end for**	5.　**end for**

Step 3: For each element in initialList, repeat step 2

We repeat *step 2* to get a *cCList* for each element in *initialList,* shown in Table 3. Since the element in *cCList* mined from $p(d_2t_2)$ is covered by element $<\{d_1t_1, d_2t_2, d_1t_3\}, \{u_1, u_3\}>$, it will be deleted. Actually, if the users in element *p* is the subset of the users in *p'* and element *p'* appears before element *p*, the *cCList* mined from *p* is redundant. We could mine $<\{d_2t_2, d_1t_3\}, \{u_1, u_3\}>$ by *step 2*. However, the user set of $p(d_2t_2)$ is the subset of $p(d_1t_1)$, therefore we even need not to mine *cCList* of $p(d_2t_2)$. Finally, we add all elements in *cCList* to *CPGList*.

Algorithm 5: CreateCPGList	Algorithm 6: RemoveRedundent
Input: *The user preference relation Λ*	**Input:** *cCList1, cCList2*
Output: *CPGList*	**Output:** *cCList1, cCList2*
1.　*CPGList = {};*	**Description:** *As input, an element from cCList2 may covered by an element from cCList1. As output, any element from cCList2 wouldn't be covered by any element from cCList1.*
2.　*initialList =* **UserClustering**(*Λ*);	
3.　**for** each element *p* ∈ *initialList* **do**	
4.　　*cCList(p)　=　CreateCandidateCPGList(p);*	
5.　　add *cCList* to *tempList;*	
6.　**end for**	1.　**for** each element $<set\{dt\}, set\{u\}>$ ∈ *cCList1*
7.　**for** each *cCList(p_i)* ∈ *tempList*	2.　　**for** each element $<set'\{dt\}, set'\{u\}>$ ∈ *cCList2*
8.　　**for** each *cCList(p_j)* ∈ *tempList*	3.　　　**if** *($set'\{dt\} ⊆ set\{dt\}$ & $set'\{u\} ⊆ set\{u\}$)*
9.　　　**if** $p_j.t$ after $p_i.t$	4.　　　　**remove** element $<set'\{dt\}, set'\{u\}>$ **from** *cCList2;*
10.　　　　*RemoveRedundent(cCList(p_i), cCList(p_j));*	
11.　　　**end if**	5.　　　**endif**
12.　　**endfor**	6.　　**end for**
13.　**endfor**	7.　**end for**
14.　*CPGList = tempList;*	
15.　**Return** *CPGList;*	

As shown in different shape regions in Fig.1 (b), three *CPG* are mined. $g_1 = <\{t_1d_1, t_2d_2, t_3d_1\}, \{u_1, u_3\}>$, $g_2 = < \{t_1d_1, t_4d_3\}, \{u_2, u_3, u_4\}>$ and $g_3 = <\{t_2d_3, t_3d_3\}, \{u_4, u_5\}>$. We can easily generate *GPR* and *UGR* by *CPGList*. For example, as shown in Fig.1(c)(d), the *CPG* g_1 has $GPR_1 = <g_1, \{t_1d_1, t_2d_2, t_3d_1\}>$, $UGR_1 = <g_1, \{u_1, u_3\}>$.

Table 3. All cCList

P	$cCList$
$p(d_1t_1)$	$\{<\{d_1t_1,d_2t_2,d_1t_3\}, \{u_1,u_3\}>, <\{d_1t_1,d_3t_4\},$
$p(d_2t_1)$	$\varPhi$
$p(d_2t_2)$	$\{<\{d_2t_2, d_1t_3\}, \{u_1,u_3\}>\}(Delete)$
$p(d_3t_2)$	$\{<\{ d_3t_2,d_3t_3\}, \{u_4,u_5\}>\}$
$p(d_1t_3)$	$\varPhi$
$p(d_3t_3)$	$\varPhi$

Step 5: CPG merging

Normally, when we finish **Step 4**, we've already got all *CPG* we could mine from the *UPR* table. But at this moment, all *CPG* are actually original common preference groups, called *Ori-CPG*. *Ori-CPG* is sensitive to the difference among user preferences that means all user preferences are completely consistent in the same *CPG*. In this paper, we absorb the advantages of *CF* technology and take it into *CPG* mining algorithm, called *CF-CPG*. It can tolerate the slight difference among user preferences, realize the preference inferring in certain extent and thus gain a better recommendation effect. For a more concise description to our approach, we did not distinguish *CF-CPG* from *Ori-CPG* in an elaborate and uniform used concept of *CPG*. Table.4 gives a simple example of the situation where the divergence between CPG_1 and CPG_2 is only about topic t_1 and the difference of referred users set is very small. So we merge the two groups to obtain a bigger range of share and recommendation. It is similar as the idea of *CF* technology, but the operated target is group instead of single user. As a result, the recommendation efficiency will be higher due to the number of *CPG* is far less than that of users and *CPG* actually pre-stores the preference relation among users.

Table 4. The situation of CPG merging

The principle of *CPG* merging follows the formula (4). When the similarity value $H(g,g')$ is more than a certain threshold, g and g' will be merged and the merging method is to find the minimal rectangle containing them. It means some users in the new *CPG* will be automatically recommended those preferences which they did not have ago. We suppose the user in g', namely $U(g')$, is more than that in g, namely $U(g)$, and then the comparison about the number of topic is inverse, otherwise g' will contain g, or they have not intersection at all that does not lead to *CPG* merging.

$$H(g, g') = \prod \left(\frac{Sum(\Phi \cap \Phi')}{Sum(\Phi \cup \Phi')} \right) \cdot \frac{1}{\sum_{i \in \{TUT' - T \cap T'\}} Sub(F_i, F_i')}$$

$$Sub(F_i, F_i') = | \sum_{j \in U^\wedge} F_{ij}' / Sum(U^\wedge) - F_i |$$

$$U^\wedge = U(g') - U(g) \tag{4}$$

where Φ, Φ' denote the user/topic sets of group g and g', respectively. $Sum(*)$ computes the unit amount in a set. $(TUT' - T \cap T')$ represents the difference of topics for g and g', in other words, it is a set of topics which have different affection degrees for the same topic. The intuitional means of formula (4) is to calculate the similarity between two *CPG* which is proportional to their common user, topic and affection degree. Although it seems that the computation complexity of formula (4) is high, we actually only compute a part of it every time instead of all parts together. E.g. we can directly stop merging if the intermediate result has been lower than the threshold, because each part, such as $\frac{1}{\sum_{i \in \{TUT' - T \cap T'\}} Sub(F_i, F_i')}$, $\frac{Sum(\Phi \cap \Phi')}{Sum(\Phi \cup \Phi')}$ or only $\Phi \in T$, is a decimals and their product can be only smaller.

5   Data Object Recommendation

As mentioned in section 3, a data object contains several semantic topics. o_{topic} is a set of semantic topics t_i implied in the object o. we calculate the similarity between the o_{topic} of a data object o and g_{topic} of each *CPG* and then determine into which *CPG* the object o will be recommended. We give a similarity calculating method in this subsection.

The similarity of o_{topic} and g_{topic} is determined by *Positive Score (P_{score})* and *Negative Score (N_{score})*. The formula of similarity score is given below.

$$Sim(o_{topic}, g_{topic}) = \frac{P_{score}}{\delta + N_{score}} \tag{5}$$

As mentioned in previous section, g_{topic} is a set of weighted semantic topics $\{d_i t_i\}$, the weight d_i indicate *CPG* member's affection degrees for semantic topic t_i. g_{topic} also include a set of semantic topics, we use g_t indicate the semantic topic set in g_{topic}. The matched topics between o_{topic} and g_t contribute to P_{score} and the mismatch topics contribute to N_{score}. The semantic topic set t_{Match} includes all the semantic topics in both o_{topic} and g_t ($t_{Match} = o_{topic} \cap g_t$), which contributes to P_{score}. In formula (6), α is the P_{score} value, which is equal to the sum of affection degrees of all matched topics. δ is a constant to avoid division by zero.

$$\alpha = \sum_{t_m \in t_{Match}} d_m \tag{6}$$

When o_{topic} and g_t is fully matched, that is $o_{topic} = g_t$, we calculate negative score β_1 with formula (7). β_1 indicates the differential of weight d between matched semantic topics contribute N_{score} to the similarity. If $o_{topic} \subset g_t$, $N_{score} = \beta_1 + \beta_2$. Negative score β_2 is caused by the mismatch of *CPG* topics. Formula (8) shows that β_2 equals to the sum of affection degrees of all mismatched topics in g_t. If $o_{topic} \supset g_t$, N_{score} $=\beta_1 + \beta_3$. Negative score β_3 is caused by the mismatch of object topics. $avg\,d$ is the average value of affection degrees in matched semantic topics. $\beta_k (k = 1,2,3)$ are all N_{score}, which indicate factors in inverse proportion to the value of $Sim(o_{topic}, g_{topic})$. If $o_{topic} \cap g_t \neq \emptyset$, $o_{topic} \not\subset g_t$ and $o_{topic} \not\supset g_t$, $N_{score} = \beta_1 + \beta_2 + \beta_3$.

$$\beta_1 = \sum_{t_i, t_j \in t_{Match} \ and \ i \neq j} |d_i - d_j| \qquad (7)$$

$$\beta_2 = \sum_{t_c \in g_{Mismatch}} d_c \qquad (8)$$

$$\beta_3 = \sum_{t_d \in o_{Mismatch}} avg\,d \qquad (9)$$

With the similarity score between o_{topic} and g_{topic}, we can choose a threshold τ for social media object recommendation. If the similarity score $Sim(o_{topic,}\ g_{topic}) > \tau$, the data object o will be recommended to the *CPG g*. We can also rank the recommended objects in *CPG* based on the similarity scores.

6   Experiments and Analysis

The *CPG* automatic mining algorithm and data recommendation algorithm are implemented in Java. We design and conduct a series of experiments to evaluate the efficiency and precision of the proposed approach. All experiments are conducted on a PC with 2.8 GHz CPU and 3 GB memory, running the Windows 7 operating system.

Discovering users' preference is the foundation of our *CPG* automatic mining and data recommendation approaches. *Flickr* which is a popular social media site for image sharing and recommendation provides a function which allows users express their interests by marking "favorite" to images. Each user's "favorite" images set imply his/her interests. Therefore, we can utilize this function in *Flickr* for *CPG* mining and recommendation evaluation, i.e., the image in the "favorite" set is the correct recommendation.

Based on the above considerations, we first initialize a user set. The user set is determined by those users who marks one or many of Top-k most interesting images of each day from 2010.1 to 2011.12.We eliminate the users who have favorite images less than 100 and larger than 1,000. Finally, we download all favorite images and relative tags, and totally collect 10,695 users, 10,928,510 images and 61,944,695 tags. In fact, we are concerned with the semantic topics contained in images so that we extract semantic topics by using classical *LDA* (*Latent Dirichlet Allocation*) method on those tags annotating images. Of course, the tag preprocessing is necessary before *LDA* computing because most of tags are repeated and meaningless. The processed tag set will be transformed into concept set including 52,366 concepts at last. We use 1/4 user favorite images, namely 2010.1 to 2010.6, to extract semantic topics and model the user's preference, and the rest images are used for recommendation evaluation.

6.1 Efficiency Evaluation

Due to no similar work about mining potential group, this section only shows the efficiency experiment results of our own approach. The experiment conducts based on two different *CPG* mining strategies: *Ori-CPG* and *CF-CPG*. The algorithms given in Section 4.2 introduced the *Ori-CPG* strategy which mines the original CPG while the *CF-CPG* strategy performs additional merging process on *CPG* with *CF*'s idea. Both memory inverted index on mined *CPG* and hash method are used to further improve the efficiency of our approach.

There are 4 parameters $<U, T, F, minUser>$ affecting the efficiency of the *CPG* mining algorithm, where U is the number of users, T is the number of semantic topics, F is the quantification levels of interestingness, *minUser* is minimal numbers of users to form a *CPG*. We conduct a series of experiments to evaluate the efficiency of our *CPG* mining algorithm by two strategies. All experiments in this subsection are repeated 5 times for each case and the averaged values are employed.

Fig.2 (a) shows how the parameter *minUsers* affect the time efficiency. We set the parameter $T = 100$, $F = 3$ and user sizes $U = 8,000$. The experimental result shows

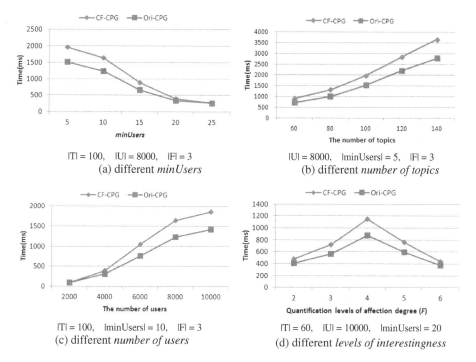

|T| = 100, |U| = 8000, |F| = 3
(a) different *minUsers*

|U| = 8000, |minUsers| = 5, |F| = 3
(b) different *number of topics*

|T| = 100, |minUsers| = 10, |F| = 3
(c) different *number of users*

|T| = 60, |U| = 10000, |minUsers| = 20
(d) different *levels of interestingness*

Fig. 2. Efficiency Evaluation

that: *Ori-CPG* strategy has a higher mining efficiency than *CF-CPG* strategy because the latter has additional computation task and the reason is the same for the following experiments.

The larger *TopicSpace* will lead to the richer semantic topic implied in *CPG* set and more accurate user preference expression. However, it will also consume more computing time for *CPG* mining and data object recommendation. With the number of topics increasing, the efficiency variation of the *CPG* mining algorithm is shown in Fig.2 (b). It shows the computing time is in proportion to the number of topics. Therefore, it is necessary to determine an appropriate range of *TopicSpace*, namely the number of representative topics which will be discussed in section 6.2.

Fig.2 (c) shows the experiment result when $T = 100$, $minUsers = 10$, $F = 3$ and user number varies from 2,000 to 10,000. The two curves which presents slight upward parabola indicates our approach has a good performance for the increasing of user number.

The experiment results in Fig.2 (d) are different with others. There is respectively a peak in the two curves. The reason of emerging peak value is that if quantification level is low, the personality of each user will be weakened and most users tend to be same. As a result, both the number of mined *CPG* and calculating time are low. When the quantification level of affection degree is increased, the original difference among users is exposed so that *CPG* number and calculating time consuming rapidly rises. However, after the quantification level continues to increase, the difference among users is magnified. It is difficult to find those users with the same preferences in this status and both *CPG* number and calculating time decline sharply.

6.2   Precision Evaluation

We consider that users' interests were expanding, so that we suppose those new interests of each user emerged in the later one and a half years were the result of data recommendation and would contribute to the precision of corresponding recommendation approach. Of course, the defect of the assumption is that the final precision evaluation for our approach would be lower than the actual result because it is impossible to get real user feedback information in Flickr.

Table 5. Precision comparing of different approaches

	Ori-CPG	*CF-CPG*	*CSP*
2010.7-2010.12	*33.1%*	*37.3%*	*41.2%*
2010.7-2011.6	*33.2%*	*42.4%*	*40.8%*
2010.7-2011.12	*33.5%*	*48.3%*	*41.4%*

In this section, we compare the precision of data recommendation based on our two *CPG* mining strategies and *CSP*[2] for 3 different time range of dataset, shown in Table 5. We can see that the recommendation precision of *CF-CPG* strategy has the most significant increases and gradually gain the highest value as time goes on because the users possibly spend a long time on expanding their interests which are just the inferring result of *CF-CPG*. On the contrary, *Ori-CPG* and *CSP* have little change as time goes on because of lacking any inferring. Due to the specialty of our *CPG* mining approach, it discovers the essentially common preference groups among users while the traditional groups used in *CSP* are built by subjective consciousness. In our experiment, we ignore the feedback learning in *CSP* in consideration of fairness. *Ori-CPG* is based on the real situation completely, so that it could reflect the fact of common user preference. But it has not any inferring function and can't predict the improving of user preference. As a result, *Ori-CPG* obtains the lower precision than *CF-CPG* when we take new data as verification where *CF-CPG* absorbs the idea of *CF* technology. Because the reason of dataset, we can't evaluate the real inferring effect of *CF-CPG* which relies on the user feedback for the recommended objects. In fact, some data are possibly recommended to those potential users who don't reveal their whole preferences and are not even aware of some preferences. However, it will affect the precision evaluation results of *CF-CPG* approach if those data recommended are not in users' favorite lists.

In Fig.3, we compare the precision variation of data recommendation for different numbers of topics. Two strategies obtain similar trend of curves. Their precision growth rates have clearly slowed down after the horizontal ordinate comes near 160. It means we should select about 160 semantic topics, namely $k=160$, to mine *CPG* and realize data recommendation.

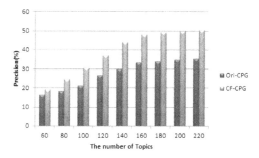

Fig. 3. Precision comparing for different numbers of topics

7   Conclusion

In the circumstance of massive user generated unstructured data, the data sharing and recommendation approaches take more important role than information retrieval approaches for data diffusion. In the paper, we analyse the disadvantages of current data sharing and recommendation methods and propose automatic group mining approach based on user's common preferences, which lead to sufficient data diffusion and improve the sociability between users. The experimental results indicate that our approach can efficiently discover potential groups and gain preferable recommendation effect.

References

1. Negoescu, R.A., Gatica-Perez, D.: Analyzing Flickr groups. In: Proceedings of the 7th ACM International Conference on Image and Video Retrieval, pp. 417–426 (2008)
2. Yu, J., Jin, X., Han, J., Luo, J.: Collection-based sparse label propagation and its application on social group suggestion from photos. Journal ACM Transactions on Intelligent Systems and Technology (TIST) 2(2) (2011)
3. Zheng, N., Li, D., Liao, S., Zhang, L.: Flickr group recommendation based on tensor decomposition. In: Proceedings of the 33nd ACM SIGIR Conference, pp. 737–738 (2010)
4. Negoescu, R.A., Gatica-Perez, D.: Topickr: flickr groups and users reloaded. In: Proceedings of the 16th ACM International Conference on Multimedia, pp. 857–860 (2008)
5. Cai, J., Zha, Z.-J., Tian, Q., Wang, Z.: Semi-automatic Flickr Group Suggestion. In: Lee, K.-T., Tsai, W.-H., Liao, H.-Y.M., Chen, T., Hsieh, J.-W., Tseng, C.-C. (eds.) MMM 2011 Part II. LNCS, vol. 6524, pp. 77–87. Springer, Heidelberg (2011)
6. Cui, B., Tung A.K.H., Zhang, C., Zhao, Z.: Multiple feature fusion for social media applications. In: Proceedings of ACM SIGMOD Conference, pp. 435–446 (2010)
7. Chen, H., Chang, M., Chang, P., et al.: SheepDog: group and tag recommendation for flickr photos by automatic search-based learning. In: Proceedings of the 16th ACM International Conference on Multimedia, pp. 737–740 (2008)
8. Bu, J., et al.: Music recommendation by unified hypergraph: combining social media information and music content. In: Proceedings of the 18th ACM International Conference on Multimedia, pp. 391–400 (2010)
9. Lerman, K., Jones, L.: Social Browsing on Flickr. In: International Conference on Weblogs and Social Media (2007)
10. Zwol, R.: Flickr: Who is Looking? In: Proceedings of ACM International Conference on Web Intelligence, pp. 184–190 (2007)
11. Datta, R., Joshi, D., Li, J., Wang, J.: Image retrieval: Ideas, influences, and trends of the new age. ACM Computing Surveys 40(2) (2008)
12. Konstas, I., Stathopoulos, V., Jose, J.M.: On social networks and collaborative recommendation. In: Proceedings of the 32nd ACM SIGIR Conference, pp. 195–202 (2009)
13. Goto, M., Komatani, K., et al.: Hybrid Collaborative and Content-based Music Recommendation Using Probabilistic Model with Latent User Preferences. In: Proceedings of the 7th International Conference on Music Information Retrieval, pp. 296–301 (2006)
14. Sarwar, B.M., Karypis, G., Konstan, J.A., Riedl, J.: Item-based collaborative filtering recommendation algorithms. In: Proceedings of the 10th ACM International World Wide Web Conference, pp. 285–295 (2001)
15. Zhu, M., Badii, A.: Semantic-associative visual content labeling and retrieval: A multimodal approach. Signal Processing: Image Communication 22(6), 569–582 (2007)
16. Barnard, K., Duygulu, P., Forsyth, D.A., Freitas, N., Blei, D.M., et al.: Matching Words and Pictures. Journal of Machine Learning Research 3, 1107–1135 (2003)
17. Zheng, N., Li, Q., Liao, S., Zhang, L.: Which photo groups should I choose? A comparative study of recommendation algorithms in Flickr. Journal of Information Science 36(5), 733–750 (2010)
18. Smeulders, A., Worring, M., Santini, S., Gupta, A., Jain, R.: Content-Based Image Retrieval at the End of the Early Years. IEEE Transactions on Pattern Analysis and Machine Intelligence (TPAMI) 22(12), 1349–1380 (2000)

A Framework for Time-Aware Recommendations

Kostas Stefanidis[1], Irene Ntoutsi[2], Kjetil Nørvåg[1], and Hans-Peter Kriegel[2]

[1] Department of Computer and Information Science, Norwegian University of Science
and Technology, Trondheim, Norway
{kstef,kjetil.norvag}@idi.ntnu.no
[2] Institute for Informatics, Ludwig Maximilian University, Munich, Germany
{ntoutsi,kriegel}@dbs.ifi.lmu.de

Abstract. Recently, recommendation systems have received significant attention. However, most existing approaches focus on recommending items of potential interest to users, without taking into consideration how temporal information influences the recommendations. In this paper, we argue that time-aware recommendations need to be pushed in the foreground. We introduce an extensive model for time-aware recommendations from two perspectives. From a *fresh-based* perspective, we propose using a suite of aging schemes towards making recommendations mostly depend on fresh and novel user preferences. From a *context-based* perspective, we focus on providing different suggestions under different temporal specifications. The proposed strategies are experimentally evaluated using real movies ratings.

1 Introduction

Recommendation systems provide users with suggestions about products, movies, videos, pictures and many other items. Many systems, such as Amazon, NetFlix and MovieLens, have become very popular nowadays. One popular category of recommendation systems are the collaborative filtering systems (e.g., [11,8]) that try to predict the utility of items for a particular user based on the items previously rated by similar users. That is, users similar to a target user are first identified, and then, items are recommended based on the preferences of these users. Users are considered as similar if they buy common items as in case of Amazon or if they have similar evaluations as in case of MovieLens.

The two typical types of entities that are dealt in recommendation systems, i.e., users and items, are represented as sets of ratings, preferences or features. Assume, for example, a restaurant recommendation application (e.g., ZAGAT.com). Users initially rate a subset of restaurants that they have already visited. Ratings are expressed in the form of preference scores. Then, a recommendation engine estimates preference scores for the items, i.e., restaurants in this case, that are not rated by a user and offers him/her appropriate recommendations. Once the unknown scores are computed, the k items with the highest scores are recommended to the user.

Although there is a substantial amount of research performed in the area of recommendation systems, most of the approaches produce recommendations by

S.W. Liddle et al. (Eds.): DEXA 2012, Part II, LNCS 7447, pp. 329–344, 2012.
© Springer-Verlag Berlin Heidelberg 2012

ignoring the temporal information that is inherent in the ratings, since ratings are given at a specific point in time. Due to the fact that a huge amount of user preferences data is accumulated over time, it is reasonable to exploit the temporal information associated with these data in order to obtain more accurate and up to date recommendations. In this work, our goal is to use the time information of the user ratings towards improving the predictions in collaborative recommendation systems. We consider two different types of time effects based upon the recency/freshness and the temporal context of the ratings and consequently, we propose two different time-aware recommendation models.

The *fresh-based recommendations* assume that the most recent user preferences better reflect the current trends and thus, they contribute more in the computation of the recommendations. To account for the recency of the ratings we distinguish between the *damped window model* that gradually decreases the importance of ratings over time and the *sliding window model* that counts only for the most recent data and ignores any previous historical information. For example, consider a movie recommendation system that gives higher priority to new releases compared to other old seasoned movies (damped window model) or focuses only on new releases (sliding window model).

From a different perspective, *context-based recommendations* offer different suggestions for different time specifications. The main motivation here, is that user preferences may change over time but have temporal repetition, i.e., recur over time. As an example consider a tourist guide system that should provide different suggestions during summer than during winter. Or, a restaurant recommendation system that might distinguish between weekdays (typically business lunches) and weekends (typically family lunches).

It is the purpose of this paper to provide a framework for studying various approaches that handle different temporal aspects of recommendations. To deal with the sparsity of the explicitly defined user preferences, we introduce the notion of support in recommendations to model how confident the recommendations of an item for a user is. We also consider different cases for selecting the appropriate set of users for producing the recommendations of a user.

The rest of the paper is organized as follows. A general, time-invariant recommendations model is presented in Sect. 2. Time is introduced in Sect. 3, where we distinguish between the aging factor (Sect. 3.1) and the temporal context factor (Sect. 3.2). The computation of recommendations under different temporal semantics is discussed in Sect. 4, while in Sect. 5, we present our experiments using a real dataset of movie ratings. Related work is presented in Sect. 6. Finally, conclusions and outlook are pointed out in Sect. 7.

2   A Time-Free Recommendation Model

Assume a set of items $\mathcal{I}$ and a set of users $\mathcal{U}$ interacting with a recommendation application $\mathcal{A}$. Each user $u \in \mathcal{U}$ may express a preference for an item $i \in \mathcal{I}$, which is denoted by $preference(u, i)$ and lies in the range $[0.0, 1.0]$. We use $\mathcal{Z}_i$ to denote the set of users in $\mathcal{U}$ that have expressed a preference for item i. The cardinality of the items set $\mathcal{I}$ is usually high and typically users rate only a few of these items, that is, $|\mathcal{Z}_i| << |\mathcal{U}|$ for a specific item i. For the items unrated by the users, a *relevance score* is estimated by invoking a recommendation strategy.

In this section, we first present a model for time-free recommendations (Sect. 2.1) and then define the top-k recommendations (Sect. 2.2). The time–free recommendations model is the generally used recommendations model where the notion of time is completely ignored.

2.1 Defining Time-Free Recommendations

There are different ways to estimate the relevance of an item for a user. In general, the recommendation methods are organized into three main categories: (i) *content-based*, that recommend items similar to those the user has preferred in the past (e.g., [17,13]), (ii) *collaborative filtering*, that recommend items that similar users have liked in the past (e.g., [11,8]) and (iii) *hybrid*, that combine content-based and collaborative ones (e.g., [5]).

Our work falls into the *collaborative filtering* category. The key concept of collaborative filtering is to use, for a given user $u \in \mathcal{U}$, the preferences of other users in $\mathcal{U}$ in order to produce relevance scores for the items unrated by u. But, *which is the appropriate set of users, hereafter called peers, for computing the recommendations of u?* Due to the inherent fuzziness associated with this question, there exists no single definition for locating the peers of u. In our model, we assimilate three different aspects of peers: (i) *close friends*, (ii) *domain experts* and (iii) *similar users*.

The *close friends* of a user u are explicitly selected by u. Computing recommendations using only close friends is based on the assumption that these users would have similar tastes for most things, because of the closeness of relationship.

CLOSE FRIENDS: Let $\mathcal{U}$ be a set of users. The *close friends* $\mathcal{C}_u$, $\mathcal{C}_u \subseteq \mathcal{U}$, of a user $u \in \mathcal{U}$ are explicitly defined by u.

From a different perspective, *domain experts* can be used for producing recommendations for specific queries, since they are considered to be knowledgeable on a specific topic or domain. Several methods deal with the problem of finding experts (e.g., [6]); the focus of this paper though is on how to exploit experts preferences to recommend interesting items to other users and not on how to identify these experts. So, we consider that the set of experts for a given query are predefined, e.g., experts in tablet pcs.

DOMAIN EXPERTS: Let $\mathcal{U}$ be a set of users and Q be a query. The *domain experts* $\mathcal{D}_Q$, $\mathcal{D}_Q \subseteq \mathcal{U}$, are the users considered as experts for the domain of Q.

We denote this set as $\mathcal{D}_Q$, so, not dependent on the user, since typically experts are associated with specific queries, subjects or domains rather than with certain users.

Alternatively, a user can opt to employ the preferences of the users that exhibit the most similar behavior to him/her in order to produce relevance scores for the items unrated by him/her, even if other friendship or expert relationships exist. *Similar users* are located via a *similarity function* $simU(u, u')$ that evaluates the proximity between u and u'. Several methods can be applied for selecting the similar users of a user u. A direct method is to locate those users u' with similarity $simU(u, u')$ greater than or equal to a threshold value.

SIMILAR USERS: Let $\mathcal{U}$ be a set of users. The *similar users* $\mathcal{S}_u$, $\mathcal{S}_u \subseteq \mathcal{U}$, of a user $u \in \mathcal{U}$ is a set of users, such that, $\forall u' \in \mathcal{S}_u$, $simU(u, u') \geq \delta$ and $\forall u'' \in \mathcal{U} \backslash \mathcal{S}_u$, $simU(u, u'') < \delta$, where δ is a threshold similarity value.

Clearly, one could argue for other ways of selecting $\mathcal{S}_u$, e.g., by taking the m most similar users to u. Our main motivation here is that we opt for selecting only highly connected users even if the resulting set of users $\mathcal{S}_u$ is small.

We define now the general notion of peers for a user by taking into account the three different cases.

Definition 1 (Peers). *Let $\mathcal{U}$ be a set of users, u be a user in $\mathcal{U}$ and Q be a query posed by u. The peers $\mathcal{P}_{u,Q}$, $\mathcal{P}_{u,Q} \subseteq \mathcal{U}$, of u for Q are either: (i) the close friends $\mathcal{C}_u$ of u, (ii) the domain experts $\mathcal{D}_Q$ for Q, or (iii) the similar users $\mathcal{S}_u$ of u.*

Based on the peers of a user for a query, we define formally the relevance of an item for a user as follows:

Definition 2 (Time-free Relevance). *Let u be a user in $\mathcal{U}$, Q be a query posed by u and $\mathcal{P}_{u,Q}$ be the peers of u for Q. If u has not expressed any preference for an item i, the time-free relevance of i for u under Q is:*

$$relevance^f(u, i, Q) = \frac{\sum_{u' \in (\mathcal{P}_{u,Q} \cap \mathcal{Z}_i)} contribution(u, u') \times preference(u', i)}{\sum_{u' \in (\mathcal{P}_{u,Q} \cap \mathcal{Z}_i)} contribution(u, u')}$$

$$where\ contribution(u, u') = \begin{cases} 1, & if\ \mathcal{P}_{u,Q}\ is\ \mathcal{C}_u\ or\ \mathcal{D}_Q \\ simU(u, u'), & if\ \mathcal{P}_{u,Q}\ is\ \mathcal{S}_u \end{cases}$$

The relevance score of user u for an item i depends on the peers of u that have given a rating for i, i.e., those in $\mathcal{P}_{u,Q} \cap \mathcal{Z}_i$. The *contribution*$(u, u')$ reflects the importance of each *preference*(u', i) for u; this importance depends on how "reliable" u' is for u. When close friends or domain experts are used, contribution is set to 1, since we are certain about the importance of the preferences of the selected users. For the similar users case, the contribution of each user u' depends on the similarity between u and u'.

As already mentioned, due to the abundance of items in a recommendation application, users, even the expert ones, rate only a small portion of them. So, the following question usually arises: *How confident are the relevance scores associated with the recommended items?* To deal with this problem, we introduce the notion of *support* for each candidate item i for user u, which defines the fraction of peers of u that have expressed preferences for i.

Definition 3 (Time-free Support). *Let u be a user in $\mathcal{U}$, Q be a query posed by u and $\mathcal{P}_{u,Q}$ be the peers of u for Q. The time-free support of i for u under Q is:*

$$support^f(u, i, Q) = |\mathcal{P}_{u,Q} \cap \mathcal{Z}_i| / |\mathcal{P}_{u,Q}|$$

Intuitively, the notion of support expresses how reliable is our estimation for i.

To estimate the worthiness of an item recommendation for a user, we propose to combine the *relevance* and *support* scores in terms of a *value* function.

Definition 4 (Time-free Value). *Let $\mathcal{U}$ be a set of users and $\mathcal{I}$ be a set of items. For $\sigma \in [0,1]$, the time-free value of an item $i \in \mathcal{I}$ for a user $u \in \mathcal{U}$ under a query Q, such that, $\nexists preference(u,i)$, is:*

$$value^f(u,i,Q) = \sigma \times relevance^f(u,i,Q) + (1-\sigma) \times support^f(u,i,Q)$$

We take a generic approach for computing the *time-free value* of an item for a user. More sophisticated functions can be designed. However, this general form of weighted summation is simple and easy to implement. Moreover, when $\sigma = 1$, *value* maps to *relevance*, which is the typically used recommendation score.

2.2 Top-k Time-Free Recommendations

Given a query Q submitted by a user u and a restriction k on the number of the recommended items, the goal is to provide u with k suggestions for items that are highly relevant to u and exhibit high support.

Definition 5 (Top-k Time-free Recommendations). *Let $\mathcal{U}$ be a set of users and $\mathcal{I}$ be a set of items. Given a query Q posed by a user $u \in \mathcal{U}$, recommend to u a list of k items $\mathcal{I}_u =< i_1, \ldots, i_k >$, $\mathcal{I}_u \subseteq \mathcal{I}$, such that:*

(i) $\forall i_j \in \mathcal{I}_u$, $\nexists preference(u,i_j)$,
(ii) $value^f(u,i_j,Q) \geq value^f(u,i_{j+1},Q)$, $1 \leq j \leq k-1$, $\forall i_j \in \mathcal{I}_u$, and
(iii) $value^f(u,i_j,Q) \geq value^f(u,x_y,Q)$, $\forall i_j \in \mathcal{I}_u$, $x_y \in \mathcal{I} \backslash \mathcal{I}_u$.

The first condition ensures that the suggested items do not include already evaluated items by the user (for example, do not recommend a movie that the user has already watched). The second condition ensures the descending ordering of the items with respect to their value, while the third condition defines that every item in the result set has value greater than or equal to the value of any of the non–suggested items.

3 Time-Aware Recommendations

The general time-free recommendation model assumes that all preferences are equally active and potentially they can be used for producing recommendations. This way though the temporal aspects of the user ratings are completely ignored. However, the information needs of a user evolve over time, especially if we consider a long period of time, either smoothly (i.e., drift) or more drastically (i.e., shift). This fact makes the recent user preferences to reflect better the current trends than the older preferences do. From a different point of view, user interests differ from time to time which means that users may have different needs under different temporal circumstances. For example, during the weekdays one might be interested in reading IT news whereas during the weekends he/she might be interested in reading about cooking, gardening or doing other hobbies.

To handle such different cases, we propose a framework for time-aware recommendations that incorporates the notion of time in the recommendations process with the goal of improving their accuracy. We distinguish between two types of time-aware recommendations, namely the fresh-based and the context-based ones. The *fresh-based recommendations* pay more attention to more recent user ratings thus trying to deal with the problems of drift or shift in the user information needs over time. The *context-based recommendations* take into account the temporal context under which the ratings were given (e.g., weekdays, weekends).

In our time-aware recommendation model, the rating of a user u for an item i, i.e., $preference(u, i)$, is associated with a timestamp $t_{u,i}$, which is the time that i was rated by u. So, this timestamp declares the freshness or age of the rating. Below, we first define the fresh-based recommendation model (Sect. 3.1) and then, the temporal context-based recommendation model (Sect. 3.2). We also present a variant of the top-k recommendations problem by defining the top-k time-aware recommendations (Sect. 3.3).

3.1   Fresh-Based Recommendations

Generally speaking, the popularity of the items of a recommendation application changes with time; typically, items lose popularity while time goes on. For example, a movie, a picture or a song may lose popularity because they are too seasoned. Motivated by the intuition that the importance of preferences for items increases with the popularity of the items themselves, fresh-based recommendation approaches care for suggesting items taking mainly into account recent and novel user preferences.

Driven by the work in stream mining [10], we use different types of aging mechanisms to define the way that the historical information (in form of ratings) is incorporated in the recommendation process. Aging in streams is typically implemented through the notion of windows, which define which part of the stream is active at each time point and thus could be used for the computations. In this work, we use the *damped window model* that gradually decreases the importance of historical data comparing to more recent data and the *sliding window model* that remembers only the preferences defined within a specific, recent time period. We present these cases in more detail below. Note that the static case (Sect. 2), corresponds to the *landmark window model* where the whole history from a given landmark is considered.

Damped Window Model. In the damped window model, although all user preferences are active, i.e., they can contribute to produce recommendations, their contribution depends upon their arrival time. In particular, the preference of a user u for an item i is weighted appropriately with the use of a temporal decay function. Typically, in temporal applications, the exponential fading function is employed, so the weight of $preference(u, i)$ decreases exponentially with time via the function: $2^{-\lambda(t-t_{u,i})}$, where $t_{u,i}$ is the time that the preference was defined and t is the current time. Thus, $t - t_{u,i}$ is actually the age of the preference. The parameter λ, $\lambda > 0$, is the decay rate that defines how fast the past history is forgotten. The higher λ, the lower the importance of historical preferences compared to more recent preferences.

Under this aging scheme, the so-called *damped relevance* of an item i for a user u with respect to a query Q in a given timepoint t is given by:

$$relevance^d(u,i,Q) = \frac{\sum_{u' \in (\mathcal{P}_{u,Q} \cap \mathcal{Z}_i)} 2^{-\lambda(t-t_{u',i})} \times contribution(u,u') \times preference(u',i)}{\sum_{u' \in (\mathcal{P}_{u,Q} \cap \mathcal{Z}_i)} contribution(u,u')}$$

So, all user item scores are weighted by the recency $2^{-\lambda(t-t_{u',i})}$.

Since all preferences are active, the *damped support* of i for u under Q is equal to the corresponding time-free support, that is:

$$support^d(u,i,Q) = support^f(u,i,Q)$$

Finally, the *damped value* of i for u under Q is computed as in the time-free case by aggregating the relevance and support scores ($\sigma \in [0,1]$):

$$value^d(u,i,Q) = \sigma \times relevance^d(u,i,Q) + (1-\sigma) \times support^d(u,i,Q)$$

Sliding Window Model. The sliding window model employs an alternative method which exploits only a subset of the available preferences, and in particular, the most recent ones. The size of this subset, referred to as window size, might be defined in terms of timepoints (e.g., use the preferences defined after Jan 2011) or records (e.g., use the 1000 most recent preferences). We adopt the first case. The preferences within the window are the active preferences that participate in the recommendations computation. Let t be the current time and W be the window size. Then, a preference of a user u for an item i, $preference(u,i)$, is active only if $t_{u,i} > t - W$.

In the sliding window model, the *sliding relevance* of an item i for a user u under a query Q is defined with regard to the active preferences of the peers of u for i. More specifically:

$$relevance^s(u,i,Q) = \frac{\sum_{u' \in (\mathcal{P}_{u,Q} \cap \mathcal{X}_i)} contribution(u,u') \times preference(u',i)}{\sum_{u' \in (\mathcal{P}_{u,Q} \cap \mathcal{X}_i)} contribution(u,u')}$$

where $\mathcal{X}_i$ is the set of users in $\mathcal{Z}_i$, such that, $\forall u' \in \mathcal{X}_i$, $t_{u',i} > t - W$.

The *sliding support* of i for u under Q is defined as the fraction of peers of u that have expressed preferences for i that are active at time t. That is:

$$support^s(u,i,Q) = |\mathcal{P}_{u,Q} \bigcap \mathcal{X}_i|/|\mathcal{P}_{u,Q}|$$

Finally, the *sliding value* of i for u under Q, for $\sigma \in [0,1]$, is:

$$value^s(u,i,Q) = \sigma \times relevance^s(u,i,Q) + (1-\sigma) \times support^s(u,i,Q)$$

3.2   Temporal Context-Based Recommendations

In contrast to fresh-based recommendations, the context-based ones assume that although the preferences may change over time, they display some kind of temporal repetition. Or in other words, users may have different preferences under different temporal contexts. For instance, during the weekend a user may prefer to watch different movies from those in the weekdays. So, a movie recommendation system should provide movie suggestions for the weekends that may differ from the suggestions referring to weekdays.

As above, the rating of a user for an item, $preference(u, i)$, is associated with the rating time $t_{u,i}$. Time is modeled here as a multidimensional attribute. The dimensions of time have a hierarchical structure, that is, time values are organized at different levels of granularity (similar to [16,18]). In particular, we consider three different levels over time: $time\_of\_day$, day_of_week and $time\_of\_week$ with domain values { "morning", "afternoon", "evening", "night" }, { "Mon", "Tue", "Wed", "Thu", "Fri", "Sat", "Sun" } and { "Weekday", "Weekend" }, respectively. It is easy to derive such kind of information from the time value $t_{u,i}$ that is associated with each user rating by using SQL or other programming languages. More elaborate information can be extracted by using the WordNet or other ontologies.

Let Θ be the current temporal context of a user u. We define the *context-based relevance* of an item i for u under a query Q expressed at Θ based on the preferences of the peers of u for i that are defined for the same context Θ. Formally:

$$relevance^c(u, i, Q) = \frac{\sum_{u' \in (\mathcal{P}_{u,Q} \cap \mathcal{Y}_i)} contribution(u, u') \times preference(u', i)}{\sum_{u' \in (\mathcal{P}_{u,Q} \cap \mathcal{Y}_i)} contribution(u, u')}$$

where $\mathcal{Y}_i$ is the set of users in $\mathcal{Z}_i$, such that, $\forall u' \in \mathcal{Y}_i$, $t_{u',i} \mapsto \Theta$, that is, the user rating has been expressed for a context equal to Θ. For example, if the temporal context of a user query is "Weekend", only the user preferences given for context "Weekend" would be considered.

The *context-based support* of i for u under Q is defined with respect to the number of peers of u that have expressed preferences for i under the same to the query context. That is:

$$support^c(u, i, Q) = |\mathcal{P}_{u,Q} \cap \mathcal{Y}_i| / |\mathcal{P}_{u,Q}|$$

Similar to the fresh-based recommendations, the *context-based value* of i for u under Q is calculated taking into account the context-based relevance and support. For $\sigma \in [0, 1]$:

$$value^c(u, i, Q) = \sigma \times relevance^c(u, i, Q) + (1 - \sigma) \times support^c(u, i, Q)$$

3.3   Top-k Time-Aware Recommendations

Next, we define the time-aware variation of the top-k recommendations applicable to both fresh-based and context-based approaches.

Definition 6 (Top-k Time-aware Recommendations). *Let $\mathcal{U}$ be a set of users and $\mathcal{I}$ be a set of items. Given a query Q posed by a user $u \in \mathcal{U}$ at time t mapped to Θ, recommend to u a list of k items $\mathcal{I}_u = < i_1, \ldots, i_k >$, $\mathcal{I}_u \subseteq \mathcal{I}$, such that:*

    (i) $\forall i_j \in \mathcal{I}_u$, $\nexists preference(u, i_j)$, for the fresh-based recommendations, and $\forall i_j \in \mathcal{I}_u$, $\nexists preference(u, i_j)$ that is associated with context equal to Θ, for the context-based recommendations,

    (ii) $value^o(u, i_j, Q) \geq value^o(u, i_{j+1}, Q)$, $1 \leq j \leq k-1$, $\forall i_j \in \mathcal{I}_u$, and

    (iii) $value^o(u, i_j, Q) \geq value^o(u, x_y, Q)$, $\forall i_j \in \mathcal{I}_u$, $x_y \in \mathcal{I} \backslash \mathcal{I}_u$,

where o corresponds to the same d (for the damped window model), s (for the sliding window model), or c (for the context-based model).

The first condition ensures that the suggested items do not include already evaluated items by the user either in general or under a specific context, while the second and the third conditions resemble those of Definition 5.

4 Time-Aware Recommendations Computation

Assume a user that submits a query presenting his information needs. Each query is enhanced with a contextual specification expressing some temporal information. This temporal information of the query may be postulated by the application or be explicitly provided by the user as part of his query. Typically, in the first case, the context implicitly associated with a query corresponds to the current context, that is, the time of the submission of the query. As a query example, for a restaurant recommendation application, consider a user looking for restaurants serving chinese cuisine during the weekend. As part of his/her query, the user should also provide the aging scheme that will be used.

Then, we locate the peers of the user (Sect. 4.1) and employ their preferences for estimating the time-aware recommendations (Sect. 4.2). Recommendations are presented to the user along with explanations on the reasons behind them (Sect. 4.3). In following, we overview the details of each step.

4.1 Selecting Peers

Our model assumes three different kinds of peers, namely close friends, domain experts and similar users. For each submitted query Q of a user u, u specifies the peers that will be used for producing his/her recommendations. This selection step of the peers is, in general, application dependent. For example, when a user is asking for advice for a personal computer, the domain experts may fit well to the user needs, while when asking for a suggestion about a movie, the user's close friends may provide good answers. In a similar manner, when using a trip advisor, the choice of users with similar tastes seems appropriate.

For the close friends case, the set of peers of u consists of the close friends of u, while for the domain experts case, the set of peers of u consists of the users that are considered to be experts for Q. We assume that this information is already known. For the similar users case, we need to calculate all similarity measures $simU(u, u')$ for all users $u' \in \mathcal{U}$. Those users u' with similarity $simU(u, u')$ greater than or equal to the threshold δ represent the similar users of u.

4.2   Computing Recommendations

Having established the methodology for finding the peers of a user, we focus next on how to generate valued recommendations for him/her. Given a user $u \in \mathcal{U}$ and his/her peers $\mathcal{P}_{u,Q}$, the procedure for estimating the value score of an item i for u requires the computation of the relevance and support of i. Note that we do not compute value scores for all items in $\mathcal{I}$, but only for the items $\mathcal{I}'$, $\mathcal{I}' \subseteq \mathcal{I}$, that satisfy the query selection conditions. To do this, we perform a pre-processing step to select the relevant to the query data by running a typical database query. For example, for a query about destinations in Greece posed to a travel recommendation system, we ignore all the rest destinations.

For computing the scores of the items in $\mathcal{I}'$, pairs of the form $(i, value^o(u, i, Q))$ are maintained in a set $\mathcal{V}_u$, where o corresponds to d, s or c for the damped window, sliding window and context-based approach, respectively. As a post-processing step, we rank all pairs in $\mathcal{V}_u$ on the basis of their value score. To provide the top-k recommendations to u, we report the k items with the highest scores, i.e., the k first items in $\mathcal{V}_u$.

Next, we discuss separately the particulars of each time-aware recommendation approach. For the damped window approach, all the preferences of the peers of u are employed for computing recommendations. However, this is not the case for the other two approaches, where only a subset of the peers preferences are taken into consideration. More specifically, for the sliding window approach, only the most recent preferences are used, while for the context-based approach, the preferences that are defined for a temporal context equal to the query context. This can be seen as a preference pre-filtering step. It is worth noting that, since some preferences are ignored, some of the peers may not contribute finally to the recommendation list construction.

Moreover, for the context-based approach, the associated set of preferences for a specific query may be empty, that is, there may be no preferences for the query. In this case, we can use for the recommendation process these preferences whose context is more general than the query context. For example, for a query with context "Sat", we can use a preference defined for context "Weekend". The selection of the appropriate preferences can be made more efficient by deploying indexes on the context of the preferences. Such a data structure that exploits the hierarchical nature of context, termed profile tree, is introduced in [18].

As a final note, consider that the two approaches for computing time-aware recommendations can be applied together. For instance, we can apply the context-based approach first. Then, we can apply the damped window approach. This way, the importance of the preferences that are defined for the query context decreases with time.

4.3   Presenting Recommendations

After identifying the k items with the highest value scores for a user u, u is presented with these items. Recently, it has been shown that the success of recommendations relies on explaining the cause behind them [19]. This is our motivation for providing an explanation along with each suggested item, i.e., for explaining why this specific recommendation appears in the top-k list.

To do this, we present recommendations along with their explanations as text by using a simple template mechanism. Since explanations depend on the employed approach, different templates are associated with the two different approaches.

For the fresh-based approach, the reporting results have the following form:

ITEM @i IS RECOMMENDED BY THE SYSTEM
BECAUSE OF ITS HIGH VALUE SCORE, @$value^o$,
COMPUTED USING THE RECENT PREFERENCES OF YOUR @$peers$.
@$|\mathcal{P}_{u,Q} \cap \mathcal{T}_i|$ USERS OUT OF YOUR @$|\mathcal{P}_{u,Q}|$ PEERS HAVE RATED THIS ITEM.

In this case, o corresponds to d or s and $\mathcal{T}_i$ to $\mathcal{Z}_i$ or $\mathcal{X}_i$ for the damped window or the sliding window approach, respectively.

Similarly, for the context-based approach, the results are presented as follows:

ITEM @i IS RECOMMENDED BY THE SYSTEM
BECAUSE OF ITS HIGH VALUE SCORE, @$value^c$,
COMPUTED USING THE PREFERENCES, DEFINED FOR A TEMPORAL CONTEXT
EQUAL TO THE QUERY ONE, OF YOUR @$peers$. @$|\mathcal{P}_{u,Q} \cap \mathcal{Y}_i|$ USERS OUT OF
YOUR @$|\mathcal{P}_{u,Q}|$ PEERS HAVE RATED THIS ITEM.

We use the symbol @ to mark parameter variables. Variables are replaced with specific values at instantiation time. For a movie recommendation system, an example of a reported result with its explanation is the following.

ITEM *Dracula* IS RECOMMENDED BY THE SYSTEM
BECAUSE OF ITS HIGH VALUE SCORE, 0.9,
COMPUTED USING THE PREFERENCES, DEFINED FOR A TEMPORAL CONTEXT
EQUAL TO THE QUERY ONE, OF YOUR *close friends*.
27 USERS OUT OF YOUR 68 PEERS HAVE RATED THIS ITEM.

5   Experiments

In this section, we evaluate the effectiveness of our time-aware recommendation system using a real movie ratings dataset [1], which consists of 100,000 ratings given from September 1997 till April 1998 by 1,000 users for 1,700 items. The monthly split is shown in Fig. 1(a), while the split per weekends and weekdays is shown in Fig. 1(b).

Since there is no information about actual friends and experts in this dataset, we employ as the peers of a given user his/her similar users. To this end, the notion of user similarity is important. We use here a simple variation; that is,

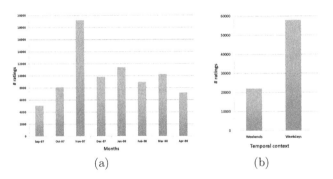

Fig. 1. (a) Ratings per month and (b) ratings per temporal context

we use distance instead of similarity. More specifically, we define the distance between two users as the Euclidean distance over the items rated by both. Let $u, u' \in \mathcal{U}$ be two users, $\mathcal{I}_u$ be the set of items for which $\exists preference(u, i)$, $\forall i \in \mathcal{I}_u$, and $\mathcal{I}_{u'}$ be the set of items for which $\exists preference(u', i)$, $\forall i \in \mathcal{I}_{u'}$. We denote by $\mathcal{I}_u \cap \mathcal{I}_{u'}$ the set of items for which both users have expressed preferences. Then, the distance between u, u' is:

$$distU(u, u') = \sqrt{\sum_{i \in \mathcal{I}_u \cap \mathcal{I}_{u'}} (preference(u, i) - preference(u', i))^2 / |\mathcal{I}_u \cap \mathcal{I}_{u'}|}$$

To evaluate the quality of the recommendations, we use a predictive accuracy metric that directly compares the predicted ratings with the actual ones [12]. A commonly used metric in the literature is the *Mean Absolute Error* (MAE), which is defined as the average absolute difference between predicted ratings and actual ratings: $MAE = \sum_{u,i} |preference(u, i) - value^o(u, i, Q)|/N$, where N is the total number of ratings in the employed dataset and o corresponds to d, s or c. Clearly, the lower the MAE score, the better the predictions.

Next, we report on the results for the sliding window model, the damped window model and the context-based model compared to the time-free model.

Sliding window model. To illustrate the effectiveness of the sliding window model, we use windows of different sizes W. The window size $W = 1$ stands for the most recent month, i.e., April 1998, the window size $W = 2$ stands for both April 1998 and March 1998, and so forth. The window size $W = 8$ includes the whole dataset, from April 1998 till September 1997. We denote the resulting dataset as D_W, where $W = [1 - 8]$ is the window size. For each dataset D_W, we compute the recommendations for each user by considering the user ratings within the corresponding window W. We compare the predicted values with the actual values given by the user within the same window W and report the average results.

The results for different windows are presented in Fig. 2(a). In the same figure, we also show the effect of the user distance threshold. In general, recommendations present better quality for small windows (this is not the case for the smallest window size ($W = 1$) because of the small amount of ratings used for predictions). For example, for a user distance threshold equal to 0.03 and $W = 2$, the predictions are improved around 2.5% compared to $W = 8$ (i.e., compared to the time-free recommendations model). Or, for a threshold equal to 0.06 and $W = 3$, the predictions are improved on average 0.5% compared to $W = 8$. Moreover, the larger the window, the smaller the improvement. Regarding the effect of the user similarity thresholds, as expected, for larger user distance thresholds, the MAE scores increase for all window sizes, since more dissimilar users are considered for the suggestions computation.

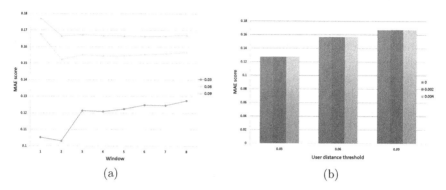

(a)   (b)

Fig. 2. MAE scores for (a) the sliding window and (b) the damped window model

Damped window model. Next, we evaluate the effect of the decay rate λ in the recommendations accuracy. We use different values for λ; the higher the λ is, the less the historical data count. The value $\lambda = 0$ corresponds to the time-free model. We downgrade the original ratings based on the decay factor λ and the time difference between the end of the observation period (22/04/1998) and the ratings timestamp.

The results of this experiment are shown in Fig. 2(b). Practically, this aging model seems to not offer any (or offer a very small) improvement in this setting, i.e., for the employed dataset. As above, larger distance thresholds lead to larger MAE scores.

Context-based recommendations. In this set of experiments, we demonstrate the effect of temporal context on producing recommendations. We consider two different temporal contexts "Weekends" and "Weekdays". For the "Weekends" context, we base our predictions only on ratings defined for weekends ($D_{weekends}$), whereas for the "Weekdays" context, we consider ratings from Monday to Friday ($D_{weekdays}$). The predicted values are compared to the actual values given by the user within the same temporal context through the MAE metric.

Fig. 3 displays the results. Except for the two temporal contexts, "Weekends" and "Weekdays", we also present the scores for the time-free model, i.e., when the whole dataset is used. Generally speaking, the temporal context affects the recommendations accuracy. In particular, for both contexts, "Weekends" and "Weekdays", the quality of the recommendations is improved compared to the time-free approach that completely ignores the temporal information of the ratings. For example, for a user distance threshold equal to 0.03, the predictions for "Weekends" are improved on average 0.95% when using ratings for "Weekends" instead of using the whole rating set. Similarly, for a distance equal to 0.06, the predictions for "Weekdays" are improved around 0.5%. Also, larger distance thresholds values result in larger MAE scores, that is, the quality of the recommendations decreases with the user distance threshold.

In overall, time plays an important role towards improving the quality of the proposed recommendations. The sliding window and the context-based approaches increase the recommendations accuracy. However, a mere decay model seems to be not adequate. Our goal is to design a more elaborate aging scheme that considers not only the age of the ratings but also other parameters, such as the recency and popularity of the recommended items and the context under which the ratings were given. We expect that the time effect will be more evident for datasets that span a larger period of time. Further experimentation with other kinds of peers provided by the application dataset will also be interesting.

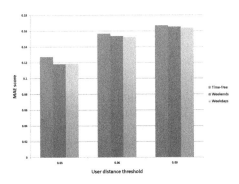

Fig. 3. MAE scores for context-based approach

6   Related Work

The research literature on recommendations is extensive. Typically, recommendation approaches are distinguished between: *content-based*, that recommend items similar to those the user previously preferred (e.g., [17,13]), *collaborative filtering*, that recommend items that users with similar preferences liked (e.g., [11,8]) and *hybrid*, that combine content-based and collaborative ones (e.g., [5]). Several extensions have been proposed, such as employing multi-criteria ratings (e.g., [2]) and defining recommendations for groups (e.g., [4,15,14]).

Recently, there are also approaches focusing on enhancing recommendations with further contextual information (e.g., [3,16]). In these approaches, context is defined as a set of dimensions, or attributes, such as location, companion and time, with hierarchical structure. While a traditional recommendation system considers only two dimensions that correspond to users and items, a context-aware recommendation system considers one additional dimension for each context attribute. In our approach, we focus on a particular case of this model, that is, the three-dimensional recommendations space among users, items and time, since our specific goal is to study how the time effects contribute to the improvement of predictions.

Moreover, there are some approaches which incorporate temporal information to improve recommendations effectiveness. [21] presents a graph-based recommendation system that mixes long-term and short-term user preferences to improve predictions accuracy, while [20] considers how time can be used into matrix factorization models by examining changes in user and society tastes and habits, and items popularity. [9] uses a strategy, similar to our damped window model, that decreases the importance of known ratings as time distance from recommendation time increases. However, the proposed algorithm uses clustering to discriminate between different kinds of items. [7] introduces the idea of micro-profiling, which splits the user preferences into several sets of preferences, each representing the user in a particular temporal context. The predictions are computed using these micro-profiles instead of a single user model. The main focus of this work is on the identification of a meaningful partition of the user preferences using implicit feedback. In our paper, the goal is to examine time from different perspectives. This way, we use a general model for time, considering time either as specific time instances or specific temporal conditions, in order to define a unified time-aware recommendation model.

7 Conclusions

In this paper, we studied different semantics to exploit the time information associated with user preferences to improve the accuracy of recommendations. We considered various types of time effects, and thus, proposed different time-aware recommendation models. Fresh-based recommendations care mainly for recent and novel preferences, while context-based recommendations are computed with respect to preferences with temporal context equal to the query context. Finally, we demonstrated our approach using a real dataset of movie ratings. There are many directions for future work. One is to extend our framework so as to support a novel mode of interaction between users and recommendation systems; our goal is to exploit the whole rating history to produce valued recommendations and, at the same time, use the fresh ratings to assist users in database exploration.

References

1. Movielens data sets, http://www.grouplens.org/node/12 (visited on November 2011)
2. Adomavicius, G., Kwon, Y.: New recommendation techniques for multicriteria rating systems. IEEE Intelligent Systems 22(3), 48–55 (2007)

3. Adomavicius, G., Sankaranarayanan, R., Sen, S., Tuzhilin, A.: Incorporating contextual information in recommender systems using a multidimensional approach. ACM Trans. Inf. Syst. 23(1), 103–145 (2005)

4. Amer-Yahia, S., Roy, S.B., Chawla, A., Das, G., Yu, C.: Group recommendation: Semantics and efficiency. PVLDB 2(1), 754–765 (2009)

5. Balabanovic, M., Shoham, Y.: Content-based, collaborative recommendation. Commun. ACM 40(3), 66–72 (1997)

6. Balog, K., Bogers, T., Azzopardi, L., de Rijke, M., van den Bosch, A.: Broad expertise retrieval in sparse data environments. In: SIGIR, pp. 551–558 (2007)

7. Baltrunas, L., Amatriain, X.: Towards time-dependant recommendation based on implicit feedback. In: CARS, pp. 1–5 (2009)

8. Breese, J.S., Heckerman, D., Kadie, C.M.: Empirical analysis of predictive algorithms for collaborative filtering. In: UAI, pp. 43–52 (1998)

9. Ding, Y., Li, X.: Time weight collaborative filtering. In: CIKM, pp. 485–492 (2005)

10. Gama, J.: Knowledge Discovery from Data Streams. CRC Press (2010)

11. Konstan, J.A., Miller, B.N., Maltz, D., Herlocker, J.L., Gordon, L.R., Riedl, J.: Grouplens: Applying collaborative filtering to usenet news. Commun. ACM 40(3), 77–87 (1997)

12. Melville, P., Sindhwani, V.: Recommender systems. In: Encyclopedia of Machine Learning, pp. 829–838 (2010)

13. Mooney, R.J., Roy, L.: Content-based book recommending using learning for text categorization. In: ACM DL, pp. 195–204 (2000)

14. Ntoutsi, I., Stefanidis, K., Nørvåg, K., Kriegel, H.-P.: Fast group recommendations by applying user clustering. In: ER (2012)

15. O'Connor, M., Cosley, D., Konstan, J.A., Riedl, J.: Polylens: A recommender system for groups of user. In: ECSCW, pp. 199–218 (2001)

16. Palmisano, C., Tuzhilin, A., Gorgoglione, M.: Using context to improve predictive modeling of customers in personalization applications. IEEE Trans. Knowl. Data Eng. 20(11), 1535–1549 (2008)

17. Pazzani, M.J., Billsus, D.: Learning and revising user profiles: The identification of interesting web sites. Machine Learning 27(3), 313–331 (1997)

18. Stefanidis, K., Pitoura, E., Vassiliadis, P.: Managing contextual preferences. Inf. Syst. 36(8), 1158–1180 (2011)

19. Tintarev, N., Masthoff, J.: Designing and evaluating explanations for recommender systems. In: Recommender Systems Handbook, pp. 479–510. Springer (2011)

20. Xiang, L., Yang, Q.: Time-dependent models in collaborative filtering based recommender system. In: Web Intelligence, pp. 450–457 (2009)

21. Xiang, L., Yuan, Q., Zhao, S., Chen, L., Zhang, X., Yang, Q., Sun, J.: Temporal recommendation on graphs via long- and short-term preference fusion. In: KDD, pp. 723–732 (2010)

A Hybrid Time-Series Link Prediction Framework for Large Social Network

Jia Zhu, Qing Xie, and Eun Jung Chin

School of ITEE, The University of Queensland, Australia
{jiazhu,felixq,eunjung}@itee.uq.edu.au

Abstract. With the fast growing of Web 2.0, social networking sites such as Facebook, Twitter and LinkedIn are becoming increasingly popular. Link prediction is an important task being heavily discussed recently in the area of social networks analysis, which is to identify the future existence of links among entities in the social networks so that user experiences can be improved. In this paper, we propose a hybrid time-series link prediction model framework called DynamicNet for large social networks. Compared to existing works, our framework not only takes timing as consideration by using time-series link prediction model but also combines the strengths of topological pattern and probabilistic relational model (PRM) approaches. We evaluated our framework on three known corpora, and the favorable results indicated that our proposed approach is feasible.

1   Introduction

With the fast growing of Web 2.0, social networking web sites such as Facebook, Twitter and LinkedIn are becoming increasingly popular. For example, people can use Facebook to find their friends and share similar interests. These systems can provide sufficient information to users from collaboration. Therefore, it is important to capture information effectively from social networks.

Link prediction is a sub-field of social networks analysis. It is concerned with the problem of predicting the future existence of links among entities in the social networks. Most of the existing works like [12,13] are mainly designed for homogeneous network, and the attribute values of entities are difficult to gain and measure. Thus, the use of topological features among entities in a social networks is critical.

In this paper, we propose a hybrid time-series link prediction framework called DynamicNet, using both attribute and structural properties, by carefully considering the timing of publication, in order to compute accurate link prediction in social networks. Timing is an important factor which has not fully considered by existing works and we can retrieve useful information like periodic patterns and temporal trends from time-series analysis [2,9]. For example, if two authors have cooperated together in the past three years, they have higher possibility to cooperate in the next three years than if they only have cooperated ten years ago. In addition, our framework combines both topological pattern and probabilistic

S.W. Liddle et al. (Eds.): DEXA 2012, Part II, LNCS 7447, pp. 345–359, 2012.

relational model approaches. The topological pattern approach is to put whole social networks into a graph and exploit local and global topological patterns from the observed part of the network, and then make decisions for link existence by calculating weight for pair of nodes [17,18]. On the other hand, probabilistic relational model (PRM) is a framework [6] to abstract observed network into a graph model. PRM represents a joint probability distribution over the attributes of a relation data set possibly with heterogeneous entities [15].

However, according to existing researches, both approaches can not get ideal outcomes in some cases. For instances, it is difficult for PRM to get the well-established data patterns in large graphs such as paths and cycles if descriptive attributes, link structure information, and potentially descriptive attributes of the links are not available [8]. On the other hand, topological pattern is not flexible enough to factorize the entire graph and it is strict by predefined threshold [18].

In order to minimize these issues, our hybrid link prediction framework is to first use topological pattern to get descriptive information for entities in the graph to construct a model graph, and then apply probabilistic relational model to learn so that both vertices and edges in the graph can be regenerated and the patterns in the link structure can be discovered.

Extensive experiments have been performed on three real data sets: DBLP, Wikipedia and IMDB. We show that our approach performs significantly better than several selected state-of-the-art methods in different scenarios. The rest of this paper is organized as follows. In Section 2, we discuss related works in link prediction. In Section 3, we formulate our problem and describe the details of our approach including problem formulation. In Section 4, we present our experiments, evaluation metrics, and results. We conclude this study in Section 5.

2   Related Work

The previous studies of link prediction methods differ significantly with respect to three types of approaches: node-wise similarity, topological pattern and probabilistic model based approaches. The following review of some previous works is presented in chronological order.

Hoff et al. [7] introduced a class of latent class models from the perspective of social networks analysis which try to project all the networked objects to a latent space, and the decision for link existence is based on their spatial positions. Their method provides a model based spatial representation of social relationships.

Taskar et al. [21] proposed a model by applying the Relational Markov Network framework to define a joint probabilistic model over the entire link graph. The application of the RMN algorithm to this task requires the definition of probabilistic patterns over sub graph structures and provided significant improvements in accuracy over flat classification.

Newman [16] proposed based on empirical evidence that the probability of co-authorship x and y is correlated with the product of the number of collaborators of x and y, and used it to infer the probability of their collaboration in the future.

Bilenko and Mooney [4] represented an original instance by a feature vector where each feature corresponds to whether a word in a vocabulary appears in the string representation of the instance. They made a pair instance by representing it as a pairwise vector in an N dimensional space to compare similarity.

Basilico and Hofmann [3] proposed to use the inner product of two nodes and their attributes as similarity measure for collaborative filtering. Their method showed how generalization occurs over pairs with a kernal function generated from either attribute information or user behaviors.

Martin et al. [14] proposed a solution for protein interaction social networks. They extracted some signatures from consecutive amino acids which have the hydrophilic characteristics from whole sequence that in fact pairs of amino acid sequences for link prediction.

Bilgic et al. [5] demonstrated that jointly performing collective classification and link prediction in an iterative way can be effective for both tasks.

Huang et al [9] introduced the time-series link prediction model problem and took into consideration temporal evolutions of link occurrences to predict link occurrence probabilities at a particular time. Their experiments demonstrated that time-series prediction models of link occurrences achieve comparable performance with static graph link prediction models.

3   Our Approach

3.1   Problem Formulation

There are five types of objects in our graph, namely Persons, Keywords, Entities, Venues and Time. We use DBLP[1], Wikipedia[2] and IMDB [3] as our test datasets. The explanations of these five objects are given in Table 1.

Table 1. Objects References

Objects	DBLP	Wikipedia	IMDB
Persons	Authors	Contributors	Actors
Keywords	Keywords in Titles	Keywords in Description	Keywords in Movie Titles
Entities	Papers	Wikipedia Page	Movies
Venues	Venue Names	Wikipedia Regions	Movie Locations
Time	Published Time	Contribution Time	Showed Time

Given a set of instances $v_i \in V$, which is normalized as a graph $G(V, E)$, where E is the set of links, the task is to predict an unobserved link $e_{ij} \in E$ existing between a pair of nodes v_i, v_j, which are a pair of persons, in the graph. The qualified link should satisfy a list of criteria both on the attribute and structural properties in the graph networks. We formulate a list of criteria for an unobserved link between two persons, which should satisfy as below:

---

[1] http://www.informatik.uni-trier.de/~ley/db/

[2] http://en.wikipedia.org/wiki/Wiki

[3] http://www.imdb.com

1) *Timing* - Two persons who have cooperated each other recently are more likely to be linked than those who only have cooperated long times ago.
2) *Frequency* - Two persons who have cooperated many times before are more likely to be linked than those who only have cooperated few times.
3) *Node based Topological Patterns* - Two persons are likely to be linked if they have a number of common neighbors in the graph or the product of their collaborators is high [16] or their shared attributes are very rare [1].
4) *Path based Topological Patterns* - Two persons are likely to be linked if there are many paths indirectly connecting between them or the number of steps between them are low or their neighbours are similar [10].
5) *Probabilistic Relational Model* - Our framework uses probabilistic rational model, and in a typical relational network, two persons are likely to be linked if they have similar distribution according to a set of clique templates [20].

3.2 DynamicNet Framework

Graph Structure of DynamicNet. As discussed in the previous sections, there are five types of objects in the graph of our model. Persons object is a concrete node and other four objects are attribute nodes to represent attribute properties of the graph. Entities, Keywords and Venues are also connection nodes where Entities is to connect Persons and other two is to connect Entities. Fig. 1 is an sample of the graph. There are two different scenarios, and one is direct link between persons. For example, Person_1 has collaborated with Person_2 on Entity_1, therefore they have possibility to cooperate again and similar situation to Person_4 and Person_5. The other is indirect link. Person_1 might work with Person_5 in the future because their works Entity_1 and Entity_2 have connections via Keywords and Venues which means their work topics are similar and have been released in the same location before. Person_3 also has high chance to work with Person_5 because they have same collaborator Person_4. We give a list of definitions for our basic graph $G(V, E)$ as below:

Definition 1 *Path: We call $p(u, v)$ is a path donating a sequence of vertices from vertex u to vertex v through a set of edges $e_i(i = 1, 2, ..., n), e \in E$ and $E \in G$.*

Definition 2 *Path Set: We call $PATH(u, v)$ is the set of all possible paths from vertex u to v.*

Definition 3 *Cycle: We call $c(k)$ is a cycle donating a length k cycle where k vertices connected in a closed chain. The number of vertices in a cycle equals the number of edges.*

Definition 4 *Cycle Set: We call $C(k)$ is the set of all possible cycles in graph G in the length k.*

Edge Weighting. Like most of existing graph approaches, edge weighting is very important for link prediction. In this section, we propose our edge weighting methods for the edges from concrete node to attribute node and from attribute node to attribute node. We define the weight of edge from concrete node p to attribute node a in Equation.(1):

$$\omega(p, a) = \frac{d \cdot Score(a)}{N(p)} \tag{1}$$

where $Score(a)$ is the importance score of attribute node a in the graph, $N(p)$ is the number of links to p and $d \in (0, 1]$ is a decay factor.

The weight of edge from attribute node a to attribute node b is defined in Equation.(2):

$$\omega(a, b) = \frac{d \cdot (Score(a) + Score(b))}{\gamma \cdot N_a(a, b) + (1 - \gamma) \cdot N_p(a, b)} \tag{2}$$

where $Score(a)$ and $Score(b)$ are the importance scores of attribute nodes a and b in the graph, $d \in (0, 1]$ is a decay factor, $N_a(a, b)$ is the number of links from other attribute nodes to a and b, $N_p(a, b)$ is the number of links from concrete nodes to a and b. Since the link from concrete node to attribute node is different to the link from attribute node to attribute node, γ is the parameter to control the weight of two different types of links.

Regards to the importance score of an attribute, we define it in Equation.(3):

$$Score(a) = \frac{N(a)}{N(G)} \tag{3}$$

where $N(a)$ is the total number of links to a and $N(G)$ is the total number of links in graph G.

Topological Pattern Based Probability Model. In this section, we propose a probability model to calculate the probability of a link with the weighting methods we have introduced in the last section based on topological pattern information in the graph, such as path and cycle formation.

Simply speaking, to predict an unobserved link in a graph is actually to calculate the occurrence probability of this link. In our model, the probability is determined by the number of paths and cycles with different lengths and degrees.

We first start from our sample graph in Fig. 1. The key is that the path from Person_3 to Person_4 and the path from Person_4 to Person_5 should be treated as a combination determinants for the cooperation of Person_3 and Person_5.

Therefore, an ideal link between two concrete nodes v_i, v_j should remain paths or cycles with other concrete nodes from topological pattern of view. We define the probability of a link $e_{(i,j)}, e \in E$ based on path and cycle patterns in Equation.(4) and (5):

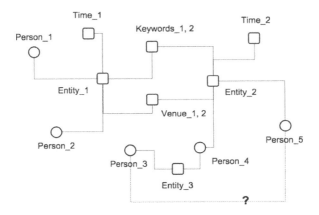

Fig. 1. Sample Graph of DynamicNet with Unobserved Link

$$PR_p(v_i, v_j) = d_i \cdot d_j \Sigma_{p \in P} PR(v_i|u_p) PR(u_p|v_j) \qquad (4)$$

$$PR_c(v_i, v_j) = d_i \cdot d_j \Sigma_{c \in C} PR(v_i|u_c) PR(u_c|v_j) \qquad (5)$$

where $PR_p(v_i, v_j)$ is the transition probability from v_i to v_j via all possible paths P, and u_p is the set of nodes between v_i and v_j in path p. Similarly, $PR_c(v_i, v_j)$ is the transition probability between v_i and v_j in all possible cycles C_k, and u_c is the set of nodes between v_i and v_j in cycle c. k is the length of cycle and d_i, d_j are the degrees of v_i, v_j Then we have the probability of link occurrence of $e_{(i,j)}$ defined in Equation.(6):

$$PR(v_i, v_j) = PR_p(v_i, v_j) \cdot PR_c(v_i, v_j) \qquad (6)$$

Time-Series Link Prediction Model. In this section, we apply time-series model to our probability model for link prediction because time-series models of link occurrences achieve comparable link prediction performance with commonly used static graph link prediction model [9].

As we mentioned before, timing is important for link prediction. Much richer information could be extracted from the frequency time series of the link occurrences, such as the periodic patterns and temporal trends of the graph [9]. For each link $e_{(i,j)}$, assume the probability in time $T+1$ is $PR_{(T+1)}(v_i, v_j)$ according to the period from 1 to T, then we have Equation.(7) by applying autoregressive model [19]:

$$PR_{(T+1)}(v_i, v_j) = \alpha_1 PR_1(v_i, v_j) + ... + \alpha_T PR_T(v_i, v_j) + \varepsilon_t \qquad (7)$$

where the term ε_t is the source of randomness and is called white noise. α_T is the weighting parameter for each time, and we give higher weight to the year closer to the future year.

Further Improvements with Probabilistic Relational Model and DynamicNet Wrap-Up. We have introduced our model based on topological pattern with time-series link prediction model in previous sections. However, like other traditional graph models, the model uses a single graph to model the relationship among the attributes of entities and it is difficult to measure entity attributes over entire graph because the graph is static and limited by predefined threshold.

Though applying time-series can improve the performance in certain degrees, the graph in each period is still static which can not fully reflect the dynamic change of data in the graph. In this section, we involve probabilistic relational model to represent a joint probability distribution over the attributes of relational data.

We take the original data as training graph G_D, and the basic graph in our model with attribute properties we discussed before as model graph G_M. Then we have a new inference graph G_I denoting the corresponding conditional probability distribution (CPD) [20] as shown in Fig. 2.

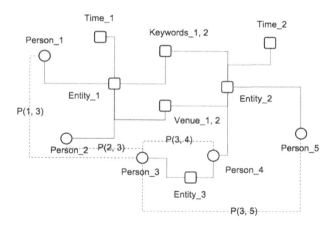

Fig. 2. Sample Inference Graph of DynamicNet

In this sample inference graph, we take Person_3 as example and the link occurrence conditional probability distribution between this person and other persons are shown as $P(1,3)$, $P(2,3)$, $P(3,4)$ and $P(3,5)$. We can then determine if the existence of these links according to their CPD value.

We use Relational Markov Networks (RMN) [20] to generate the inference graph. It uses a set of relational clique templates C to define all the clique. A clique c is a set of nodes V_c in graph G, such that each $V_i, V_j \in V_c$ are connected. Each clique c is associated with a clique potential $\phi_c(V_c)$ to define a joint distribution. So the joint probability for each potential link is computed as a product over the clique templates C and is defined in Equation.(8):

$$P(v_i, v_j) = \frac{1}{Z} \prod_{C_i \in C} \prod_{C_j \in C} \phi_{C_i}(v_i)\phi_{C_j}(v_j) \tag{8}$$

where Z is a normalization constant.

Regards to the clique templates C, $C = (F, W, S)$ consists of three components [20]:

1) $F = F_i$ - a set of entity variables F_i which is of type $E(F_i)$.
2) $W(F.R)$ - a boolean formula using conditions of the form $F_i.R_j = F_k.R_l$.
3) $F.S \subseteq F.X \bigcup F.Y$ - a selected subset of content and label attributes in F.

A clique template specifics a set of cliques in an instantiation L:

$$C(L) = \{c = f.S : f \in L(F) \wedge W(f.r)\} \tag{9}$$

where f is a tuple of entities f_i in which each f_i is type of $E(F_i)$; $L(F) = \{L(E(F_1)) \times ... \times L(E(F_n))\}$ denotes the cross-product of entities in the instantiation; the clause $W(f.r)$ ensures that the entities are related with each other. Therefore, using the log-linear representation of potentials, $\phi_{C_i}(v_c) = exp\{W_C.f_C(V_C)\}$, we have:

$$logP(v_i, v_j) = \Sigma_{C \in \mathbf{C}} w_C.f_C(L.v_i, L.v_j) - logZ(L.v_i, L.v_j) \tag{10}$$

where $f_C(L.v_i, L.v_j) = \Sigma_{c \in C(L)} f_C(L.v_i, L.v_j)$ is the sum over all appearances of the template $C(L)$ in the instantiation and f is the vector of all f_C.

We follow the rules in RMN, and for the input data, graph G_D is actually a single instantiation, where the same parameters are used multiple times and once for each different entity that uses a feature. We select certain period to construct model graph G_M with the probability calculation methods we introduced earlier and take the results as boolean formula for each template to generate inference graph G_I. For every link between two nodes, there is an edge between the labels of these nodes. We use maximum a posteriori (MAP) estimation to avoid overfitting and assume that different parameters are a priori independent during learning stage. More details about RMN learning can be found at [20].

With probabilistic relational model, we can dynamically modelling data over the entire graph from different periods. We wrap-up the whole DynamicNet framework as shown in Fig. 3 to give an overview of this framework. From this diagram we can see after taking the original data graph as input, DynamicNet constructs a new graph with attribute properties and calculate the link occurrence probability according to the edge weighting and topological pattern based probability model components. The performance can be improved by applying time-series link prediction model. The process can stop here, however this graph can be a model graph with probabilistic relational model to produce inference graph for better outcomes. More details are given in the next sections.

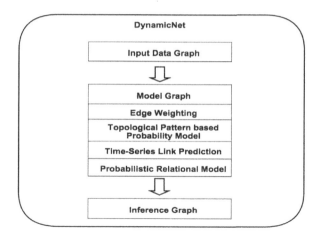

Fig. 3. Overview of DynamicNet Framework

4   Evaluations

In this section, we show our experiments on real data sets to demonstrate the performance of DynamicNet.

4.1   Corpora and Data Preparation

In our experiment, we use three popular corpora: DBLP, Wikipedia and IMDB. For DBLP corpus, we select top three conferences of six categories in computer science according to Microsoft Academic Search website[4], Databases, Data Mining, World Wide Web, Bioinformatics, Multimedia and Information Retrieval. We extract 400 entries for each category from year 2004 to 2009.

For Wikipedia corpus, we extract 1400 Wikipedia pages from 4 different domains, Sports, Entertainment, Science and Geography. We take those contributors who updated the same Wikipedia page as collaborators. For IMDB corpus, we select 800 movies from four types of movies, Action, Comedy, War and Romance, from year 2004 to 2009. We stem select keywords from paper titles, page main description and movie titles using the Lovins stemmer[5]. Table 2 shows the details of our corpora.

4.2   Evaluation Metrics and Baseline Methods

We first select 400 pairs of persons from DBLP, Wikipedia and IMDB each who have cooperated in 2010. 100 pairs are in the case of direct link which means they have worked together before between 2004 and 2009 and 100 pairs are in the

---

[4] http://academic.research.microsoft.com/
[5] http://en.wikipedia.org/wiki/Stemming

Table 2. Data Corpora

	DBLP	Wikipedia	IMDB
Num. of Entries	2400	1400	800
Num. of Categories	6	4	4
Num. of Persons	5388	8760	3021
Num. of Keywords	13462	23370	4380
Num. of Links	38530	92455	21965

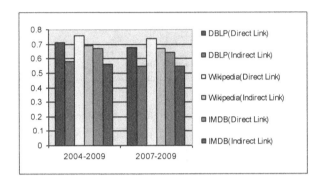

Fig. 4. Results of Topological Pattern based Probability Model

case of indirect link which means they have not worked together between 2004 and 2009. The other 200 pairs are the same context as above but the period is only between 2007 and 2009. The number of data from each category are equal.

We calculate the accuracy as $Accuracy = \frac{N}{100}$, where N is the number of pairs of persons being predicted correctly.

For baseline methods, to prove our framework can achieve the criteria we formulated in Section 3, we select CommonNeighbors and Preferential Attachment [16] as current node based topological pattern based approaches; Katz [11] and SimRank [10] as current path based topological pattern based approaches to compare.

4.3   Evaluation Results

Results of Topological Pattern Based Probability Model. To evaluate the topological pattern based probability model, we set different values to the parameters $\gamma = 0.2$ and $d = 0.5$ in our edge weighting method. As the results are shown in Fig. 4, it is not surprising that the accuracy of direct link is higher than indirect link because it makes sense that two persons should have higher probability to work together in the future if two persons have cooperated before. On the other hand, the results show that the model focuses purely on the static link structure which can not handle very well in the case of indirect link. There is no much difference between different periods.

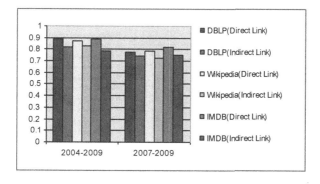

Fig. 5. Results of Time-Series Link Prediction Model

Results of Time-Series Link Prediction Model. In the topological pattern based probability model, the time attribute nodes are being calculated as number of attributes but it did not have the impacts on the results because there are no nodes connected via time attribute nodes.

For time-series link prediction model evaluation, we set $\varepsilon_t = 0$ to simplify the process and give different value to the weighting parameter α_T for different period as shown in Table 3 and make sure $\alpha_1 + \alpha_2 + ... + \alpha_T = 1$. As the results shown in Fig. 5, with time-series link prediction model, the accuracy has been improved a lot especially for the period between 2004 and 2009 because more probability distribution information over whole periods are involved.

Table 3. α_T Value in Time-Series Link Prediction Model

α_T	2004-2009	2007-2009
α_1	0.05	0.1
α_2	0.1	0.3
α_3	0.15	0.6
α_4	0.2	
α_5	0.23	
α_6	0.27	

Results of DynamicNet. In this section, we show the performance of DynamicNet with probabilistic relational model. We use the model graph we discussed before with attribute properties. The model graph was learned based on our corpora and labelled by using our topological pattern based probability model with and without time-series link prediction model (TLP) to get best parameters before used to generated inference graph. From the inference graph, we pick the link with the highest conditional probability distribution as predicted results to verify our framework performance. The results are shown in Fig. 6.

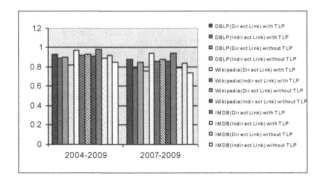

Fig. 6. Results of DynamicNet

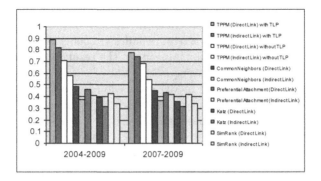

Fig. 7. Comparisons with Baseline Methods on DBLP Corpus

As we can see from the results, DynamicNet can improve accuracy around average 9% to other models because it involves probabilistic relational model so that we can dynamically modelling data over the entire graph from different periods. Additionally, from all three corpora we know that whatever which corpus, the model with time-series link prediction model is higher than the model without time-series link prediction model, and data over more periods can provide more information to the model to produce better outcomes.

Overall Comparisons with Baseline Methods. In this section, we compare the performance of DynamicNet with other baseline methods, CommonNeighbors, Preferential Attachment, Katz and SimRank. Due to the time restriction, we did not implement these methods with time-series link prediction. We first compare them with our topological pattern based probability model (TPPM) with and without time-series link prediction model (TLP). The results are shown in Fig. 7, 8 and 9.

As we can see from the results, our topological pattern based probability model performs better than baseline methods in all cases while we found that most of

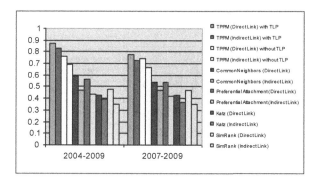

Fig. 8. Comparisons with Baseline Methods on Wikipedia Corpus

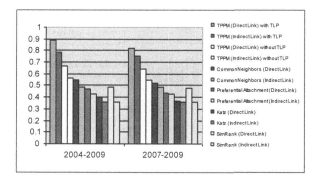

Fig. 9. Comparisons with Baseline Methods on IMDB Corpus

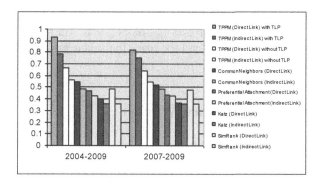

Fig. 10. DynamicNet VS Baseline Methods on DBLP Corpus

baseline methods are not flexible enough to handle the case of indirect link which proves the flexibility of our framework. To more precisely evaluate DynamicNet, we also use the baseline methods to label and learn the model graph to generate inference graph. The results can be found at Fig. 10, 11 and 12.

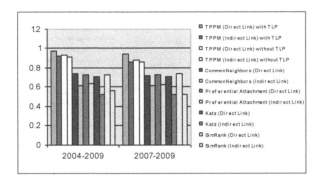

Fig. 11. DynamicNet VS Baseline Methods on Wikipedia Corpus

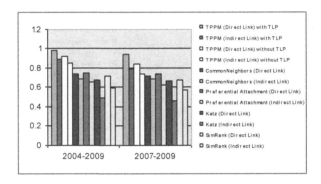

Fig. 12. DynamicNet VS Baseline Methods on IMDB Corpus

As we can see from the results, though the performance of baseline methods improves a lot with probabilistic relational model, it is still far more lower than DynamicNet especially for indirect link situation when taking time-series link prediction model into account.

5   Conclusions and Future Work

In this paper, we propose a hybrid time-series link prediction framework called DynamicNet for large social networks. Our framework not only takes timing as consideration by using time-series link prediction model but also combines the strengths of topological pattern based probability model and probabilistic relational model (PRM). We evaluated our framework on three known corpora with several baseline methods in the case of direct link and indirect link. The experiments indicated that our framework is feasible and flexible.

Next, we will focus on investigating the task of link recommendations, which is a task not only to determine the existence of a link but also to provide ranking for a list of candidate links.

References

1. Adamin, L.A., Adar, E.: Friends and neighbors on the web. Social Networks (2003)
2. Alwan, L.C., Roberts, H.V.: Time-series modeling for statistical process control. J. Bus. Econom. Statist. (1988)
3. Basilico, J., Hofmann, T.: Unifying collaborative and content-based filtering. In: ICML (2004)
4. Bilenko, M., Mooney, R.J.: Adaptive duplicate detection using learnable string similarity measures. In: KDD (2003)
5. Bilgic, M., Namata, G., Getoor, L.: Combining collective classification and link prediction. In: ICDM Workshops (2007)
6. Friedman, N., Getoor, L., Koller, D., Pfeffer, A.: Learning probabilistic relational models. In: IJCAI (1999)
7. Hoff, P., Raftery, A., Handcock, M.S.: Latent space appraches to social network analysis. Journal of the American Statistical Association (2002)
8. Huang, Z.: Link prediction based on graph topology: The predictive value of the generalized clustering coefficient. In: Twelfth ACM SIGKDD International Conference on Knowledge Discovery and Data Mining (2006)
9. Huang, Z., Lin, D.K.J.: Time-series link prediction problem with applications in communication surveillance. INFORMS Journal on Computing (2009)
10. Jeh, G., Widom, J.: Simrank: a measure of structural-context similarity. In: KDD (2002)
11. Katz, L.: A new status index derived from sociometric analysis. Psychometrika (1953)
12. Leroy, V., Cambazoglu, B.B., Bonchi, F.: Cold start link prediction. In: KDD 2010 (2010)
13. Lichtenwalter, R.N., Lussier, J.T., Chawla, N.V.: New prespectives and methods in link prediction. In: KDD 2010 (2010)
14. Martin, S., Roe, D.C., Faulon, J.L.: Predicting protein-protein interactions using signature products. Journal of Bininformatics (2005)
15. Neville, J.: Statistical models and analysis techniques for learning in relational data. PhD thesis (2006)
16. Newman, M.E.J.: The structure and function of complex networks. SIAM Review (2003)
17. Nowell, D.L., Kleinberg, J.M.: The link prediction problem for social networks. In: CIKM 2003 (2003)
18. Nowell, D.L., Kleinberg, J.M.: The link prediction problem for social networks. JASIST (2007)
19. Shumway, R.H.: Applied statistical time series analysis. Prentice Hall, Englewood Cliffs (1988)
20. Taskar, B., Abbeel, P., Koller, D.: Discriminative probabilistic models for relational data. In: UAI (2002)
21. Taskar, B., Wong, M.F., Abbeel, P., Koller, D.: Link prediction in relational data. In: NIPS (2003)

A Comparison of Top-k Temporal Keyword Querying over Versioned Text Collections

Wenyu Huo and Vassilis J. Tsotras

Department of Computer Science and Engineering
University of California, Riverside
Riverside, CA, USA
{whuo,tsotras}@cs.ucr.edu

Abstract. As the web evolves over time, the amount of versioned text collections increases rapidly. Most web search engines will answer a query by ranking all known documents at the (current) time the query is posed. There are applications however (for example customer behavior analysis, crime investigation, etc.) that would need to efficiently query these sources as of some past time, that is, retrieve the results as if the user was posing the query in a past time instant, thus accessing data known as of that time. Ranking and searching over versioned documents considers not only keyword constraints but also the time dimension, most commonly, a time point or time range of interest. In this paper, we deal with top-k query evaluations with both keyword and temporal constraints over versioned textual documents. In addition to considering previous solutions, we propose novel data organization and indexing solutions: the first one partitions data along ranking positions, while the other maintains the full ranking order through the use of a multiversion ordered list. We present an experimental comparison for both time point and time interval constraints. For time-interval constraints, different querying definitions, such as aggregation functions and consistent top-k queries are evaluated. Experimental evaluations on large real world datasets demonstrate the advantages of the newly proposed data organization and indexing approaches.

1 Introduction

Versioned text collections are textual documents that retain multiple versions as time evolves. Numerous such collections are available today and a well-known example is the collaborative authoring environment, such as Wikipedia [1], where textual content is explicitly version-controlled. Similarly, web archiving applications such as the Internet Archive [2] and the European Archive [3] store regular crawls (over time) of web pages on a large scale. Other time-stamped textual information such as, weblogs, micro-blogs, even feeds and tags, as also create versioned text collections.

If a text collection does not retain past documents, then a search query ranks only the documents as of the most current time. If the collection contains versioned documents, a search typically considers each version of a document as a separate document and the ranking is taken over all documents independently to the document's version (creation time). There are applications however, where this approach is not

S.W. Liddle et al. (Eds.): DEXA 2012, Part II, LNCS 7447, pp. 360–374, 2012.

adequate. Consider the following example: in order for a company to analyze consumer comments on a specific product before some event occurred (new product, advertisement campaign etc.), a temporal constraint may be very useful. For example, to view opinions on iphone4, a time-window within 06/07/2010 (announce date) and 10/04/2011 (announce date of iphone4s) could be a fair choice. Many investigation scenarios also require combining the keyword search with a time-window of interest. For example, while considering a financial crime, an investigator may need to identify what information was available to the accused as of a specific time instant in the past.

Providing "as-of" queries is a challenging problem. First is the data volume. Document collections like Wikipedia and Internet Archive, are already huge even if only their most recent snapshot is considered. When searching in their evolutionary history, we are faced with even larger data volumes. Moreover, how to quickly return the top-k temporally ranked candidates is another new challenge. Note that returning all qualified results without temporal constraints would not be efficient since two extra steps are required: (i) filtering out results later than the query specified time constraint, and, (ii) ranking the remaining results so as to provide the top-k answers.

In this paper we present an experimental evaluation of the top-k query over versioned text collections, comparing previously proposed as well novel approaches. In particular the key contributions can be summarized as:

1. Previous methods related to versioned text keyword search are suitably extended for top-k temporal queries.
2. Novel approaches are proposed in order to accelerate top-k temporal queries. The first approach partitions the temporal data based on their ranking positions, while the other maintains the full rank order using a multiversion ordered list.
3. In addition to top-k time-point keyword based search, we also consider two time-interval (or time-range) variants, namely "aggregation ranking" and "consistent" top-k querying.
4. Experimental evaluations with large-scale real-world datasets are performed on both the previous and newly proposed methods.

The rest of the paper is organized as follows. Preliminaries and related work are introduced in section 2. Our novel approaches appear in section 3. Different query definitions of time-interval top-k queries are presented in section 4. All techniques are comprehensively evaluated and compared in a series of experiments in section 5 while the conclusions appear in section 6.

2   Preliminaries and Related Work

2.1   Definitions

The data model for versioned document collections was formally introduced in [10], and used by later works [9, 5, 6]. Let D be a set of n documents $d_1,d_2,...,d_n$ where each document d_i is a sequence of m_i versions: $d_i = \{d_i^1, d_i^2, ..., d_i^{m_i}\}$. Each version has a semi-closed validity time-interval (or lifespan) $life(d_i^j) = [t_s, t_e)$. Moreover, it is

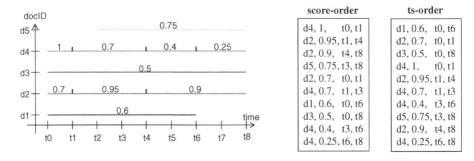

Fig. 1. Example of versioned documents with scores for one term

assumed that different versions of the same document   have disjoint life spans. An example of five documents and their versions appears in Fig. 1; each document corresponds to a colored line, while segments represent different versions of a document.

The inverted file index is the standard technique of text indexing for keyword queries, deployed in many search engines. Assuming a vocabulary V, for each term v in V, the index contains an inverted list L_v consisting of postings of the form (d, s) where d is a document-identifier and s is the so-called payload score. There are numerous existing relevance scoring functions, such as tf-idf [7], language models [13] and Okapi BM25 [14]. The actual scoring function is not important for our purposes; for simplicity we assume that the payload score contains the term frequency of v in d.

In order to support temporal queries, the inverted file index must also contain temporal information. Thus [10] proposed adding the temporal lifespan explicitly in the index postings. Each posting includes the validity time-interval of the corresponding document version: (d_i, s, t_s, t_e) where the document d_i had payload score s during the time interval $[t_s, t_e)$.

If the document evolution contains few changes over time, the associated score of most terms is unchanged between adjacent versions. In order to reduce the number of postings in an index list, [10] coalesces temporally adjacent postings belonging to the same document that have identical (or approximate identical) scores.

A general keyword search query Q consists of a set of x terms $q = (v_1, v_2, \ldots, v_x)$ and a temporal interval $[lb, rb]$. Without loss of generality, we use the aggregated score of a document version for keyword query q is the sum of the scores from each term v. The time-interval $[lb, rb]$ restricts the candidate document versions as a subset of the original collection: $D^{[lb, rb]} = \{d_i^j \in D \mid [lb, rb] \cap life(d_i^j) \neq \varnothing\}$. When $lb = rb$ holds, the query time interval collapses into a single time point t. For simplicity we first concentrate on time-point query and more complex time-interval queries are discussed in section 4 with related variations.

The answer R to a Top-K Time-Point keyword query **TKTP** $= (q, t, k)$ over collection D is a set of k document versions satisfying: $\{d_i^j \in R \mid (\exists v \in q : v \in d_i^j) \wedge (d_i^j \in D^t) \wedge$

$(\forall d' \in (D^t - R) : s(d_i^j) \geq s(d'))\}$ where $D^t = \{d_i^j \in D \mid t \in life(d_i^j)\}$. The first condition presents the keyword constraint, the second condition the temporal constraint, while the third implies that the top-k scored document versions are returned. Now we present how to answer query TKTP using previous methods based on temporal inverted indexes.

2.2 Previous Methods

The straightforward way (referred to as **basic**) to solve query TKTP uses exactly one inverted list for each vocabulary term v with the posting (d_i, s, t_s, t_e). To answer the top-k queries, corresponding inverted lists are traversed and postings are fetched. When a posting is scanned, it is also verified for the time point specified in TKTP.

The sort-order of the index lists is also important. One natural choice is to sort each list in score order. This method (score-order) enables the classical top-k algorithms [11] to stop early after having identified the k highest scores with qualified lifespan. Another suitable sorting choice is to order the lists first by the start time t_s and then by score (t_s-order) which is beneficial for checking the temporal constraint. However, this approach is not efficient for top-k querying, especially when the query includes multiple terms. Fig. 1 shows the score-order and ts-order lists for a specific term.

Note that the efficiency of processing a top-k temporal query is influenced adversely by the wasted I/O due to read but skipped postings. We proceed with various materialization ideas of the slice the whole list of a term into several sub-lists or partitions thus improving processing costs.

Interval Based Slicing splits each term list along the time-axis into several sub-lists, each of which corresponds to a contiguous sub-interval of the time spanned by the full list. Each of these sub-lists contains all coalesced postings that overlap with the corresponding time interval. Note that index entries whose validity time-interval spans across the slicing boundaries are replicated in each of the spanned sub-lists.

The selection of the corresponding time-intervals where the slices are created is vital as discussed in [9, 5]. One obvious strategy is to eagerly slice sub-lists for all possible time instants (and adjacent identical lists can be merged). This will create one sub-list per time instant; this will provide ideal query performance for a TKTP query since only the postings in the sub-list for the query time point will be accesses. We refer to this method as **elementary**.

Note that the **basic** and **elementary** methods are two extremes: the former requires minimal space but requires more processing at query time since many entries irrelevant to the temporal constraint are accessed; the latter provides the best possible performance (for time-point query) but is not space-efficient (due to copying of entries among sub lists). To explore the trade-off between space and performance, [5] employs a simple but practical approach (referred to as **Fix**) in which a partition boundary is placed after a fixed time window. The window size can be a week, a month, a year, or other flexible choices. Fig. 2 shows the Fix-2 and Fix-4 sub-lists of our running example from Fig. 1, with the partition time window size as 2 and 4 time instants respectively. Nevertheless, all variations of the interval based slicing suffer from an index-size blowup since entries whose valid-time interval spans across the slicing boundaries are replicated.

Fix-2:

[t0, t2):	[t2, t4):	[t4, t6):	[t6, t8):
d4, 1, t0, t1	d2, 0.95, t1, t4	d2, 0.9, t4, t8	d2, 0.9, t4, t8
d2, 0.95, t1, t4	d5, 0.75, t3, t8	d5, 0.75, t3, t8	d5, 0.75, t3, t8
d2, 0.7, t0, t1	d4, 0.7, t1, t3	d1, 0.6, t0, t6	d3, 0.5, t0, t8
d4, 0.7, t1, t3	d1, 0.6, t0, t6	d3, 0.5, t0, t8	d4, 0.25, t6, t8
d1, 0.6, t0, t6	d3, 0.5, t0, t8	d4, 0.4, t3, t6	
d3, 0.5, t0, t8	d4, 0.4, t3, t6		

Fix-4:

[t0, t4):	[t4, t8):
d4, 1, t0, t1	d2, 0.9, t4, t8
d2, 0.95, t1, t4	d5, 0.75, t3, t8
d5, 0.75, t3, t8	d1, 0.6, t0, t6
d2, 0.7, t0, t1	d3, 0.5, t0, t8
d4, 0.7, t1, t3	d4, 0.4, t3, t6
d1, 0.6, t0, t6	d4, 0.25, t6, t8
d3, 0.5, t0, t8	
d4, 0.4, t3, t6	

Fig. 2. Time Interval Based Slicing sub-list examples

Stencil Based Partitioning. Another index partitioning method along the time-axis was proposed in [12]. It is distinguished from the interval based slicing by using a multi-level hierarchical (vertical) partitioning of the lifespan. The inverted list of term v, at level L_0 contains the entire lifespan of this list, while level L_{i+1} is obtained from L_i by partitioning each interval in L_i into b sub-intervals. Such a partitioning is called a **stencil**; each index posting is placed into the deepest interval in the multi-level partitioning that fits its range. A stencil-based partition of three levels with $b = 2$ for the running example (from Fig. 1) is shown in Fig. 3.

Comparing to the time interval based slicing, the stencil based partitioning has significant advantage in space because each posting falls into a single list, the deepest sub-interval that it fits. Nevertheless, for a time-point query stencil based partitioning has to fetch multiple sub-lists, one from each level.

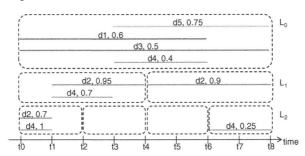

Fig. 3. Stencil-based partitioning with 3 levels and $b = 2$

The sort-order of each sub-list is again important. Since the temporal partitioning already shreds one full list into several sub-lists along the time-axis, a more appropriate choice for top-k queries is score-ordering.

Temporal Sharding. The approach proposed in [6] is to **shard** (or horizontally partition) each term list along the document identifiers instead of time. Entries in a term list are thus distributed over disjoint sub-lists called shards, and entries in a shard are ordered according to their start times t_s. So as to eliminate wasteful reads, within a shard g_i, entries satisfy a staircase property: $\forall p, q \in g_i, ts(p) \leq ts(q) \Rightarrow te(p) \leq te(q)$.

An optimal greedy algorithm for creating this partitioning is given in [6]; an example of temporal sharding for the term list from Fig. 1, is shown in Fig. 4.

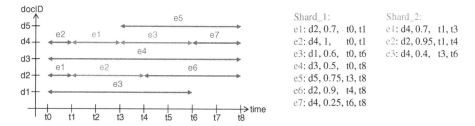

Fig. 4. Temporal sharding example

As with the stencil based approach, the space usage for temporal sharding is optimal since there are no replications of index entries. However, for query processing, all shards for each term need to be accessed, resulting in multiple sub-list readings. Moreover, the entries in each shard can only be time-ordered (based on start time t_s). Thus the benefit of score-ordering for ranked queries cannot be achieved, because all temporal valid entries have to be fetched.

3   Novel Approaches

A common characteristic of existing works is that they only consider the versioned documents on the time- and docID-axes, and try to partition the data along either direction. Instead, we view the index entries from a new angle -- namely, their score over time, and create index organizations to improve the performance of top-k querying. The *score-time* view of the example from Fig. 1 is shown in Fig. 5.

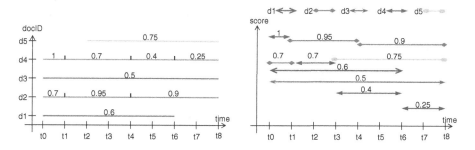

Fig. 5. The Score-Time view of the versioned documents

Recall that to answer a TKTP query we should be able to quickly find the top-k scores of a term at a given time instant. The main idea behind the score-time view is to maintain an index that will provide the top scores per term at each time instant.   For example, at time t_0, the term depicted in Fig.5 had scores 1 (from d_4), 0.7 (from d_2), 0.6 (from d_1) and 0.5 (from d_3). These orderings change as time proceeds; for example at time t_2, the top score is 0.95 from d_2, etc. In *rank-based* partitioning (section 3.1), we first discuss a simplistic approach (SPR) where an index is created for each rank position of a term. For example, there is an index that maintains the top score over time, then one for the second top score, etc. More practical is the group ranking approach (GR) where an index is created to maintain the group of the top-g scores (g is a constant), then the next top-g etc. We also consider temporal indexing methods (section 3.2). One solution is to use the Multiversion B-tree and maintain the whole ranked list in order over time. We realize however that these ranked lists are always accessed in order, so a better solution is provided (multiversion list) that links appropriately the data pages of the temporal index, without overhead of the index nodes.

3.1   Rank Based Partitioning

The **Single Position Ranking (SPR)** approach creates a separate temporal index for each ranking position of a term. Thus, for the i-th ranking position ($i = 1,2,...$), a sublist is maintained that contains all the entries that ever existed on position i over time. Together with each entry we maintain the time interval during which this entry occupied that position. All sub-list entries are ordered based on their recorded starting time; a B+tree built on the start times can easily locate the appropriate entry at a given time. The SPR of our running example (from Fig. 1) is shown in Fig. 6(a). Space can be saved by using only the start time of each entry but for simplicity we show the end times as well (the end time is needed only if there is no entry in a particular position, but this is true only at the last position).

Using the SPR approach, to process a TKTP query about time t, the first k sub-lists have to be accessed for each relevant term; from each sublist the B+tree will provide the appropriate score (and document id) of this term at time t. If each sub-list has m items on average, the estimated time complexity is $O(k \cdot log_B m)$ (here B corresponds to the page size in records). Many sub-list accesses can degrade querying performance; moreover, in this simple SPR method the same posting can be duplicated in multiple ranking position sub-lists. This unavoidable replication may result in storage overhead.

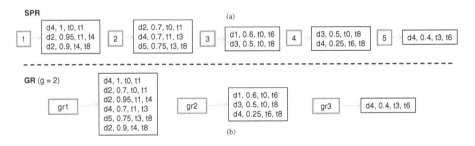

Fig. 6. Ranking position based partitioning

Group Ranking (GR). In order to save space and improve querying performance, GR maintains an index not for a single ranking position, but for a group of positions. Let the group size be g. For example, the first g ranked elements are in group gr_1, the next g ranked elements are in group gr_2, etc. Thus, compared to the n sub-lists maintained in SPR for n ranking positions, GR uses instead n/g sub-lists. With respect to the I/O of top-k querying, we only need k/g random accesses (each of them still logarithmic).

As with the SPR each member within a group also records the time interval that the member was in the group. For example, assume that group gr_i maintains ranking positions $(i-1)g+1$ through ig. If at time t_s the score of a particular term falls within these positions, this score is added to the group, with an interval starting at t_s. As long as this score falls within the ranking positions of this group, it is considered part of the group; if at time t_e it falls out of the group, the end time of its interval is updated to t_e.

To save on update time, within each group we do not maintain the rank order. That is, each group is treated as an unordered set of scores that evolves over time. To answer a TKTP query that involves a particular group gr at time t, we need to identify what members group gr had at time t. Since the size of the group is fixed, we can easily sort these member scores and provide them to the TKTP result in rank order. However, it is guaranteed that given time t, the members in gr_i have no lower scores than those in group gr_j where $1 \leq i < j \leq (n/g)$. The GR approach for the above example (from Fig. 1) with $g = 2$ is shown in Fig. 6(b).

An interesting question is what index to employ for maintaining each group over time. Different than SPR, each group at a given time may contain multiple entries; thus a B+index on the temporal start times is not enough. Instead, temporal index structures that maintain and reconstruct efficiently an evolving set over time, like the snapshot index [15] can be used to accelerate temporal querying.

Note that when implementing GR in practice, each group may have a different size g. It is preferable to use smaller g for the top groups and larger g for the lower groups (since the focus is on top-k, the few top groups will be accessed more frequently and thus we prefer to give faster access). For simplicity however, we use the same g for all groups.

3.2   Using a Multiversion List

Consider the ordered list of scores that a term has over all documents at time t; as time evolves, this list changes (new scores are added, scores are promoted, demoted or even removed, etc). Temporal indexing methods have addressed a more general problem: how to maintain an evolving set of keys over time. This set is allowed to change by adding, deleting or updating keys; the main temporal query supported is the so called: temporal-range query: "given t, provide the keys that were in the set at time t, and are within key range r". The Multiversion B-tree (**MVBT**) proposed in [8], is an asymptotically optimal (in terms of I/O accesses under linear space) solution to the temporal range query. Assuming that there were a total of n changes that occurred in the set evolution, then the MVBT uses linear space ($O(n/B)$). Consider a range temporal query that specifies range r and time t, and let a_t denote the number of keys that were within range r at time t (i.e., the number of keys that satisfy the query); the MVBT answers the above query using $O(log_B n + a_t/B)$ page I/Os, which is optimal in linear space [8].

In order for the MVBT to maintain order among the keys, it uses a B+tree to index the set. As the set evolves, so does the B+-tree. Conceptually the MVBT contains all B+-trees over time; for a given query time t the MVBT provides access to the root of the appropriate B+-tree, etc. Of course, the MVBT does not copy all B+-trees (as this would result in quadratic space). Instead it uses clever page update policies.

In particular, when a key k is added to the evolving set at time t a record is inserted in the (leaf) data page whose range contains k; this record stores key k and a time interval of the form: $[t, *)$. The '*' denotes that key k has not been updated yet. If later at time t' this key is removed from the set, its record is not physically deleted. Instead this change is represented by changing the '*' to t' in this record's interval. A record is called "alive" for all time instants in its interval. Given a query about time t, the MVBT tree identifies all data pages that contain alive records for that time t. In contrast to a regular B+-tree that deals with pages that get underutilized due to record deletions, the MVBT pages cannot get underutilized because no record is ever deleted. Like the B+-tree pages can get full of records and need to be split (*page overflow*). However, the MVBT needs to also guarantee that the number of "alive' records in a page do not fall below a lower threshold l (*weak version underflow*) and also do not go over an upper threshold u (*strong version overflow*)- note that l and u are $O(B)$. If a page overflows, a time-split occurs, that copies the alive records of the overflown page (at the time of the overflow) to a new page. If there are too few alive records, the page is merged with a sibling page that is also first time-split. If there are too many alive records, a key split is first applied (among the alive records) [8].

Using the MVBT for our purposes means that the scores play the role of "keys". That is, the MVBT will maintain the order of scores over time. Since however term records are accessed by the docID they belong to, a hashing index is also needed that, for a given docID, it provides the leaf page that holds the record with this term's current score. This hashing scheme need only maintain the most current scores (i.e., it does not need to maintain past positions).

Nevertheless, the above MVBT approach has a significant overhead. In particular, it is built to answer queries about any range of scores. This is achieved by starting from an appropriate root of the MVBT and follow index nodes until the leaf data pages in the query range are accessed. For top-k processing however, we only access scores in decreasing order, starting with the largest score at a particular time instant. Thus, what we actually need, is a way to access the leaf page that has the highest scores at a particular time, and then follow to its sibling leaf page (with the next lower scores) at that time, etc. We still maintain the split policies among the leaf pages, but we do not use the MVBT's index nodes. Effectively we maintain a **multiversion list** (**MList**), i.e., of the leaf data pages over time.

To access the leaf data page that has the highest scores at a given time, we maintain an array A with records of the form (t, p) where t is a time instant and p is a pointer to the leaf page with the highest scores at time t. If later at time t' another page p' becomes the leaf page with the highest scores, array A is updated with a record (t', p'). If this array becomes too large for main memory, it can easily be indexed by a B+-tree on the (ordered) time attribute.

For the above "list of leaf pages" idea to work, each leaf page needs to "remember" the next sibling leaf page (with lower scores) at each time. (Note: the MVBT does not require the sibling pointers, since access to siblings is done through the parent index nodes). One could still use the array approach (one array responsible to keep access to the second leaf page, one for the third etc.) but this would require many array lookups at query time (each such lookup taking $O(log_B n)$ page I/Os. Instead, we propose to embed these arrays within the page structure. That is, within each leaf page, we allocate a space of c records (where c is a constant) for the sibling page pointer records (also of the form (t,p)). As a result, each leaf page has now space for B-c score records. Since however, the sibling page can change over time, it is possible that for a leaf page p the sibling will change more than c times. If this happens at time t, page p is "time split", that is, a new leaf page p' is created containing only the currently alive records of page p and with an empty array for sibling pointers. Moreover, p' replaces p in the list. If before t, the list of leaf pages contained pages (in that order) $m \rightarrow p \rightarrow v$, a new record (t, p') is added in the array of page m, and the array of page p' is initialized with a record (t,v). If p was the first page, the record (t,p') is added to array A.

The advantage of the **Mlist** approach is apparent at query processing time. A search is first performed within array A for time t (in $O(log_B n)$ page I/Os). This will provide access to the page with the highest scores at time t. Find the next sibling page at time t however will be provided by looking among the c records of this page, etc. That is, the top-k scores at time t will be accessed in $O(log_B n + k/B)$ page I/Os. The justification is that after the access to array A, each leaf page (except possibly the last one) will provide $O(B)$ of the top-k scores (since we are using the MVBT splitting policies within the B-c space of each leaf page and c is a constant, each page is guaranteed to provide at least $l=O(B)$ scores that were valid at the query time t.

4   Top-k Time Interval Queries

Until now we focused on the top-k time point (TKTP) querying, and analyzed different index structures for solving it. We proceed with the time interval top-k query. The main difference is that in the TKTP, each document has at most one valid version at the given time point t; while for an interval querying, each document may have multiple versions valid during the given time interval $[lb, rb]$. As a result, there are different variations, depending on how the top-k is defined (which of the valid scores per document participate in the top-k computation). Here, we summarize the different definitions of top-k time-interval queries and discuss how to process them efficiently within the proposed index structures.

4.1   Classic Top-k Time-Interval Query

This query definition is a straight forward extension from the top-k time point query. For a Top-K Time Interval keyword query **TKTI** = (q, lb, rb, k) over collection D, we require the answer R be a set of k document versions satisfying: $\{d_i^j \in R \mid (\exists v \in q : v \in d_i^j)$

$\wedge (d_i^j \in D^{[lb,rb]}) \wedge (\forall d' \in (D^{[lb,rb]} - R) : s(d_i^j) \geq s(d'))\}$ where $D^{[lb,rb]} = \{d_i^j \in D \mid [lb,rb] \cap life(d_i^j) \neq \varnothing\}$. This definition only changes the time constraints from a time point t to a time range $[lb, rb]$. The returned top-k answers are different versions, which may be from the same document, that is, we consider each document version as an independent object.

Processing a TKTI query is similar to processing a TKTP query. For some of the described index methods, multiple sub-lists have to be accessed instead of one.

For example in time interval based slicing and stencil based partitioning, all the sub-lists (or stencils) overlapping with the query time-interval should be checked in order to find the correct top-k results. The multiple parallel sub-lists can be accessed in a round-robin fashion which is compatible with top-k algorithms.

4.2   Document Aggregated Top-k Time-Interval Query

Another possibility is to treat each document as one object, that is, a document appears at most once in the result. There are various approaches in aggregating relevance scores of the document versions that existed at any point in the temporal constraint $[lb, rb]$ to obtain a document relevance score $drs(d_i, lb, rb)$. Three aggregation relevance models are mentioned in [10]:

MIN. This model judges the relevance of a document based on the minimum score. It is formally defined as: $drs(d_i, lb, rb) = \min\{s(d_i^j) \mid [lb, rb] \cap life(d_i^j) \neq \varnothing\}$. The MIN scores of our five-document example for interval $[t_0, t_8)$ are $d_2=0.7$, $d_3=0.5$, $d_4=0.25$, $d_1=0$, $d_5=0$.

MAX. In contrast, this model takes the maximum score as an indicator. It is formally defined as: $drs(d_i, lb, rb) = \max\{s(d_i^j) \mid [lb, rb] \cap life(d_i^j) \neq \varnothing\}$. MAX scores of our five-document example for interval $[t_0, t_8)$ are $d_4=1$, $d_2=0.95$, $d_5=0.75$, $d_1=0.6$, $d_3=0.5$.

TAVG. Finally, the TAVG model assigns the score to each document using a temporal average among all its valid versions. Since score $s(d_i^j)$ is piecewise-constant in time, $drs(d_i, lb, rb)$ can be efficiently computed as a weighted summation of these segments. TAVG scores of our five-document example for interval $[t_0, t_8)$ are $d_2=0.89$, $d_4=0.51$, $d_3=0.5$, $d_5=0.47$, $d_1=0.45$.

After the aggregation mechanism has been defined, one can consider the Aggregated Top-K Time-Interval keyword query $\mathbf{TKTI}^A = (q, lb, rb, k)$ over collection D, that finds the top k documents with aggregated scores over all their valid document versions. To process the aggregated top-k time-interval query, we need to extend the traditional top-k algorithms (such as TA and NRA) by recording the bookkeeping information and computing the scores and thresholds with candidates at document-level. The relevance score of a document in the query temporal-context depends on the scores of its version that are valid during this period.

4.3   Consistent Top-k Time Range Query

The consistent top-k search finds a set of documents that are consistently in the top-k results of a query throughout a given time interval. The result of this query has size 0 to k; queries can have empty results if k is small or the rankings change drastically. A relaxing consistent top-k query utilizes a relax factor r, $0 < r <= 1$, and seeks for documents that are in the top-k for at least $r \times (rb - lb)$ time in the $[lb, rb]$ interval. For a Consistent Top-K Time Interval keyword query $\mathbf{TKTI}^C = (q, lb, rb, k)$ over collection D, the documents in the answer R are in the top-k for at least $r \times (rb - lb)$ time in the $[lb, rb]$ interval. The consistent top-3 query of our five-doc example for time-interval $[t_0, t_8)$ has only one result as d_2 if $r = 1$, and has three results as d_1, d_2 and d_5 if $r = 0.6$.

In [16] several algorithms were introduced to answer the consistent top-k query; the most efficient ones are based on the assumption that there is a list containing all versions satisfying the keyword and time interval constraints and the list is ordered by score. This assumption coincides with the purpose of our proposed index structures, thus we can access the qualified entries and execute the consistent top-k time interval query using the proposed approaches in [16].

5   Experimental Evaluations

Dataset Description and Methods Implemented: We used news-like articles as our primary versioned document collection. We collected US and world-wide English newspaper websites and treated each URL as a single document. Then their historical homepage versions were retrieved by crawling the Internet Archive [2] from 1997.1.1 until 2011.12.31. We created two different datasets with daily unit time granularity. The US based news had many frequent updates. The size of raw data is about 0.2 TB, with 12,649 documents and 1,542,893 versions; thus on average there are 122 versions per document in the US dataset. For the world-wide news websites, the size of raw data is about 50 GB, with 5,046 documents and 275,981 versions, so on average there are 55 versions per document. Previous related works create query workloads by extracting frequent queries from the AOL query logs. In addition to this traditional query workload, we use popular keywords (such as "twitter", "iphone", "lady gaga" etc.) from the Google Zeitgeist [4] annual reports from 2001 until 2011. Overall, we formed 200 queries with 265 terms for both classic and popular keywords.

We organize the data into term inverted list(s) using the previous and novel approaches. In the basic method with score-ordering (referred to as **Basic-s**) we create one inverted list per term. The second method is elementary time-interval slicing with a merging of adjacent identical sub-lists (**Ele**). For the Fix approach we used a time-window length of 30 days (**Fix-30**). The stencil based partitioning was implemented with 3 levels and $b = 4$ (**Stencil**). Temporal sharding is referred as **Shard**, while the single position ranking model appears as **SPR**. Two group ranking methods were implemented with group sizes of 25 and 50 (**GR-25** and **GR-50**). For comparison purposes we also included the **MVBT** index (with the appropriate hashing secondary index).The multiversion list approach (**MList**) uses a factor $a = c / B$ to present the ratio of the number of pointer records to the number of all records in a page.

Comparison Results: First, the space usage for all implemented methods on both the US-news and World-news datasets is presented in Table 1. The page size is 4 Kbytes while $B = 100$ records. The table presents the space consumed (in GB) to implement the index methods for the 256 terms used in our experiments. Clearly, the elementary time-interval slicing has a huge space overhead while the Stencil and Shard methods present substantial space savings. As expected, the Basic-s approach has the minimal space requirements. Fix-30 uses more space since a record may appear in more partitions while in Stencil and Shard, each record appears once. The additional space that Stencil and Shard use wrt Basic-s is due to the additional structures they utilize. Among the rank-based partitioning methods, SPR uses more space than the GR approaches; this is because the SPR approach has to maintain one index per ranked position. GR-25 uses more space than GR-50 since it uses more index structures (one per group). For the MList method, we show the results of $a = 7\%$ and $a = 10\%$ (referred to MList-7 and MList-10). The MList approaches also use linear space (but due to the copying of records at page splits, the space is more than the Stencil and Shard approaches). MList uses slightly more space than the MVBT because of the use of sibling pointers and the splits they create.

Table 1. The space usage (in GB) for the 256 terms used in the queries

Methods	Basic-s	Ele	Fix-30	Stencil	Shard	SPR	GR-25	GR-50	MVBT	MList-7	MList-10
US	1.93	213.34	4.26	2.24	2.31	6.65	6.12	5.93	3.79	4.02	3.95
World	0.35	38.6	0.78	0.41	0.43	1.21	1.12	1.06	0.69	0.75	0.78

The top-k temporal queries include both time-point (in our dataset this corresponds to one day) and time-interval queries. For each temporal keyword query, we randomly choose 50 time constraints from the 15-year lifespan from 1997 to 2011, and record the average performance. For TKTI[A], we use TAVG scoring; for TKTI[C], we use $r = 1$. The page I/O costs for top-20 queries using the US-news dataset are shown in Table 2 (the best performance for each query is shown in bold). For time interval queries, the time-interval lengths used were 15 days, 30 days, and 60 days. We also present the I/O costs for top-100 queries on both US-news and World-news datasets in Table 3 for both time-point query and 30-day time-interval queries.

Table 2. The page I/O cost of top-20 temporal keyword queries for US news

Methods	Basic-s	Ele	Fix-30	Stencil	Shard	SPR	GR-25	GR-50	MVBT	MList-7	MList-10
TKTP	49.16	3.74	7.44	11.16	87.5	33.24	6.26	7.8	5.34	5.58	5.02
TKTI-15	65.32	56.22	13.9	16.5	90.64	40.74	14.92	17.26	11.46	11.7	11.18
TKTI-30	81.76	108.7	16.48	20.42	93.58	45.9	19.32	22.88	15.74	16.22	15.28
TKTI-60	105.6	195.82	31.26	35.8	95.22	49.66	23.06	26.14	22.38	22.9	21.7
TKTIA-15	74.16	67.84	20.8	24.12	96.54	48.38	20.42	22.8	18.68	19.54	16.92
TKTIA-30	89.84	126.4	23.18	27.84	98.3	50.1	25.78	26.2	21.9	23.84	21.42
TKTIA-60	112.96	209.56	41.06	46.76	103.86	60.22	31.14	33.84	30.32	30.82	29.68
TKTIC-15	68.48	60.6	17.42	19.48	92.82	44.34	16.68	19.12	14.04	14.58	13.74
TKTIC-30	83.52	110.58	19.5	22.38	96.04	47.48	21.5	24.04	18.18	20.36	17.44
TKTIC-60	108.34	201.42	35.74	39.22	98.72	53.82	26.7	27.98	25.6	26.24	24.18

Table 3. The page I/O cost of top-100 temporal keyword queries for US and World news

US	Basic-s	Ele	Fix-30	Stencil	Shard	SPR	GR-25	GR-50	MVBT	MList-7	MList-10
TKTP	93.4	10.14	25.74	38.7	102.68	162.4	29.64	21.18	20.72	21.84	19.12
TKTI-30	157.84	315.3	48.62	70.22	114.2	233.94	92.82	62.94	46.92	49.38	46.24
TKTIA-30	171.8	336.44	53.5	79.18	118.24	241.48	115.74	75.32	52.1	55.92	51.48
TKTIC-30	163.52	324.86	50.26	73.42	115.7	236.5	101.36	67.28	49.06	52.06	48.2
World	Basic-s	Ele	Fix-30	Stencil	Shard	SPR	GR-25	GR-50	MVBT	MList-7	MList-10
TKTP	85.44	9.96	23.36	37.52	98.8	156.44	27.5	20.84	20.12	18.22	18.84
TKTI-30	143.32	306.58	45.74	69.62	110.28	228.36	83.34	54.62	45.32	44.78	45.1
TKTIA-30	152.7	322.36	50.82	77.84	115.66	237.02	103.6	70.7	51.16	49.82	50.56
TKTIC-30	147.24	311.92	47.78	72.16	112.72	231.84	91.76	62.58	47.24	46.3	46.82

The elementary time-interval slicing has the best snapshot querying performance for both top-20 and top-100 queries. This is to be expected since the answer is basically prepared for each time instant (at the cost of huge storage requirements). Among the other methods, the newly proposed approaches (GR, MList) outperform the previous methods (Stencil and Shard). The best performance is provided by the MList-10 method. It has better performance than the MVBT given it accesses the answer faster (by avoiding the MVBT index traversal). Considering its low space requirements, this approach provides the overall best performance for TKTP queries.

For time-interval queries, the Ele method's performance degrades drastically, especially for longer time-interval. The group ranking method's performance is related to its group size g as it relates to k. For top-20 querying, a group size of 25 works better than a group size of 50 (the answer can be found by accessing the first group only); while for top-100 querying, GR-50 is a better choice (only two groups need to be accessed instead of four for GR-25, thus less index accesses). For top-20 interval queries, the MList-10 had consistently the best performance for each query.

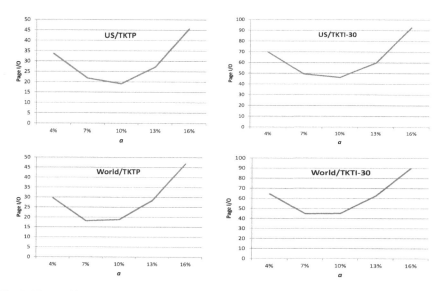

Fig. 7. The Multiversion list method for different ratio a using the US and World news datasets

Interestingly, for the top-100 interval queries the MList-7 shows better performance for the World-news dataset. The reason for that is that this dataset has fewer updates. As a result, there will be fewer pointer changes in the ordered list, thus a smaller a will provide enough space to hold the pointer structure. This can also be seen in the space requirements for this dataset: the fewer pointer splits mean that MList-7 uses less space than MList-10 (and thus the lists are shorter and the query performance better). The above observation implies that the performance of the multiversion list method is related to the value of a. There are two opposing factors affecting the query performance with respect to a. For a given page size, a small a implies that the area allocated to sibling pointers is small; thus few sibling page changes can cause the page to split. More splits use more space and the query time increases. On the other hand, a large a implies that the space allocated for the regular records in a page is small, thus the page can split faster due to the record updates. This also increases space and query time. The optimized value of a depends on the dataset characteristics. Figure 7 depicts the page I/O for the top-100 results returned by point

(TKTP) and interval (TKTI-30) queries for the US and World-news datasets. For the US-news dataset, $a = 10\%$ has the best average performance for both time-point querying (TKTP) and 30-day time-interval querying (TKTI-30) while for the World-news dataset (which has less update frequency), the performance is optimized for $a = 7\%$.

6   Conclusion

We presented an experimental comparison of indexing methods over versioned text collections for top-k temporal keyword queries. In addition to previous methods, we proposed novel solutions that partition the data along the score-time axes. Among all methods, the multiversion list provided the most robust performance considering space usage and query time efficiency for both time-point and time-interval queries. We examined variations of the time-interval queries, including the document-level aggregated top-k queries and consistent top-k queries. The performance of the multiversion list is affected by the value of a, the percentage of a data page allocated to hold sibling pointers. As future work, we plan to devise a model that can optimize the value of a based on the frequency of updates, the size of the page and other factors.

Acknowledgements. This work is partially supported by NSF grant IIS-0910859.

References

[1] Wikipedia, http://en.wikipedia.org/
[2] Internet Archive, http://www.archive.org/
[3] European Archive, http://www.europarchive.org/
[4] Google Zeitgeist, http://www.google.com/zeitgeist/
[5] Anand, A., Bedathur, S., Berberich, K., Schenkel, R.: Efficient Temporal Keyword Queries over Versioned Text. In: CIKM (2010)
[6] Anand, A., Bedathur, S., Berberich, K., Schenkel, R.: Temporal Index Sharding for Space-Time Efficiency in Archive Search. In: SIGIR (2011)
[7] Baeza-Yates, R., Ribeiro-Neto, B.: Modern Information Retrieval. Addison-Wesley (1999)
[8] Becker, B., Gschwind, S., Ohler, T., Seeger, B., Widmayer, P.: An asymptotically optimal multiversion B-tree. VLDB Journal (1996)
[9] Berberich, K., Bedathur, S., Neumann, T., Weikum, G.: A Time Machine for Text Search. In: SIGIR (2007)
[10] Berberich, K., Bedathur, S., Weikum, G.: Efficient Time-Travel on Versioned Text Collections. In: BTW (2007)
[11] Fagin, R., Lotem, A., Naor, M.: Optimal Aggregation Algorithms for Middleware. J. Comput. Syst. Sci. 66(4), 614–656 (2003)
[12] He, J., Suel, T.: Faster Temporal Range Queries over Versioned Text. In: SIGIR (2011)
[13] Ponte, J.M., Croft, W.B.: A Language Modeling Approach to Information Retrieval. In: SIGIR (1998)
[14] Robertson, S.E., Walker, S.: Okapi/keenbow at TREC-8. In: TREC (1999)
[15] Tsotras, V.J., Kangelaris, N.: The Snapshot Index: an I/O Optimal Access Method for Snapshot Queries. Information System 20(3), 237–260 (1995)
[16] U, L.H., Mamoulis, N., Berberich, K., Bedathur, S.: Durable Top-k Search in Document Archives. In: SIGMOD (2010)

An Efficient SQL Rewrite Approach for Temporal Coalescing in the Teradata RDBMS

Mohammed Al-Kateb, Ahmad Ghazal, and Alain Crolotte

Teradata Labs, El Segundo CA, USA
{mohammed.al-kateb,ahmad.ghazal,alain.crolotte}@teradata.com

Abstract. The importance of temporal data management is manifested by a considerable attention from the database research community. This importance is becoming even more evident by the recent increasing support of temporal features in major commercial database systems. Among these systems, Teradata offers a native support to a wide range of temporal analytics. In this paper, we address the problem of temporal coalescing in the Teradata RDBMS. Temporal coalescing is a key temporal query processing operation, which merges adjacent or overlapping timestamps of value-equivalent rows. From existing approaches to implement temporal coalescing, pursuing an SQL-based approach is perhaps the most feasible and the easiest applicable. Along this direction, we propose an efficient SQL rewrite approach to implement temporal coalescing in the Teradata RDBMS by leveraging runtime conditional partitioning – a Teradata enhancement to ANSI ordered analytic functions – that enables to express the coalescing semantic in an optimized join-free single-scan SQL query. We evaluated our proposed approach over a system running Teradata 14.0 with a performance study that demonstrates its efficiency.

1 Introduction

Time is an integral part of each and every real-world concept and application. This aspect is realized in many database applications that deal with data whose values may change over time (e.g., financial applications [4]). In response to this reality, database research community has contributed a standing history of research proposals to enable and facilitate temporal data management over relational databases [3] [5] [7] [10] [11] [12].

Commercial database providers also subscribe to the pressing necessity of temporal data management as we recently witnessed a boost in supporting temporal functionalities and analytics in major commercial database systems. One example is Oracle's initiative of flashback queries [8], which allow issuing transaction-time range queries and retrieving a former snapshot of the data. Another example is IBM DB2's support for many temporal elements including business-time (i.e., valid-time), system-time (i.e., transaction-time) and temporal uniqueness constraints [6]. Yet another example is Teradata recently revealing its native support to a wider range of temporal features [14], including valid-time tables, transaction-time tables, bi-temporal tables, temporal constraints, temporal predicates and functions, and partitioned temporal tables.

S.W. Liddle et al. (Eds.): DEXA 2012, Part II, LNCS 7447, pp. 375–383, 2012.

In this paper, we address the problem of temporal coalescing in the Teradata RDBMS. Temporal coalescing [1] is a fundamental operation for temporal databases [2]. It merges adjacent or overlapping timestamps of value-equivalent rows[1] [2]. That is, coalescing two value-equivalent rows with adjacent or overlapping time intervals generates a single row whose time interval is bounded between the start timestamp of the first row and the end timestamp of the second row.

Implementing temporal coalescing efficiently is a challenging issue [2]. One approach is to implement coalescing as a native functionality. But, a native implementation usually requires complicated modifications to the database low-level internals. Another approach is to coalesce a temporal table in-memory. This approach, however, is not practical in many situations in which temporal tables are too large to load in memory. Another common approach is to express the coalescing semantic via SQL constructs. This approach, however, usually comes about as a rather complex query. Such a query typically involves both a self-join on the base table to keep track of the start and the end points of coalesced rows and exclusion joins with temporary tables (created as, and populated from, the base table being coalesced) to detect and handle the presence of temporal gaps [11]. Anther form of this complex query replaces exclusion joins with a grouped COUNT aggregation on top of a three-way self-join over the base table [11].

To overcome this complexity, Zhou et al. propose to express coalescing using ordered analytic functions [15]. While this approach indeed simplifies coalescing to a join-free SQL query, it has two main drawbacks. First, it requires two retrieve steps from the base table. With large table sizes, accessing base table twice can have a negative impact on the overall query performance. Second, the technical idea of this approach presumes the open-ended time interval model which effectively means that the end timestamp of a preceding temporal row has the same value of the begin timestamp of the succeeding temporal row. Limiting the time interval model to this assumption is essential to correctly keep track of the count of each timestamp, which is a necessary aggregation step, among other intermediate steps, in the coalescing mechanism proposed in [15].

Considering the aforementioned drawbacks, we propose an efficient SQL rewrite approach for implementing temporal coalescing in the Teradata RDBMS. While our approach is built on using ordered analytic functions as proposed in [15], its unique key idea is to leverage run-time conditional partitioning – a Teradata enhancement to the ordered analytic functions. Run-time partitioning extends the basic SQL partitioning functionality offered by the SQL "PARTITION BY" construct. It provides the means to impose dynamic data-dependent conditions on window aggregate processing using the "RESET WHEN" construct. If and when the "RESET WHEN" condition is satisfied during query evaluation, a new virtual partition is instantly constructed. Then, window functions are evaluated on each virtual partitions. Using the "RESET WHEN" construct enables to express the coalescing semantic in an *optimized* join-free *single-scan* SQL query and, at the same time, maintains the *flexibility* of handling either open-ended or

---

[1] This paper assumes *tuple timestamping*. Attribute timestamping is generally more complex [13], and can be sought for future work.

closed-ended *time interval models* equally. Our proposed approach is simple to implement since it leverages existing Teradata functionality, and is expected to perform efficiently given that Teradata has a robust and mature query rewrite engine. Moreover, our proposed approach can be easily adopted by other database systems with temporal features to extend their temporal implementation.

We present the results of a performance study conducted to compare our approach against the approach proposed in [15] with respect to elapsed time. The performance metric is reported for different number of coalescing attributes and for varying sizes of the base table. The experimental results show that our coalescing scheme outperforms the scheme in [15], particularly for reasonably larger number of coalescing attributes and larger size of the base table.

The rest of this paper is organized as follows. Section 2 discusses the importance and challenges of temporal coalescing. Section 3 presents our temporal coalescing approach. Section 4 presents the performance study. Finally, Section 5 outlines concluding remarks and suggests future work.

2   Temporal Coalescing

Temporal coalescing [1], which merges adjacent or overlapping timestamps of value-equivalent rows, is essential to temporal query processing because queries evaluated on un-coalesced data may generate incorrect answers [2].

Table 1. David's employment records

Name	Dept	Title	Validity
David	PS	Engineer	(2001, 2004)
David	PS	Sr. Engineer	(2004, 2006)
David	PM	Sr. Engineer	(2006, UNTIL_CHANGED)

Example 1 (Importance of coalescing). Consider the employment history of the employee David shown in Table 1. The first row reflects when David was first hired in 2001 in Professional Support (PS) as an Engineer. The second row shows that, in 2004, David was promoted to Senior Engineer. The third row reflects that, in 2006, David moved from PS to Product Management (PM) and still holds this position to date[2]. Suppose that the organization's director needs to know the employees who worked for PS for at least five "consecutive" years. Such a query can be expressed in TSQL[3] [9] as follows.

```
SELECT E.Name, E.Department, VALID(E)
FROM   Employee (Name, Department) as E
WHERE  CAST(VALID(E) AS INTERVAL YEAR) >= 5
```

---

[2] In Teradata, UNTIL_CHANGED represents an open-ended value for the end of a period validity in case such a value is not known beforehand.

[3] In this TSQL syntax, Employee (Name, Department) as E specifies coalescing on Name and Department, VALID(E) returns the validity interval of rows in the result, and CAST converts the validity interval to the specified granularity, i.e., year. The query returns the pairs of the values of Name and Department along with the interval during which the values have been the same.

Without temporal coalescing, the history of David's employment (see Table 2) represents his employment in PS in two separate rows with two different temporal validities. Given this un-coalesced data, each of the two rows has a time interval that is less than five years. Therefore, an erroneous conclusion will be returned excluding David from the query result. With coalescing (see Table 3), however, the two rows with PS are merged into one row because their temporal validities are adjacent. With the length of the new time interval being greater than five years, David will correctly qualify for inclusion in the query result.

Table 2. *Without* coalescing

Name	Dept	Validity
David	PS	(2001, 2004)
David	PS	(2004, 2006)
David	PM	(2006, UNTIL_CHANGED)

Table 3. *With* coalescing

Name	Dept	Validity
David	PS	(2001, 2006)
David	PM	(2006, UNTIL_CHANGED)

Example 2 (Coalescing with temporal gaps). Temporal coalescing becomes more challenging in the presence of temporal gaps. Recall David's employment history in Example 1 and assume that after spending only one year in PM, David moved back to PS on 2007 as shown in Table 4.

Table 4. David's employment records with *temporal gaps*

Name	Dept	Title	Validity
David	PS	Engineer	(2001, 2004)
David	PS	Sr. Engineer	(2004, **2006**)
David	PM	Sr. Engineer	(2006, 2007)
David	PS	Sr. Engineer	(**2007**, UNTIL_CHANGED)

On one hand, if coalescing overlooks the temporal gap between 2006 and 2007 during which David was in PM, the result of coalescing will erroneously consider that David has been an employee in PS all the way from 2001 to date as shown in Table 5. On the other hand, with coalescing being able to detect and handle this temporal gap, the correct result will be as shown in Table 6.

Table 5. *Ignoring* temporal gaps

Name	Dept	Validity	
David	PS	(2001, UNTIL_CHANGED)	✗
David	PM	(2006, 2007)	✓

Table 6. *Considering* temporal gaps

Name	Dept	Validity	
David	PS	(2001, **2006**)	✓
David	PM	(2006, 2007)	✓
David	PS	(**2007**, UNTIL_CHANGED)	✓

3   Proposed Approach

Our proposed approach for temporal coalescing in Teradata is motivated by three design objectives. First, it should avoid self-joins and nested queries. Second, it should reduce the number of accesses to the base temporal table. Third, it should be flexible to work equally with open-ended and closed-ended interval models.

To meet these objectives, we propose to implement temporal using ordered analytic functions and runtime partitioning. Ordered analytic functions are SQL constructs used to apply aggregate functions to a partition of the data. Runtime partitioning is a Teradata functional enhancement to ordered analytic functions that provides the means to specify data-dependent partitioning conditions on window aggregate processing using "RESET WHEN" construct. When the "RESET WHEN" condition is satisfied during query evaluation, a new virtual partition is instantly constructed and the window aggregate is evaluated on each partition. "RESET WHEN" itself encapsulates simple CASE expressions to dynamically form IF-THEN logic, which is used to construct run-time partitions.

We describe our solution using the example of David's employment records shown in Table 7. Herein, for simplicity and without loss of generality, we assume that the validity period is represented by ValidFrom and ValidUntil attributes.

Table 7. David's employment records

Name	Dept	Title	ValidFrom	ValidUntil
David	PS	Engineer	2001	2004
David	PS	Sr. Engineer	2004	2006
David	PM	Sr. Engineer	2006	2007
David	PS	Sr. Engineer	2007	2012

Table 8. Coalescing on dept. affiliation

Name	Dept	StartDate	EndDate
David	PS	2001	2006
David	PM	2006	2007
David	PS	2007	2012

Assume we need to know David's affiliation history. Our approach to generate the result of such a coalescing request (see Table 8) is to use ordered analytic functions and "RESET WHEN" construct as in the following SQL query:

```
WITH E (Name, Department, dummy1, dummy2, mn, mx) AS
(SELECT  Name
        ,Department
        ,MAX(ValidUntil)
            OVER (PARTITION BY Name, Department ORDER BY ValidFrom
            ROWS BETWEEN 1 PRECEDING AND 1 PRECEDING)
            as preceding_end_validity
        ,ValidFrom as current_begin_validity
        ,MIN(ValidFrom)
            OVER (PARTITION BY Name, Department ORDER BY ValidFrom
            RESET WHEN (current_begin_validity > preceding_end_validity)
            ROWS BETWEEN UNBOUNDED PRECEDING AND CURRENT ROW)
        ,ValidUntil
  From Employee)
SELECT Name, Department, MIN(mn) as StartDate , MAX(mx) as EndDate
FROM E
GROUP BY Name, Department, mn
ORDER BY Name, StartDate
```

In the above example, coalescing is achieved mainly by "PARTITION BY" and "RESET WHEN" constructs. "PARTITION BY" groups rows with the same values of coalescing attributes (e.g., *Name* and *Department*) in one partition. "RESET WHEN" detects the presence of temporal gaps in a given partition (when the *start* timestamp of a *current* row is *greater* than the *end* timestamp of a *preceding* row) and further divides it into sub-partitions. Aggregate functions are then applied to each of the individual partitions. Note that the change in the department affiliation resulted in a temporal gap during which David left PS for PM before returning back to PS. This case emphasizes an actual need for the "RESET WHEN" functionality to detect and handle such temporal gaps.

In light of this example, we outline our coalescing procedure as follows. We assume that the input to coalescing is a coalescing clause in the form of $\mathcal{C}(\mathcal{T}, \mathcal{A})$, where $\mathcal{T}$ denotes the temporal table and $\mathcal{A}$ denotes the set of coalescing attributes. Given this input, we maintain a temporary table (in the above example, it is E) that uses SQL ordered analytic functions and Teradata "RESET WHEN" functionality to select the following elements from each row $\mathcal{R}_i$ in $\mathcal{T}$:

1. $\mathcal{R}_i.\mathcal{A}$
2. The end timestamp (i.e., the value of ValidUntil) of the preceding row $\mathcal{R}_{i-1}$.
3. The start timestamp (i.e., the value of ValidFrom) of the current row $\mathcal{R}_i$.
4. The cumulative minimum value of start timestamp up to the current row $\mathcal{R}_i$, *reseting when* there is a *temporal gap* between the preceding row $\mathcal{R}_{i-1}$ and the current row $\mathcal{R}_i$ (i.e., value of element 3 is greater than the value of element 2.).
5. The end timestamp (i.e., ValidUntil value) of the current row.

Note that at this point, the data in E, excluding dummy attributes, looks as shown in Table 9. As can be seen in Table 9, the rows that should coalesce together share the same cumulative minimum start timestamp mn (since the start timestamp is monotonically increasing), while each row holds the value of its own end timestamp (i.e., ValidUntil) in mx. Hence, finally, we select the minimum of mn and the maximum of mx to return the coalesced temporal interval of the corresponding row. In Table 9, for instance, the two rows sharing the same mn value of 2001 are coalesced (i.e., merged) together with the validity from 2001 to 2006. Each of the other two rows represents a single coalesced row. Therefore, the final result consists of three rows as shown in Table 10.

It can be easily seen that our coalescing approach achieves its design objectives. First, coalescing is done in a *join-free* query. Second, it *accesses* the base

Table 9. Projected content of E

Name	Dept	mn	mx
David	PM	2006	2007
David	PS	**2001**	2004
David	PS	**2001**	2006
David	PS	2007	2012

Table 10. Final result of coalescing

Name	Dept	StartDate	EndDate
David	PS	2001	2006
David	PM	2006	2007
David	PS	2007	2012

table only *once*. Third, it can handle either *open-ended or closed-ended* interval models by simply controlling the condition of the "RESET WHEN" construct.

4 Experiments

We compare the performance of our coalescing approach (CWRW: **C**oalescing **W**ith **R**eset **W**hen) to that of coalescing using window analytic functions only (CWAF: **C**oalescing **W**ith **A**nalytic **F**unctions) as proposed in [15]. We measure the performance in terms of elapsed time – a performance metric automatically captured by Teradata DBQL (DataBase Query Log). Elapsed time is the duration between the start time of a query (i.e., the time the query is submitted for execution) and its first response time (i.e., the time it starts to return results). The experiments were conducted on a system running latest Teradata release 14.0 with a single-node machine of 4G memory, 2 CPUs, and 4 AMPs[4].

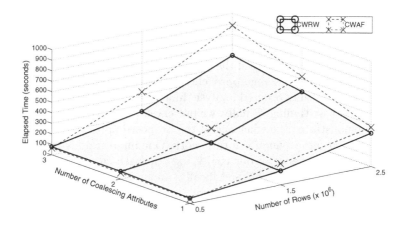

Fig. 1. Elapsed Time of CWRW and CWAF

We synthesized a dataset of employment records that portrays temporal changes in employment history simulating the following scenario. An employment record stores values of time-invariant attributes (e.g., *Name* and *SSN*) and time-varying attributes (e.g., *Dept.* and *Grade*). For some time-varying attributes, old values can be future values, thus, causing temporal gaps (e.g., an employee was first hired in a department, then moved to another department, and later returned back to the first department). The number of changes in an employee's history is generated at random and the changing time-varying attribute is also selected at random. A change in one time-varying attribute may result in a change in another time-varying attribute (e.g., upgrading an

---

[4] AMP (Access Module Processor) is a Teradata virtual component responsible for receiving query plans, performing actual database steps, and producing query results.

employee's grade is assumed to result in a pay raise). The syntactic dataset models temporal changes starting from 1966 to 2012 with the possibility that an employee's employment may either end before 2012 (due to termination or retirement) or continue until 2012. We loaded the generated data using Teradata Fastload – a utility for loading large amounts of data in a rapid manner.

We report the results with respect to two factors – table size (the number of rows in a table) and number of coalescing attributes (the number of time-varying attributes subject to coalescing). We vary the number of rows from 500K to 2.5M rows[5] and change the number of coalescing attributes from one to three[6].

Figure 1 shows that our CWRW beats CWAF. On one end, the performance of CWRW is similar to CWAF for smaller table size regardless of the number of coalescing attributes. On the other end, CWRW outperforms CWAF for larger table sizes and the magnitude of improvement increases further as the number of coalescing attributes increases – nearly 10% improvement for coalescing on one attribute and 50% improvement for coalescing on three attributes.

5   Conclusion and Future Work

In this paper, we examined the implementation of temporal coalescing in the Teradata RDBMS. First, we discussed the challenges of temporal coalescing and the drawbacks of existing coalescing approaches. Second, we proposed our temporal coalescing approach using ordered analytic functions jointly with Teradata runtime conditional partitioning. Finally, we presented the results of a performance study that demonstrates the efficiency of the proposed approach.

This work opens an avenue for future work. One interesting issue is to investigate how the performance of our coalescing approach can be improved using other existing Teradata features. Specifically, we envision vertical (i.e., column) partitioning – a physical database design choice made available in the recent Teradata release, 14.0 – as a potential way to further optimize the coalescing performance since coalescing is naturally applied to a small set of columns from the entire table. Another direction is to explore the utility of temporal coalescing with respect to other temporal analytic features already supported in Teradata.

References

1. Böhlen, M.H.: Temporal coalescing. In: Encyclopedia of Database Systems, pp. 2932–2936 (2009)
2. Böhlen, M.H., Snodgrass, R.T., Soo, M.D.: Coalescing in temporal databases. In: VLDB, pp. 180–191 (1996)
3. Dyreson, C., et al.: A consensus glossary of temporal database concepts. SIGMOD Rec. 23(1), 52–64 (1994)
4. Chandra, R., Segev, A.: Managing temporal financial data in an extensible database. In: VLDB 1993, pp. 302–313 (1993)

---

[5] The exact number of rows is 494,041, 1,484,467, and 2,471,405, respectively.

[6] Temporal applications normally target a small number of attributes for coalescing.

5. Date, C., Darwen, H.: Temporal Data and the Relational Model. Morgan Kaufmann Publishers Inc., San Francisco (2002)
6. IBM. A matter of time: Temporal data management in DB2 for z/OS (2010), http://www.ibm.com/developerworks/data/library/techarticle/dm-1204db2temporaldata/dm-1204db2temporaldata-pdf.pdf
7. Jensen, C.S., Snodgrass, R.T.: Temporal data management. IEEE TKDE 11(1), 36–44 (1999)
8. Oracle. Oracle flashback technologies (2010), http://www.oracle.com/technetwork/database/features/availability/flashback-overview-082751.html
9. Snodgrass, R.T., et al.: TSQL2 language specification. SIGMOD Rec. 23(1), 65–86 (1994)
10. Shoshani, A., Kawagoe, K.: Temporal data management. In: VLDB 1986, pp. 79–88 (1986)
11. Snodgrass, R.T.: Developing Time-Oriented Database Applications in SQL. Morgan Kaufmann (1999)
12. Tansel, A.U.: Temporal databases theory, design, and implementation. The Benjamin/Cummings Publishing Company, Inc. (1993)
13. Tansel, A.U.: Temporal relational data model. IEEE TKDE 09(3), 464–479 (1997)
14. Teradata. Teradata temporal analytics (2010), http://www.teradata.com/database/teradata-temporal/
15. Zhou, X., Wang, F., Zaniolo, C.: Efficient Temporal Coalescing Query Support in Relational Database Systems. In: Bressan, S., Küng, J., Wagner, R. (eds.) DEXA 2006. LNCS, vol. 4080, pp. 676–686. Springer, Heidelberg (2006)

HIP: *I*nformation *P*assing for Optimizing Join-Intensive Data Processing Workloads on *H*adoop

Seokyong Hong and Kemafor Anyanwu

North Carolina State University, Department of Computer Science
Raleigh, USA
{shong3,kogan}@ncsu.edu

Abstract. Hadoop-based data processing platforms translate join intensive queries into multiple "jobs" (MapReduce cycles). Such multi-job workflows lead to a significant amount of data movement through the disk, network and memory fabric of a Hadoop cluster which could negatively impact performance and scalability. Consequently, techniques that minimize sizes of intermediate results will be useful in this context. In this paper, we present an information passing technique (*HIP*) that can minimize the size of intermediate data on Hadoop-based data processing platforms.

1 Introduction

MapReduce [1] has become the de facto standard for large scale data processing frameworks because of its simple programming model, ease of parallelization, and scalability. Its programming model allows users to describe their desired tasks by implementing two functions (*map* and *reduce*). Each of map and reduce functions is called in a separate phase (*map phase* and *reduce phase*, respectively) and a pair of each function constitute a single MapReduce job. When a job executes, Hadoop [2], an open source MapReduce implementation, schedules multiple instances of the map function (*mappers*) and the reduce function (*reducers*) over the cluster. Each mapper is assigned a partition of input data named *split* and applies the map function to its split to generate intermediate key-value pairs. The intermediate pairs are sorted and partitioned by a key and the partitions are transferred to appropriate reducers to apply the reduce function. Non-trivial tasks (e.g., tasks consisting of several join operations) could result in multiple job cycles across which the execution of the join operations is distributed. However, lengthy MapReduce workflows can lead to poor performance due to the sorting, transferring, and merging overheads between map and reduce phases [3, 4]. The costs associated with MapReduce cycles are heavily dependent on the sizes of the intermediate results (including those generated after both the map and reduce phases). Consequently, an important problem is how to keep those sizes minimal.

In traditional database systems, one family of promising techniques for keeping intermediate results minimal is *Sideways Information Passing (SIP)* [7–11].

S.W. Liddle et al. (Eds.): DEXA 2012, Part II, LNCS 7447, pp. 384–391, 2012.

Such techniques generate summary information from certain operators and make such information available to other associated operators in a same query plan. By exploiting summary information, the recipient operators can prune "unnecessary" (not ultimately relevant to the final output) records from their inputs. However, while existing techniques assume parallelism techniques such as inter-operator and/or pipelined parallelism where multiple operators execute concurrently, MapReduce processing is based on partitioned parallelism where a single operator is executed at a time on different data partitions. A consequence of this is that the complete summary for each operator is also partitioned across different nodes. Further, this partitioning may be based on a different set of partitioning keys than that used for subsequent operations that need to utilize the summary. Therefore, there is a challenge of keeping track which subsets of summary partitions are relevant to which future data partitions. A second issue is that existing techniques assume a flexible communication framework for passing information between operators whereas MapReduce platforms such as Hadoop use a very limited communication model between nodes involved in processing. This presents challenges with respect to how the summaries can be communicated across operators.

In this paper, we present an *Information Passing* (*HIP*) approach suitable for the MapReduce computing model, which addresses the unnecessary data movement problem. Specifically, we propose (i) an extended MapReduce-based data processing model that enables information passing to ameliorate the join processing performance, and (ii) an information passing approach for integrating the technique into Apache Hive. Results of an empirical evaluation are presented that demonstrate the superiority of information passing-based approach for multi-join workloads.

2   Background and Related Work

2.1   Sideways Information Passing

The SIP technique proposed in [8] adaptively uses information passing for processing RDF data. At compile-time, the query optimizer calculates the entire equivalence classes in each of which operators that share a variable in predicates are grouped together. During the query execution, operators such as index scan operators and merge join operators in a group generate filters and share them via shared memory. Operators in each group use the shared filters to prune unnecessary records. An adaptive query processing technique in [10] exploits a special operator named *Eddy*. This operator is responsible for routing tuples between relational operators so that it maximizes the query processing performance. *Magic-set rewriting* technique generates and propagates a filter set between nested query blocks. An inner query block can prune irrelevant records that do not meet its outer block's predicates. Techniques in [7, 9] support information passing among operators in distributed environments. Semi-join approach [9] generates information about join columns of a relation and passes it to a remote site. Such information is used for pruning irrelevant records and the resulting records are shipped to the other site and joined. In [7], distributed operators generate summary information about their intermediate records. Such summary information is stored in a centralized summary repository so that those operators can exchange summary information and prune unnecessary records.

2.2   Hadoop-Based Data Processing Systems

In extended Hadoop platforms such as Apache Pig [13] and Apache Hive [12], a user query is ultimately compiled into a sequence of MapReduce cycles in which each MapReduce cycle hosts the execution of some subset of operators in the query plan. The left part in Fig 1a shows the general architectural structure using Hive as an example. Such frameworks typically consist of a *compiler/optimizer* component and a *MapReduce job executor* on top of Hadoop. The former component produces an optimized execution plan consisting of several operations and assigning the operations to one or more MapReduce cycles. The latter component is responsible for submitting each job to Hadoop. Multiple operators can be assigned to execute in a map or reduce phase. We discuss the dataflow in a Hadoop job using abstract operators as shown in Fig 1b. *MapsidePrimary* (P_M) is logically the first operator executed in map phase and is responsible for scanning the records of an input table. *MapsideTerminal* (T_M) is the final operator in map phase, which passes intermediate records to a Hadoop object, called *OutputCollector*. The OutputCollector stores intermediate results to local disk. The reduce phase has similar operators, *ReducesidePrimary* (P_R) that receives input records from the map phase and *ReducesideTerminal* (T_R) that passes the resulting records to the OutputCollector to materialize them on the HDFS.

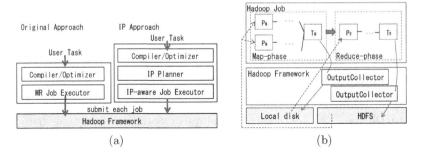

(a)   (b)

Fig. 1. (a) Data Processing on Hadoop, (b) Abstract Operators in a Single Job

2.3   Join Processing Techniques on Hadoop

In MapReduce-based data processing systems, the most widely adopted join approach is *Standard Repartitioning Join* [5, 6]. Processing each join operation in reduce phase involves data sorting and transferring overhead between map- and reduce-phase. *Fragment-Replication Join* and *Map-Merge Join* are alternative join strategies that can reduce the data sorting and shuffling overhead by processing join in map phase [5, 6]. However, the fragment-replication join assumes that one of the input tables is small enough to fit in memory. In case of map-merge join, an additional pre-processing phase is required to sort and co-partition input tables. Moreover, this approach may not be effective in scenarios involving intermediate tables or when join key is different from the partitioning key. Recently, [6] proposed a *Semi-Join* strategy for use on MapReduce platforms. Performing a semi-join operation requires three MapReduce cycles. The first two cycles are responsible for reducing the size of an input table by creating a list of distinct join key values from a table and filtering out unmatched records in the other table. In the last cycle, the reduced table is joined with the other table. However, the semi-join strategy achieves performance improvement only if the first two steps are performed in pre-processing phase.

3   Integrating Information Passing into Hadoop Data Processing Platforms

Our discussion will proceed in the context of the Apache Hive platform [12] but can be generalized to other similar platforms. In order to integrate information passing, support for combining an information passing plan into a query plan at compile-time and run-time support for generating and transferring summaries for use by appropriate jobs are required. These are achieved in our extended Hive implementation by the *IP Planner* and *IP-aware Job Executor* respectively (Fig 1a right). The IP planner receives a query plan (Hadoop job plan) from the Hive compiler and augments it with a corresponding IP plan. An IP plan consists of three data structures: *dataflow graph*, *dependency graph*, and *IP descriptor table*. A dataflow graph describes all the relationships based on data dependencies between Hadoop jobs in a job plan. The data dependencies induces ordering on the execution of MapReduce jobs in a job plan. Such information is captured in a job dependency graph. An IP descriptor table contains IP descriptors for all jobs in a job plan. Each IP descriptor includes the *job name, the type of* P_R (e.g., join or groupby), *a reference* to a Hive's job descriptor instance, and *IP input* and *output descriptors* as shown in Fig 2c. The main purpose of input or output descriptors is to keep track of source and destination paths of summary information files for each job.

3.1   Run-Time Execution of IP Plans

Given an IP plan, our IP-aware job executor determines whether the currently scheduled job should generate summary information ($IP\_CREATION$) and/or exploit summaries generated by previously executed jobs (IP_USAGE). Decisions are encoded and distributed to mappers and reducers through *JobConf*. JobConf is a facility provided by the Hadoop framework to propagate system-wide and application-specific configurations to every node. Once the IP-aware job executor makes the decision to enable information passing for a particular job, it replaces the default P_M and T_R with relevant IP-aware P_M and T_R operators.

Creation of Summary Information: As described in Section 2.1, T_R is the final logical operator in a job, which is responsible for gathering final data and

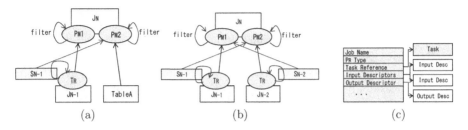

Fig. 2. (a) Join a table and an intermediate table, (b) Join two intermediate tables, (c) IP Descriptor

passing them into the MapReduce framework. Therefore, if the current job requires the generation of summary information on a column(s) of its output, our IP-aware T_R operator produces a list of distinct hash values on the next join column and stores the list into HDFS in a zip-compressed format. Whenever a record is passed to T_R, it locates the target column, calculates a hash value for it, and puts the hash value in an in-memory buffer. When the operator is closed, it stores all the hash values into HDFS.

Utilization of Summary Information: Once the IP-aware job executor decides to enable *IP_USAGE* for a job (J_C), it propagates its summary to the *DistributedCache*, facility provided by Hadoop to distribute files in an efficient manner. The DistributedCache reduces communication overhead since it transfers a single copy of data to each node's local disk rather than to each mapper or reducer even when multiple mappers/reducers execute on the same node. There are two cases to consider. Fig 2a shows the first case where the binary join operator receives an intermediate table generated by a previous job. In this case, P_M2 in J_N is coordinated to load summary information S_{N-1} so that it prunes irrelevant records in *TableA* before being sorted and transferred to reducers. In the second case, a join operator receives its all input from previous jobs as shown in Fig 2b. In this case, P_M1 and P_M2 are coordinated to load the summary information generated by J_{N-2} and J_{N-1} respectively. When an IP-enabled P_M operator is initialized, the corresponding summary information is loaded into in-memory buffer. Whenever a record is passed, the operator calculates the hash value for the column(s) to be joined in reduce phase and retrieves the hash value from the in-memory buffer. If the value does not exist, then it prevents the record from being passed to the next operator. Otherwise, it passes the record to its neighboring operator.

Decision-Making Process: if information passing has been enabled, the executor looks up the dataflow graph to check whether the result from the execution of the currently scheduled job, J_C will be fed into its neighbor and whether the neighboring job has a join operator as P_R. If true, the executor enables *IP_CREATION* so that J_C generates summary information during its execution. Next, in order to determine whether to enable the current job J_C, to load and exploit summary information files P_i $(1 \leq i \leq r)$, generated from a previous job, J_P, the job executor checks whether the IP feature is activated in the Hive configuration file. If true, the job executor checks if J_C has a join operator as P_R, looks up the dependency graph to find any neighbor, J_P. Then, the job executor measures the total size of summary information files generated by J_P and compares the size with a user-configured threshold C. This is necessary since a large amount of summary may reduce available heap memory causing memory shortage and decrease processing performance. If the size does not exceed the threshold, the job executor activates the IP feature so that the the summary files are loaded and used by each corresponding IP-aware P_M operator. Otherwise, the job executor launches J_C according to the original MapReduce job plan.

4   Experimental Evaluation

4.1   Evaluation Framework

Experiment Setup: Experiments were conducted on a 20-node Hadoop cluster where each node has a 3.0 GHz dual core Xeon processor and 4GB main memory, and Redhat Enterprise Linux 5 runs as an operating system. Apache Hadoop 0.20 [2] with 512MB block size, 1 replication factor, 1280MB heap size for mappers/reducers, and no speculative execution was used. We present a performance evaluation of the HIP approach that was implemented by extending Hive 0.5 [12]. We compared the performance of three join approaches, the Hive's repartitioning-join, and map-join approaches, and the semi-join approach proposed in [6] against the HIP technique. The first two steps of the semi-join were implemented as vanilla MapReduce applications. For the last phase, we used Hive's map-side join.

Workload and Datasets: We evaluated the different join processing approaches under different join selectivities to assess the impact of the amount of intermediate data. To achieve this, we generated several relation instances over the same schema (*Table B*). The different instances varied in size and the distribution of join values in the join column. Another dataset (*TableA*) was generated with unique join keys in the join column and the set of keys contained in the set of join values in *TableB*. The test workload consists of 2 jobs in which *JobA* executes a self join on *TableA* generating a summary that is equivalent to the set of join values in the table. The second job (*JobB*) joins the output records of *JobA* with *TableB*. The properties of datasets are shown in Table 1.

Table 1. Input Properties in Scenario 1 and 2

	Table	Table Size	Key Density
Scenario1	TableA (Key: 25B, ColumnA: 10B, ColumnB: 65B)	30MB	1
	TableB (Key: 25B, ColumnA: 100B, ColumnB: 75B)	1-40GB	1
Scenario2	TableA (Key: 25B, ColumnA: 10B, ColumnB: 65B)	20MB-50GB	100
	TableB (Key: 25B, ColumnA: 100B, ColumnB: 75B)	40GB	1

4.2   Experimental Results

The HIP approach improves performance due to early pruning of unnecessary data before the sorting and transferring steps. Performance improvements are expected to be proportional to the degree of unnecessary data reduction. The amount of data reduction in shuffle phase is shown in Table 2. In scenario1, a small input relation (*TableA*) is joined with a small fraction of a large relation *TableB*. As the size of *TableB* increases, the amount of unnecessary data that are eventually not joined also does. As presented in Fig 3a, our information passing approach shows better performance than the other three approaches in the first scenario. Particularly, as the degree of data reduction increases, the HIP approach shows better execution time. For Hives map-join, a significant amount of time was consumed to load input tables. Consequently, the map-join did not show better performance than the repartitioning join. The semi-join approach shows the worst performance since the second and third stages require large input data to be scanned and load. The performance of the repartitioning join and the map-side join were significantly degraded as the size of input dataset increases while impact on the information passing approach was less. In the second scenario, the semi-join, map-join, and our information passing approaches show relatively better performance than repartitioning join approach when the size of *TableA* is not more than 0.2GB (Fig 3b). However, the semi-join and the map-join techniques could not run with table sizes greater than 0.2GB in our experiment setting. This is because all of *TableA* records cannot fit into available heap memory for the map-join to work. In the semi-join case, the third stage

Table 2. Data Reduction Size

Scenario1	TableB Size(GB)	1	10	20	30	40
	Reference (%)	5.86	0.59	0.29	0.2	0.15
	Reduction (GB)	0.77	8.15	16.35	24.55	28.65
Scenario2	TableA Size(GB)	0.0195	0.2	10	30	50
	Reference (%)	0.001	0.01	0.5	1.5	2.5
	Reduction (GB)	33.09	33.19	32.96	32.63	32.54

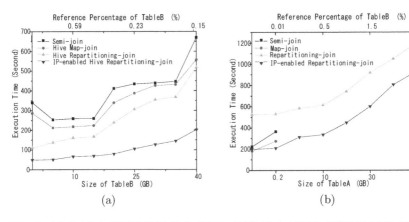

(a)   (b)

Fig. 3. (a) TableA:30MB,TableB:1-40GB, (b)TableA:20MB-50GB,TableB:40GB

requires map-join and the reduced *TableB* through the first two stages does not fit into the available memory. Compared to the map-join and the semi-join, the HIP approach executes even with quite larger sizes of *TableA* since our technique requires less information: summary vs. the entire table which contains a lot of duplicate join key values.

5 Conclusions

The HIP approach effectively adapts information passing to MapReduce resulting improved join performance. Since MapReduce frameworks do not support features such as indexing and cost-based optimization, information passing provides an effective optimization alternative for multi-job workloads. Experimental evaluations demonstrate its advantages with respect to performance.

Acknowledgments. This work was partially funded by NSF grant IIS-0915865.

References

1. Dean, J., Ghemawat, S.: MapReduce: Simplified Data Processing on Large Clusters. Commun. ACM 51(1), 107–113 (2008)
2. Apache Hadoop, http://hadoop.apache.org
3. Gates, A., Natkovich, O., Chopra, S., Kamath, P., Narayanam, S., Olston, C., Reed, B., Srinivasan, S., Srivastava, U.: Building a HighLevel Dataflow System on top of MapReduce: The Pig Experience. PVLDB 2(2), 1414–1425 (2009)
4. Dittrich, J., Quiané-Ruiz, J., Jindal, A., Kargin, Y., Setty, V., Schad, J.: Hadoop++: Making a Yellow Elephant Run Like a Cheetah. PVLDB 3(1), 518–529 (2010)
5. Lin, Y., Agrawal, D., Chen, C., Ooi, B.C., Wu, S.: Llama: Leveraging Columnar Storage for Scalable Join Processing in the MapReduce Framework. In: ACM SIGMOD, pp. 961–972. ACM, Athens (2011)
6. Blanas, S., Patel, J.M., Ercegovac, V., Rao, J., Shekita, E.J., Tian, Y.: A Comparison of Join Algorithms for Log Processing in MapReduce. In: ACM SIGMOD, pp. 975–986. ACM, Indianapolis (2010)
7. Ives, Z.G., Taylor, N.E.: Sideways Information Passing for Push-Style Query Processing. In: 24th International Conference on ICDE, pp. 774–783. IEEE, Cancún (2008)
8. Neumann, T., Weikum, G.: Scalable join processing on very large RDF graphs. In: ACM SIGMOD, pp. 627–640. ACM, Providence (2009)
9. Bernstein, P.A., Chiu, D.W.: Using Semi-Joins to Solve Relational Queries. J. ACM 28(1), 25–40 (1981)
10. Avnur, R., Hellerstein, J.M.: Eddies: Continuously Adaptive Query Processing. In: ACM SIGMOD, pp. 261–272. ACM, Dallas (2000)
11. Mumick, I.S., Pirahesh, H.: Implementation of Magic-sets in a Relational Database System. In: ACM SIGMOD, pp. 103–114. ACM, Minneapolis (1994)
12. Apache Hive, http://hive.apache.org
13. Apache Pig, http://pig.apache.org

All-Visible-*k*-Nearest-Neighbor Queries

Yafei Wang[1], Yunjun Gao[1], Lu Chen[1], Gang Chen[1], and Qing Li[2]

[1] College of Computer Science, Zhejiang University
{wangyf,gaoyj,chenl,cg}@zju.edu.cn
[2] Department of Computer Science, City University of Hong Kong
itqli@cityu.edu.hk

Abstract. The All-*k*-Nearest-Neighbor (A*k*NN) operation is common in many applications such as GIS and data analysis/mining. In this paper, for the first time, we study a novel variant of A*k*NN queries, namely *All-Visible-k-Nearest-Neighbor* (AV*k*NN) query, which takes into account the impact of obstacles on the *visibility* of objects. Given a data set P, a query set Q, and an obstacle set O, an AV*k*NN query retrieves for each point/object in Q its *visible* k nearest neighbors in P. We formalize the AV*k*NN query, and then propose efficient algorithms for AV*k*NN retrieval, assuming that P, Q, and O are indexed by conventional data-partitioning indexes (e.g., R-trees). Our approaches employ *pruning techniques* and introduce a *new pruning metric* called VMDIST. Extensive experiments using both real and synthetic datasets demonstrate the effectiveness of our presented pruning techniques and the performance of our proposed algorithms.

1 Introduction

The *All-Nearest-Neighbor* (ANN) operation plays an important role in a wide range of applications such as GIS [26], data analysis/mining [2, 10], VLSI layout design [13, 19], and multimedia retrieval [12, 15]. Given a data set P and a query set Q, an ANN query finds for each point/object in Q its nearest neighbor in P. A natural generalization is *All-k-Nearest-Neighbor* (A*k*NN) search which, for every point in Q, retrieves its k nearest neighbors (NNs) in P. Formally, A*k*NN$(P, Q) = \{\langle q, p \rangle \mid \forall q \in Q, \exists p \in (k\text{NN}(q) \subseteq P), \neg \exists p' \in (P - k\text{NN}(q))$ such that $dist(p', q) < dist(p, q)\}$, in which A*k*NN$(P, Q)$ represents the result set of the A*k*NN query w.r.t. P and Q, $k\text{NN}(q)$ denotes the set of k NNs for q, and $dist(\ )$ refers to the Euclidean distance metric. An example of A1NN ($k = 1$) retrieval is depicted in Figure 1(a), with $P = \{p_1, p_2, p_3, p_4, p_5, p_6\}$ and $Q = \{q_1, q_2, q_3\}$. A1NN$(P, Q) = \{\langle q_1, p_2 \rangle, \langle q_2, p_3 \rangle, \langle q_3, p_5 \rangle\}$.

Traditional A*k*NN search does not take obstacles into consideration. Nevertheless, physical obstacles (e.g., buildings, blindages, etc.) are *ubiquitous* in the real world, and their existence may affect the *visibility/distance* between objects and hence the result of queries. Moreover, in some applications, the users might be only interested in the objects that are *visible* or *reachable* to them. Recently, spatial clustering and spatial queries with obstacle constraints have been explored in the literature [6, 7, 8, 11, 14, 17, 18, 20, 22, 24, 26]. To the best of our knowledge, however, there is *no* prior work on the A*k*NN retrieval in the presence of obstacles.

S.W. Liddle et al. (Eds.): DEXA 2012, Part II, LNCS 7447, pp. 392–407, 2012.

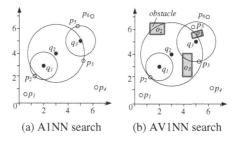

(a) A1NN search   (b) AV1NN search

Fig. 1. Illustration of A*k*NN and AV*k*NN queries for $k = 1$

In this paper, we study a novel variant of A*k*NN queries, namely *All-Visible-k-Nearest-Neighbor* (AV*k*NN) query, which takes into account the impact of obstacles on the *visibility* of objects. Given a data set P, a query set Q, and an obstacle set O, an AV*k*NN query finds for each point in Q its *visible* k nearest neighbors in P. Consider, for instance, Figure 1(b), where $P = \{p_1, p_2, p_3, p_4, p_5, p_6\}$, $Q = \{q_1, q_2, q_3\}$, and $O = \{o_1, o_2, o_3\}$ (in this paper we assume that an obstacle is represented by a rectangle, although it might be in any shape such as triangle and pentagon). The result of an AV1NN ($k = 1$) query is $\{\langle q_1, p_2 \rangle, \langle q_2, p_2 \rangle, \langle q_3, p_3 \rangle\}$, which is *different* from the result of previous A1NN search.

We focus on the AV*k*NN retrieval because, it not only is a *new* problem from the research point of view, but also is *useful* in many applications. For example, suppose that the transportation department of Hong Kong wants to install traffic surveillance cameras to monitor m different accident-prone sites. Obviously, each site s should be *visible* to at least one camera c for safety. Furthermore, the distance between the site s and its corresponding camera(s) c is expected to be as *small* as possible in order to improve the video quality. In this scenario, an AV*k*NN query, which takes as inputs a set of monitoring cameras (denoting a data set P), a set of accident-prone sites (representing a query set Q), and a set of buildings/blindages (denoting an obstacle set O), may help the decision-maker to find out for each site in Q its k visible nearest cameras in P. In addition, the AV*k*NN query, as a stand-alone tool or a stepping stone, could also be applied in complex spatial data analysis/mining (e.g., clustering, outlier detection, etc.) involving obstacles.

A naive solution to tackle AV*k*NN retrieval is to perform visible k NN search [14] for each point in a data set P. Unfortunately, this approach is *very inefficient* since it needs to traverse the data set P and the obstacle set O *multiple times* (i.e., $|P|$ times), resulting in *high* I/O and CPU costs, especially for *larger* P. Motivating by this, in this paper, we propose two efficient algorithms, i.e., *Grouping-based Multiple-access Obstacle Tree algorithm* (GMOT) and *Grouping-based Single-access Obstacle Tree algorithm* (GSOT), for AV*k*NN query processing, assuming that P, Q, and O are indexed by conventional data-partitioning indexes (e.g., R-trees [1]). In particular, our methods utilize *pruning techniques* and introduce a new *pruning metric* called VMDIST to improve the query performance. In brief, the key contributions of this paper are summarized as follows:

- We formalize the AVkNN query, a *new* addition to the family of spatial queries in the presence of obstacles.
- We develop effective *pruning techniques* and a *new pruning metric* to facilitate the pruning of unqualified obstacles.
- We propose two efficient algorithms to answer *exact* AVkNN retrieval.
- We conduct extensive experimental evaluation using both real and synthetic datasets to verify the effectiveness of our presented pruning techniques and the performance of our proposed algorithms.

The rest of the paper is organized as follows. Section 2 briefly surveys the related work. Section 3 presents the problem statement. Section 4 elaborates AVkNN query processing algorithms. Extensive experiments and our findings are reported in Section 5. Finally, Section 6 concludes the paper with some directions for future work.

2   Related Work

One area of related work concerns ANN/AkNN queries. Existing approaches can be classified into two categories, depending on whether the data set P and/or the query set Q are indexed. The first one [25] involves solutions that do not assume any index on the underlying datasets, but they handle ANN/AkNN retrieval by scanning the entire database at least once, resulting in expensive overhead. In particular, Zhang et al. [25] propose a hash-based method using spatial hashing for processing ANN search. Methods of the other category [3, 5, 15, 25] incur significantly lower query cost by performing the ANN/AkNN retrieval on appropriate index structures (e.g., R-trees [1]). Zhang et al. [25] present two approaches using R-trees, i.e., *multiple Nearest Neighbor search* (MNN) and *Batched Nearest Neighbor* (BNN), to tackle ANN queries. Chen and Patel [3] develop an ANN query processing algorithm based on an enhanced bucket quadtree index structure termed MBRQT, and introduce a pruning metric called NXNDIST to further improve the search performance. Sankaranarayanan et al. [15] investigate the AkNN algorithm (using R-tree) for applications involving large point-clouds in image processing. Recently, Emrich et al. [5] introduce *trigonometric pruning* to shrink the search space, and propose a novel AkNN query processing algorithm which is based on the trigonometric pruning and employs an SS-tree. In the literature, the AkNN query is also referred to as *kNN join* [2, 21, 23]. Bohm and Krebs [2] first explore *k*NN join. Their solution is based on a specialized index structure termed *multipage index* (MuX), and hence is inapplicable for general-purpose index structures (e.g., R-trees). Xia et al. [21] propose GORDER to tackle *k*NN join. It sorts input datasets into the *G-order* (an order based on grid), and then applies the *scheduled block nested loop join* on the G-ordered data to obtain the result. Yu et al. [23] present a *k*NN join algorithm using iDistance. In addition, Hjaltason and Samet [9] and Corral et al. [4] utilize closest pairs (CP) algorithms to answer ANN queries. However, they incur expensive cost since the termination condition for the CP-based algorithm is the identification of the NN for all points in Q. It is worth pointing out that all the aforementioned approaches do not take into account the physical obstacles that are *ubiquitous* in the real world and may affect the *visibility* between objects, and thus cannot be (directly) applicable to deal with AVkNN search efficiently.

Another area of related work is spatial queries in the presence of obstacles. As mentioned in [6], the existence of obstacles could affect the *distance* or/and *visibility*. First, in terms of distance, Zhang et al. [26] propose a suite of algorithms for processing several common spatial queries (e.g., range query, NN retrieval, etc.) with obstacle constraints. Xia et al. [20] present a more detailed study of *obstructed* NN search. Recently, *continuous* NN and *moving* k-NN queries in the presence of obstacles are also explored in [6, 11]. Second, in terms of *visibility*, Nutanong et al. [14] investigate *visible* NN retrieval. In the sequel, other studies along this line include *visible* reverse NN search [7], *continuous visible* NN search [8], and *group visible* NN search [22], respectively. Note that, different from the above works, we aim to tackle the AkNN query in the presence of obstacles. To our knowledge, this paper is the first attempt on this problem.

3   Problem Statement

In this section, we formally define the AVkNN query. Given a data set $P = \{p_1, p_2, ..., p_m\}$, a query set $Q = \{q_1, q_2, ..., q_n\}$, and an obstacle set $O = \{o_1, o_2, ..., o_r\}$ in a 2-dimensional space, the *visibility* between two points is defined in Definition 1 below.

Definition 1 (Visibility [7]). *Given two data points* $p_i, p_j \in P$ *and O, p_i and p_j are visible to each other iff there is* no *any obstacle o_i in O such that the straight line connecting p_i and p_j, denoted as $[p_i, p_j]$, crosses o_i, i.e.,* $\forall o_i \in O, [p_i, p_j] \cap o_i = \varnothing$.

Based on Definition 1 above, we formulate *visible k nearest neighbors* and *all-visible-k-nearest-neighbor query* in Definition 2 and Definition 3 below, respectively.

Definition 2 (Visible k Nearest Neighbors). *Given P, O, and a query point q, a data point $p_i \in P$ is one of the* visible k *nearest neighbors (VkNNs) of q if and only if: (i) p_i is* visible *to q, i.e.,* $\forall o_i \in O, [p_i, q] \cap o_i = \varnothing$; *and (ii) there are* at most $(k - 1)$ *data points $p_j \in P - \{p_i\}$ such that p_j is* visible *to q and meanwhile has its distance to q smaller than that from p_i to q, i.e.,* $|\{p_j \in P - \{p_i\} \mid \forall o_i \in O, [p_j, q] \cap o_i = \varnothing \wedge dist(p_j, q) < dist(p_i, q)\}| < k$.

Definition 3 (All-Visible-k-Nearest-Neighbor Query). *Given P, Q, O, and an integer k (≥ 1), an all-visible-k-nearest-neighbor (AVkNN) query retrieves for each point $q \in Q$ its VkNNs in P. Formally,* AVkNN$(P, Q, O) = \{\langle q, p \rangle \mid \forall q \in Q, \exists p \in (VkNN(q) \subseteq P), \neg\exists p' \in (P - VkNN(q))$ *such that $dist(p', q) < dist(p, q)$ if p' is* visible *to q}, in which* AVkNN(P, Q, O) *represents the result set of the AVkNN query w.r.t. P, Q, and O, VkNN(q) denotes the set of VkNNs for q, and dist( ) refers to the Euclidean distance metric.*

In this paper, we study the problem of efficiently processing AVkNN retrieval.

4   AVkNN Query Processing

In this section, we propose two algorithms, i.e., *Grouping-based Multiple-access Obstacle Tree algorithm* (GMOT) and *Grouping-based Single-access Obstacle Tree algorithm* (GSOT), for answering AVkNN queries, assuming that the data set P, the

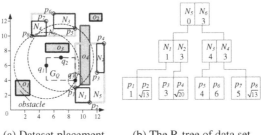

(a) Dataset placement   (b) The R-tree of data set

Fig. 2. A running example

query set Q, and the obstacle set O are indexed by three *separate* R-trees, and then provide an analysis of these algorithms.

On the whole, our solutions are based on *grouping techniques* in order to minimize the number of visi*ble k nearest neighbor* (VkNN) queries. Specifically, we splits the points from Q into *disjoint* groups $G_{Q1}, G_{Q2}, ..., G_{Qt}$, such that $\cup_{1 \le i \le t} G_{Qi} = Q$ and $\forall\ i, j \in [1, t]$ $(i \ne j)$, $G_{Qi} \cap G_{Qj} = \varnothing$. Then, for each group G_Q, we find the VkNNs of each point in G_Q by traversing the data R-tree T_P on P. Note that the number and size of grouping may affect the performance of algorithms. A small grouping incurs more VkNN queries, while a large grouping leads to high storage and computation costs. In general, a good grouping method must consider several criteria: (i) each grouping should be *small* enough to fit in main-memory; (ii) the points in each grouping should be *clustered* to reduce the computation cost; and (iii) the number of points in each grouping should be *maximized* to decrease the number of VkNN queries. Nevertheless, as mentioned in [25], an *optimal trade-off* between these criteria above is *difficult* to obtain, i.e., finding a good grouping is not easy. For simplicity, in this paper, we take every leaf node in the query R-tree T_Q on Q as one grouping. Surely, it is challenging and interesting to develop effective grouping for AVkNN retrieval, but we would like to leave it to our future work due to the limitation of space. In addition to grouping techniques, our solutions also utilize effective *pruning heuristics* to further improve the search performance. In the sequel, a running example, as illustrated in Figure 2, is employed to facilitate the understanding of different AVkNN query processing algorithms. Here, a data point set $P = \{p_1, p_2, ..., p_8\}$, organized in the R-tree of Figure 2(b) with node capacity = 2; a grouping $G_Q = \{q_1, q_2, q_3\}$ ($\subseteq Q$); and an obstacle set $O = \{o_1, o_2, o_3, o_4\}$. Note that, in Figure 2(b), the number in every entry refers to the MINMINDIST [4] (for intermediate entries) or the MINDIST (for data points) between the *minimum bounding rectangle* (MBR) of each group G_Q and the corresponding MBR of the entry. These numbers are not stored in R-tree previously but computed *on-the-fly* during query processing.

4.1   Grouping-Based Multiple-Access Obstacle Tree Algorithm

In this subsection, we present the *Grouping-based Multiple-access Obstacle Tree algorithm* (GMOT). In order to check the visibility of each query point q in the current group G_Q, we need to find all the obstacles that may affect the visibility of q and compute its visible region. However, it is observed that the cost of qualified obstacle retrieval is expensive. Therefore, we develop two pruning heuristics below to discard those unqualified obstacles which cannot affect the visibility of any point in order to improve the query performance.

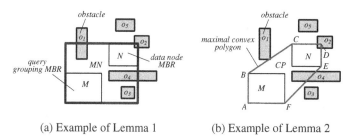

(a) Example of Lemma 1   (b) Example of Lemma 2

Fig. 3. Illustration of Lemma 1 and Lemma 2

Lemma 1. *Given the MBR M of a query grouping G_Q and the MBR N of a data node, the obstacles may affect the visibility of a certain data point p in N w.r.t. a certain query point in G_Q if they overlap the* MBR *formed by M and N, denoted as MN.*

Proof. The proof is obvious since the straight line that connects any query point q in M and any data point p in N, i.e., $[p, q]$, is bounded by the MBR *MN*.   □

An illustrative example is depicted in Figure 3(a), with a query grouping MBR M, a data node MBR N, and an obstacle set $O = \{o_1, o_2, o_3, o_4, o_5\}$. The MBR formed by M and N is the rectangle *MN* (highlighted in blue). According to Lemma 1 above, the obstacles o_1, o_2, o_3, and o_4 may affect the visibility, whereas the obstacle o_5 can be pruned away directly.

Although Lemma 1 is straightforward, it can discard a large amount of unqualified obstacles that may not affect the visibility of any data point in the current data node MBR w.r.t. any query point in the current grouping. However, the pruning power of Lemma 1 can be further enhanced, which is described in Lemma 2 below.

Lemma 2. *Given the MBR M of a query grouping G_Q and the MBR N of a data node, the obstacles may affect the visibility of a certain data point p in N w.r.t. a certain query point in G_Q if they overlap the* maximal convex polygon *formed by M and N, denoted as CP.*

Proof. Similar to the proof of Lemma 1, and thus omitted.   □

For instance, we apply Lemma 2 on the same example of Lemma 1 (shown in Figure 3(a)), and the result is illustrated in Figure 3(b). To be more specific, the maximal convex polygon *CP* formed by M and N is the hexagon *ABCDEF* (highlighted in red). Hence, according to Lemma 2 above, the obstacles o_1, o_2, and o_4 may affect the visibility, whereas the obstacles o_3 and o_5 can be discarded directly. Note that the obstacle o_3 cannot be pruned by Lemma 1.

Our first approach, namely GMOT, employs the aforementioned pruning heuristics to discard unqualified obstacles. The pseudo-code of GMOT is depicted in Algorithm 1. GMOT follows the *best-first* traversal paradigm. It maintains a heap H to store the data entries visited so far, sorted by ascending order to their MINMINDIST to the query group G_Q; and a list l_o that preserves all qualified obstacles during the search. As defined in [3, 4], the MINMINDIST between two MBRs is the *minimum* possible distance between any point in the first MBR and any point in the second MBR. It can be used as the *lower bound metric* for pruning. As an example, the MINMINDIST between MBRs M and N is illustrated in Figure 5(a).

Algorithm 1. Grouping-based Multiple-access Obstacle Tree Algorithm (GMOT)

Input: a data R-tree T_P, an obstacle R-tree T_O, a grouping G_Q $(\subseteq Q)$, an integer k
Output: the result set S_r of an AVkNN query
/* H: a min-heap sorted in *ascending order* of the MINMINDIST (to G_Q); $T_P.root$: the root node of the data R-tree T_P; $q.vknnD$: the *maximal* distance among the distances from q to its every VkNN; $G_Q.vknnD$: the *maximal* distance among the distances between each query point $q \in G_Q$ and its every VkNN; l_o: a list storing obstacles. */
1:   initialize $H = l_o = \varnothing$ and $q.vknnD = G_Q.vknnD = \infty$
2:   insert all entries of $T_P.root$ into H
3:   **while** $H \neq \varnothing$ **do**
4:       de-heap the top entry (e, key) of H
5:       **if** MINMINDIST$(G_Q, e) \geq G_Q.vknnD$ **then**   // early termination condition
6:           break
7:       **else**
8:           **if** e is a leaf node **then**
9:               form the convex polygon CP w.r.t. G_Q and e, and add all obstacles crossing
                    CP to a list l_o by traversing the obstacle R-tree T_O // use Lemmas 1 and 2
10:              **for** each $q \in G_Q$ **do**
11:                  **if** MINDIST$(q, e) \geq q.vknnD$ **then**
12:                      continue
13:                  VRC (l_o, q) // compute q's visible region using algorithm of [7]
14:                  choose the dimension *dim* with the largest projection of e in axis, and then
                        sort the points in e by ascending order of their distances to q along *dim*
15:                  **for** each point $p \in e$ **do**
16:                      **if** $dist_{dim}(q, p) \geq q.vknnD$ **then**
17:                          break
18:                      **if** p is visible to q and $dist(q, p) < q.vknnD$ **then**
19:                          add p to q's VkNN set $S_{vknn} \in S_r$ and update $q.vknnD$ if necessary
20:                  update $G_Q.vknnD$ if necessary and $l_o = \varnothing$ // for the next round
21:          **else**   // e is an intermediate (i.e., a non-leaf) node
22:              **for** each child entry $e_i \in e$ **do**
23:                  insert $(e_i, \text{MINMINDIST}(G_Q, e_i))$ into H
24:   return S_r

Initially, GMOT inserts all entries in the root of the data R-tree T_P into a min-heap H. Then, the algorithm recursively retrieve the VkNNs of each query point in G_Q until H is *empty* or the *early termination condition* satisfies (lines 3-23). Specifically, for every head entry e of H, GMOT first determines the MINMINDIST between G_Q and e, i.e., MINMINDIST(G_Q, e), is *no smaller than* current $G_Q.vknnD$ which maintains the *maximal* distance among the distances from each query point in G_Q to its VkNN (line 5). If yes, the algorithm terminates as the remaining data points cannot be one of actual answer points. Otherwise, it distinguishes two cases: (i) If e is a leaf node, GMOT uses Lemma 1 and Lemma 2 to obtain all qualified obstacles by traversing the obstacle R-tree T_O (line 9). Thereafter, for each query point q in G_Q, the algorithm first determines whether the minimal distance between q and e is *no smaller than* the current $q.vknnD$ which stores the *maximal* distance among the distances from q to its every VkNN (line 11). If yes, e cannot be one of the VkNNs of q. Otherwise, GMOT invokes our previously proposed VRC algorithm in [7] to compute the *visible region* of q, and then updates q's VkNN set S_{vknn} as well as $q.vknnD$ if necessary (lines 13-19). (ii) If e is an intermediate node, GMOT inserts its child entries into H (lines 22-23). Finally, the result set S_r of an AVkNN query is returned (line 24). Note that the operation involved in line 20 is necessary for the evaluation later.

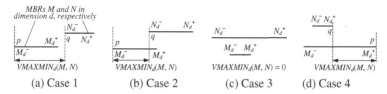

Fig. 4. Illustration of $VMAXMIN_d$ metric

Back to our running example shown in Figure 2, and suppose the integer $k = 1$. In the first place, GMOT visits the root of the data R-tree T_P and inserts its entries N_5, N_6 into a heap $H = \{N_5, N_6\}$. Then, the algorithm de-heaps the top entry N_5 of H, accesses its child nodes, and en-heaps the entries into $H = \{N_1, N_2, N_6\}$. Next, N_1 is visited. As N_1 is a leaf node, GMOT forms the convex polygon w.r.t. G_Q and N_1, and then obtains qualified obstacle o_4 and updates the obstacle list l_o to $\{o_4\}$ based on Lemmas 1 and 2. Since points p_1 and p_2 in N_1 are *visible* to each point q_i $(1 \leq i \leq 3)$ in G_Q and p_1 is *closer* to q_i than p_2, p_1 is added to the current V1NN set S_{v1nn} of q_i, and $q_i.vknnD$ is updated to $dist(q_i, p_1)$. Also, $G_Q.vknnD$ is updated to 5. The algorithm proceeds in the same manner until the result set $S_r = \{\langle q_1, p_8 \rangle, \langle q_2, p_1 \rangle, \langle q_3, p_1 \rangle\}$ is returned.

4.2   Grouping-Based Single-Access Obstacle Tree Algorithm

Although GMOT employs pruning heuristics (i.e., Lemmas 1 and 2) to prune away unqualified obstacles, it needs to traverse the obstacle R-tree T_O *multiple times* to get qualified obstacles, resulting in high I/O and CPU costs. Motivated by this, in this subsection, we propose the *Grouping-based Single-access Obstacle Tree algorithm* (GSOT), which traverses T_O *only once* during each group search, reducing the computation cost significantly. Moreover, a *new pruning metric* called VMDIST (as an upper bound for pruning) is introduced to further improve the query performance.

In a d-dimensional space, an arbitrary point p could be denoted as $\langle p_1, p_2, ..., p_d \rangle$. Based on this, we formally define the $VMAXMIN_d$ metric.

Definition 4 ($VMAXMIN_d$). *Given two d-dimensional MBRs M and N, an arbitrary point q in M, and an arbitrary point p in N, the $VMAXMIN_d(M, N)$ in dimension d is defined as*:

$$\text{VMAXMIN}_d(M, N) = \text{MAX}_{\forall q \in M}(\text{MIN}_{\forall p \in N}|q_d - p_d|)$$

In fact, $\text{VMAXMIN}_d(M, N)$ is the *maximal* distance among the *minimal* distances of a point in M to any point in N on dimension d. Figure 4 illustrates all cases of VMAXMIN_d metric, in which MBRs M and N in dimension d are represented as $\langle M_d^-, M_d^+ \rangle$ and $\langle N_d^-, N_d^+ \rangle$, respectively. For example, in Figure 4(a), $M_d^- < M_d^+ < N_d^- < N_d^+$, $\text{VMAXMIN}_d(M, N) = N_d^- - M_d^-$. Note that, other cases can be transferred to one of them. Based on Definition 4 above, we formalize VMAXMIN metric below.

Definition 5 (VMDIST). *Given two MBRs M and N, VMDIST(M, N) is defined as*:

$$\text{VMDIST}(M, N) = \begin{cases} 0 & \text{if } M \text{ and } N \text{ overlaps} \\ \sqrt[d]{\sum_{d=1}^{d} \text{VMAXMIN}_d^2(M, N)} & \text{otherwise} \end{cases} \tag{1}$$

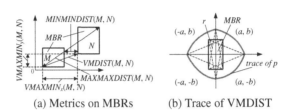

(a) Metrics on MBRs   (b) Trace of VMDIST

Fig. 5. Illustration of VMDIST properties

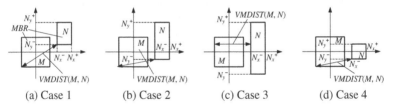

(a) Case 1   (b) Case 2   (c) Case 3   (d) Case 4

Fig. 6. Illustration of VMDIST metric

Figure 5(a) illustrates various distance metrics between two MBRs M and N. Using the VMDIST metric, we develop two pruning heuristics below to discard those unqualified obstacles in order to improve the query performance.

Lemma 3. *Given the MBR M of a query grouping G_Q, the MBR N of an obstacle node, a data point p, and let $|diagonal(M)|$ be the diagonal length of M, if $\text{VMDIST}(M, N) > \text{MAX}(\text{MAXDIST}(p, M), |diagonal(M)|)$, N does not affect the visibility of p w.r.t. M.*

Proof. We use $[(-a, a), (-b, b)]$, $[(N_x^-, N_x^+), (N_y^-, N_y^+)]$, (p_x, p_y) to represent M, N, and p respectively, and assume that the maximal distance from p to M, i.e., $\text{MAXDIST}(p, M)$ is r. If p is located inside the first quadrant, point $(-a, -b)$ will reach the largest distance from p to M. Other quadrants are similar to the first quadrant. The trace of p is the red line depicted in Figure 5(b), and the function of p is as follows:

$$(\text{MAXDIST}(M, p))^2 = \left(p_x + \frac{p_x}{|p_x|} \times a \right)^2 + \left(p_y + \frac{p_y}{|p_y|} \times b \right)^2 \tag{2}$$

As *VMDIST(M, N)* > r, any point p' in N, denoted as $(p_x'\, p_y')$, does not fall into the area bounded by p's trace (i.e., red area in Figure 5(b)). For example, as shown in Figure 6(a), if $N_x^- \geq a$ and $N_y^- \geq 0$, $(\text{MAXDIST}(M, p'))^2 = (p_x' + a)^2 + (p_y' + b)^2 \geq (N_x^- + a)^2 + (N_y^- + b)^2 > (\text{VMDIST}(M, N))^2 > r^2$. If $\text{MAXDIST}(p, M) < |diagonal(M)|$, partial MBR M is *outside* p's trace. Thus, if $\text{VMDIST}(M, N) > \text{MAX}(\text{MAXDIST}(p, M), |diagonal(M)|)$, p' does not in p's trace, and M and p' do not affect the visibility of p w.r.t. M. Similarly, other cases in Figures 6(b), 6(c), and 6(d), can be also proved. $\square$

Lemma 4. *Given the MBR M of a query grouping G_Q, the MBR N of an obstacle node, the MBR N_p of a data node, and let $|diagonal(M)|$ be the diagonal length of M, if $\text{MAX}(|diagonal(M)|, \text{MAXMAXDIST}(N_p, M)) < \text{VMDIST}(M, N)$, N does not affect the visibility of any data point in N_p w.r.t. M.*

Proof. According to the aforementioned Lemma 3, N does not overlap the largest area bounded by trace p in N_p. Consequently, N does not affect the visibility of any data point in N_p w.r.t. M. $\square$

Algorithm 2. Obstacle Pruning Algorithm (OP)

Input: a grouping G_Q ($\subseteq Q$), a leaf node e', a required obstacle list l_r, a candidate obstacle list l_c
Output: the updated list l_r of required obstacles
1:   initialize $l_r = \varnothing$
2:   **while** $l_c \neq \varnothing$ **do**
3:      remove the top entry (e, key) of l_c
4:      **if** $\text{VMDIST}(G_Q, e) > \text{MAXMAXDIST}(G_Q, e') > |diagonal(G_Q)|$ **then**
5:         add $(e, \text{VMDIST}(G_Q, e))$ to l_c // early termination
6:         break
7:      **else**
8:         **if** e is an obstacle **then**
9:            add $(e, \text{VMDIST}(G_Q, e))$ to l_r
10:        **else**   // e is a node entry
11:           **for** each entry $e_i \in e$ **do**
12:             **if** e_i is an obstacle and $\text{VMDIST}(G_Q, e_i) \leq \text{MAXMAXDIST}(G_Q, e')$ **then**
13:                add $(e_i, \text{VMDIST}(G_Q, e_i))$ to l_r
14:             **else**
15:                add $(e_i, \text{VMDIST}(G_Q, e_i))$ to l_c
16:   return l_r

Based on Lemma 4 above, we present the *Obstacle Pruning Algorithm* (OP) whose pseudo-code is shown in Algorithm 2. OP takes as input a grouping G_Q ($\subseteq Q$), a leaf node e', a list l_r storing the obstacles that are required for obtaining the final query result, and a candidate obstacle list l_c, and outputs the updated list l_r of required obstacles. It is worth noting that, the elements in l_c and l_r are sorted in ascending order of their VMDIST to G_Q. For every top entry e of l_c, OP inserts e into l_r if e is an actual obstacle and its VMDIST to G_Q is *no larger than* the MAXMAXDIST from G_Q to e', i.e., $\text{VMDIST}(G_Q, e) \leq \text{MAXMAXDIST}(G_Q, e')$ holds; otherwise, e is added to either l_c or l_r according to $\text{VMDIST}(G_Q, e)$ (lines 8-15). As defined in [3, 4], the

MAXMAXDIST between two MBRs is the *maximum* possible distance between any point in the first MBR and any point in the second MBR. For instance, the MAXMAXDIST between MBRs M and N is shown in Figure 5(a).

Having explained OP algorithm, we are ready to present our second approach, namely GSOT, for processing AVkNN search. The pseudo-code of GSOT is depicted in Algorithm 3. Like GMOT, GSOT follows the *best-first* traversal paradigm. Unlike GMOT, however, it employs our proposed pruning metric, i.e., VMDIST, to discard unqualified obstacles during the search. In particular, if current top entry e of H is a leaf node, GSOT calls OP to incrementally obtain qualified obstacles (preserved in l_r) as well as prune away those unqualified obstacles.

Back to the running example of Figure 2 again, and suppose the integer $k = 1$. First of all, GSOT visits the root of the data R-tree T_P and inserts its entries N_5, N_6 into a min-heap $H = \{N_5, N_6\}$. Then, the algorithm de-heaps the top entry N_5 of H, accesses its child nodes, and en-heaps the entries into $H = \{N_1, N_2, N_6\}$. Next, N_1 is visited. As N_1 is a leaf node, GSOT invokes OP, rather than utilizes Lemmas 1 and 2, to obtain qualified obstacles and updates the required obstacle list l_r to $\{o_1, o_3, o_4\}$. Since points p_1 and p_2 in N_1 are *visible* to each point q_i $(1 \le i \le 3)$ in G_Q and p_1 is *closer* to q_i than p_2, p_1 is added to the current V1NN set S_{v1nn} of q_i, and $q_i.vknnD$ is updated to $dist(q_i, p_1)$. Also, $G_Q.vknnD$ is updated to 5. The algorithm proceeds in the same manner until the result set $S_r = \{\langle q_1, p_8 \rangle, \langle q_2, p_1 \rangle, \langle q_3, p_1 \rangle\}$ is returned.

Algorithm 3. Grouping-based Single-access Obstacle Tree Algorithm (GSOT)

Input: a data R-tree T_P, an obstacle R-tree T_O, a grouping G_Q $(\subseteq Q)$, an integer k
Output: the result set S_r of an AVkNN query
/* l_r: a list storing the obstacles that are required for obtaining the final query result; l_c: a list storing candidate obstacles; $T_O.root$: the root node of the obstacle R-tree T_O. */
1: initialize $H = l_r = l_c = \varnothing$ and $q.vknnD = G_Q.vknnD = \infty$
2: insert all entries of $T_P.root$ into H and all entries of $T_O.root$ into l_c
3: **while** $H \ne \varnothing$ **do**
4:      de-heap the top entry (e, key) of H
5:      **if** MINMINDIST$(G_Q, e) \ge G_Q.vknnD$ **then**
6:          break   // early termination
7:      **else**
8:          **if** e is a leaf node **then**
9:              OP (GQ, e, l_r, l_c) // Algorithm 2
10:             **for** each $q \in G_Q$ **do**
11:                 **if** MINDIST$(q, e) \ge q.vknnD$ **then**
12:                     continue
13:                 VRC (l_r, q) // compute q's visible region using algorithm of [7]
14:                 choose the dimension *dim* with the largest projection of e in axis, and then sort the points in e by ascending order of their distances to q along *dim*
15:                 **for** each point $p \in e$ **do**
16:                     **if** $dist_{dim}(q, p) \ge q.vknnD$ **then**
17:                         break
18:                     **if** p is visible to q and $dist(q, p) < q.vknnD$ **then**
19:                         add p to q's VkNN set $S_{vknn} \in S_r$ and update $q.vknnD$ if necessary
20:                 update $G_Q.vknnD$ if necessary and $l_r = \varnothing$ // for the next round
21:         **else**   // e is an intermediate node
22:             **for** each child entry $e_i \in e$ **do**
23:                 insert $(e_i,$ MINMINDIST$(G_Q, e_i))$ into H
24: return S_r

4.3   Discussion

In this subsection, we reveal some characteristics of our proposed algorithms.

Lemma 5. *For each query grouping G_Q, the GMOT algorithm traverses the data R-tree T_P only once and the obstacle R-tree T_O multiple times, whereas the GSOT algorithm traverses both T_P and T_O only once.*

Proof. As shown in Algorithm 1, the GMOT algorithm traverses T_P only once based on the best-first fashion to evaluate every data point in P. In addition, once a leaf node e is visited, it uses Lemmas 1 and 2 to obtain all qualified obstacles via traversing T_O (see line 9 in Algorithm 1). We omit the proof of GSOT property since it is similar to that of GMOT property. $\qquad\square$

Lemma 6. *Both the GMOT algorithm and the GSOT algorithm can retrieve exactly the VkNNs of each point in every query grouping G_Q.*

Proof. This is guaranteed by Lemmas 1 through 4. $\qquad\square$

5   Experimental Evaluation

In this section, we experimentally evaluate the effectiveness of our developed pruning heuristics and the performance of our proposed algorithms for AVkNN search, using both real and synthetic datasets. All algorithms were implemented in C++, and all experiments were conducted on an Intel Core 2 Duo 2.93 GHz PC with 3GB RAM.

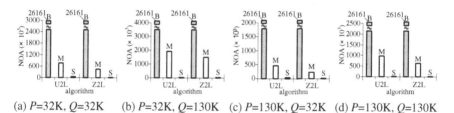

(a) *P*=32K, *Q*=32K   (b) *P*=32K, *Q*=130K   (c) *P*=130K, *Q*=32K   (d) *P*=130K, *Q*=130K

Fig. 7. Heuristic efficiency vs. problem size (*O=LA*)

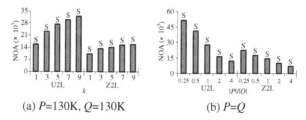

(a) *P*=130K, *Q*=130K   (b) *P=Q*

Fig. 8. Heuristic efficiency vs. *k* and |*P*|/|*O*| respectively (*O=LA*)

We employ a real dataset *LA*, which contains 131,461 2D rectangles/MBRs of streets in Los Angeles. We also create several synthetic datasets following the uniform and zipf distribution, with the cardinality varying from $0.25 \times |LA|$ to $4 \times |LA|$.

The coordinate of each point in *Uniform* datasets is generated uniformly along each dimension, and that of each point in *Zipf* datasets is generated according to Zipf distribution with skew coefficient $\alpha = 0.8$. All the datasets are mapped to a [0, 10000] × [0, 10000] square. Since AVkNN retrieval involves a data set P, a query set Q, and an obstacle set O, we deploy two different combinations of the datasets, namely, *U2L* and *Z2L*, representing $(P, Q, O) = (Uniform, Uniform, LA)$ and $(Zipf, Zipf, LA)$, respectively. Note that the data points in P are allowed to lie on the boundaries of the obstacles but not in their interior. All data and obstacle sets are indexed by R*-trees [1], with a page size of 4K bytes.

We investigate several factors, involving dataset size, k, |P|/|O|, and buffer size. Note that, in each experiment, only one factor varies, whereas the others are fixed to their default values. The *query time* (i.e., the sum of I/O cost and CPU time, where the I/O cost is computed by charging 10ms for each page access, as with [16]) and *the number of obstacles accessed* (*NOA*) are used as the major performance metrics. In the figures below, we denote B, M, S for Baseline, GMOT and GSOT algorithms, respectively. Similar to [25], unless specifically stated, the size of LRU buffer is 0.6M in the experiments.

5.1   Effectiveness of Pruning Techniques

This set of experiments aims at verifying the effectiveness of our proposed pruning heuristics (i.e., Lemmas 1, 2, and 4). Recall that GMOT integrates Lemmas 1 and 2, whereas GSOT uses Lemma 4. We measure the effectiveness of a heuristic by how much it accesses the qualified obstacles per group during the search. The Baseline algorithm presents the number of obstacles that AVkNN operation requires. Figure 7 plots the efficiency of different heuristics for different problem size instances: (i) both P and Q are small, (ii) P is small and Q is large, (iii) P is large and Q is small, and (iv) both P and Q are large. Compared with Baseline, all heuristics prune away a large number of unqualified obstacles, validating their usefulness. The obstacles accessed by GMOT and GSOT are much less than that of Baseline. Observe that GSOT (using Lemma 4) is better than GMOT (using Lemmas 1 and 2) in all cases. This is because GMOT needs to discard unqualified obstacles by traversing the obstacle T_O multiple times. In the sequel, we only report the experimental results on GSOT since GSOT always exceeds the other algorithms in all cases. Figure 8 illustrates the prune efficiency of Lemma 4 w.r.t. k and |P|/|O|, respectively, using *U2L* and *Z2L*. The diagrams confirm the observations and corresponding explanations of Figure 7.

5.2   Results on AVkNN Queries

The second set of experiments studies the performance of our proposed algorithms (i.e., Baseline, GMOT, and GSOT) in answering AVkNN queries. First, we explore the impact of different problem size, and the results are shown in Figure 9. Clearly, GSOT performs the best in all cases, as demonstrated by the subsequent experiments. The reason behind is that, GSOT only incrementally traverses the obstacle R-tree T_O only once, but GMOT needs to traverse T_O multiple times (as pointed out in

Lemma 5), and for each query point, Baseline only finds its VkNNs and does not discard any obstacle. Furthermore, a comparison between case (a) and case (d) reveals that the improvement of GSOT over GMOT and Baseline is independent of the problem size, which is also confirmed by subsequent experiments. Therefore, for clarity of experimental figures, in the rest of experiments, we only illustrate the results that shed light on the efficiency of GSOT algorithm.

Then, we study the impact of k on the performance of GSOT algorithm, using *U2L* and *Z2L*. Figure 10(a) depicts the query cost as a function of k which varies from 1 to 9. As expected, the cost of GSOT grows slightly with k. This is because, when k increases, more VkNNs of every query point need to find, incurring more node/page accesses, more obstacle retrieval, and larger search space.

Next, we investigate the effect of $|P|/|O|$ ratio on the GSOT algorithm. Towards this, we fix $k = 5$ (i.e., the median value in Figure 10(a)), and vary $|P|/|O|$ from 0.25 to 4. Figure 10(b) shows the performance of GSOT algorithm w.r.t. different $|P|/|O|$. It is observed that the total cost of GSOT demonstrates a *stepwise* behavior. Specifically, it increases slightly as $|P|/|O|$ changes from 0.25 to 1, but then ascends much faster as $|P|/|O|$ grows further. The reason is that, as the density of data set P and query set Q grows, the number of query grouping G_Q increases as well, leading to more traversals of the R-tree T_O, more visibility check, and more distance computation.

As mentioned earlier, all the above experiments are conducted with a 0.6M LRU buffer. Thus, the last set of experiments examines the efficiency of GSOT algorithm under various LRU buffer sizes, by fixing $k = 5$. Figure 10(c) plots the experimental results. Observe that the total query cost of GSOT algorithm drops with the growth of buffer size. This is because a larger buffer maximizes the probability of accessing the entries in the same main-memory, considerably reducing the total query cost.

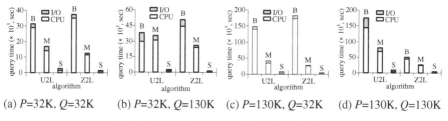

(a) P=32K, Q=32K   (b) P=32K, Q=130K   (c) P=130K, Q=32K   (d) P=130K, Q=130K

Fig. 9. AVkNN cost vs. problem size (O=LA)

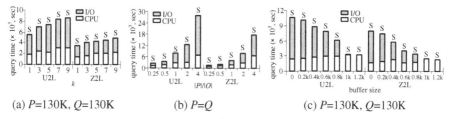

(a) P=130K, Q=130K   (b) P=Q (c) P=130K, Q=130K

Fig. 10. AVkNN cost vs. k, $|P|/|O|$, and buffer size respectively (O=LA)

6   Conclusions

This paper, for the first time, identifies and solves a novel form of all-k-nearest-neighbor queries, namely *All-Visible-k-Nearest-Neighbor* (AVkNN) query, which takes into account the impact of obstacles on the *visibility* of objects. AVkNN search is not only interesting from a research point of view, but also useful in many applications such as GIS and data analysis/mining. We carry out a systematic study of the AVkNN retrieval. First, we present a formal definition of the problem. Then, we propose two efficient algorithms, i.e., GMOT and GSOT, for *exact* AVkNN query processing, assuming that a data set P, a query set Q, and an obstacle set O are indexed by three separated conventional data-partitioning indexes (e.g., R-trees). Our approaches employ *pruning techniques* and a *new pruning metric* called VMDIST to improve the query performance. Finally, we conduct extensive experiments using both real and synthetic datasets to confirm the effectiveness of our developed pruning techniques and the performance of our proposed algorithms. In the future, we intend to extend our methods to handle other variants of AVkNN queries, e.g., *constrained* AVkNN retrieval, etc. In addition, we plan to investigate visibility queries for moving objects and moving obstacles.

Acknowledgements. This work was supported by NSFC 61003049, ZJNSF Y110278, Fundamental Research Funds for the Central Universities under grant 2010QNA5051, 2012QNA5018, the Key Project of Zhejiang University Excellent Young Teacher Fund, and the Returned Scholar Fund for the Personnel Office of Zhejiang Province.

References

1. Beckmann, N., Kriegel, H.-P., Schneider, R., Seeger, B.: The R*-tree: An Efficient and Robust Access Method for Points and Rectangles. In: SIGMOD, pp. 322–331 (1990)
2. Bohm, C., Krebs, F.: The k-Nearest Neighbour Join: Turbo Charging the KDD Process. Knowl. Inf. Syst. 6(6), 728–749 (2004)
3. Chen, Y., Patel, J.: Efficient Evaluation of All-Nearest-Neighbor Queries. In: ICDE, pp. 1056–1065 (2007)
4. Corral, A., Manolopoulos, Y., Theodoridis, Y., Vassilakopoulos, M.: Closest Pair Queries in Spatial Databases. In: SIGMOD, pp. 189–200 (2000)
5. Emrich, T., Graf, F., Kriegel, H.-P., Schubert, M., Thoma, M.: Optimizing All-Nearest-Neighbor Queries with Trigonometric Pruning. In: Gertz, M., Ludäscher, B. (eds.) SSDBM 2010. LNCS, vol. 6187, pp. 501–518. Springer, Heidelberg (2010)
6. Gao, Y., Zheng, B., Chen, G., Chen, C., Li, Q.: Continuous Nearest-Neighbor Search in the Presence of Obstacles. ACM Trans. Database Syst. 36(2), 9 (2011)
7. Gao, Y., Zheng, B., Chen, G., Lee, W.C., Lee, K.C.K., Li, Q.: Visible Reverse k-Nearest Neighbor Query Processing in Spatial Databases. IEEE Trans. Knowl. Data Eng. 21(9), 1314–1327 (2009)
8. Gao, Y., Zheng, B., Chen, G., Li, Q., Guo, X.: Continuous Visible Nearest Neighbor Query Processing in Spatial Databases. VLDB J. 20(3), 371–396 (2011)
9. Hjaltason, G.R., Samet, H.: Incremental Distance Join Algorithms for Spatial Databases. In: SIGMOD, pp. 237–248 (1998)

10. Jain, A., Murthy, M., Flynn, P.: Data Clustering: A Review. ACM Comput. Surv. 31(3), 264–323 (1999)
11. Li, C., Gu, Y., Li, F., Chen, M.: Moving k-Nearest Neighbor Query over Obstructed Regions. In: APWeb, pp. 29–35 (2010)
12. Lowe, D.: Object Recognition from Local Scale-Invariant Features. In: ICCV, pp. 1150–1157 (1999)
13. Nakano, K., Olariu, S.: An Optimal Algorithm for the Angle-Restricted All Nearest Neighbor Problem on the Reconfigurable Mesh, with Applications. IEEE Trans. Parallel Distrib. Syst. 8(9), 983–990 (1997)
14. Nutanong, S., Tanin, E., Zhang, R.: Incremental Evaluation of Visible Nearest Neighbor queries. IEEE Trans. Knowl. Data Eng. 22(5), 665–681 (2010)
15. Sankaranarayanan, J., Samet, H., Varshney, A.: A Fast All Nearest Neighbor Algorithm for Applications Involving Large Point-Clouds. Comput. Graph. 31(2), 157–174 (2007)
16. Tao, Y., Papadias, D., Lian, X., Xiao, X.: Multidimensional Reverse kNN Search. VLDB J. 16(3), 293–316 (2007)
17. Tung, A.K.H., Hou, J., Han, J.: Spatial Clustering in the Presence of Obstacles. In: ICDE, pp. 359–367 (2001)
18. Wang, X., Hamilton, H.J.: Clustering Spatial Data in the Presence of Obstacles. International Journal on Artificial Intelligence Tools 14(1-2), 177–198 (2005)
19. Wang, Y.-R., Horng, S.-J., Wu, C.-H.: Efficient Algorithms for the All Nearest Neighbor and Closest Pair Problems on the Linear Array with a Reconfigurable Pipelined Bus System. IEEE Trans. Parallel Distrib. Syst. 16(3), 193–206 (2005)
20. Xia, C., Hsu, D., Tung, A.K.H.: A Fast Filter for Obstructed Nearest Neighbor Queries. In: Williams, H., MacKinnon, L.M. (eds.) BNCOD 2004. LNCS, vol. 3112, pp. 203–215. Springer, Heidelberg (2004)
21. Xia, C., Lu, H., Ooi, B.C., Hu, J.: GORDER: An Efficient Method for KNN Join Processing. In: VLDB, pp. 756–767 (2004)
22. Xu, H., Li, Z., Lu, Y., Deng, K., Zhou, X.: Group Visible Nearest Neighbor Queries in Spatial Databases. In: Chen, L., Tang, C., Yang, J., Gao, Y. (eds.) WAIM 2010. LNCS, vol. 6184, pp. 333–344. Springer, Heidelberg (2010)
23. Yu, C., Cui, B., Wang, S., Su, J.: Efficient Index-based KNN Join Processing for High-dimensional Data. Information and Software Technology 49(4), 332–344 (2007)
24. Zaiane, O.R., Lee, C.-H.: Clustering Spatial Data in the Presence of Obstacles: A Density-based Approach. In: IDEAS, pp. 214–223 (2002)
25. Zhang, J., Mamoulis, N., Papadias, D., Tao, Y.: All-Nearest-Neighbors Queries in Spatial Databases. In: SSDBM, pp. 297–306 (2004)
26. Zhang, J., Papadias, D., Mouratidis, K., Zhu, M.: Spatial Queries in the Presence of Obstacles. In: Bertino, E., Christodoulakis, S., Plexousakis, D., Christophides, V., Koubarakis, M., Böhm, K. (eds.) EDBT 2004. LNCS, vol. 2992, pp. 366–384. Springer, Heidelberg (2004)

Algorithm for Term Linearizations
of Aggregate Queries with Comparisons

Victor Felea[1] and Violeta Felea[2]

[1] CS Department, Al.I. Cuza University, Iasi, Romania
`felea@infoiasi.ro`
[2] FEMTO-ST, Franche-Comté University, Besançon, France
`violeta.felea@femto-st.fr`

Abstract. We consider the problem of rewriting queries based exclusively on views. Both queries and views can contain aggregate functions and include arithmetic comparisons. To study the equivalence of a query with its rewriting query, the so called "linearizations of a query" need to be computed. To find the linearizations of a query, the linearizations of terms from the query need to be generated. We propose an algorithm to find these term linearizations and give a bound for its time-complexity.

1 Introduction

The problem of rewriting queries using views (RQV) is expressed, informally, as follows: given Q a query over a database schema, and $\mathcal{V}$ a set of view definitions, is it possible to answer the query Q using only the answers to the views from $\mathcal{V}$? One of the application fields of RQV concerns extracting and gathering information on the environment, based on sensor networks. Small inexpensive devices, namely sensors, communicate information to collecting points, called sinks or base stations. In this context, of information collection out of several sources (stored in views), located on sensor nodes, expressing queries using views. In [7], the authors apply the notions of query interface for data integration to sensor networks.

In case of finding a query Q', corresponding to a given query Q, and expressed by the views from $\mathcal{V}$, such that Q is equivalent to Q', the problem of (RQV) for the query Q has a solution. This query Q' is called a rewriting query of Q.

The equivalence problem of two aggregate queries with comparisons was investigated in [4], [1], [2], [5]. The equivalence problem of the two queries is achieved using the equivalence of their reduced queries. A reduced query associated to a query Q uses the so-called "linearizations" (or complete ordering) of the set of all terms from Q.

In [3], query equivalence between view-based query and schema query, using linearizations, is proven; no linearization computation has been given. This is the main aim of this work: to propose an algorithm to compute linearizations. By our best knowledge, an algorithm to obtain all linearizations of terms from a query has not been proposed in the literature.

S.W. Liddle et al. (Eds.): DEXA 2012, Part II, LNCS 7447, pp. 408–415, 2012.
© Springer-Verlag Berlin Heidelberg 2012

In the paper, we consider the problem of generating all linearizations corresponding to a query that uses arithmetic comparisons and aggregate functions; we consider a motivated example (see section 2) and give an algorithm that computes these linearizations, and its time complexity in section 3. We finally conclude.

2   Motivating Example

Let us give an example about a schema regarding sensor networks used in statistical applications concerning the weather. The network nodes are organized in clusters [6], one common architecture used in sensor networks.

Example 1. We assume there exist sensors for the following type of data: humidity, mist and temperature. One sensor can retrieve unique or multiple environmental data, e.g. temperature and humidity, or only temperature values. Sensors are placed in several locations, and have an identifier, a type, and they are grouped by regions. The type of a sensor characterizes the set of data types that it senses. A location is identified by two coordinates *lat, long*, where *lat* is the latitude and *long* is the longitude. A region has an identifier and a unique name. Let us consider the relational schema S consisting of {*REG, POINT, SENSOR, DATA*}, where *REG* represents regions and has *idReg* as region identifier, and *regName* as region name; *POINT* represents the locations from the regions, and uses the *idReg* attribute as the region identifier, and the *lat, long* attributes as the coordinates of a location from the region identified by *idReg*; *SENSOR* represents information about sensors with *idS* the sensor identifier, *lat, long* - the sensor coordinates, *typeS* its type related to the sensing capacity (e.g. *typeS* equals 1 for temperature and humidity, 2 for temperature and mist, 3 for humidity and mist, and so on), and *chS* designs if the sensor is a cluster head (its value equals 1 or 2). We assume here that the type of the sensors that can sense temperature equals either 1 or 2; *DATA* represents the values registered by sensors, and has the following attributes: *idS* - the sensor identifier, *year, month, day, period* - the date of the sensed information, the daily frequency of sensed data, expressed in seconds, and *temp, humid, mist* - the temperature, humidity, and mist values, respectively.

Let us consider the following views:

V_1: Compute all pairs of the form (*idReg, regName*), where *idReg* is the region identifier, and *regName* is the name of the region identified by *idReg*, such that in the region *idReg* at least a cluster head exists.
V_2: Compute the maximum temperature for each region, each network location, and for every year.

Let us consider the following queries:

Q_1: Find the maximum of temperature values for the *"NORTH"* region, between 2003 and 2010, considering data sensed only by cluster head sensors.
Q_2: Find the maximum temperature corresponding to the *"NORTH"* region, on September 22nd, 2009.

Example 2. Let us consider the schema, queries, and views from Example 1. We denote the attributes $idReg$, $regName$ from REG by x, y_1, respectively, $idReg$, lat, $long$ from $POINT$ by y_2, y_3, y_4, and idS, lat, $long$, $typeS$, chS from $SENSOR$ by z_1, z_2, z_3, t_1, t_2, and the attributes from the schema $DATA$ by $z_5, z_6, z_7, z_8, z_9, t, z_{11}, z_{12}$. To be short, we denote by D, R, S, P the relational symbols $DATA$, REG, $SENSOR$, $POINT$, respectively. Let $c_1 = "NORTH"$, $c_2 = 2003$ and $c_3 = 2010$. To obtain a condition associated to Q_1, we replace the variables y_1, t_2 by $c_1, 1$, respectively. Since $y_2 = x$, $z_2 = y_3$, $z_3 = y_4$, and $z_5 = z_1$, the query Q_1 is expressed as the following condition:
$Q_1 : h(x, max(t)) \leftarrow D(z_1, z_6, z_7, z_8, z_9, t, z_{11}, z_{12}) \wedge R(x, c_1) \wedge S(z_1, y_3, y_4, t_1, 1) \wedge P(x, y_3, y_4) \wedge C$, where $C \equiv ((t_1 = 1) \vee (t_1 = 2)) \wedge (z_6 \geq c_2) \wedge (z_6 \leq c_3)$.

The query Q_2, and the views V_1, V_2 defined in Example 1, have similar expressions.

3   Term Linearizations of Aggregate Queries

In this section we specify some notions about the *term linearizations* and give an algorithm to compute these *linearizations* constructed by the terms from the conditions contained in the aggregate query. For other notions, see [3].

Two aggregate queries Q_1 and Q_2 having identical heads are said *equivalent*, if their answers are equal, i.e. $Q_1(D) = Q_2(D)$, for any database D defined on Dom. Let f_i be the body of Q_i. Since each f_i is a disjunction of relational atoms and of an expression constructed from comparison atoms using the conjunction and the disjunction operators, it can contain comparisons of terms. The comparisons of an expression f_i induce, in general, a partial order among the constants and variables of the query. To study the equivalence of two queries, we need to consider linear orderings such that a given partial ordering of the set of terms is compatible with a set of linear orderings of the same set of terms.

The set of all terms from a query can be considered as a set of groups, where all attributes of a group correspond to the same value domain. Let C be a set of comparison atoms, and $t_1 < t_2$ be a comparison atom. We say that C implies $t_1 < t_2$, denoted $C \models t_1 < t_2$, if for each substitution τ, the statement ($\tau\delta=$ *true* for each $\delta \in C$) implies $\tau t_1 < \tau t_2$. For a comparison atom having the form $t_1 = t_2$ or $t_1 > t_2$ the implication relation of a comparison atom from a set of comparison atoms is similar. Let T be a set of terms (variables or constants) that are semantically equivalent. A *linearization* of T is a set of comparisons L having the terms in T such that for any two different elements t_1, t_2 from T, exactly one of the following comparisons: $t_1 < t_2$, $t_2 < t_1$, or $t_1 = t_2$ is implied by L. If T has $t_1, \ldots, t_p$ as elements, let $(t'_1, \ldots, t'_p)$ be a permutation of the elements t_j, $1 \leq j \leq p$. Then we can represent a linearization L of T as $t'_1 \rho_1 t'_2 \rho_2 \ldots t'_{p-1} \rho_{p-1} t'_p$, where the operators ρ_i are '<' or '='. We say that a substitution τ from $\{t'_1, \ldots, t'_p\}$ into Dom, that preserves constants, satisfies the linearization L, denoted $\tau(L) = true$, if for each $i, 1 \leq i < p$, $\tau t'_i = \tau t'_{i+1}$, when ρ_i is '=', and $\tau t'_i < \tau t'_{i+1}$, when ρ_i is '<'. We denote by $t_1 = t_2 = \ldots = t_h$, the set of comparisons $t_i = t_{i+1}$ contained in L, for any $i, 1 \leq i < h$. Intuitively, a linearization is a list of different variables and constants such that between two elements of the list there exists one of the '=' or '<' operators. In this way, a linearization produces a partition of all terms into *equivalence classes*; an equivalence class contains all terms $t_1, \ldots, t_h$ such that $t_1 = t_2 = \ldots = t_h$.

We say that a set of comparisons C is *satisfiable*, if there exists a substitution τ from terms into Dom such that τ satisfies all comparisons from C. We say that τ satisfies C or $\tau(C) = true$, if $\tau t_1 < \tau t_2$ if $C \equiv t_1 < t_2$, and $\tau t_1 = \tau t_2$ if $C \equiv t_1 = t_2$. The value of $\tau(C)$ is extended naturally to expressions C obtained from basic comparisons using the conjunction and disjunction operators. We say that a linearization L of T, and a set of comparisons C are compatible, if $L \cup C$ is satisfiable.

In the following we give a method to obtain all linearizations of T, compatible with C. Let C be the conjunction of the comparisons $t_i \sigma_i t'_i$, i.e. $C = (t_1 \sigma_1 t'_1) \wedge \ldots \wedge (t_h \sigma_h t'_h)$, where the operators σ_i are relational operators. Let C' be the formula obtained from C, by replacing $(t_i \leq t_j)$ with $(t_i < t_j) \vee (t_i = t_j)$, and $(t_i \geq t_j)$ with $(t_j < t_i) \vee (t_i = t_j)$. Using for the expression C', the distributivity of the conjunction versus the disjunction, we obtain a formula C'' having the form: $C'' = E_1 \vee E_2 \vee \ldots \vee E_p$ (1), where E_j is a conjunction of comparisons of the form $t_i < t_j$ or $t_i = t_j$. In this disjunction, we take only consistent conjunctions by eliminating those conjunctions that are inconsistent, i.e. which contain at least two comparisons of the form: $t_1 < t_2$ and $t_1 = t_2$, or two comparisons of the form $t_1 < t_2$ and $t_2 < t_1$. Let E be an arbitrary conjunction from C''. Associated to E, we construct a forest, denoted G_E. Firstly, for the set of all comparisons S_E of the form $t_i = t_j$ from E, we take the transitive and symmetrical closure of S_E. This closure produces a set of equivalence classes $Cl_1, \ldots, Cl_q$. Let us denote by $Term1 = \{t_1, \ldots, t_p\}$, the remainder of the variables and the constants from E. The forest G_E associated to E is defined as follows: its nodes are labeled with $Cl_j, 1 \leq j \leq q$ and $t_i, 1 \leq i \leq p$. Let $Var(C)$ be the set of variables from C and $Const(C)$ the set of constants from C.

The edge set $\mathcal{N}$ of the forest G_E is specified as follows:

- Let t_1 and t_2 be variables or constants. If $t_1 < t_2$ appears in E, and t_1 and t_2 occur in $Term1$, then $(t_1, t_2) \in \mathcal{N}$.
- Let Cl_j be a class and t_2 from $Term1$. If there exists $t' \in Cl_j$ such that $t' < t_2$ occurs in E, then $(Cl_j, t_2) \in \mathcal{N}$.
- Let t_1 be from $Term1$, and Cl_h a class. If there exists $t' \in Cl_h$ such that $t_1 < t'$ occurs in E, then $(t_1, Cl_h) \in \mathcal{N}$.
- Let Cl_j, Cl_h be two different classes. If there exist the terms t_1, t_2, where $t_1 \in Cl_j$ and $t_2 \in Cl_h$ such that $t_1 < t_2$ appears in E, then $(Cl_j, Cl_h) \in \mathcal{N}$.

Since the labels associated to different nodes contain disjoint sets of terms, we identify the nodes with their labels. If a class Cl_j consists of the elements $t_1, \ldots, t_s$, we denote it as $\{t_1, \ldots, t_s\}$ or $\{t_1 = \ldots = t_s\}$.

Example 3. Let $C \equiv (x_1 \leq x_2) \wedge (x_2 > x_3) \wedge (y_1 \leq x_2)$. The formula corresponding to C has the form: $C' = [(x_1 < x_2) \vee ((x_1 = x_2)] \wedge (x_3 < x_2) \wedge [(y_1 < x_2) \vee (y_1 = x_2)]$. The expression C'' has the form: $C'' = E_1 \vee E_2 \vee E_3 \vee E_4$, where

$E_1 = (x_1 < x_2) \wedge (x_3 < x_2) \wedge (y_1 < x_2)$, $E_2 = (x_1 < x_2) \wedge (x_3 < x_2) \wedge (y_1 = x_2)$,
$E_3 = (x_1 = x_2) \wedge (x_3 < x_2) \wedge (y_1 < x_2)$, $E_4 = (x_1 = x_2) \wedge (x_3 < x_2) \wedge (y_1 = x_2)$.

The forests associated to each of the four expressions are given in the following figure, where nodes are labeled β_i.

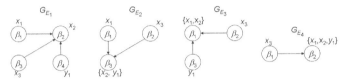

3.1 A Method to Obtain the Linearizations of C

In this subsection, we specify a method to obtain all linearizations of $Var(C) \cup Const(C)$ compatible with C and corresponding to a forest G_E associated to an expression E from C''. The construction of this set of linearizations will be done recursively in function of the number of edges from G_E.

- The base step: Let G_E having zero edges, and $\beta_1, \ldots, \beta_p$ its nodes. Let $i_1, \ldots, i_p$ be a permutation of the indexes $1, 2, \ldots, p$. Let us denote by $\mathcal{M}_{G_E}^L$ the set of linearizations defined by G_E. For this base step, we take:
  $\mathcal{M}_{G_E}^L = \{\beta_{i_1} \sigma_1 \beta_{i_2} \sigma_2 \ldots \beta_{i_{p-1}} \sigma_{p-1} \beta_{i_p} | (i_1, \ldots, i_p)$ is a permutation of $\{1, 2, \ldots, p\}, \sigma_i \in \{'<', '='\}\}$.
  We make the following convention: $\{t_1 = t_2 = \ldots = t_s\} = \{t_1' = t_2' = \ldots = t_r'\}$ becomes: $\{t_1 = t_2 = \ldots = t_s = t_1' = t_2' = \ldots = t_r'\}$.
- The inductive step: Assume that we have computed the set of linearizations corresponding to any forest G having at most n edges, where n is a natural number. Let G be a forest having $n+1$ edges. Let t_0 be an initial node from G, and $\gamma_1, \ldots, \gamma_k$ the immediate successors of t_0. Let G' be the forest obtained from G by deleting the edges (t_0, γ_j), $1 \le j \le k$, and deleting the node t_0. Since G' has at most n edges, by induction we have computed $\mathcal{M}_{G'}^L$. Using the linearizations of G', we compute the linearizations for G. Let $L \equiv s_1 < s_2 < \ldots < s_m$ be an element of $\mathcal{M}_{G'}^L$, where m is the number of the nodes of G'. An element s_i contains a set of terms from $Var(C) \cup Const(C)$. The label γ_j belongs to a unique s_i, for each $j, 1 \le j \le k$ ($\gamma_j \subseteq s_i$). Let us denote by $ind(j)$ this natural number i. Let i_0 be the minimum of $ind(j)$, for each $j, 1 \le j \le k$. Now, we define a set of linearizations for G corresponding to L by inserting t_0 in L before s_{i_0}. Let us denote by $Insert(t_0, L, l)$ the linearizations obtained from L by inserting t_0 between the positions l and $l + 1$. There are two linearizations in $Insert(t_0, L, l)$ in case $l \ge 1$, denoted L_1, L_2, where $L_1 \equiv s_1 < s_2 < \ldots < s_l < t_0 < s_{l+1} < \ldots < s_m$, and $L_2 \equiv s_1 < s_2 < \ldots < \{s_l = t_0\} < s_{l+1} < \ldots < s_m$. If $l = 0$ the insertion takes place before s_1 and $Insert(t_0, L, l)$ consists of one sequence, namely L_1, where $L_1 \equiv t_0 < s_1 < s_2 < \ldots < s_m$. For the linearization L and the node t_0, we consider the union of the sets $Insert(t_0, L, l)$, for each $l, 0 \le l < i_0$. Let us denote this set by $\mathcal{M}(t_0, L)$.

Now, we define the set of linearizations corresponding to G as follows: $\mathcal{M}_G^L = \{L'|(\exists L)(L \in \mathcal{M}_{G'}^L)$ such that $L' \in \mathcal{M}(t_0, L)\}$.

Let E_j be an expression from relation (1), $1 \leq j \leq p$. Let G_{E_j} be the forest corresponding to E_j. Let $\mathcal{M}_{G_{E_j}}^L$ be the set of linearizations computed for the forest G_{E_j}. Let $\mathcal{M}(C)$ be the union of the $\mathcal{M}_{G_{E_j}}^L$ sets, for each $j, 1 \leq j \leq p$. Regarding these notations, we have the following results.

Lemma 1. *Let E be a consistent conjunction of comparisons of the form $t_i < t_j$ or $t_i = t_j$. Let L be a linearization compatible with E, and G_E the forest associated to E. Then, we have $L \in \mathcal{M}_{G_E}^L$.*

Proof. Using the induction on the number of the comparisons of the form '$<$' from E (this number is equal to the number of the edges from G_E).

Theorem 1. *Let L be a linearization of $Var(C) \cup Const(C)$ and $\mathcal{M}(C)$ the set of linearizations computed for C as above. Then, we have: L is compatible with C iff $L \in \mathcal{M}(C)$.*

Proof. Let L be a linearization from $\mathcal{M}(C)$. There exists an integer $j, 1 \leq j \leq p$ such that $L \in \mathcal{M}_{G_{E_j}}^L$. The linearization L can be represented as $Cl(t_1) < Cl(t_2) < \ldots < Cl(t_m)$, where $Cl(t_i)$ are equivalence classes corresponding to L, and t_i is an element from that class. A class $Cl(t_i)$ consists of a single variable, a single constant, or a set of terms $\{t_1', \ldots, t_h'\}$, $h \geq 2$, and this set contains at most one constant. The union of the classes $Cl(t_i)$ is $Var(C) \cup Const(C)$. Let us define the substitution τ as follows: if $Cl(t_i) = \{t_1', \ldots, t_h'\}$, then $\tau t_j' = \tau t_l'$, $1 \leq j < l \leq h$. The substitution τ is extended to classes: $\tau Cl(t_j) = \tau t_j$. For two classes, we take the values for τ such that if $Cl(t_i) < Cl(t_j)$, then $\tau t_i < \tau t_j$. In this manner, we have $\tau(L)$ is true. Using the method to construct $\mathcal{M}_{G_{E_j}}^L$, we obtain $\tau(E_j)$ is true, hence $\tau(C)$ is true. That means $L \cup C$ is satisfiable.

Conversely, let L be a linearization compatible with C. It results that there exists an integer $j, 1 \leq j \leq p$ such that L is compatible with E_j. Using Lemma 1, we obtain $L \in \mathcal{M}_{G_{E_j}}^L$, hence $L \in \mathcal{M}(C)$.

Using Theorem 1, we obtain an algorithm that computes the set of all linearizations for a conjunction of comparisons.

3.2   Algorithm to Compute Linearizations of $\mathcal{M}(C)$

Firstly, let us denote by $LinearizationsComp(G,\mathcal{L})$ an algorithm that computes the set $\mathcal{L}$ of all linearizations corresponding to the forest G. Let us consider some notations used in this algorithm. We denote by h the number of levels from G. The terminal nodes of G are considered on level 1. Let N_j be the set of all nodes on level j, $1 \leq j \leq h$. We denote by $\mathcal{L}_j$ the set of all linearizations for nodes situated on the levels l in G, where $l \leq j$. We use the notation $\mathcal{M}(\gamma, L)$ for a set of linearizations obtained by inserting the node γ into a linearization L, as specified in subsection 3.1. By $\mathcal{L}_{j,l}$ we denote a set of linearizations of nodes from G, that is a subset of $\mathcal{L}_j$. The algorithm proceeds level by level, beginning with the first level.

Algorithm *LinearizationsComp* (input G, output $\mathcal{L}$)
compute h the number of levels from G
for all $j=1$ **to** h **do** compute N_j, the set of all nodes on level j **endfor**
compute $\mathcal{L}_1$ the set of linearizations for N_1 (see base step from Subsection 3.1)
if $h=1$ **then** $\mathcal{L} = \mathcal{L}_1$ exit **endif**
for all $j=2$ **to** h **do**
    $\mathcal{L}_j = \emptyset$
    let $N_j = \{\gamma_1, \ldots, \gamma_k\}$
    $\mathcal{L}_{j,0} = \mathcal{L}_{j-1}$
    for all $l=1$ **to** k **do**
        $\mathcal{L}_{j,l} = \emptyset$ let $\mathcal{L}_{j,l-1} = \{L_1, \ldots, L_p\}$
        for all $s = 1$ **to** p **do**
            compute $\mathcal{M}(\gamma_l, L_s)$ (see Subsection 3.1)
            $\mathcal{L}_{j,l} = \mathcal{L}_{j,l} \cup \mathcal{M}(\gamma_l, L_s)$
        endfor
    endfor
    $\mathcal{L}_j = \mathcal{L}_{j,h}$
endfor
$\mathcal{L} = \mathcal{L}_h$

Secondly, we specify the computing of the linearizations corresponding to the expression C that has comparison atoms as base expressions, and is formed of base atoms using the conjunction or disjunction operators. The expression C is equivalent to a disjunction of conjunctions, i.e. $E \equiv E_1 \vee \ldots \vee E_p$. For each expression E_j, we construct the forest G_{E_j}, and call the algorithm *LinearizationsComp* with the parameters G_{E_j} and $\mathcal{L}_j$. The set of the linearizations for C, denoted $\mathcal{M}(C)$, is the union of all $\mathcal{L}_j$, i.e. $\mathcal{M}(C) = \cup_{j=1}^p \mathcal{L}_j$.

Example 4. Let us consider Example 2. Let C be the comparison expression from Q_1. Since $z_6 \geq c2$ is equivalent to $(z_6 > c2) \vee (z_6 = c_2)$, and $z_6 \leq c_3$ is equivalent to $(z_6 < c_3) \vee (z_6 = c_3)$, the expression C is equivalent to $C'' \equiv \bigvee_{i=1}^3 [(t_1 = 1) \wedge E_i]$ $\bigvee_{i=1}^3 [(t_1 = 2) \wedge E_i]$, where $E_1 \equiv (c_2 < z_6) \wedge (z_6 < c_3)$, $E_2 \equiv (c_2 < z_6) \wedge (z_6 = c_3)$, $E_3 \equiv (z_6 = c_2) \wedge (z_6 < c_3)$. The expression $(z_6 = c_2) \wedge (z_6 = c_3)$ is non satisfiable because all constants are considered different. Concerning the constants c_2 and c_3, we assume $c_2 < c_3$, otherwise the expression C is not satisfiable. As we specified in Section 3, we represent a linearization as a set of groups, where a group is a linearization of all terms corresponding to the same value domain. For the group $\{z_6, c_2, c_3\}$, and the first conjunction from C'', we obtain one linearization: $c_2 < z_6 < c_3$. Hence, to the first conjunction of C'', we have two groups denoted g_1, g_2, where $g_1 \equiv (t_1 = 1)$, $g_2 \equiv c_2 < z_6 < c_3$. We consider an order of linearizations as follows: L_1 for the first conjunction of C'', L_2 for the fourth, L_3 for the second, L_4 for the fifth, L_5 for the third, and L_6 for the sixth. Thus, the linearizations L_i have the following forms:

$L_1 \equiv (t_1 = 1, c_2 < z_6 < c_3)$, $L_2 \equiv (t_1 = 2, c_2 < z_6 < c_3)$,
$L_3 \equiv (t_1 = 1, z_6 = c_3)$, $L_4 \equiv (t_1 = 2, z_6 = c_3)$,
$L_5 \equiv (t_1 = 1, z_6 = c_2)$ and $L_6 \equiv (t_1 = 2, z_6 = c_2)$.

3.3 Complexity Issues

Let us compute the time complexity of algorithm 1. Let G be a forest, with n nodes and p edges and h its number of levels (corresponding to the depth of the forest) with $h \leq p + 1$. For every level j, $1 \leq j \leq h$, let N_j be the number of nodes on level j. We have $\sum_{j=1}^{h} N_j = n$. The set L_j from the algorithm gives all the linearizations obtained using nodes from levels 1 to j. The number of terms for each linearization from L_j is $length_j = \sum_{k=1}^{j} N_k$. Let us denote by $|L_j|$ the cardinal of the L_j set. We have $|L_1| = N_1!$ because there is no edge between the nodes on the first level. For the other levels, we have: $|L_{j+1}| \leq N_{j+1} * length_j * |L_j|$, $1 \leq j \leq h - 1$.

We obtain $|L_h| \leq N_h * N_{h-1} * \ldots * N_1 * (N_{h-1} + \ldots + N_1) * \ldots * (N_2 + N_1) * N_1!$.

The set of terms can be grouped in classes, a class contains all terms having the same value domain. If m is the maximum number of elements in these classes, then m is the number of the forest nodes in the given algorithm.

4 Conclusion

In the paper, we propose an algorithm to compute linearizations and evaluate its complexity. These are useful in the process of constructing and proving the query equivalence between query expressed on the database schema and its rewritings using views. Queries and views considered here contain aggregate functions and arithmetic comparisons. Future work will focus on the aspects concerning using linearizations in the problem of queries equivalence when queries and views contain negations.

References

1. Afrati, F., Li, C., Mitra, P.: Rewriting queries using views in the presence of arithmetic comparisons. Theoretical Computer Science 368, 88–123 (2006)
2. Cohen, S.: Containment of aggregate queries. ACM SIGMOD 34(1), 77–85 (2005)
3. Cohen, S., Nutt, W., Serebrenik, A.: Rewriting aggregate queries using views. In: PODS, pp. 155–166 (1999)
4. Cohen, S., Nutt, W., Sagiv, Y.: Deciding equivalences among conjunctive aggregate queries. Journal of the ACM 54(2), 1–50 (2007)
5. Grumbach, S., Rafanelli, M., Shurin, S.: On the equivalence and rewriting of aggregate queries. Acta Informatica 4(8) (2004)
6. Heinzelman, W.R., Chandrakasan, A., Balakrishnan, H.: Energy-efficient Communication Protocol for Wireless Microsensor Networks. In: 33rd Annual Hawaii International Conference on System Sciences (HICSS-33), pp. 3005–3014 (2000)
7. Madden, S., Szewczyk, R., Franklin, M., Culler, D.: Supporting aggregate queries over ad-hoc wireless sensor networks. In: Proc. of 4th IEEE Workshop on Mobile Computing Systems and Applications, pp. 49–58 (2002)

Evaluating Skyline Queries on Spatial Web Objects

Alfredo Regalado, Marlene Goncalves, and Soraya Abad-Mota

Universidad Simón Bolívar, Departamento de Computación, Apartado 89000
Caracas 1080-A, Venezuela
{mgoncalves,abadmota}@usb.ve

Abstract. GoogleMaps, Google Earth and Bing Maps have stimulated the visual exploration of maps on the Web and have allowed users to query spatial web objects. A spatial web object has a geographical location and usually has associated a textual description or a link to a web document. To select the objects in response to a query the data contained in their textual description or document must be inspected. Skyline is a query processing paradigm which offers queries over multiple criteria, where each criterion is equally important. In this paper we define *Location-based Textual Skyline* queries as those which use a spatial function and a function over descriptive text as criteria in a Skyline query. For example, a *Location-based Textual Skyline query* may use the distance and the relevance of keywords in the text describing a spatial web object, as criteria to retrieve objects. We define a technique to evaluate this kind of query and develop an experimental study to evaluate the proposed technique.

Keywords: Skyline, R-Tree, Information Retrieval, Nearest Neighbor, Relevance of Keywords, Spatial Databases.

1 Introduction

Continuous improvements on technologies such as mobile devices, GPS systems, and telecommunication networks, have facilitated an increase on location-based web searches. A great percentage of the web searches are related to closeness to a current spatial location. Recent studies reveal that the volume of location-based web searches per month is high and increasing. For example, location-based web searches on Google constitute 20% of all PC searches and 40% on mobile devices [1]. These percentages represent billions of location-based web searches that are being performed on Google each month.

Skyline [2] provides a way of selecting objects of interest based on several criteria where each one is equally relevant. For example, a tourist may look for inexpensive restaurants with a nice view. If we add a spatial criterion, like the distance from a hotel, the query will combine categorical and spatial criteria.

In this paper we present an algorithm that combines a spatial function, distance, with a function on the textual description, as the criteria for computing a skyline query. We define this as a special kind of skyline query. Section 2 contains the motivation for this work. Section 3 presents the related work. In Section 4 the proposed solution is presented. In section 5 we present an experimental study of the algorithm and section 6 contains the conclusions and future work.

S.W. Liddle et al. (Eds.): DEXA 2012, Part II, LNCS 7447, pp. 416–423, 2012.

2 Motivation

Users who perform spatial searches look for objects close to them; closeness is a relevant selection criterion. However, the closest items may not satisfy other user preferences. Consider a tourist in Margarita Island, Venezuela, interested in beaches close to her hotel where she can practice windsurf and kitesurf. Figure 1(a) illustrates a map containing the location of beaches A, B, C, D, E, F, G and I near the tourist's hotel H.

Figure 1(c) lists the locations of the beaches in the map of Figure 1(a) and the Euclidean distance [3] from each beach to point H, shown in Figure 1(b), which is a 2D representation of the beaches. For simplicity, we consider that latitude and longitude values of each point representing a beach vary between 1 and 8.

Each beach can be described by a text document published on the Web, e.g. a website, a blog or some other source. Thus, spatial data can be geotagged or associated with a textual description. This description can be used as a criterion within a web search. In this

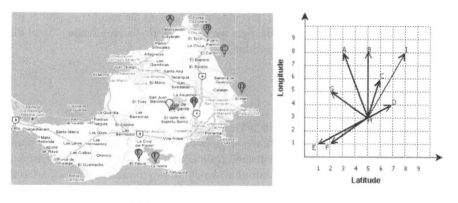

(a) Map   (b) Representation in R^2.

Beach	Location	Euclidean Distance
A	(3,8)	5.4
B	(5,8)	5.0
C	(6,6)	3.2
D	(7,4)	2.2
E	(1,1)	4.5
F	(2,1)	3.6
G	(2,5)	3.6
H	(5,3)	0.0
I	(8,8)	5.8

(c) Location and Distance between each beach and the Hotel

Beach	windsurf	kitesurf	Frequency
A	4	3	3.5
B	2	2	2.0
C	5	2	3.5
D	0	0	0.0
E	6	6	6.0
F	2	2	2.0
G	2	2	2.0
I	2	2	2.0

(d) Frequency of keywords in each beach description

Fig. 1. Map, Distance and Keyword Frequency

context, Information Retrieval techniques may be applied [4,5,6,7,8]. In Figure 1(d) the number of keywords in the description and the keyword frequency function are shown.

In spite of beach D being the closest to the hotel, its description does not contain the keywords windsurf and kitesurf. Beach E has the highest keyword frequency, but it is one of the farthest beaches. Thus, there is no beach closer to hotel H with the highest keyword frequency.

Skyline is an approach that can be used for this kind of query [2]. The Skyline [2] is the set of points where each point in the set is not dominated by some other point; point A dominates point B if it is better or equal in all criteria, and is better in at least one criterion. For example, beach E is better than beach A because it has a shorter distance and a higher frequency of keywords; while beaches D and E are incomparable because none of them is better than the other in all criteria. The Skyline resulting from this example contains beaches C, D, and E.

We call this kind of query, Location-based Textual Skyline (LTS) query, it identifies the best results in terms of preferences on distance to user's location and relevancy of keywords specified by the user. In this work, we propose an algorithm to evaluate LTS queries, which uses R-Trees [9] for spatial data management and inverted files [10] for textual data management. Moreover, this algorithm reduces the number of probes per dominance between objects of the dataset. We also perform an empirical study of the quality and execution time of the algorithm.

3   Related Work

Spatial Skyline queries have gained increased interest. In [11] a Spatial Skyline set is defined in terms of multiple distance functions; a point spatially dominates another point if it has a better or equal value in all distance functions, and if it is better in at least one distance value. The authors of [12] introduce an algorithm to evaluate Location-based Skyline (LS) which is a special case of Spatial Skyline on categorical data. In [13] a keyword-matched skyline query is defined, this is the set of objects whose textual descriptions contain all keywords of the query. Nevertheless, a keyword-matched skyline query does not deal with a distance function as part of the Skyline criteria and in [11] and [12] a function on the keywords is not a Skyline criterion.

The use of keyword frequency in a textual description as a Skyline criterion relates to the area of *Information Retrieval (IR)* [16]; in IR a set of documents is selected based on the relationship between a query and the documents. IR techniques may be used to evaluate the frequency function or keyword relevancy in the web spatial object description.

The works [14,15] propose algorithms for low-dimensional Skyline. These algorithms evaluate Skyline queries without dynamic functions as their criteria; since we are interested in distance and keyword frequency which are dynamic functions, we would have to precompute their values in order to use these algorithms.

Top-k is another strategy which identifies the best k objects, but unlike Skyline, it uses a score function [17]. In [4,6,18], a top-k query returns the best k objects in terms of a score function that combines two criteria using a weight parameter. The criteria used may include a normalized Euclidean distance to the user location, and a normalized similarity function of the keywords. A weight parameter specifies the importance of one criterion over the other. In Skyline, a criterion is not more important than other and there is no weight parameter to be computed. There is a relationship between Skyline and top-1 object, i.e. for any Skyline there is a monotone ranking function for which it is the best.

4 Proposed Solution

A naive algorithm to evaluate Location-based Textual Skyline queries is to precalculate the distance function and orderly access the data by these precalculated values. Conversely, the relevancy function may be precalculated first and the data ordered by these values. Suppose that the distance function is precalculated, the process continues by accesing an object, computing the frequency of the required keyword for this object and comparing both values against its predecessor object in the Skyline, to determine if the new object is dominated by its predecessor. To avoid the distance precomputation but still access the objects by distance order, an R-Tree structure may be used. This structure organizes the objects by the closeness to each other, determined by a nearest neighbor function.

We developed two algorithms, the naive which uses an R-tree structure with indexes and the $Predecessor^+$ algorithm, which uses the same structures but has a stop condition, to reduce the number of dominance comparisons. In this paper we describe the $Predecessor^+$ algorithm in subsection 4.2, before that description, we present the IR-tree and the additional structures used by both algorithms in subsection 4.1.

4.1   Data Structures

Our algorithms use a data structure known as IR-Tree [19]. An IR-Tree is an R-Tree [9] extended with a pointer to an inverted file [10]. The IR-Tree combines the advantages of an R-Tree with the notion of inverted files for better performance.

In an R-tree, the internal nodes contain regions (bounding rectangles) in which the space represented in the tree is divided. Each region may contain other regions. The leaf nodes have individual spatial objects contained in the region of the parent node. Figure 2 presents an example of an IR-Tree. The root node, R_4, represents a region (the whole space) that includes regions R_1, R_2, and R_3. The right-most leaf node contains the spatial objects E and F, which are contained in region $R3$.

Each node of the tree has an inverted file which summarizes the word occurrence frequency in the documents describing the spatial objects or regions contained in that tree node. Each entry of an inverted file corresponds to a word and has a list of pairs; for leaf nodes, each pair contains the identifier of a spatial object and the value of the word occurrence frequency in the document associated with that object. For internal nodes, each entry of the inverted file for a word, contains the summary frequency value (maximum in our case) for all the objects in each region of the internal node. The table of Figure 2 shows four inverted files for the IR-Tree in Figure 2. For example, Inverted File 1 has word frequencies for each document A, B, and I, while Inverted File 4 has word frequencies for each region R_1, R_2, and R_3.

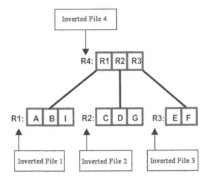

Words	Inverted File 1	Inverted File 2	Inverted File 3	Inverted File 4
windsurf	(A,4),(B,2),(I,2)	(C,5),(G,2)	(E,6),(F,2)	$(R_1,4),(R_2,5),(R_3,6)$
kitesurf	(A,3),(B,2),(I,2)	(C,2),(G,2)	(E,6),(F,2)	$(R_1,3),(R_2,2),(R_3,6)$
food	(A,3),(B,2),(I,3)	(C,3),(G,3)	(E,3),(F,2)	$(R_1,3),(R_2,3),(R_3,3)$
sports	(A,5),(I,5)	(C,5),(G,5)	(E,6),(F,0)	$(R_1,5),(R_2,5),(R_3,6)$
surf	(B,3),(I,5)	(C,2),(D,5),(G,1)	(E,1),(F,3)	$(R_1,5),(R_2,5),(R_3,3)$

Fig. 2. An example IR-Tree

Table 1. An example Inverted File

Word	Frequency
windsurf	(A,4),(B,2),(C,5),(E,6),(F,2),(G,2),(I,2)
kitesurf	(A,3),(B,2),(C,2),(E,6),(F,2),(G,2),(I,2)
food	(A,3),(B,2),(C,3),(E,3),(F,2),(G,3),(I,3)
sport	(A,5),(C,5),(E,6),(G,5),(I,5)
surf	(B,3),(C,2),(D,5),(E,1),(F,3),(G,1),(I,5)

Additionally, $Predecessor^+$ requires an inverted file to index the textual data by word. Table 1 shows an example inverted file, where for each word that describes a beach there is a collection of pairs, each pair contains the word frequency and the beach identifier. For example, the word *windsurf* appears in seven documents and four times in the document associated with beach A.

4.2 $Predecessor^+$ Algorithm

This algorithm uses the IR-tree to access the objects by increasing order of distance and the inverted files to examine the objects in word frequency order. An object may be dominated by objects with the same value in one attribute or contained in the predecessor Skyline (see [20] for a proof). Therefore, the function $skyline$ probes equal-distance nearest neighbors that are not dominated by themselves or their predecessor Skyline. An analogous comparison is done with word frequency values.

As can be verified in tables 1(c) and 1(d) from section 2, the first two objects accessed in steps 3 and 7 of the Algorithm 1 are D and E (minimum distance and maximum frequency); they are inserted into L_{s_d} and L_{s_r}, respectively. In the next iteration,

---

Algorithm 1. Algorithm $Predecessor^+$

Require: IR: *IR-Tree*; I: *Inverted File*
1: $L_{s_d} \leftarrow \emptyset; L_{s_r} \leftarrow \emptyset; pred_{dist} \leftarrow \emptyset; pred_{relevance} \leftarrow \emptyset;$
2: **while** $L_{s_d} \cap L_{s_r} = \emptyset$ and there exist objects in IR or I **do**
3:     $L_{o_d} \leftarrow nextNNEquals(IR); L_d \leftarrow skyline(L_{o_d}, pred_{dist});$
4:     **if** $L_d <> \emptyset$ **then**
5:         $L_{s_d} \leftarrow L_{s_d} \cup L_d; pred_{dist} \leftarrow L_d$
6:     **end if**
7:     $L_{o_r} \leftarrow nextNNEquals(I); L_r \leftarrow skyline(L_{o_r}, pred_{relevance})$
8:     **if** $L_r <> \emptyset$ **then**
9:         $L_{s_r} \leftarrow L_{s_r} \cup L_r; pred_{relevance} \leftarrow L_r$
10:    **end if**
11: **end while**
12: $S \leftarrow merge(L_{s_d}, L_{s_r});$
13: **return** S

$nextEquals(IR)$ returns C. Since D and C are incomparable, Object C is a Skyline and is inserted into L_{s_d}. Later, $nextEquals(I)$ produces the objects A and C, these are the next objects to be considered which have the same frequency value. C dominates A which is discarded. The predecessor Skyline, E, does not dominate C, therefore, C is Skyline. Object C belongs to $L_{s_d} \cap L_{s_r}$, this is the stop condition for this algorithm, as was proven in [20]. L_{s_d} has the objects C and D, and L_{s_r} has the objects E and C. These lists are merged into S in step 12. Finally, the Location-Based Textual Skyline is C, D, and E which is returned in step 13.

5   Experimental Study

Due to space restrictions in this paper we do not describe the naive algorithm, which is the same as the $Predecessor^+$ algorithm but without the special stop condition that allows it to perform less comparisons to find all the objects which belong to the Skyline. We present the empirical evaluation results for both algorithms. In all experiments, we measured the total execution time and the number of dominance probes. The algorithms were implemented in Java and were executed on a server *SunFire V440* with *SunOS 5.10*, two processors *sparcv9* of $1,281$ MHZ, 16 GB RAM and 4 HD *Ultra320 SCSI* of 73 GB.

Synthetic and real datasets were used in the experiments. Datasets comprise objects with a spatial location and a textual description. The synthetic datasets were created with a Java program, varying the distribution and cardinality of the data. Uniform distribution and correlated datasets were used. The number of words per object varied in the range $[100, 300]$ and the frequency of words was between 10% and 20%.

The real dataset was provided by CANTV, a phone company. This dataset has network elements with their spatial location and a textual description. To build and manipulate the IR-Tree structures used by the algorithms, we selected the open source library presented in [21]. This library was adapted to make each node point to an inverted file that describes the objects in each region. For the inverted files, we employed a nested hash of pairs (key, collection), where key is a word, and collection is a set of pairs (id, value), id is the document identifier and value represents the number of repetitions of a word in that document. Thirty queries were used for the experiments, with spatial coordinates and 1, 2, and 3 keywords.

The results of the experiments showed that the average execution time for the $Predecessor^+$ algorithm is the lowest, because of the stop condition which avoids scanning all the data, as the naive algorithm does. For all datasets, the average time of $Predecessor^+$ is between 1.6% and 10.2% of the average time of the naive algorithm. Moreover, the behavior of the two algorithms is similar when they are executed on both independent and correlated datasets. Since the naive algorithm scans all data, its execution time is similar on the three datasets. Regarding the average number of probes executed, the experiments showed that as the dataset size increases, the number of probes also increases. The number of dominance probes executed by $Predecessor^+$ is between 18.2% and 45.1% of the probes performed by the naive algorithm.

6   Conclusions and Future Work

In this paper we define a new type of query, called *location-based textual Skyline*. We also define two algorithms to evaluate these queries. The algorithms use index structures and retrieve objects based on their proximity to the user location and the keyword occurrence frequency in the textual description of the objects. An experimental study showed that $Predecessor^+$ reduces the number of dominance probes because it is able to stop earlier than the naive algorithm. Consequently, the performance of $Predecessor^+$ is better and only checked 20% of the dominated objects.

In the work described here, we use a keyword frequency function on a single document, but this could be extended by using other functions over text. There is a single document describing the object and it exists in a centralized site, several documents located in different sites could be considered.

The issue that remains open is the extension of the LTS queries on more than two dimensions.

References

1. Comscore releases december 2011 U.S. search engine rankings,
   http://www.comscore.com/Press_Events/Press_Releases/2012/1/
   comScore_Releases_December_2011_U.S._Search_Engine_Rankings
2. Börzsönyi, S., Kossmann, D., Stocker, K.: The skyline operator. In: ICDE, pp. 421–430 (2001)
3. Breu, H., Gil, J., Kirkpatrick, D., Werman, M.: Linear time Euclidean distance transform algorithms. IEEE Transactions on Pattern Analysis and Machine Intelligence, 529–533 (2002)
4. De Felipe, I., Hristidis, V., Rishe, N.: Keyword search on spatial databases. In: ICDE, pp. 656–665 (2008)
5. Zaragoza, H.: Information retrieval: Algorithms and heuristics. Inf. Retr. 5(2-3), 271–274 (2002)
6. Cong, G., Jensen, C.S., Wu, D.: Efficient retrieval of the top-k most relevant spatial web objects. PVLDB 2(1), 337–348 (2009)

7. Hariharan, R., Hore, B., Li, C., Mehrotra, S.: Processing spatial-keyword (sk) queries in geographic information retrieval (gir) systems. In: SSDBM, pp. 16–25 (2007)
8. Sanderson, M., Kohler, J.: Analyzing geographic queries. In: SIGIR Workshop on Geographic Information Retrieval, pp. 1–2 (2004)
9. Guttman, A.: R-trees: A dynamic index structure for spatial searching. In: SIGMOD Conference, pp. 47–57 (1984)
10. Zobel, J., Moffat, A.: Inverted files for text search engines. ACM Comput. Surv. 38(2), 1–56 (2006)
11. Sharifzadeh, M., Shahabi, C.: The spatial skyline queries. In: VLDB, pp. 751–762 (2006)
12. Kodama, K., Iijima, Y., Guo, X., Ishikawa, Y.: Skyline queries based on user locations and preferences for making location-based recommendations. In: GIS-LBSN, pp. 9–16 (2009)
13. Choi, H., Jung, H., Lee, K.Y., Chung, Y.D.: Skyline queries on keyword-matched data. Information Sciences (in press, corrected proof) (2012)
14. Lu, H.-X., Luo, Y., Lin, X.: An Optimal Divide-Conquer Algorithm for 2D Skyline Queries. In: Kalinichenko, L.A., Manthey, R., Thalheim, B., Wloka, U. (eds.) ADBIS 2003. LNCS, vol. 2798, pp. 46–60. Springer, Heidelberg (2003)
15. Köhler, H., Yang, J.: Computing large skylines over few dimensions: The curse of anti-correlation. In: APWeb, pp. 284–290 (2010)
16. Baeza-Yates, R.A., Ribeiro-Neto, B.A.: Modern Information Retrieval - the concepts and technology behind search, 2nd edn. Pearson Education Ltd., Harlow (2011)
17. Carey, M.J., Kossmann, D.: On saying "enough already!" in sql. In: SIGMOD Conference, pp. 219–230 (1997)
18. Rocha-Junior, J.B., Vlachou, A., Doulkeridis, C., Nørvåg, K.: Efficient processing of top-k spatial preference queries. PVLDB 4(2), 93–104 (2010)
19. Li, Z., Lee, K.C.K., Zheng, B., Lee, W.-C., Lee, D.L., Wang, X.: Ir-tree: An efficient index for geographic document search. IEEE Trans. Knowl. Data Eng. 23(4), 585–599 (2011)
20. Regalado, A.: Evaluación de consultas skyline espacial textual. M.S. thesis, Universidad Simón Bolívar. Trabajo de Grado. Magister en Ciencias de la Computación (2011)
21. Deegree - java framework for geospatial solutions,
http://www.java2s.com/Open-Source/Java-Document/GIS/
deegree/org.deegree.io.rtree.html

Alternative Query Optimization
for Workload Management[*]

Zahid Abul-Basher[1], Yi Feng[1], Parke Godfrey[1], Xiaohui Yu[3,1],
Mokhtar Kandil[2], Danny Zilio[2], and Calisto Zuzarte[2]

[1] York University, Toronto, ON, Canada
{zahidur,yfeng,godfrey,xhyu}@yorku.ca
[2] IBM Toronto Lab, Markham, ON, Canada
{mkandil,zilio,calisto}@ca.ibm.com
[3] Shandong University, Jinan, Shandong, China

Abstract. Systems with heavy workloads run many queries concurrently. Modern database workloads—as those incurred by business intelligence applications—involve ad-hoc, highly complex, expensive queries. While query plans are optimized individually, the workload overall is not. Plans running together incur resource contention, resulting in suboptimal performance. To address this, we introduce the idea of *alternative-objective query optimization*. Multiple query plans for the same query are generated, each optimized for an alternative resource usage. At runtime, the workload manager then can choose the plan for the query that works best for runtime conditions. This balances the system load, reducing contention, to increase overall workload throughput.

Keywords: query optimization, workload management.

1 Introduction

Key to the success of relational database management systems has been high quality query optimization. For a given query, a cost-based query optimizer finds a plan that is best with respect to some cost *objective*. A common objective used is the *time to completion* of the query plan. During optimization, the completion time of partial query plans is estimated based on the estimated usage of different resources(e.g., CPU and I/O usage)and other factors . These estimates are then used to compare and prune partial query plans during plan enumeration, to arrive at the plan with the best estimated cost.

Our motivation for considering alternative optimization objectives is that systems with heavy workloads run many queries concurrently. Modern database workloads, such as those incurred by business intelligence applications, involve

---

[*] This work was supported by NSERC Discovery Grants, IBM CAS Fellowships, the National Natural Science Foundation of China Grants (No. 61070018, No. 60903108), the Program for New Century Excellent Talents in University (NCET-10-0532), the Independent Innovation Foundation of Shandong University (2012ZD012, 2009TB016), and the SAICT Experts Program.

S.W. Liddle et al. (Eds.): DEXA 2012, Part II, LNCS 7447, pp. 424–431, 2012.

ad-hoc, highly complex, expensive queries. Query plans running together can lead to resource contention, resulting in sub-optimal performance. Different execution plans for the same query, however, often involve quite different resource consumption profiles. So query plans that are optimized for different resource usages could be mixed at runtime by an intelligent workload manager to reduce contention to balance system load, thereby increasing workload throughput.

We introduce the concept of *alternative-objective query optimization* to address these issues. The key idea is to retool the optimizer so that a given query can be optimized for multiple objectives with a *single* run of the optimizer. The results of optimization will be different query plans, each optimized for a different objective. Such plans can then be used to support intelligent workload management and various applications.

To this end, we identify useful, practical alternative optimization objectives to the standard time-to-completion. In this work, we investigate the objectives of *buffer-pool width*,[1] *CPU usage* and *rate*, and *I/O usage* and *rate*. We illustrate how to optimize queries with respect to alternative objectives. We develop an approach that balances overall query performance (time-to-completion) with the alternative objectives. We modify bottom-up query plan generation to generate plans with respect to different optimization criteria, along with the best plan for time to completion. Our approach has been implemented in IBM DB2 (Viper) for the alternative objectives of *buffer-pool width*, *CPU usage*, and *I/O usage*.[2]

We can summarize the main contributions of this paper as follows.

1. We introduce the concept of alternative-objective query optimization and propose an algorithm that optimizes a given query for the multiple objectives, within a single run of the optimizer.
2. We devise *alternative optimization objectives* that are effective, *and* that can be employed by our alternative-objective optimization algorithm.
3. We implement alternative-objective optimization within IBM DB2, and demonstrate its effectiveness via experiments over TCP-H.

2   Background and Related Work

There has been significant research on workload management, to improve throughput (to execute more queries within the same period of time). Most prior efforts fall into two categories: (1) approaches that focus on resource allocation for concurrent queries (e.g., [2,7]), and (2) proposals that specifically deal with problem queries (e.g., unexpectedly long-running queries) in the workloads (e.g., [1,3]). Previous work in the first category considers how to allocate resources for multiple concurrent queries with widely varying resource consumption profiles. Since problem queries can cause serious performance degradation in the DBMS,

---

[1] Buffer-pool *width*, or *footprint*, is a measure of how much of the buffer pool the query plan occupies at runtime (if it is to achieve its predicted performance).

[2] The implementation is a code branch-off of the commercial code base of IBM DB2. We hope to fold these techniques into the commercial release in the future.

existing work in the second category proposes approaches that are specifically targeted at managing those queries.

Our work shares a focus with prior work on resource allocation in that we also aim to best utilize available resources to achieve high system throughput. However, all existing approaches assume that query plans are fixed at runtime. In our approach, the workload manager enjoys the freedom to choose from a set of different plans for each query, based upon the availability of each resource.

Multi-query optimization (MQO) strives to exploit common sub-expressions to reduce the query evaluation cost [5,6,8]. Although MQO bears some resemblance to our proposal in that they are both motivated by improving the system performance in the presence of multiple concurrent queries, they employ totally different methodologies. MQO focuses on reducing the cost of queries via the reuse of common intermediate results, where the cost is still measured in terms of time to completion. Our approach, on the other hand, optimizes each query for different objectives, leading to different candidate query plans, which allows the workload manager to improve the system performance by piecing together the right plans (chosen from the candidate plans) for the queries. In addition, our approach could accommodate the preferences of users and/or applications through the use of different objectives.

In IBM DB2, an alternative optimization objective called *time to first tuple* is already implemented. This measures how much time the query plan takes to present the first answer tuple (as opposed to having generated all the answer tuples). Note, however, this alone does not suffice for our more general goal for alternative-objective optimization, nor does it provide a general approach to generate plans with respect to alternative objectives, as we do in this paper.

3   Alternative Optimization

Important system resources for query evaluation are I/O, CPU, and buffer-pool. Temporary space (*tempspace*) is another critical, limited resource, which is reserved disk space where temporary results during query processing are written (*spilled*), when too large to keep in the buffer pool. There are numerous ways one can define optimization objectives (cost formulas) with respect to these resources. Part of our work has been to devise optimization objectives both that are effective and that can be used efficiently within the optimization procedure.

1. **CPU.**
   (a) **raw CPU.** How many CPU cycles are consumed over the life of the plan's execution.
   (b) **CPU load.** How many CPU cycles per unit time are consumed, on average, by the plan.
2. **I/O.**
   (a) **raw I/O.** How many I/O's—page reads and writes—are consumed over the life of the plan's execution.
   (b) **I/O load.** How many I/O's per unit time are consumed, on average, by the plan.

3. **bufferpool width.** The maximum footprint in the *buffer pool* that the query plan occupies—to achieve the plan's predicted performance—at any given point during its execution.

4. **tempspace width.** The maximum footprint in the *tempspace* that the query plan occupies—to achieve the plan's predicted performance—at any given point during its execution.

On the one hand, *CPU load* and *I/O load*, would be quite useful in scheduling to reduce resource contention. However, these are *not* monotonic in bottom-up planning. A sub-plan may have high CPU load, while the overall plan does not. So one cannot efficiently optimize with respect to these. On the other hand, *raw CPU* and *raw I/O* are monotonic in bottom-up planning, and so can be used in dynamic programming as in Selinger's join-enumeration algorithm. However, they are not directly useful from a resource contention point of view.

One can find good plans *indirectly* for *CPU load* and *I/O load*, though, with respect to *raw CPU* and *raw I/O*. Any best alternative-objective plan has a longer running time than the best time-to-completion (*base*) plan. (Else, the plan would be the best base plan itself.) Let the best raw-CPU plan A use A_c CPU cycles overall and A_t time, and let the base plan B use B_c cycles and B_t time. Since $A_c < B_c$ but $A_t > B_t$, it follows $A_c/A_t < B_c/B_t$, so A's *CPU load* is better too.[3] Likewise, *raw I/O* indirectly finds better plans for *I/O load*.

The objective *bufferpool width* is not the same as *I/O*. A plan may have heavy I/O load, but still have small bufferpool width. Consider a large index nested loop join. It probes often the index of the inner table, so it may have high I/O load. Since an individual probe's pages do not need to remain in the buffer pool after the probe, though, its bufferpool width is small. If the join were done via a hash join instead, all pages of each partition of the outer table would need to reside simultaneously, resulting in a large bufferpool width. *Temp-space width* is the analogous secondary-memory resource over which concurrently executing plans may contend. This is also a monotonic measure in bottom-up planning.

A System R style optimizer, using Selinger's join enumeration algorithm, can be retooled to find alternative-objective plans simultaneously, along with finding the best time-to-completion (*base*) plan as before. We call our approach *bounded-alternative optimization*. During the bottom-up planning and pruning, we find the sub-plans that are *best* by each of the alternative cost formulas that are within a multiplicative constant (the *threshold coefficient*) in time cost of the best *base* sub-plan for that same "sub-query". This ensures the final best alternative plan found—if any—is no worse than the threshold cost (the threshold coefficient times the time-to-completion cost of the best base plan found). Thus, any alternative plan that is better on its objective (than is the base plan), is within the threshold cost, and is no worse than another alternative plan of the same objective (that is, worse on *both* time and alternative costs) is kept.

---

[3] This does not say that the best plan with respect to *raw CPU* is the *best* possible plan with respect to *CPU load*. A better CPU-load plan could exist. Finding it would be prohibitively expensive, however.

Additional query plans with *interesting orders* must already be kept while building the plan in a bottom-up fashion. They are kept to benefit the generation of potentially better super-plans. Consider sort-merge join, which consists of two steps: sorting and merging. If any of the input relations / streams is already sorted on the join attributes, the sorting operation can be skipped for that stream (as the merging can be done directly). Just as Selinger's algorithm also keeps these additional plans that generate *interesting orders*, we keep the additional plans that are "interesting" with respect to our alternative objectives, as discussed above. Keeping these extra plans during enumeration does add some overhead to the optimization time, of course, but we do not expect it to be substantial. This is confirmed by experimentation, discussed next.

4 Experimentation

We have implemented alternative-objective optimization for the alternative objective of bufferpool width, CPU raw cost, and I/O raw cost as a prototype branch off of IBM DB2 9.2 (Viper). In the current DB2 query cost-based optimizer, after an operator generates the alternative query sub-plans, it prunes the alternative sub-plans with the same properties by comparing their costs measured in timerons, and keeps the best base plan. We extended the optimizer to generate and carry additional plans for the alternative objectives during plan generation.

Our extensions to the optimizer include two components: first, it carries along multiple candidate query plans with respect to the various objectives during the optimization stage. Second, it uses a specialized cost function to determine which alternative query plans for the various objectives should be carried.

We employ the TPC-H benchmark for our experiments [4]. For this, we installed TPC-H on DB2 9.2 running on IBM AIX [4]. The size of the database we generated is one gigabyte. The number of the physical nodes of the database was set to a single node. We set the query plan execute in explain mode to use the command *db2exfmt* for extracting the query plan trees and their details.

We ran our prototype over the TPC-H database with a threshold coefficient of four to find alternative plans with respect to *buffer-pool width*, *CPU usage*, and *I/O usage*. Of the 22 standard queries, *five* alternative plans were found for *buffer-pool width*, *two* for *CPU usage*, and *eleven* for *I/O usage*. (In other cases, the alternative plan was the same as the base plan.) Some of the alternative plans offer just marginal improvement. We next set a cut-off of 95%—so the alternative plan consumes 95% or less of its targeted resource than the base plan—and target *load* instead of *usage* for *CPU* and *I/O* as the measure. Then *four* alternative plans for *buffer-pool width*, *two* for *CPU load*, and *eight* for *I/O load* remain.

Figure 1 illustrates the results. The upper left chart shows that the buffer-pool plans's *times* are longer than the base plans's, but their buffer-pool *widths* are less than the base plans's. The *upper bars* show the ratio of the buffer-pool plans's times to the base plans's times. The *lower bars* show the ratios of the

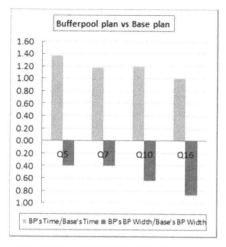

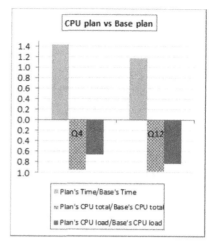

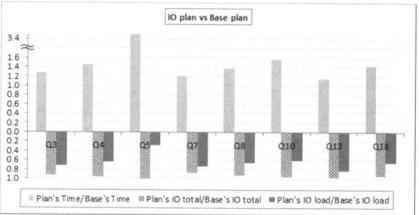

Fig. 1. Alternative plan vs. Base plan

buffer-pool plans's buffer-pool widths to the base plans's buffer-pool widths. For both the upper right and lower charts, three bars are shown. Again, the *upper bars* show the ratio of alternative's times to the base plans's times. The *left lower bar* shows the ratio of the alternative plans's alternative measures—CPU usage or I/O usage—to the base plans's. The *right lower bars* in these cases show the ratio of the alternative plans's CPU load or I/O load, respectively, to base plans's.

Figure 2 shows the difference between the base plan (left) and the buffer-pool plan (right) for query Q_5. For the base plan (left), the total time (in timerons) to completion is 688,995, and its maximum buffer width is 6,358 pages. As observed, the maximum buffer pool usage of this plan occurs at operator #7, a HSJOIN (hash join). A hash join has a large buffer-pool footprint because it needs an entire hash partition of one of the two inputs in memory at any given time, in addition to the input width for the other input.

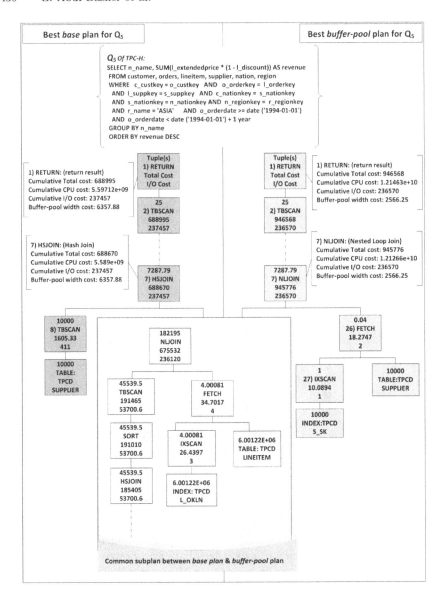

Fig. 2. Best base and buffer-pool plan for Q_5

As for the buffer-pool plan (right), the total time (in timerons) to completion for this plan is 946,568, and its maximum buffer width is 2,566 pages. The main change with respect to the base plan is in operator #7, which is now a NLJOIN (a nested-loop join). This needs only a quite small buffer-pool footprint; it needs two input widths for joining the two input streams. This reduces the size of the result set flowing upward, so the rest of the operations occupy less buffer pool. Thus, the buffer-pool plan for Q_5 runs in 1.4 times the runtime as the best base plan, but in 40% less buffer-pool space than what the best base plan requires.

The time taken by the modified optimizer to produce alternative query plans plus the base plan compared against the standard optimizer to optimize the query (to produce just the base plan) was not substantially more. For the 22 queries in TPC-H that we tested, the average time overhead was 5%. (There was an outlier exception with Query #8, which experienced 50% overhead.) This overhead is quite acceptable.

These experiments over TPC-H offer strong proof of concept that our bounded technique for alternative optimization objectives in a System R style optimizer over realistic workloads can result in query plans that are significantly different and that offer worthwhile tradeoffs in system resource usage.

5   Conclusions

Alternative-objective query optimization bridges the local optimality of individual query plans with global optimality of the workload. We have devised alternative objectives—tuned for conserving different key system resources—that are effective and that can be implemented efficiently within the "standard" optimization procedure. We modified Selinger's join enumeration to find best alternative-objective plans, along with the standard best time-to-completion plan, within a single invocation of the optimizer. We implemented this within IBM DB2 (in an experimental code base). We have shown over the TPC-H benchmark the approach finds realistic alternative plans, while adding insignificant overhead. In the next stage of our work, we aim to show that alternative-optimiztion can be used effectively to increase workload throughput.

References

1. Dayal, U., Kuno, H., Wiener, J.L., Wilkinson, K., Ganapathi, A., Krompass, S.: Managing operational business intelligence workloads. SIGOPS 43, 92–98 (2009)
2. Krompass, S., Gmach, D., Scholz, A., Seltzsam, S., Kemper, A.: Quality of Service Enabled Database Applications. In: Dan, A., Lamersdorf, W. (eds.) ICSOC 2006. LNCS, vol. 4294, pp. 215–226. Springer, Heidelberg (2006)
3. Krompass, S., Kuno, H., Dayal, U., Kemper, A.: Dynamic workload management for very large data warehouses: juggling feathers and bowling balls. In: VLDB, pp. 1105–1115 (2007)
4. T. organization. Tpc-h homepage. TPC-H homepage
5. Park, J., Segev, A.: Using common subexpressions to optimize multiple queries. In: ICDE, pp. 311–319 (1988)
6. Rosenthal, A., Chakravarthy, U.S.: Anatomy of a mudular multiple query optimizer. In: VLDB, pp. 230–239 (1988)
7. Schroeder, B., Harchol-Balter, M., Iyengar, A., Nahum, E.: Achieving class-based qos for transactional workloads. In: ICDE, Washington, DC (2006)
8. Sellis, T.K.: Multiple-query optimization. ACM TODS 13, 23–52 (1988)

Online Top-k Similar Time-Lagged Pattern Pair Search in Multiple Time Series

Hisashi Kurasawa, Hiroshi Sato, Motonori Nakamura, and Hajime Matsumura

NTT Network Innovation Laboratories,
3-9-11 Midori-cho Musashino-City, Tokyo, Japan

Abstract. We extract the relation among multiple time series in which a characteristic pattern in a time series follows a similar pattern in another time series. We call this a 'post-hoc-relation'. For extracting many post-hoc-relations from a large number of time series, we investigated the problem of reducing the cost of online searching for the top-k similar time-lagged pattern pairs in multiple time series, where k is the query size. We propose an online top-k similar time-lagged pattern pair search method that manages the candidate cache in preparation for the top-k pair update and defines the upper bound distance for each arrival time of pattern pairs. Our method also prunes dissimilar pattern pairs by using an index and the upper bound distance. Experimental results show that our method successfully reduces the number of distance computations for a top-k similar pattern update.

Keywords: Multiple Time Series, Time-lagged Correlation.

1 Introduction

We extracted the relation among multiple time series in which a characteristic pattern in a time series follows a similar pattern in another time series. We call this 'post-hoc-relation'. An example of a post-hoc-relation is shown in Figure 1. Note that a post-hoc-relation is different from a correlation. While a correlation measures the dependence between two time series, a post-hoc-relation represents the order of two similar segmented time series.

We believe that a large number of post-hoc-relations help us understand the relationships among multiple time series. Ultimately, we believe that a post-hoc-relation will reveal an unknown connection among physical elements, such as situations, environmental factors, and events. As the proverb "Post hoc ergo propter hoc (after this therefore because of this)" says, the post-hoc-relation has generally not been recognized as useful for analyzing events before carefully determining its causation. Needless to say, if a post-hoc-relation between two time series occurs only a few times, we assume it an accident. However, if a post-hoc-relation occurs many times, we believe it can be used for understanding the connection of the time series pair. Thus, we extracted a large number of post-hoc-relations.

S.W. Liddle et al. (Eds.): DEXA 2012, Part II, LNCS 7447, pp. 432–441, 2012.

Post-hoc-relations have not been investigated to the best of our knowledge. A fundamental problem of the time series similarity search is the computational cost. We developed an online post-hoc-relation search method that accepts a variety of similarity functions and an undefined similarity threshold. We generally need to select the similarity function and threshold for determining post-hoc-relations before searching for them. However, it is difficult to accurately set the most appropriate similarity function and threshold because they depend on the time series. Thus, we assume we can examine and evaluate them while searching in real-time. We designed our method to be flexible over them.

In summary, we make the following three contributions. Firstly, we propose an online top-k similar pattern pair search method as the first solution for retrieving post-hoc-relations, which is open to various similarity functions. Secondly, we use the normalized Euclidean distance as a temporary measure, which normalizes the amplitudes and offsets of the patterns. Finally, this method includes a caching technique, which defines the upper bound distance for each arrival time of pattern pairs, and a low update cost index, which inserts a new pattern and searches similar indexed patterns at the same time.

2   Related Work

Many techniques have been proposed for reducing the cost of time series searches. These techniques are categorized into three types: indexing, correlation analysis, and motif discovery.

Time series representations and indexes have been studied for indexing time series. Time series representations [9,3] reduce the dimensions of time series and calculate a lower bound distance for a specific distance with low computational cost. On the other hand, indexes developed for a specific space [1] recursively group time series by their similarities and reduce the number of distance computations while searching by pruning time series dissimilar from a query by using these groups. Although time series representations and indexes effectively reduce search cost for a specific distance and a specific space, we cannot use them because we have not yet fixed the similarity function for determining post-hoc-relations. Furthermore, they do not search for similar pairs.

Correlation analysis techniques are used to find all pairs with correlations above a given threshold [17,6,12]. Since the computational and I/O costs are larger for a larger number of time series, such techniques are aimed at reducing these costs. They use dimension reduction and pruning techniques for reducing computational cost and use caching strategies for reducing the I/O cost. Although correlation analysis techniques effectively reduce the similar pattern pair search cost for a large number of time series, they mainly handle patterns that arrive at the same time. Some methods [12] handle time-lagged patterns, but they take into account a very short time-lag.

A motif is a pair of similar non-overlapping patterns in a time series. Motif discovery techniques are aimed at finding unknown repeated patterns [10,15,2]. Recent studies have attempted to find motifs from a time series in real time [11,7].

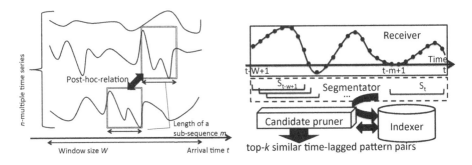

Fig. 1. Post-hoc-relation   **Fig. 2.** Overview

Since the memory requirement for storing motif candidates is larger for discovering from a longer time series, they were developed for motif candidate management. Although the concept of the motif is similar to the post-hoc-relation, these techniques search motifs from only a single time series.

As mentioned above, none of the above-mentioned techniques can search the top-k similar time-lagged pattern pairs from multiple time series. There are two issues to be addressed. One is to propose a caching strategy and the other issue is to improve the update technique of the index. In this paper, we mainly describe the caching strategy and the index update technique.

3   Problem Definition

A time series is a continuous sequence of values. Assume that we have n-multiple time series S^1, S^2, $\cdots$, S^n, the value of the ith time series S^i at time t_x is $S^i[t_x]$. We keep the most recent time series of size W, which consists of the values $S^i[t - W + 1 : t] = \{S^i[t - W + 1], S^i[t - W + 2], \cdots, S^i[t]\}$ where $S^i[t]$ is the most recent value.

A pattern $S^i_{t_x}$ of length m $(m \leq W)$ in a time series S^i is a sequence $S^i[t_x - m + 1 : t_x]$. The number of patterns in S^i is $w = W - m + 1$.

Our method uses similarity measures in metric spaces. Let $M = (D, d)$ be a metric space defined for a domain of objects D and a distance function $d : D \times D \mapsto \mathbb{R}$. This metric satisfies the four postulates [16]: non-negativity, symmetry, identity, and triangle inequality.

We use the normalized Euclidean distance as a temporary measure. Given two pattern $S^i_{t_x}, S^j_{t_y}$, the normalized Euclidean distance $d(S^i_{t_x}, S^j_{t_y})$ between $S^i_{t_x}$ and $S^j_{t_y}$ is defined as $\sqrt{\sum_{c=0}^{m-1} \left(\frac{S^i[t_x - c] - \mu_{S^i_{t_x}}}{\sigma_{S^i_{t_x}}} - \frac{S^j[t_y - c] - \mu_{S^j_{t_y}}}{\sigma_{S^j_{t_y}}} \right)^2}$ where m, $\mu_{S^a_{t_b}}$, and $\sigma_{S^a_{t_b}}$ are the length, average, and standard deviation of pattern $S^a_{t_b}$, respectively.

A top-k similar pattern pair query is defined as follows.

Definition 1. *Given n time series $S^1[t - W + 1 : t], S^2[t - W + 1 : t], \cdots, S^n[t - W + 1 : t]$, a top-$k$ similar pattern pair query with threshold k returns the pattern pair set A. A satisfies*

$$
\begin{cases}
A \subseteq \{(S_{t_x}^i, S_{t_y}^j) | S_{t_x}^i \in S, S_{t_y}^j \in S, i \neq j\} \ , \\
|A| = k \ , \\
\forall(S_{t_x}^i, S_{t_y}^j) \in A, \forall(S_{t_a}^a, S_{t_b}^b) \in (\{(S_{t_x}^i, S_{t_y}^j) | S_{t_x}^i \in S, S_{t_y}^j \in S, i \neq j\} - A) \\
\quad d(S_{t_x}^i, S_{t_y}^j) \leq d(S_{t_a}^a, S_{t_b}^b) \ ,
\end{cases}
\tag{1}
$$

where S is a set of patterns in the time series.

4   Online Top-k Similar Pattern Pair Search

This section explains our proposed method, which is designed for online top-k most similar time-lagged pattern pair search in multiple time series. Our method is aimed at reducing the distance computations for such a search.

Our method obtains multiple stream time series and incrementally outputs the k closest time-lagged pattern pairs. It consists of four components: receiver, segmentator, candidate pruner, and indexer, as shown in Figure 2. The **receiver** obtains multiple time series and keeps the latest values for each one. Given n time series $S^1, S^2, \cdots, S^n$, the receiver stores the most recent W values of each time series into memory. The **segmentator** reads a pattern from the memory stored by the receiver. Given time series ID i and time t_x, the segmentator returns the pattern $S_{t_x}^i$. The length of $S_{t_x}^i$ is m. The **candidate pruner** manages the candidate cache, defines the upper bound distances, and outputs the k closest time-lagged pattern pairs. Given candidate cache B, the candidate pruner returns the upper bound distance list U, updated B, and the result set A. The **indexer** maintains the index of the patterns and searches similar pattern pair candidates. Given the new patterns $S_t^1, S_t^2, \cdots, S_t^n, U$, and B, the indexer returns the updated B.

We explain the details of the candidate pruner and indexer in Sections 4.1 and 4.2, respectively.

4.1   Candidate Caching Techniques

The candidate pruner caches similar pattern pair candidates in preparation for immediate updates of the top-k pattern pairs. It prevents repeated distance computations of pairs during the updates. Moreover, it defines the upper bound distance for each arrival time of pattern pairs used for searching similar pattern pairs by the index.

The strategy involves deciding which pattern pairs to bring into the candidate cache and how to remove them from the cache. The k closest pattern pairs update when one of the closest pattern pairs expires, or a new closest pattern pair is detected. The former case is that in which a closest pattern pair S_{t-w}^i and $S_{t_y}^j$ $(t - w \leq t_y < t)$ is deleted from the result set, and one of the similar pattern pair candidates in the candidate cache is inserted into the result set. The latter case is that in which a new closest pattern pair $S_{t_x}^i$ and S_t^j $(t - w < t_x \leq t)$ is found by the indexer and inserted into the result set, and the $k + 1$th closest pattern pair is deleted from the result set. These observations indicate that the cached candidates are needed only for the former case.

The above observation tells us that the candidate cache should have the top-k similar pattern pairs from each arrival time to the most recent arrival time, which leads to the following lemma.

Lemma 1. *Suppose $S_{t_x,t}$ and $A_{t_x,t}$ denote the pattern pairs from t_x to t and the k closest pattern pairs in $S_{t_x,t}$, respectively. The candidate cache should include the set $A_{t-w+1,t} + A_{t-w+2,t} + \cdots + A_{t-1,t} + A_{t,t}$.*

Proof. After $S_{t-w+1,t-w+1}$ expires, the updated result set $A_{t-w+2,t+1}$ consists of the new closest pattern pairs detected at $t+1$ and a subset of $A_{t-w+2,t}$.

We can then determine the upper bound distance of the candidate cache for each arrival time of pattern pairs. The upper bound distance U_{t_x} of t_x is the maximum distance of $A_{t_x,t}$. This means that U_{t_x} does not increase as time passes because the maximum distance of $A_{t_x,t}$ is equal to or less than that of $A_{t_x,t+1}$. The upper bound distance is useful in deciding to insert a similar pattern pair into the candidate cache. Furthermore, it helps the indexer to reduce the distance computations while searching new closest pattern pairs. This is because the indexer can discard more dissimilar pattern pairs by using the smaller upper bound distance.

Let us consider the data structure of the candidate cache. It should satisfy three operations: insertion of new similar pattern pair candidates, deletion of the cached pattern pair with distances over the upper bound distance, and deletion of the cached pattern pair that has expired. We use a queue for B of w max-heaps $B_t, B_{t-1}, \cdots, B_{t-w+1}$ and use a list for U of w upper bound distances U_t, $U_{t-1}, \cdots, U_{t-w+1}$. B keeps similar pattern pair candidates on the basis of the arrival time of the older pattern in each pair candidate, so that expired similar pattern pair candidates can be easily deleted by dequeueing the expired max-heaps B_{t-w} from B. For example, a similar pattern pair candidate $S_{t_x}^i$ and $S_{t_y}^j$ ($t_x \leq t_y$) is stored in B_{t_x}.

When the new patterns $S_t^1, S_t^2, \cdots, S_t^n$ are segmented, the indexer searches their similar pattern pairs by using U and updates B. Then the candidate pruner starts to update U and output the k closest pattern pairs as follows. First, the candidate pruner makes an empty max-heap as result set A, and merges it with B from the most recent B_t to the oldest B_{t-w+1}. The upper bound distance U_{t_x} of time t_x is set as the maximum distance of A after merging B_{t_x}. As a result, the upper bound distance of the later time is larger. After merging, the candidate pruner adds an empty max-heap B_{t+1} to B and sets U_{t+1} as ∞. At the same time, the candidate pruner deletes the oldest max-heap B_{t-w+1}. Finally, it outputs A, which consists of the k closest pattern pairs.

4.2 Indexing Techniques

The indexer classifies the patterns by their arrival time and organizes them by their similarity. As a result, it can prune dissimilar pattern pairs while searching by the index and can reduce the distance computations for finding new similar pattern pairs. We developed an exact similarity search index in metric spaces for the indexer. The goal was to reduce not only the distance computations of searching but also the distance computations of pattern inserting and deleting.

Our method uses *ball partitioning* [14], which is a major indexing technique. Because coordinates are not explicitly defined in metric spaces, ball partitioning uses the distances between patterns. It recursively partitions a set of patterns into two subsets on the basis of the distance from a reference pattern to each pattern. As a result, it creates a balanced binary tree structure index. It then prunes sets of dissimilar patterns by using the distance between the reference pattern and a query. Most metric space indexes use ball partitioning. They improve the reference pattern selection [8], structure [4], and dynamic index update [5,13].

Although current dynamic indexes have been designed for the increase of indexed patterns, we believe that it is not crucial for our index because of the static number of indexed patterns. Current methods require too much computational cost to dynamically modify the reference pattern and its partitioning distance while updating. They keep patterns for a long time and are aimed at improving scalability. It is true that we have to consider scalability because the number of patterns in multiple time series increases as the number of time series or the number of patterns in a time series increases. However, the indexer manages the same number of patterns before and after the update of the top-k pattern pairs. The indexer inserts the most recent patterns; on the other hand, it deletes the oldest patterns. Furthermore, each indexed pattern pair is stored only for a short time. Thus, we do not adopt the modification techniques for reference patterns.

The above observation tells us that the index should be designed for frequent insertions and deletions rather than the optimal search. We wanted to develop an index that requires low distance computations for updating and is practical for searching. We propose two types of indexes: *tree-array* and *single multi-time tree*.

The tree-array index has w tree structure sub-index, which are created on the basis of the ball partitioning technique by using n patterns for each arrival time. It uses a queue for managing the w sub-indexes. Thus, it repeatedly dequeues the oldest sub-index; on the other hand, it constructs and queues the newest sub-index. It prunes dissimilar pattern pairs of each arrival time separately by using the w sub-indexes and the w upper bound distances. It is advantageous for adapting the index to the trend in patterns because it can select good reference patterns for each arrival time while constructing the sub-index. However, it does not scale because it has to manage the same number of indexes as that of patterns in a time series.

On the other hand, the single multi-time tree index has one tree structure index, which is created on the basis of the ball partitioning technique by using initial $w \cdot n$ patterns. It inserts and deletes patterns without changing reference patterns. That is, this index updates only leaf nodes. It classifies patterns on the basis of their arrival time at leaf nodes. A leaf node has a queue that stores patterns in order of their arrival time. An advantage of the static structure is that we can reduce distance computations of the pattern insertions by inserting a new pattern at the same time of traversing the index for searching similar patterns. It prunes a set of dissimilar pattern pairs of more than one arrival time in a leaf node at once by using the index and the upper bound distance list. It scales more than the tree-array index due to the single tree structure. However, its reference patterns in the index may be ineffective while updating. Although the number of patterns in a leaf node is balanced at the start, it may become unbalanced as time passes. The unbalanced leaf nodes lead to low search performance.

We compared the single multi-time tree index with the tree-array index and obtained better experimental results when using the former. We decided to use the single multi-time tree index for our indexer. We discuss the tree balance evaluation and the comparison of index performance in Section 5.

4.3   Cost Analysis

We estimated the cost of updating top-k similar time-lagged pattern pairs by using the number of distance computations between patterns. The number of indexed patterns is $w \cdot n$, where n is the number of time series. The initial index construction cost is $O(w \cdot n \cdot \log (w \cdot n))$. The index insert cost for a new pattern is $O(\log (w \cdot n))$. The cost of finding the patterns similar to the new pattern is from $O(\log (w \cdot n))$ to $O(w \cdot n)$. This cost depends on the balance of the leaf nodes in the index. Thus, the total update cost for the k closest pattern pair search is from $O(n \cdot \log (w \cdot n))$ to $O(w \cdot n^2)$. For reference, the update cost of the naive search with our caching, that without caching, and that of the tree-array index are $O(w \cdot n^2)$, $O(w^2 \cdot n^2)$, and from $O(n \cdot w \cdot \log n)$ to $O(w \cdot n^2)$, respectively.

The memory requirement is as follows. The number of values in the receiver is $W \cdot n$, and the number of reference patterns in the index is $O(w \cdot n \cdot \log (w \cdot n))$. However, these patterns are in the initial $w \cdot n$ patterns. Thus, the actual number of reference patterns is equal to or less than $w \cdot n$ and are in the initial $W \cdot n$ values. The number of internal nodes in the index is $O(w \cdot n \cdot \log (w \cdot n))$. Each internal node has 5 values: one time series ID, one arrival time, one partitioning distance, and two child node pointers. The number of tuples in B is equal to or less than $k \cdot w$. Each tuple has 5 values: the distance, two time series IDs, and two arrival times. Therefore, the total memory size is $O(W \cdot n + w \cdot n \cdot \log (w \cdot n) + k \cdot w)$. For reference, the memory size of the naive search with our caching is $O(W \cdot n + k \cdot w)$ and that with naive caching is $O(W \cdot n + w^2 \cdot n^2)$.

5   Performance Evaluation

We conducted experiments on a real dataset and compared our method with two others. We evaluated the balance of leaf nodes in our index and the update performance with respect to n, w, and k. The experimental results indicate that our method is good at finding similar pairs in a large number of time series and a large number of patterns in a time series.

We used three methods: naive, tree-array, and ours. The naive method is the naive search with our candidate cache. This method does not use an index and computes the distances of all pairs in the dataset. The tree-array method uses the tree-array index with our candidate cache. Our method uses the single multi-time tree index with our candidate cache. Note that we omitted the experimental

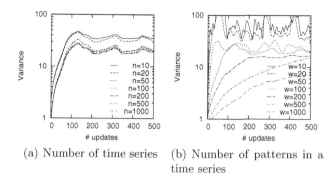

(a) Number of time series

(b) Number of patterns in a time series

Fig. 3. Balance of the leaf nodes

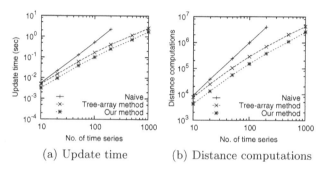

(a) Update time

(b) Distance computations

Fig. 4. Update performance with respect to the number of time series

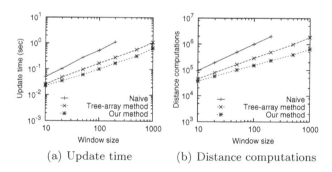

(a) Update time

(b) Distance computations

Fig. 5. Update performance with respect to the number of patterns in a time series

evaluation of a method without caching because it is clear that it requires too much computational cost, as mentioned in Section 4.3.

We used a real datasets named New York Stock Exchange, which is a collection of day stock prices on the New York Stock Exchange from the beginning of 2000 to the end of 2011. The number of time series is 2,689. We crawled this data from the web [1].

---
[1] http://finance.yahoo.com

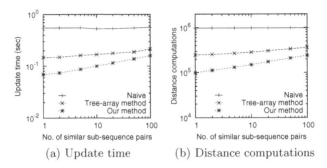

(a) Update time   (b) Distance computations

Fig. 6. Update performance with respect to the number of similar pattern pairs

We implemented the above three methods with C language. We conducted the experiment on a Linux PC equipped with an Intel(R) Quad Core Xeon(TM) X5690 3.46-GHz CPU and 144 GB of memory. Our codes were compiled with GCC 4.6. All the datasets were processed in memory for all the methods.

We used the normalized Euclidean distance as the similarity function. We randomly selected n time series from the dataset over 100 times and evaluated the performance over 500 updates. Each result was the average cost. We set k, w, n, and m to 10, 100, 100, and 32 as the standard values, respectively.

Figure 3 shows the balance of leaf nodes in our index. The horizontal axis is the variance and the vertical axis is the number of update times. We can see that the variance of the larger w and n converges to the lower value. This means that the refinement of leaf node balance is hardly needed if we search similar time-lagged pattern pairs from a large number of patterns.

Figure 4 shows the update performance with respect to the number of time series n. The horizontal axis is the number of time series, the vertical axis of Figure 4 (a) represents the update time, and that of Figure 4 (b) is the number of distance computations for an update. From these results, we can see that our method reduces the number of distance computations and achieves quick updates.

Figure 5 shows the update performance with respect to the number of time series w. The horizontal axis is the number of patterns in a time series, the vertical axis of Figure 5 (a) represents the update time, and that of Figure 5 (b) is the number of distance computations for an update. Our method outperformed the other two methods for longer patterns in a time series. We can see that the single multi-time tree structure works better than the tree-array structure, and the unbalance of leaf nodes in our method is not a critical issue for these datasets.

Figure 6 shows the update performance with respect to the number of time series k. The horizontal axis is the number of similar pattern pairs, the vertical axis of Figure 6 (a) represents the update time, and that of Figure 6 (b) is the number of distance computations for an update. The computational cost of our method and the tree-array method increased with respect to k, while the naive method required the same number of distance computations for all the numbers of k. However, we can see our method outperformed the other two methods .

6 Conclusion

We proposed an online top-k similar time-lagged pattern pair search method as the first solution for retrieving post-hoc-relations. Since we are examining similarity thresholds and functions for measuring post-hoc-relations, our method is designed to be flexible over them. Our method reduces the distance computational cost for searching by using the candidate cache and index.

We think that the current query definition has to be modified. We may get the top-k patterns belong to only two similar time series and it is not informative. We are currently investigating the query definition, similarity thresholds, and functions for port-hoc-relations.

References

1. Assent, I., Krieger, R., Afschari, F., Seidl, T.: The ts-tree: efficient time series search and retrieval. In: EDBT (2008)
2. Castro, N., Azevedo, P.: Multiresolution motif discovery in time series. In: SDM (2010)
3. Chen, Q., Chen, L., Lian, X., Liu, Y., Yu, J.X.: Indexable pla for efficient similarity search. In: VLDB (2007)
4. Chevez, E., Navarro, G.: A compact space decomposition for effective metric indexing. Pattern Recognition Letters 24(9), 1363–1376 (2005)
5. Ciaccia, P., Patella, M., Zezula, P.: M-tree: An efficient access method for similarity search in metric spaces. In: VLDB (1997)
6. Gandhi, S., Nath, S., Suri, S., Liu, J.: Gamps: compressing multi sensor data by grouping and amplitude scaling. In: SIGMOD (2009)
7. Lam, H.T., Calders, T., Pham, N.: Online discovery of top-k similar motifs in time series data. In: SDM (2011)
8. Jagadish, H.V., Ooi, B.C., Tran, K.L., Yu, C., Zhang, R.: idistance: An adaptive b+-tree based indexing method for nearest neighbor search. ACM Trans. on Database Systems (TODS) 30(2), 364–397 (2003)
9. Lin, J., Keogh, E., Lonardi, S., Chiu, B.: A symbolic representation of time series, with implications for streaming algorithms. In: DMKD (2003)
10. Lin, J., Keogh, E., Lonardi, S., Patel, P.: Finding motifs in time series. In: SIGKDD (2002)
11. Mueen, A., Keogh, E.: Online discovery and maintenance of time series motifs. In: KDD (2010)
12. Mueen, A., Nath, S., Liu, J.: Fast approximate correlation for massive time-series data. In: SIGMOD (2010)
13. Navarro, G., Reyes, N.: Dynamic spatial approximation trees. Journal of Experimental Algorithmics 12, 1.5:1–1.5:68 (2008)
14. Uhlmann, J.K.: Satisfying general proximity/similarity queries with metric trees. Information Processing Letters 40(4), 175–179 (1991)
15. Yankov, D., Keogh, E., Medina, J., Chiu, B., Zordan, V.: Detecting time series motifs under uniform scaling. In: KDD (2007)
16. Zezula, P., Amato, G., Dohnal, V., Batko, M.: Similarity Search: The Metric Space Approach (Advances in Database Systems). Springer (2005)
17. Zhang, T., Yue, D., Gu, Y., Yu, G.: Boolean representation based data-adaptive correlation analysis over time series streams. In: CIKM (2007)

Improving the Performance for the Range Search on Metric Spaces Using a Multi-GPU Platform

Roberto Uribe-Paredes[1], Enrique Arias[2],
José L. Sánchez[2], Diego Cazorla[2], and Pedro Valero-Lara[3]

[1] Computer Engineering Dept., Univ. of Magallanes, Punta Arenas, Chile
[2] Computing Systems Dept, Univ. of Castilla-La Mancha, Albacete, Spain
[3] Centro de Inv. Energéticas, Medioambientales y Tec., Madrid, Spain
roberto.uribeparedes@gmail.com,
{enrique.arias,jose.sgarcia,diego.cazorla}@uclm.es, pedro.valero@ciemat.es

Abstract. Nowadays, similarity search is becoming a field of increasing interest because these kinds of methods can be applied to different areas in science and engineering, for instance, pattern recognition, information retrieval, etc. This search is carried out over metric indexes decreasing the number of distance evaluations during the search process, improving the efficiency of this process. However, for real applications, when processing large volumes of data, query response time can be quite high. In this case, it is necessary to apply mechanisms in order to significantly reduce the average query response time. In this sense, the parallelization of the metric structures processing is an interesting field of research. Modern GPU/Multi-GPU systems offer a very impressive cost/performance ratio. In this paper, we show a simple and fast implementation of similarity search method on a Multi-GPU platform. The main contributions are mainly the definition of a generic metric structure more suitable for GPU platforms, the efficient usage of GPU memory system and the implementation of the method in a Multi-GPU platform.

Keywords: Range queries, similarity search, metric spaces, parallel processing, Multi-GPU platforms.

1 Introduction

In the last decade, the search of similar objects in a large collection of stored objects in a metric database has become a most interesting problem. This kind of search can be found in different applications such as voice and image recognition, data mining, plagiarism detection and many others. A typical query for these applications is the *range search* which consists in obtaining all the objects that are at some given distance from the consulted object. Basically, similarity is modeled in many interesting cases through metric spaces, and the search of similar objects through range search or nearest neighbors. A metric space $(\mathbb{X}, d)$ is a set $\mathbb{X}$ and a distance function $d : \mathbb{X}^2 \to \mathbb{R}$, so that $\forall x, y, z \in \mathbb{X}$ fulfills the properties of positiveness $[d(x, y) \geq 0$, and $d(x, y) = 0$ iff $x = y]$, symmetry

S.W. Liddle et al. (Eds.): DEXA 2012, Part II, LNCS 7447, pp. 442–449, 2012.

$[d(x,y) = d(y,x)]$ and triangle inequality $[d(x,y) + d(y,z) \geq (d(x,z)]$. This concept of similariy is associated to the concept of Metric space data structures, which can be grouped into two classes [1]: *clustering*-based (*BST* [2], *GHT* [3], *M-Tree* [4], *GNAT* [5], and many others), and *pivots*-based methods (*LAESA* [6], *FQT* and its variants [7], *Spaghettis* and its variants [8], *FQA* [9], *SSS-Index* [10] and others).

However, for the above mentioned real applications, when processing large volumes of data, query response time can be quite high. In this case, it is necessary to apply mechanisms in order to significantly reduce the average query response time. From parallelism point of view, currently, most of the previous and current works developed in this area are carried out considering classical distributed or shared memory platforms. However, Modern GPU/Multi-GPU systems offer a very impressive cost/performance ratio as compared to multiprocessor or multicomputer platforms that are usually more expensive gaining in significance and popularity within the scientific computing community. As far as we know, the solutions considered till now developed on GPUs are based on *kNN* queries without using data structures. This means that GPUs are basically applied to exploit its parallelism only for exhaustive search (brute force) [11,12,13,14].

In this paper three main contributions are outlined. The first one is the use of a generic metric structure instead of, for instance, Spaghettis data structure. Secondly, a GPU-based implementation is presented which improves the performance of the version presented in [15] by optimizing the memory accesses. Finally, a Multi-GPU version of the similarity search algorithm is introduced. Accordingly to that, the paper is structured as follows. In Section 2 the generic structure is explained and, in Section 3 the GPU and Multi-GPU implementations are described. The case studies, the platform and the experimental results as well as a discussion of them is carried out in Section 4. Finally, the conclusions and future work are commented in Section 5.

2   A Generic Metric Data Structure

In this work, we propose the use of a generic metric data structure (*GMS*), thinking of the posterior implementation in GPUs. This generic metric data structure consists in a bidimensional array data structure (see Figure 1(a)) whose dimensions are $Q \times P$, being Q the number of queries and P the number of pivots. In general terms, other popular metric data structure as Spaghettis [8] and SSS-Index [10] could be considered as a generic bidimensional array data structure. The difference between these structures is the way of obtaining the pivots or the way in which the structure is stored. Thus, this *GMS* avoids the sorting process (against the Spaghettis method) because this process is computationally expensive on GPUs; and it also avoids the selection of the pivots which is carried out randomly (against the SSS-Index selection).

For this generic metric data structure, the searching process, given a query q and a range r, is carried out according to the following steps:

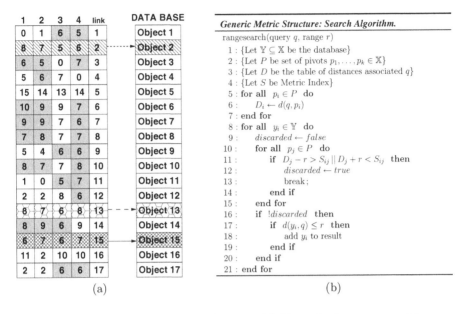

1	2	3	4	link	DATA BASE
0	1	6	5	1	Object 1
8	7	5	6	2	Object 2
6	5	0	7	3	Object 3
5	6	7	0	4	Object 4
15	14	13	14	5	Object 5
10	9	9	7	6	Object 6
9	9	7	6	7	Object 7
7	8	7	7	8	Object 8
5	4	6	6	9	Object 9
8	7	7	8	10	Object 10
1	0	5	7	11	Object 11
2	2	8	6	12	Object 12
8	7	6	8	13	Object 13
8	9	6	9	14	Object 14
6	7	6	7	15	Object 15
11	2	10	10	16	Object 16
2	2	6	6	17	Object 17

(a)

Generic Metric Structure: Search Algorithm.

rangesearch(query q, range r)

1 : {Let $\mathbb{Y} \subseteq \mathbb{X}$ be the database}
2 : {Let P be set of pivots $p_1, \ldots, p_k \in \mathbb{X}$}
3 : {Let D be the table of distances associated q}
4 : {Let S be Metric Index}
5 : for all $p_i \in P$ do
6 : $D_i \leftarrow d(q, p_i)$
7 : end for
8 : for all $y_i \in \mathbb{Y}$ do
9 : $discarded \leftarrow false$
10 :   for all $p_j \in P$ do
11 :   if $D_j - r > S_{ij} \| D_j + r < S_{ij}$ then
12 : $discarded \leftarrow true$
13 :   break;
14 :   end if
15 :   end for
16 :   if $!discarded$ then
17 :   if $d(y_i, q) \leq r$ then
18 :   add y_i to result
19 :   end if
20 :   end if
21 : end for

(b)

Fig. 1. Metric structure and search algoritm. (a) Example for query q with ranges $\{(6, 10), (5, 9), (2, 6), (4, 8)\}$ to pivots. (b) Range search algorithm.

1. From the distance between q and all pivots $p_1, \ldots, p_k$ we obtain k intervals in the form $[a_1, b_1], \ldots, [a_k, b_k]$, where $a_i = d(p_i, q)$ - r and $b_i = d(p_i, q) + r$.
2. The objects in the intersection of all intervals are candidates to the query q.
3. For each candidate object y, the distance $d(q, y)$ is calculated, and if $d(q, y) \leq r$, then the object y is a solution to the query.

Figure 1(a) represents the generic data structure. This structure is built using 4 pivots to index a database of 17 objects [15].

3   GPU and Multi-GPU Implementations

In this section we describe the GPU-based implementation of the range search algorithm 1(b). This version improves a previous one [15] by using the GPU shared memory in order to significantly reduce the memory accesses latency.

According to the three steps described in Section 2 to complete the searching process, the following data structures have to be allocated on shared memory (the fastest memory):

1. The first part consists in computing the distances between the set of queries, Q, and the set of pivots, P. A $Q \times P$ matrix is used to store all distances, which are calculated at the same time in a single call to the kernel. This kernel consists of as many threads as the number of queries. In fact, each thread calculates independently the distance from a query to all pivots.

As all threads need to access the pivots, that is, they need to access the same positions of memory, we store the pivots in shared memory, and the bidimensional structure with the distances of all the queries and pivots is allocated on global memory (the slowest memory).

2. The second part of the parallel implementation consists in determining whether each element of the database is or not a candidate for every query. This part has been implemented as an iterative process. In each iteration the candidates for a particular query are computed in one kernel call. In this kernel as many threads as the number of elements of the database are launched. Each thread determines if every data of the dataset is candidate or not. Thus, this kernel returns a list of candidates for a given query. Finally, when this process finishes we obtain one list of candidates for each query. Again, we take advantage of the GPU shared memory storing there the distances between a given query and the pivots.

3. The kernel which implements the third part determines if each candidate is really a solution. In this kernel, the number of threads corresponds to the number of candidates for each query. Each thread calculates the distance between one candidate and one query, and determines if this candidate is a valid solution. Finally, as result we obtain one list of solutions for each query. Precisely, the query is the structure to allocate on shared memory.

The Multi-GPU implementation is, a priori, a naive implementation consisting in a unique platform with several GPUs and using OpenMP to get out the databases or the queries among the GPUs. In this case, we have decided to split the queries and to replicate the database, and the processing of each query is completely independent.

4   Experimental Results

4.1   Case Studies and Platform

As case studies, we have considered two datasets: a subset of the Spanish dictionary (86,061 words) using the edit distance and for each query, was considered a range search between 1 and 4. The second space is a set of $112,682$ color histograms (112-dimensional vectors) from an image database. Any quadratic form can be used as a distance, thus we chose Euclidean distance as the simplest meaningful alternative. The radius used was that allowing to retrieve 0.01, 0.1 and 1% from the dataset.

For both databases we create the metric data structure with 90% of the data set randomly chosen, and reserve the rest 10% for queries. The pivots have been chosen using the SSS-Index method, taking into account a low and a high number of pivots, due to the fact that the GPU has hard memory constraints. The database has to fit in the shared memory of the GPU. Thus, it is out of the scope of this paper considering big databases, and it is considered on the Future work section.

For all the experiments, the execution time considered takes into account only the searching process, including CPU-GPU data transfer time. The computation of the different structures, for instance the pivots, is computed offline. In this way, all the methods are in the same conditions in order to carry out a correct comparison.

The hardware platform used is a 2 Quadcore Xeon E5530 at 2.4GHz and 48GB of main memory with 1 Nvidia Tesla C1060 240 cores at 1.3GHz and 4 GB of global memory, using CUDA SDK v3.2 (http://developer.nvidia.com/). The compilation has been done using gcc 4.3.4 compiler and OpenMP library.

4.2   Results and Discussion

Figures 2(a) and 2(b) show the execution time of Spaghettis, SSS-Index and GMS versions. The number of pivots considered for color histograms case study is 57 and 119, for the Spanish dictionary case study is 44 and 328.

In the case of Spanish dictionary, as the number of pivots increases the execution time decreases as the Figure 2(a) shows, just contrary to the vectors case study. This situation is normal because the behaviour of the method with respect to the database is unpredictable due to the space distribution. In fact, if we make a detailed study of the obtained results, when range is 1 or 2, Spaghettis is slightly better in terms of execution time, due to the benefits of the binary search. However, this process becomes an inconvenient when range is 3 or 4 because most of the data structure is covered. In general, for all the metric data structures considered in this work, the execution time for range 1 and 44 pivots is much better than using 328 pivots (almost 2.47 times better). However, due to the space distribution this changes with range 2 - 4.

In Figure 2(b) Spaghettis has the best execution time for low and high number of pivots. However, GMS is very close to it, especially when the problem size increases, that is, the percentage of information retrieved from the database is 1%. This result is evident due to Spaghettis takes benefit of the binary search. However, this advantage becomes a drawback from a GPU point of view. The worst behaviour corresponds to the SSS-Index method. Remark that the GMS behaviour is closer to the Spaghettis one.

Regarding the GPU implementation, the use of Spaghettis is not considered because the binary search is computationally expensive being a handicap to be running on a GPU platform.

The results obtained with the sequential implementation allow us to obtain some important conclusions for the GPU shared memory implementation:

– For color histograms case study (Figure 2(b)) the number of pivots to take into account is 57, where GMS implementation on the GPU has better behaviour than SSS-Index. However, using 57 pivots the structure does not fit on the shared memory of the GPU. Considering that circumstance, the number of pivots has to be reduce to 32. In this case, the GMS implementation is the best taking advantage of the GPU hierarchy memory. The behaviour of SSS-Index was worst that considering 57 pivots.

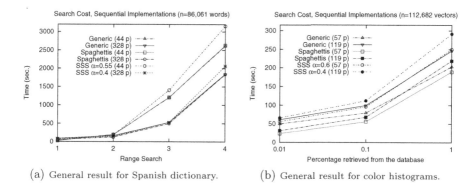

(a) General result for Spanish dictionary.   (b) General result for color histograms.

Fig. 2. Execution time for sequential implementation of the generic, Spaghettis and SSS algorithms

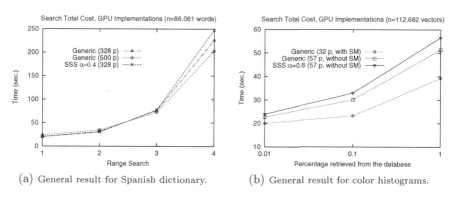

(a) General result for Spanish dictionary.   (b) General result for color histograms.

Fig. 3. Execution time for the shared memory GPU implementation (generic) and for GPU implementation without considering shared memory (generic and SSS-Index)

– For spanish dictionary case study (Figure 2(a)), the number of pivots has to be as high as possible, around 328 pivots. However, this high number of pivots has been considered because SSS-Index does not provide another bigger with good behaviour. In Figure 3(a) we can appreaciate that the behaviour of the GMS is better than the SSS-Index when the range increases. But, according to the shared memory size, and thinking in middle and high ranges, a suitable pivots number is 500. This number has been taken into account for an additional test in the case of the GMS, that allows us to increase the number of pivots.

Due to the fact that GPU experiments show that SSS-Index has a worst behaviour than GMS, it is not considered in these experiments.

Figure 4 shows the execution time of the Multi-GPU implementation. Notice that the Multi-GPU implementation is up to 12 times faster than the sequential for the color histograms case study, and up to 31 times faster than the sequential implementation for the Spanish dictionary case study.

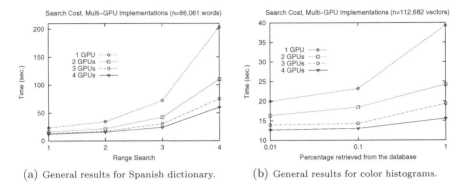

(a) General results for Spanish dictionary.   (b) General results for color histograms.

Fig. 4. Execution time for the Multi-GPU implementation (generic)

5   Conclusions and Future Work

In this work, we have presented an implementation of the similarity search using a GMS for GPU-based platforms. In order to evaluate these implementations we have used two different databases, Spanish dictionary and color histograms.

The experimental results of the sequential implementation allow us to conclude that GMS has a similar behaviour than Spaghettis, and it is much better than SSS-Index. Also, the sequential tests have guided us about the number of pivots to take into account, especially considering the GPU and Multi-GPU implementations with a limited memory capacity.

With respect to the GPU implementation, an exhaustive study of the algorithm allows to take advantage of the memory hierarchy of current GPUs, in particular, shared memory. It is possible to appreciate that, independently of the use of shared memory, the best execution time has been obtained using GMS because it fits better in a GPU platform.

With respect to the best sequential method presented in Section 2, the GPU implementation is more than 4.5 times faster for color histograms and almost 10 for Spanish dictionary. Regarding the Multi-GPU implementation, it is up to 12 times faster for color histograms and up to 31 for the Spanish dictionary.

With respect the future work, and taking into account that probably the database can not be allocated completely on main memory, a block-oriented implementation has to be carried out minimizing data transfers between secondary memory and main memory (CPU) and between main memory and GPU memory.

Also, the authors are developing MPI, OpenMP and Hybrid versions of the sequential algorithm proposed in this paper make easy to the user the use of the underlying architecture, through a user-friendly web interface.

Acknowledgments. This work has been partially supported by the Ministerio de Ciencia e Innovación, project SATSIM (Ref: CGL2010-20787-C02-02), Spain.

References

1. Chávez, E., Navarro, G., Baeza-Yates, R., Marroquín, J.L.: Searching in metric spaces. ACM Computing Surveys 33(3), 273–321 (2001)
2. Kalantari, I., McDonald, G.: A data structure and an algorithm for the nearest point problem. IEEE Transactions on Software Engineering 9(5) (1983)
3. Uhlmann, J.: Satisfying general proximity/similarity queries with metric trees. Information Processing Letters 40, 175–179 (1991)
4. Ciaccia, P., Patella, M., Zezula, P.: M-tree: An efficient access method for similarity search in metric spaces. In: The 23rd International Conference on VLDB, pp. 426–435 (1997)
5. Brin, S.: Near neighbor search in large metric spaces. In: The 21st VLDB Conference, pp. 574–584. Morgan Kaufmann Publishers (1995)
6. Micó, L., Oncina, J., Vidal, E.: A new version of the nearest-neighbor approximating and eliminating search (AESA) with linear preprocessing-time and memory requirements. Pattern Recognition Letters 15, 9–17 (1994)
7. Baeza-Yates, R., Cunto, W., Manber, U., Wu, S.: Proximity Matching Using Fixed-queries Trees. In: Crochemore, M., Gusfield, D. (eds.) CPM 1994. LNCS, vol. 807, pp. 198–212. Springer, Heidelberg (1994)
8. Chávez, E., Marroquín, J., Baeza-Yates, R.: Spaghettis: An array based algorithm for similarity queries in metric spaces. In: 6th International Symposium on String Processing and Information Retrieval (SPIRE 1999), pp. 38–46. IEEE CS Press (1999)
9. Chávez, E., Marroquín, J., Navarro, G.: Fixed queries array: A fast and economical data structure for proximity searching. Multimedia Tools and Applications 14(2), 113–135 (2001)
10. Pedreira, O., Brisaboa, N.R.: Spatial Selection of Sparse Pivots for Similarity Search in Metric Spaces. In: van Leeuwen, J., Italiano, G.F., van der Hoek, W., Meinel, C., Sack, H., Plášil, F. (eds.) SOFSEM 2007. LNCS, vol. 4362, pp. 434–445. Springer, Heidelberg (2007)
11. Kuang, Q., Zhao, L.: A practical GPU based kNN algorithm. In: International Symposium on Computer Science and Computational Technology (ISCSCT), pp. 151–155 (2009)
12. Garcia, V., Debreuve, E., Barlaud, M.: Fast k nearest neighbor search using GPU. In: Computer Vision and Pattern Recognition Workshop, pp. 1–6 (2008)
13. Bustos, B., Deussen, O., Hiller, S., Keim, D.: A Graphics Hardware Accelerated Algorithm for Nearest Neighbor Search. In: Alexandrov, V.N., van Albada, G.D., Sloot, P.M.A., Dongarra, J. (eds.) ICCS 2006, Part IV. LNCS, vol. 3994, pp. 196–199. Springer, Heidelberg (2006)
14. Barrientos, R.J., Gómez, J.I., Tenllado, C., Matias, M.P., Marin, M.: kNN Query Processing in Metric Spaces Using GPUs. In: Jeannot, E., Namyst, R., Roman, J. (eds.) Euro-Par 2011, Part I. LNCS, vol. 6852, pp. 380–392. Springer, Heidelberg (2011)
15. Uribe-Paredes, R., Valero-Lara, P., Arias, E., Sánchez, J.L., Cazorla, D.: Similarity search implementations for multi-core and many-core processors. In: 2011 International Conference on High Performance Computing and Simulation (HPCS), pp. 656–663 (July 2011)

A Scheme of Fragment-Based Faceted Image Search

Takahiro Komamizu[1], Mariko Kamie[1],
Kazuhiro Fukui[2], Toshiyuki Amagasa[2,3], and Hiroyuki Kitagawa[2]

[1] Graduate School of SIE, University of Tsukuba, Japan
[2] Faculty of Engineering, Information and Systems, University of Tsukuba, Japan
[3] Institute of Space and Astronautical Science, JAXA, Japan
{taka-coma,kamie}@kde.cs.tsukuba.ac.jp,
{kfukui,amagasa,kitagawa}@cs.tsukuba.ac.jp

Abstract. Retrieving desired images from large amounts of images has been increasingly important. Traditionally keyword search and content-based image retrieval have been used for image search. However, such retrieval methods have scarcely assumed a case when users have few terms for desired images or few images similar to the desired images. For this problem, we use *fragments* of images extracted from datasets as values of facets in faceted navigation. We extract parts of images as fragments. Then, we make a several number of groups for each part to decide representative images for faceted navigation. Our empirical user study based on an example application using face image data, FUKUWARAI, shows that our proposal successfully supports users to find desired images.

Keywords: Faceted Navigation, Image Retrieval, Faceted Image Search, Fragments, Fragment-based Faceted Search.

1 Introduction

Retrieving desired images from large amounts of images is increasingly important because of ease to generate and share images with help of image generating devices (e.g. camera and scanner) and image sharing services (e.g. Flickr and Picasa). Basically there are 2 kinds of image retrievals, one is to input terms and the other is to input images. A representative application for the former is keyword search which obtains images that match with the given keywords. On the other hand, a typical example of the other way of retrieving image is called content-based image retrieval (CBIR fro short) [2], in which users input images to obtain similar images from image database.

However, these ways overlook a more general case that users have few concrete information needs, while these ways have thriven for cases when users have concrete information needs. For instance, if you had seen a flower on the way home, how do you know the detail information about that flower? If you know the name of the flower, it seems to be no problem. Otherwise, you need to input a lot of information to identify the flower in the database, e.g. color and shape

S.W. Liddle et al. (Eds.): DEXA 2012, Part II, LNCS 7447, pp. 450–457, 2012.

of the flower, leaves and the location. However, it is not easy to express such sensuous information by text. We are tacking this problem in this paper.

Faceted navigation [6] is an exploratory search method over objects making use of attributes called facets. Ordinary faceted navigation is designed for objects with attributes which are categorical texts in general. Faceted navigation for image search has been studied by Yee et al. [11] and van Zwol et al. [12]. Both of their works depend on textual information (metadata, and query logs and outer ontologies). Here is one big issue they have not dealt with; "what can be facets if there is no textual information associated with images?" We can meditate to annotate images automatically to enable faceted navigation over images by applying their works, but it is still not easy enough to annotate images [4].

The main idea behind faceted navigation with no textual information is to employ expressiveness of images themselves due to the fact that images are more informative than texts. In our setting, users do not necessarily have concrete requirements for target images which are hard for them to express by terms. For example, let us assume a user looks for images of Michael Jackson with singing at the beginning of his retrieval. The point here is that how Michael Jackson is singing (i.e. crooning, shouting, or else) is not important for the user, because during the exploration the user is going to find sufficient images. Our strategy of defining the restrictive images is to fragment images in the dataset into several parts of images and then use them as facets. Still images of one part are different from others of the same part (in other words, exactly same images rarely exist), so we apply grouping techniques such as clustering to decide representative images in each part.

The remainder of this paper organized as follows: Section 2 introduces faceted navigation as preliminary, and Section 3 shows related work for our research i.e. faceted navigation over images. Then Section 4 discusses our proposed method, including prototype system implementation for face images. We evaluate our method through the prototype system via user study and report it in Section 5. Finally, Section 6 concludes this paper.

2   Faceted Navigation

Faceted Navigation [6] is one of the exploratory search methods [10] which enables exploration making use of attributes assigned to objects called facets. Faceted navigation has spotlighted in broad areas, such as e-commerce, library science and information retrieval. For example, amazon.com and eBay support faceted navigation to explore items for shopping. DBLP and IEEE Xplore, famous bibliographic repositories, use it to find papers, articles, journals and so on. In faceted navigation, users select facets from a list. The system then retrieves objects and returns the resulting objects along with the list of facets related to the resulting objects. This is repeated until users get satisfied from the resulting objects.

Faceted navigation is an exploration over attributed objects O. Each objects $o \in O$ has the set of attributes A which is referred as $o.A$. Each attribute $a \in A$

consists of a pair of attribute name and value, which are referred as $a.name$ and $a.value$, respectively. The set of facets F over the set of objects O is the union of attributes of all objects, i.e. $F = \bigcup_{o \in O} o.A$. Each facet $f \in F$ has the set of values $f.V$. During faceted navigation process, the resulting objects and their facets with values are shown to users in response to each user's selection. The resulting objects O_S over selected facets $S \subseteq F$ are given as follows:

$$O_S = \bigcap_{f \in S} \{o \mid o \in O, o.a.name = f.name, o.a.value = f.value\}$$

Given the resulting objects O_S, the set of facets F_{O_S} for O_S is given as follows:

$$F_{O_S} = \{f \mid f \in F, o \in O_S, o.f \neq null\}$$

Users select a pair of a facet and a value from F_{O_S} for each retrieval, result objects are calculated again for the next selection. Users can find results which satisfy their needs by repeating this process.

3   Related Work

The ideas for applying faceted navigation over image databases has been proposed in several works [11,12,3,7]. Most of them uses textual information [11,12,3] and there are few exceptions like [7]. Yee et al. [11] proposed faceted navigation over text annotated images and used WordNet to generate facets from annotations. They assume that images are all annotated. However, the assumption is too strong because images are not always annotated. Meanwhile, the work by van Zwol et al. [12] at Yahoo! Research is available for the case when annotations are not available. van Zwol et al. uses search logs of Yahoo! Image Search to mine facets using outer sources such as Wikipedia. Hare et al. [3] investigated quality of facets based on existing auto-annotation techniques and this work shows several problems over automatic annotation to images, e.g., age estimations of artifacts. Stober et al. [7] have worked on visualization of images to exploration. Their approach is, firstly computing features of images, secondly mapping them into 2 dimensional space keeping distance based on the features, and finally distorting the 2 dimensional space according to similarity for effective visualization. Their system shows (complete) images but similar images allocated nearer. Their work is different from ours because they visualize full images and users can zoom into clustered images, while our proposal uses *partial* images as facets and users can explore images by selecting the partial images. In addition, to the best of our knowledge, there has been no work for faceted navigation that acquires image fragments as facets.

4   Fragment-Based Faceted Image Search

This section introduces our proposed model, fragment-based faceted image navigation, for image databases. Our proposed model realizing faceted navigation for image database without textual metadata is based on fragmentation of images and make the fragments as facets as shown in Fig. 1. For more expressiveness of facet granularity, we build up facet hierarchy from these facets like Fig. 2. Then, facets are going to be assigned values which are fragmented images. There are a large number of fragmented images for each facet and being shown all of them is troublesome for users to find desired images. So, our model makes several groups of fragmented images and uses typical images, i.e., representative images for each group, as values of facets. Then, with these facets and their values, users can narrow down whole images into restricted number of images by selecting representatives of fragmented images. In the following section, we show definitions of our model in Section 4.1, navigation process of our model in Section 4.2, and an implementation of our model in Section 4.3.

4.1   Definitions

Firstly, we give the definition of image database D which consists of images, $D = \{o_1, o_2, ..., o_{|D|}\}$, where each image o_i is an object.

To fragment images into small parts, we define the set of fragment types, $T = \{t_1, t_2, ..., t_{|T|}\}$, which matches partially on images. Each fragment type is an attribute of an object in analogous to general faceted navigation. Fragment types can be given by any application or software to extract partial images. Fig. 1 shows an example of fragment types of human face. The human face in Fig. 1 is fragmented into forehead, eyes, nose, mouth, left hair and right hair.

In some cases, the fragment types are relatively small to the user intuitions. For example, seeing eyes only is not sufficient to find the target persons, but to see more widely, e.g., eyes with nose or eyes with nose and mouth, makes it more intuitive than eyes themselves. For this purpose, we build up fragment type hierarchy from fragment types by concatenating neighbor fragments. Fig. 2 shows an example of hierarchical fragment type construction. The arrows in the figure mean construction steps, merging neighbor fragments into one.

Fragments are corresponding partial images for fragment types (Definition 1). Developed applications for fragment types extract partial images.

Definition 1 (Fragment). *Given the image database D and the set of fragment types T, fragments S is defined as follows:*

$$S = \{extract(o, t) \mid o \in D, t \in T\}$$

where $extract(o, t)$ is the extraction function which extracts a fragment of the fragment type t from image o. Each fragment $s \in S$ has pointers to the original image, $s.original$, and to the fragment type $s.type$. □

Actually, the fragment types represent features of images in the database, consequently the fragment types are facets of the image database. However, not all of the fragment types can be facets because some of the fragment types may not appear on the images in the database. By excluding such fragment types,

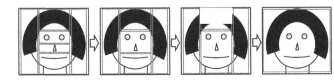

Fig. 1. Fragment types   **Fig. 2.** Examples of facet hierarchy construction

facets are defined as a set of fragment types which occur at least one image in the database. A set of facets is referred as F.

For each facet, values of the facet are the set of fragments which fragment type is the facet. Naively, the set of values of a facet f is defined as $f.V = \{s \mid s \in S, s.type = f\}$. However, this definition leads the problem. The problem is each fragment is an identical partial image among fragments because images are rarely exactly same. We thus need another way to reduce such identical fragments. We define groups of similar fragments. To use as typical values of fragments, we define *representatives* as follows:

Definition 2 (Representatives). *Given a set of groups of fragments G_f which fragments in each group are similar each other, the set of representatives for each group is referred as R_f and defined as follows:*

$$R_f = \{s \mid g \in G_f, s \in g, isCenter(s, g)\}$$

where $isCenter(s, g)$ function returns true when the fragment s is center of the group g; otherwise it returns false. Each representative $r \in R_f$ has pointers to members of group which r belongs and it is referred as $r.members$. $\square$

4.2   Navigation

For the navigation over image database D, we use the representative of a facet as the value. The following describes how users navigate over image database D:

1. (System) showcases current result images and candidate facets
2. (User) selects one facet to obtain values of the facet
3. (System) returns values of the facet for current result images
4. (User) selects one of the values of the facet
5. (User) repeat 1-4 until obtaining sufficient result images

In the following, we denote the current resulting images by $O' \subseteq D$ which match to the currently selected representatives of facets, which are referred by $Q = \{r_1, r_2, ..., r_{|Q|}\}$ where a fragment r_i is a selected representative of a facet.

Navigation is realized in 2 kinds of data accesses, (1) one is getting results for currently selected representatives and (2) the other is getting representatives of a specified facet for current results. The detail of these accesses are as follows:

(1) Given selected representatives, Q, to obtain resulting images, the system accesses the database to obtain result images, O'. The results over selected facets and representatives, Q, is defined as $O' = \bigcap_{r \in Q} r.members$.

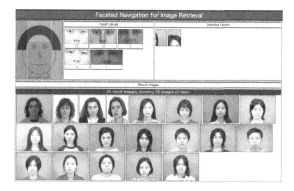

Fig. 3. FUKUWARAI interface

(2) Given a selected facet f and result images O', the representatives of f, $R'_f \subseteq R_f$ for current result images O' is given as Definition 3.

Definition 3 (Representative set). *Given a set of images O', the set of representatives R'_f for specified facet f is defined as follows:*

$$R'_f = \{r \mid r \in R_f, r, members \cap O' \neq \emptyset\} \qquad \square$$

4.3   FUKUWARAI: The Implementation

Fig. 3 shows the interface of faceted navigation for face image database. There are 4 components on the interface, upper 3 and lower 1 components. The upper left component is selection component of facets (blue one represents non-selected facet and red represents previously selected facets), the upper middle component holds representatives of a selected facet (e.g., eye and nose in the figure), the upper right component keeps selected representatives enabling deselection of selected facets by clicking one (e.g., left hair and forehead are selected), and the lower component shows a list of result images for currently selected representatives (e.g., 20 images out of 35 result images are shown in the figure). When users select one of representatives on the upper middle component, the system retrieves result images for selected representatives (already selected left hair, forehead and newly selected one of eye and nose representatives) and show the results.

5   Empirical User Study

We conduct the empirical user study to evaluate our proposed model for the case when users have few keyword inputs. We use publicly available datasets such as MUCT Database [5], FEI Face Database [1,8], and CUHK Database [9]. We use only images which faces are seeing front and less affected by lights. Totally, around 600 images are used for our experiment. The user study is held

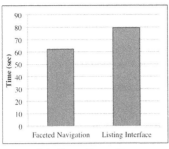
(a) Overall results.

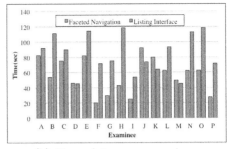
(b) Elapsed time per examinee.

Fig. 4. Experimental results

by 16 people including 10 under graduates, 5 graduates, and 1 Ph.D holder. Each examinee has been measured time in second to find a randomly chosen image.

In this experiment, we use FUKUWARAI[1], interface shown in Section 4. To compare with possible interface to retrieve face images over images, we use listing interface as counter-interface. In the counter-interface, whole image database divided into 13 pages which each page shows about 50 images in random order. The counter-interface is natural interface to assume the case that few information needs for retrieval, that is a lot of results being returned for incomplete query. The other point of counter-interface is that users are able to find desired images, while faceted navigation has some possibility that is not able to find desired images making use of facets because of intuition mismatching of facets and representatives.

Fig. 4(a) shows average retrieval time for each retrieval interface. The left bar is our implementation, labeled as `Faceted Navigation`, and the other is counter-interface, labeled as `Listing Interface`. This result shows our proposal helps users to navigate result images via facets. However, the difference is not ideally significant, because generally the user study depends on skills of examinees. Thus for further analysis we evaluate for each examinee (Fig. 4(b)). The figure shows average time for retrieval for each examinee, blue bar is our implementation (`Faceted Navigation`) and red one is counter-interface (`Listing Interface`). The horizontal axis shows examinee ids starting from A-P. By counting number of examinees which report retrieved time of our implementation is faster than that of counter-interface, there are 12 out of 16.

Also we can see the result of time for each examinee different from each other. This is because some examinees feel easy to use faceted navigation but the other do not. Once examinees miss where they are in the search space, they can hardly find desired images on both interfaces. This makes our faceted navigation better results, because faceted navigation has advantage to help examinees to narrow results down but the counter-interface does not.

---

[1] Fukuwarai is one of Japanese traditional cultural playing in the beginning of the year. There are a face plate and separated parts, e.g. eyes, nose and mouth, then a player puts them on the face plate without seeing the plate. All around enjoy the funny picture made by the player.

6   Conclusion and Future Works

We proposed a novel scheme for faceted image search which uses fragments of images in the database as facets and group fragments to abstract values used for retrieval. We have implemented an interface for face image databases, and use it for the experimental user study which shows our method works. For the future direction, automatic grouping of images is important but yet difficult. Because our method highly depends on quality of grouping (in this paper we assume perfect grouping), to improve the grouping results is our next task.

Acknowledgements. This research has been supported by the Grant-in-Aid for Scientific Research from MEXT (#21240005), and "New generation network R&D program for innovative network virtualization platform and its application(s)," the Commissioned Research of NICT.

References

1. The FEI Face Database, http://fei.edu.br/~cet/facedatabase.html
2. Datta, R., Joshi, D., Li, J., Wang, J.Z.: Image Retrieval: Ideas, Influences, and Trends of the New Age. ACM Comput. Surv. 40(2) (2008)
3. Hare, J.S., Lewis, P.H., Enser, P.G.B., Sandom, C.J.: Semantic Facets: An in-depth Analysis of a Semantic Image Retrieval System. In: CIVR, pp. 250–257 (2007)
4. Kumar, N., Belhumeur, P.N., Nayar, S.K.: FaceTracer: A Search Engine for Large Collections of Images with Faces. In: Forsyth, D., Torr, P., Zisserman, A. (eds.) ECCV 2008, Part IV. LNCS, vol. 5305, pp. 340–353. Springer, Heidelberg (2008)
5. Milborrow, S., Morkel, J., Nicolls, F.: The MUCT Landmarked Face Database. PRASA (2010), http://www.milbo.org/muct
6. Sacco, G.M., Tzitzikas, Y.: Dynamic Taxonomies and Faceted Search. Springer (2009)
7. Stober, S., Hentschel, C., Nürnberger, A.: Multi-facet exploration of image collections with an adaptive multi-focus zoomable interface. In: IJCNN, pp. 1–8. IEEE (2010)
8. Thomaz, C.E., Giraldi, G.A.: A new ranking method for Principal Components Analysis and its application to face image analysis. Image and Vision Computing 28(6), 902–913 (2010)
9. Wang, X., Tang, X.: Face photo-sketch synthesis and recognition. IEEE Trans. Pattern Anal. Mach. Intell. 31, 1955–1967 (2009)
10. White, R.W., Kules, B., Drucker, S.M., Schraefel, M.: Supporting Exploratory Search. Introduction, CommuSupporting Exploratory Senications of the ACM 49(4), 36–39 (2006)
11. Yee, K.P., Swearingen, K., Li, K., Hearst, M.A.: Faceted Metadata for Image Search and Browsing. In: Proc. CHI, pp. 401–408 (2003)
12. van Zwol, R., Sigurbjörnsson, B., Adapala, R., Pueyo, L.G., Katiyar, A., Kurapati, K., Muralidharan, M., Muthu, S., Murdock, V., Ng, P., Ramani, A., Sahai, A., Sathish, S.T., Vasudev, H., Vuyyuru, U.: Faceted Exploration of Image Search Results. In: Proc. WWW, pp. 961–970 (2010)

Indexing Metric Spaces with Nested Forests

José Martinez and Zineddine Kouahla

Lunam Université
Laboratoire d'informatique de Nantes-Atlantique (LINA, UMR CNRS 6241)
École polytechnique de l'université de Nantes – BP 50609 – F-44306 Nantes cedex 3
`name.surname@univ-nantes.fr`

Abstract. Searching for similar objects in a dataset, with respect to a
query object and a distance, is a fundamental problem for several applications that use complex data. The main difficulties is to focus the search
on as few elements as possible and to further limit the computationally-
extensive distance calculations between them. Here, we introduce a forest
data structure for indexing and querying such data. The efficiency of our
proposal is studied through experiments on real-world datasets and a
comparison with previous proposals.

1 Introduction

For several decades, indexing techniques have been developed to deal with efficient searches over large collections of data. When considering vectorial data, it
turns out that search and indexing become more and more difficult when the dimension of the vectors increases. This has been named the "dimensionality-curse
problem." The reader can find several surveys that present and compare existing
multidimensional indexing techniques [8] [2]. Nevertheless, objects to be indexed
are often more complex than mere vectors (e.g., sets, graphs). Hence, the focus
of indexing has partly moved from multidimensional spaces to metric spaces,
i.e., from exploiting the data representation itself to working on the similarities
that can be computed between objects.

This paper introduces a Metric-space Forest Indexing (MFI) technique: Section 2 introduces metric spaces and kNN queries and reviews a short taxonomy of
indexing techniques. Then, Section 3 introduces our proposal, overviews its main
characteristics and the corresponding algorithms. Section 4 discusses experimental results. Section 5 concludes the paper and introduces research directions.

2 Indexing and Querying in Metric Spaces

Definitions. Formally, a metric space is defined for a set of elements that are
comparable through a given distance.

Definition 1 (Metric space). *Let $\mathcal{O}$ be a set of elements. Let $d : \mathcal{O} \times \mathcal{O} \to \mathbb{R}^+$
be a distance function, which verifies: (i) non-negativity: $\forall (x, y) \in \mathcal{O}^2, d(x, y) \geq$*

S.W. Liddle et al. (Eds.): DEXA 2012, Part II, LNCS 7447, pp. 458–465, 2012.

0, *(ii) reflexivity:* $\forall x \in \mathcal{O}, d(x, x) = 0$, *(iii) symmetry:* $\forall (x, y) \in \mathcal{O}^2, d(x, y) = d(y, x)$, *(iv) triangle inequality:* $\forall (x, y, z) \in \mathcal{O}^3, d(x, y) + d(y, z) \le d(x, z)$. *Then* $(\mathcal{O}, d)$ *is a metric space.*

The concept of metric space is rather simple and leads to some common possibilities for querying a database. These are called similarity queries. We consider k nearest neighbour (kNN) searches.

Definition 2 (kNN query). *Let* $(\mathcal{O}, d)$ *be a metric space. Let* $q \in \mathcal{O}$ *be a query point and* $k \in \mathbb{N}$ *be the expected number of answers. Then* $(\mathcal{O}, d, q, k)$ *defines a kNN query, the value of which is* $S \subseteq \mathcal{O}$ *such that* $|S| = k$ *(unless* $|\mathcal{O}| < k$*) and* $\forall (s, o) \in S \times \mathcal{O}, d(q, s) \le d(q, o)$.

Background. Metric spaces introduce the notion of topological ball; it allows to distinguish between inside and external objects. Another useful partitioning concept is the one of generalised "hyper-plane."

Definition 3 (Closed Ball). *Let* $(\mathcal{O}, d)$ *be a metric space. Let* $p \in \mathcal{O}$ *be a pivot object and* $r \in \mathbb{R}^+$ *be a radius. Then* $(\mathcal{O}, d, p, r)$ *defines a (closed) ball, which can partition inner objects from outer objects, respectively* $I(\mathcal{O}, d, p, r) = \{o \in \mathcal{O} : d(p, o) \le r\}$ *and* $O(\mathcal{O}, d, p, r) = \{o \in \mathcal{O} : d(p, o) > r\}$ *(noted* $I(p, r)$ *and* $O(p, r)$ *for short).*

Definition 4 (Generalised hyper-plane). *Let* $(\mathcal{O}, d)$ *be a metric space. Let* $(p_1, p_2) \in \mathcal{O}^2$ *be two pivots, with* $d(p_1, p_2) > 0$. *Then* $(\mathcal{O}, d, p_1, p_2)$ *defines a generalised hyper-plane* $H(\mathcal{O}, d, p_1, p_2) = \{o \in \mathcal{O} : d(p_1, o) = d(p_2, o)\}$ *which can partition "left-hand" objects, i.e.,* $L(\mathcal{O}, d, p_1, p_2) = \{o \in \mathcal{O} : d(p_1, o) \le d(p_2, o)\}$ *from "right-hand" objects:* $R(\mathcal{O}, d, p_1, p_2) = \{o \in \mathcal{O} : d(p_1, o) > d(p_2, o)\}$ *(noted* $L(p_1, p_2)$ *and* $R(p_1, p_2)$ *for short).*

Based on these two partitioning techniques, we can introduce a short taxonomy of some related indexing techniques. The first class does not enforce space partitioning. There, we find M-tree, Slim-tree, etc. The M-tree [7] is an n-nary tree that builds a balanced index of including and overlapping balls, allows incremental updates, and performs reasonably well. Unfortunately, it suffers from the problem of overlapping that increases the number of distance calculations to answer a query. The Slim-tree [10] is an optimised version; it mainly reorganises the M-tree index in order to reduce overlaps.

The second class, the one based on space-partitioning, is richer. Two sub-approaches are included: One of them uses ball partitioning, like VP-tree, mVP-tree, etc. The other approach uses hyper-plane partitioning such as GH-tree, GNAT [4], etc.

The GH-tree [11] is a binary tree. The principle of this technique is the recursive partitioning of a space into "half-planes." Each other object is associated to the nearest pivot. It has proven its efficiency in some dimensions but it is still inefficient in large dimensions.

Similarly, the VP-tree [12] recursively partitions the space thanks to a ball. The building process is based on choosing a (random) object as the centre and finding the median distance to the other objects as the radius. The mVP-tree [3] is an m-nary version of the VP-tree where median is replaced by quantiles. Often, it behaves better but there are not enough differences to investigate further.

More recently, the MM-tree [9] uses the principle of partitioning by balls too, but it is *also* based on the exploitation of regions obtained from the *intersection* between two balls. The partitioning is done in an incremental way. When a new object arrives at leaf, containing a single value, they are used a the centre of two balls, the radii of which are equal to their interdistance. These two balls partition the space into four *disjoint* regions: their intersection, their respective differences, and the complement of their union. Therefore, it is quaternary tree. In order to improve the balancing of the tree, a semi-balancing algorithm is applied near to the leaves, which reorganises the objects in order to gain one level when possible. An extension of this technique has been developed: the onion-tree [5]. Its aim is to divide the last region into successive expanding balls. That improves the search algorithm, because the last region is particularly vast. It is a variable arity tree.

3   Metric Forest Indexing

In this section, we introduce a forest, called MFI for short, as an indexing technique in metric spaces. More specifically, it is a *family* of forests (and trees). The framework is generic and allows various implementations. It is a memory-based metric access method that divides recursively the dataset into *disjoint* regions.

Definitions. Let us introduce formally the MFI generic family.

Definition 5 (MFI). *Let $M = (\mathcal{O}, d)$ be a metric space. Let $E \subset \mathcal{O}$ be a subset of objects to be indexed. Let $1 \leq c_{\max} \leq |E|$ be the maximal cardinal that one associates to a leaf node.*

We define $\mathcal{N}$ as the nodes of a so-called MFI in the usual two-fold way: Firstly, a leaf node L consists merely of a subset of the indexed objects, all of them belonging to a closed ball: $((p, r), E_L) \in (\mathcal{O} \times \mathbb{R}^+) \times 2^{\mathcal{O}}$ with: $E_L \subseteq E$, and $|E_L| \leq c_{\max}$. The contents of the leaves partition E.

Secondly, an internal node N is a generic ordered sequence of couples of variable length: $[(P_1, N_1), \ldots, (P_n, N_n)] \in (\mathcal{P} \times \mathcal{N}^{\mathbb{N}})^{\mathbb{N}}$ where $\mathcal{P}$ is a generic data structure, and each N_i is a (recursive) sequence of (sub) nodes with $n \geq 0$. (Note that the forest becomes de facto a tree when all its internal nodes have arity at most 1.)

Next, this generic definition is instantiated hereafter based on some heuristic rules, namely: (i) avoiding too large balls, especially near the root, by relying on a forest rather than a tree (somewhat like onion-trees and M-trees), (ii) intersecting them to further limit their volume (like MM-trees), (iii) partitioning them into concentric rings (slightly like onion-trees and some VP-trees variants), (iv) reducing as much as possible concavity (like GH-trees and unlike VP-trees).

Figure 1 illustrates one such MFI instance, more precisely the structure of its inner node. This instance is rather clumsy at first sight. Let us explain it.

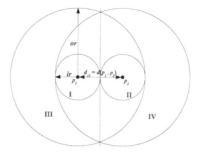

Fig. 1. The partitioning principle of a proposed instance of an MFI node

Two distinct pivots are chosen at random in a given data subset. They serve both as the centres of two balls each and as the base points for a hyperplane. The two balls create rings that help to reduce the number of children searches when the query ball has an intersection either with the ring or with the inner ball but not both.

For each pivot, the first ball has an inner radius that is *at most* equal to half the distance between the pivots. The enclosing ball has an outer radius that is *at most* equal to the third half of this distance. The goal is to avoid too large balls since in "multidimensional" spaces the volume grows exponentially with respect to the radius.

Thanks to these balls and to the hyper-plane, we can divide the space into four *disjoint* regions, namely (I) and (II) the inner balls, (III) and (IV) the intersection between the outer balls and their corresponding "half-plane." There remains a fifth part that is not to be treated as part of the node. The rationale behind that idea is again to avoid very large balls, especially near the root of the "tree" because in that case no pruning could occur during a search.

An almost formal definition is given through Algorithm 1. It presents the batch version of building that MFI instance. Notice that the full definition is slightly

---

Algorithm 1. Building an MFI instance in batch-mode

---

function Build $\left(E \subseteq \mathcal{O}, d \in \mathcal{O} \times \mathcal{O} \to \mathbb{R}^+, c_{\max} \in \mathbb{N}^*\right) \in \mathcal{N}^{\mathbb{N}}$
if $E = \emptyset$ **then return** $[\,]$ [i.e., an empty sequence]
elsif $|E| \leq c_{\max}$ **then let** $p = \mathrm{any}\{(o \in E\}$
$\qquad\qquad r = \max\{d(p, o) : o \in E\}$
$\qquad\qquad$ **return** $[((p, r), E)]$ [i.e., a singleton sequence with one leaf]
else let $(p_1, p_2) = \mathrm{any}\{(o_1, o_2) \in E^2 : d(o_1, o_2) > 0\}$ [two *random* and *disjoint* pivots]
$\qquad d_{12} = d(p_1, p_2)$; $ir = \frac{1}{2}d_{12}$; $or = \frac{3}{2}d_{12}$
$\qquad E_1 = \{o \in E : d(p_1, o) \leq ir \wedge d(p_1, o) \leq d(p_2, o)\}$ $[o \in I(p_1, ir) \cap L(p_1, p_2)]$
$\qquad E_2 = \{o \in E : d(p_2, o) \leq ir \wedge d(p_1, o) > d(p_1, o)\}$ $[o \in I(p_2, ir) \cap R(p_1, p_2)]$
$\qquad E_3 = \{o \in E : ir < d(p_1, o) \leq or \wedge d(p_1, o) \leq d(p_2, o)\}$
$\qquad E_4 = \{o \in E : ir < d(p_2, o) \leq or \wedge d(p_1, o) > d(p_2, o)\}$
$\qquad$ **return** $[((p_1, p_2), (\mathrm{Build}(E_1, d, c_{\max}), \ldots, \mathrm{Build}(E_4, d, c_{\max})))]$
$\qquad\qquad + \mathrm{Build}(E \setminus E_1 \setminus E_2 \setminus E_3 \setminus E_4, d, c_{\max})$

---

more complex in order to optimise searches; the generic parameter $\mathcal{P}$ contains redundant information (e.g., d_{12}) as well as additional information (e.g., exact bounds, not only ir and or), rather than just p_1 and p_2. Also, note that this proposal is just one among several others. We shall exemplify this in Section 4 since previous proposals have been implemented under our framework.

Building an MFI. For cases where the whole collection of data to index is known before hand, a batch-mode algorithm is adequate. However, in most situations, an incremental version would be required. This version is more intricate, therefore, due to space limitation, we describe it only in literary.

The insertions are done in a top-down way. Initially, the forest is empty.

The very first insertion in an empty sequence creates a new leaf with a single object. The following insertions in a leaf make it grow until a maximum number of elements, i.e., $c_{\max}$, is attained. Due to time complexity considerations, its value cannot be larger than $\sqrt{n}$ where $n = |E|$ is the cardinal of the whole population of objects to be inserted in the tree (actually only known in batch mode, but it can be lower-bounded and even increased as the populations increases). The ball is updated whenever a new object is inserted in the leaf extension.

When the cardinal limit is reached, a leaf is replaced by an internal node and four new leaves are obtained by splitting the former set of objects into four subsets according to the same conditions as of Algorithm 1. Possibly, a sibling is created too. In fact, this step reuses Algorithm 1 on a local subset. During a split, the choice of the pivots plays an important role along with the $c_{\max}$ parameter. The goal is to balance, as much as possible, the tree. Without clear guidelines, we "decided" to choose two objects at random. Indeed, the forest tends to be rather balanced, hence inserting a new object is a logarithmic operation, in amortised cost.

Inserting an object into an internal node amounts to selecting the subtree that has to contain it with respect to the very same conditions and applying the insertion recursively. (As a side-effect, some redundant node variables may be modified.) However, the new object does not necessarily belong to regions I to IV. In that case, we create again a brother node, i.e., the "tree" becomes an actual forest.

Let us note that, at each internal node, only two distances are calculated in order to insert a new object.

Similarity Queries. Although quite intricate, we developed a "standard" kNN search that runs a "branch-and-bound" algorithm where the query radius, initialised to infinity, is monotonically decreased down to the distance to the forthcoming k^{th} answer. The algorithm accepts an additional parameter, a solution "so-far," initialised to an empty answer.

Basically, the forest is traversed in pre-order. When arriving at a leaf node, the currently known sub-solution is merged with the local sub-solution. Note that we rely on "k-sort" and "k-merge" variants of the sort and merge algorithms respectively where the size of the answer is limited at most to the first k values. These variants are much faster.

When arriving on a sequence of internal nodes, the various regions are checked, the remaining candidates are ordered and hopefully some of them pruned; finally the search is pursued in the corresponding elected sub-nodes in expected order of relevance.

4   Case Study and Analysis

In order to explore the efficiency of our approach, we run some experiments. Firstly, we choose a few real-world datasets and the accompanying queries, then run our proposal along previously introduced ones, and finally evaluate the performances of kNN searches.

Datasets. Two used real-world datasets are described in Table 1. The so-called intrinsic dimension of a dataset gives a measure of the complexity of indexing it. We use Chávez et al. formula [6]: $d = \frac{1}{2} \frac{\mu^2}{\sigma^2}$ where μ is the average of the observed distances between pairs of objects and σ^2 is the variance. French cities coordinates provides a base line, since 2D coordinates are easy to index. By contrast, MPEG-7 Scalable Colour Descriptors (SCD) are multimedia descriptors with 64 dimensions. They are not suitable, in this form, for common multidimensional indexing techniques such as R-trees, X-trees [1], etc. However, their intrinsic dimensionality is not that high, "only" 8.

Algorithms. To run the search algorithms, we used 1,001 different objects as queries. In all the experiments, we run kNN searches with k equal to 20. Actually, we developed and used five algorithms: (i) *Naive*. This is the basic search algorithm, which consists of an improved sort limited to the first k^{th} values, the complexity of which is $O(n.D) + O(n. \log_2 k)$ where D is the complexity of the used distance. It serves as a (worst) comparison stallion. (ii) *FMI*. This is our proposal. (iii) *GH, VP, and MM*. These are *not* the original versions but three bucketed implementations of them, developed under our framework. This allows a fair comparison, all the versions being implemented, somewhat optimised, and instrumented with a largely common code. Hence, their respective merits depend mainly on the way they partition the space.

The $c_{\max}$ parameter was chosen either as a constant, 8, or the logarithm in base 2 (respectively 15 and 13 for the datasets at hand), or the square root of the size of the collection (respectively 189 and 100). For our MM-tree-like implementation, only the constant 8 is used; it corresponds to the original semi-balancing algorithm (and it is a tight constraint to limit the time complexity for building the MM-tree).

Table 1. Real-world datasets

Dataset	#Elements	Distance	Dimensions Apparent	Intrinsic	Description
French cities	35,183	L_2	2	1.912	2D Coordinates
Colour histo.	10,000	L_1	64	8.314	MPEG-7 SCD

Table 2. French cities experimental measures

VP	8	log	sqrt
min	1,660	**1,170**	2,170
average	10,017	9,306	20,318
stddev	7,056	5,933	13,270
median	8,049	8,573	18,282
max	32,570	28,103	58,736

MM	8	k-sort	
min	1,311	min	317,007
average	4,635	average	318,300
stddev	**1,124**	stddev	553
median	4,579	median	318,292
max	9,482	max	319,877

GH	8	log	sqrt
min	1,695	1,322	1,449
average	15,346	12,099	10,788
stddev	8,480	6,444	6,110
median	13,458	10,940	9,677
max	35,396	31,658	30,458

MFI	8	log	sqrt
min	1,576	1,444	1,800
average	4,223	**4,205**	8,130
stddev	1,316	2,022	3,905
median	4,018	**3,601**	7,312
max	**8,845**	11,778	20,662

Table 3. MPEG-7 SCD experimental measures

VP	8	log	sqrt
min	61,763	116,563	588,529
average	1,102,326	1,083,944	1,368,951
stddev	410,149	375,947	**264,875**
median	1,188,690	1,160,832	1,371,284
max	1,967,888	1,860,595	1,887,112

MM	8	k-sort	
min	152,170	min	1,950,465
average	1,224,772	average	1,950,767
stddev	449,584	stddev	100
median	1,292,343	median	1,950,770
max	2,079,803	max	1,951,045

GH	8	log	sqrt
min	271,929	233,864	198,431
average	1,465,409	1,320,001	1,254,771
stddev	511,888	443,818	402,526
median	1,573,054	1,429,399	1,367,281
max	2,500,494	2,204,342	1,884,863

MFI	8	log	sqrt
min	59,093	**47,476**	182,950
average	983,814	**976,551**	1,055,460
stddev	372,059	376,884	363,224
median	**1,046,715**	1,052,645	1,132,867
max	1,724,726	**1,716,615**	1,792,628

Measures on kNN Searches. When running the queries, we extracted several measures. Hereafter, we present only the overall performance, i.e., the *sum* of (i) the number of accessed objects, (ii) the number and the cost of distance computations, and (iii) the number of distance comparisons including the sort and merge phases. Tables 2 and 3 summarise the measures. We highlighted the smallest values of each statistical parameter, excluding k-sort, the baseline.

Experiments for the French cities dataset, the easy one, show clearly that the performances are largely better than a k-sort (See Table 2), between 1.3 and 6.4% on the averages with respect to it. Although there is no clear winner between MM and MFI, we note that performances tend to degrade as the size of the leaves increases, but this degradation is less important for our proposal than for GH- and VP-like trees. Also, a logarithmic size of the leaves seems to be the better compromise for the three related candidates. Ours is the best both on the average and median values.

Experiments for the multimedia dataset show clearly one point: the overall performances largely degrade for all the proposals, becoming in some cases worse than the naïve algorithm! On the averages, the ratios are between 50.4 and 75.1%... This is just another illustration of the so-called "curse of dimensionality" problem. In this difficult situation, our proposal is the one that resists the best and, again a logarithmic size of the leaves seems to be the best choice.

5 Conclusion

In this paper, we have extended the hierarchy of indexing methods in metric spaces with a family of indices consisting of a generalisation of metric trees: metric forests. We provided a first instantiation that combines several heuristics that seem *a priori* interesting when partitioning a search space. Experiments are encouraging and we shall work on determining the variants and/or parameters that best contribute to reducing the search effort in general. Then, dedicated sequential and/or parallel optimisations could be applied.

References

1. Berchtold, S., Keim, D., Kriegel, H.: The X-tree: An index structure for high-dimensional data. In: VLDB, pp. 28–39 (1996)
2. Böhm, C., Berchtold, S., Keim, D.: Searching in high-dimensional spaces: Index structures for improving the performance of multimedia databases. ACM Comput. Surv. 33(3), 322–373 (2001)
3. Bozkaya, T., Özsoyoglu, M.: Indexing large metric spaces for similarity search queries. ACM TODS 24, 361–404 (1999)
4. Brin, S.: Near neighbor search in large metric spaces. In: VLDB, pp. 574–584 (1995)
5. Carélo, C., Pola, I., Ciferri, R., Traina, A., Traina Jr., C., Ciferri, C.A.: Slicing the metric space to provide quick indexing of complex data in the main memory. Inf. Syst. 36(1), 79–98 (2011)
6. Chávez, E., Navarro, G., Baeza-Yates, R., Marroquín, J.: Searching in metric spaces. ACM Comput. Surv. 33(3), 273–321 (2001)
7. Ciaccia, P., Patella, M., Zezula, P.: M-tree: An efficient access method for similarity search in metric spaces. In: VLDB, pp. 426–435 (1997)
8. Gaede, V., Günther, O.: Multidimensional access methods. ACM Comput. Surv. 30(2), 170–231 (1998)
9. Pola, I.R.V., Traina Jr., C., Traina, A.: The MM-Tree: A Memory-Based Metric Tree Without Overlap Between Nodes. In: Ioannidis, Y., Novikov, B., Rachev, B. (eds.) ADBIS 2007. LNCS, vol. 4690, pp. 157–171. Springer, Heidelberg (2007)
10. Traina Jr., C., Traina, A., Seeger, B., Faloutsos, C.: Slim-Trees: High Performance Metric Trees Minimizing Overlap Between Nodes. In: Zaniolo, C., Grust, T., Scholl, M.H., Lockemann, P.C. (eds.) EDBT 2000. LNCS, vol. 1777, pp. 51–65. Springer, Heidelberg (2000)
11. Ulhmann, J.: Satisfying general proximity/similarity queries with metric trees. Information Processing Letters 40, 175–179 (1991)
12. Yianilos, P.: Data structures and algorithms for nearest neighbor search in general metric spaces. In: ACM-SIAM Symposium on Discrete Algorithms, Philadelphia, PA, pp. 311–321 (1993)

Navigating in Complex Business Processes[*]

Markus Hipp[1], Bela Mutschler[2], and Manfred Reichert[3]

[1] Group Research & Advanced Engineering, Daimler AG, Germany
markus.hipp@daimler.com
[2] University of Applied Sciences Ravensburg-Weingarten, Germany
bela.mutschler@hs-weingarten.de
[3] Institute of Databases and Information Systems, University of Ulm, Germany
manfred.reichert@uni-ulm.de

Abstract. In order to provide information needed in knowledge-intense business processes, large companies often establish intranet portals, which enable access to their process handbook. Especially, for large business processes comprising hundreds or thousands of process steps, these portals can help to avoid time-consuming access to paper-based process documentation. However, business processes are usually presented in a rather static manner within these portals, e.g., as simple drawings or textual descriptions. Companies therefore require new ways of making large processes and process-related information better explorable for end-users. This paper picks up this issue and presents a formal navigation framework based on linear algebra for navigating in large business processes.

Keywords: Process Navigation, Process Visualization.

1 Introduction

Large, knowledge-intense business processes, like the ones for engineering the electric-/electronic components in a car [1], may comprise hundreds or thousands of process steps. Usually, each process step is associated with task-related information, like engineering documents, development guidelines, contact information, or tool instructions—denoted as *process information*. To handle such a large information space (cf. Fig. 1), companies use web-based intranet portals. The goal is to provide a central point of access for their staff members enabling them to quicker find and access the process information needed. Process information, however, are often manually linked within these portals and hard-wired navigation structures are used to explore them. Process information not linked at all is not directly accessible for users.

In these portals, business processes and process information are usually visualized in a rather static manner [2,3], e.g., in terms of simple document lists (cf. Fig. 1). Van Wijk has shown that such visualizations often result in an information overload, rather disturbing than supporting the user [4]. Furthermore, process participants may have different perspectives on a business process and

[*] This research has been done in the niPRO project funded by the German Federal Ministry of Education and Research (BMBF) under grant number 17102X10.

S.W. Liddle et al. (Eds.): DEXA 2012, Part II, LNCS 7447, pp. 466–480, 2012.
© Springer-Verlag Berlin Heidelberg 2012

Fig. 1. Providing process information in intranet portals

related process information [5,6]. For example, consider the development of an ABS[1] control unit:

- **Requirement Engineers (Use Case 1):** Requirement engineers write a general specification for the ABS control unit. For this purpose, they need detailed instructions, templates, and contact persons. Needed information are also logic relations between process steps and process information.
- **Project Managers (Use Case 2):** Project managers must be able to identify the reasons for missed project deadlines, which negatively affect overall project goals. In this context, they need information about the status of all process steps as well as an abstract view on process steps and associated process information (e.g. due dates and duration of process steps).

These two use cases illustrate the diversity of process tasks and related process information needed. Obviously, users may have different roles and hence follow different navigation goals. Requirement engineers, for example, need very detailed information, whereas project managers ask for information on a more abstract level.

This paper introduces a process navigation framework that allows different users to intuitively and effectively navigate in and explore complex business process models. Our approach provides both processes and process information on different levels of abstraction. In particular, users can dynamically reach their navigation goal independent of their specific role. To provide a sound foundation of this navigation framework, linear algebra is used. We further demonstrate the applicability of the framework along scenarios from the automotive domain.

The remainder of this paper is organized as follows: Section 2 introduces basic notations and our core navigation model. Section 3 describes advanced concepts of our process navigation framework in detail. Section 4 applies these concepts to a real-world scenario. Section 5 discusses related work. Finally, Section 6 concludes the paper with a summary and an outlook.

2   Basic Notations and Core Navigation Model

We first introduce basic notation needed for the understanding of our navigation framework. Specifically, we reuse an existing navigation concept for complex information spaces to business processes—Google Earth [7].

---

[1] Antilock Braking System

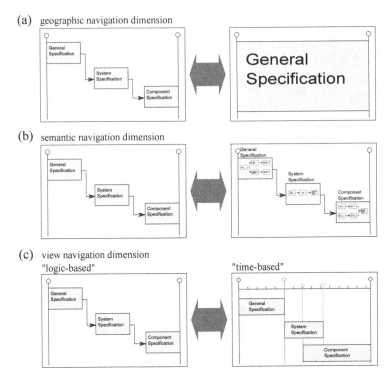

Fig. 2. Different navigation dimensions

Navigation Dimensions. We distinguish between geographic, semantic, and view navigation dimension. The *geographic dimension* allows for visual zooming without changing the level of detail. Regarding a process model with three process steps, for example, the user may zoom into the first step labeled "General Specification" (cf. Fig. 2a). A metaphor reflecting this dimension is using a magnifier, while reading a newspaper. In the *semantic dimension*, process information may be displayed at different levels of detail. Assuming that process steps comprise multiple activities, these activities may be additionally displayed by increasing the value of the semantic dimension (cf. Fig. 2b). Finally, the *view dimension* allows users to emphasize specific information, while reducing other, e.g., the duration of process steps. For example, the view may change from a logic-based to a time-based one (cf. Fig. 2c). Overall, these three dimensions define the *navigation space.*

Navigation State (NS). A navigation state corresponds to a specific point within the navigation space. The three navigation dimensions of this space are scaled in different values of which each represents a relative scale. For the sake of simplicity, we use natural numbers for this. Hence, in our context, we can define a navigation state as a triple. Let g be the value of the geographic dimension, s the one of the semantic dimension, and v the value of the view dimension. A specific navigation state NS can then be represented as:

$$NS = (g, s, v) \text{ with } g, s, v \in \mathbb{N} \tag{1}$$

Note that g, s, and v may be manually selected by the user. The set of all possible navigation states NS_{total} is as follows:

$$NS_{total} = \{(g, s, v) | g, s, v \in \mathbb{N}\} \tag{2}$$

Some of these navigation states may not make sense from a semantical point of view, i.e., they disturb the user or are forbidden by definition. Think of Google Earth and assume the user wants to see the whole globe (geographic dimension) and all city names at the same time (semantic dimension). In such a navigation state, labels would significantly overlap due to limited screen space. Hence, such a navigation state should be not reachable and be added to the set of forbidden navigation states $NS_{forbidden}$. In turn, we denote the set of allowed navigation states as *basis model BM*.

Basis Model (BM). The basis model corresponds to the set of allowed navigation states within the navigation space:

$$BM = NS_{total} \backslash NS_{forbidden} \tag{3}$$

Process Interaction. Changing values of the three navigation dimensions corresponds to a state transition within the navigation space. Since such state transitions are *user-driven*, we denote them as *process interactions*. In our framework, process interactions are represented by vectors. Changing the view from "logic-based" to "time-based" (cf. Fig. 2c) constitutes an example of such an interaction. A *one-dimensional* process interaction, in turn, is an activity transforming a given navigation state into another one by changing the value of exactly one navigation dimension. We assume that $g, s, v \in \mathbb{N}$ and $e = (\tilde{e}_1, \tilde{e}_2, \tilde{e}_3)$. A one-dimensional process navigation Int_{oneDim} can then be defined as follows:

$$Int_{oneDim} = \{(\tilde{e}_1, \tilde{e}_2, \tilde{e}_3) | \tilde{e}_1, \tilde{e}_2, \tilde{e}_3 \in \{0, 1, -1\} \text{ and } \|e\| = 1\} \tag{4}$$

In turn, a *multi-dimensional* process interaction can be defined as an activity transforming one navigation state into another by changing the value of several navigation dimensions at the same time (e.g., both the geographic dimension and the semantic dimension may be changed at once). Google Earth, for example, implicitly uses multi-dimensional interactions when the user applies the scroll wheel to zoom. If the geographic dimension is changed, the semantic dimension is changed accordingly. Since this functionality is well known and accepted by users, we apply it to process navigation as well. We define multi-dimensional process interaction as follows:

$$Int_{multiDim} = \{(\tilde{e}_1, \tilde{e}_2, \tilde{e}_3) | \tilde{e}_1, \tilde{e}_2, \tilde{e}_3 \in \{0, 1, -1\}\} \tag{5}$$

Navigation Model (NM). In our framework, a navigation model corresponds to a pre-defined set of allowed process interactions. This set may contain one-dimensional as well as multi-dimensional process interactions. According to (4) and (5), and due to the subset relation between one- and multi-dimensional process interactions (6a), the set of all possible process interactions Int_{total} can be defined as follows:

$$Int_{oneDim} \subset Int_{multiDim} \tag{6a}$$

$$\Rightarrow Int_{total} = Int_{multiDim} \tag{6b}$$

This set of allowed process interactions can be further reduced by manually discarding the set of forbidden process interactions $Int_{forbidden}$. Thus, NM can be defined as follows:

$$NM = Int_{total} \backslash Int_{forbidden} \tag{7}$$

Navigation Sequence (NavSeq). A navigation sequence is a sequence of process interactions. It describes the path along which the user navigates from a start navigation state NS_0 to an end navigation state NS_n:

$$NavSeq = (a_1, \ldots, a_n, NS_0, NS_n)$$
$$\text{with } a_1, \ldots, a_n \in NM \wedge NS_0, NS_n \in BM \tag{8}$$

Process Navigation (PN). Finally, process navigation can be defined as 4-tuple consisting of the basis model, the navigation model, a start state NS_0, and a navigation sequence defined by the user:

$$PN(BM, NM, NS_0, NavSeq) \tag{9}$$

Navigation sequence $NavSeq$ can be further investigated by applying linear algebra to the process navigation 4-tuple in (9) (cf. Section 3.1).

3   Process Navigation Framework

Basically, our process navigation framework comprises two main components (cf. Fig. 3): the *navigation* and *presentation layers*. The navigation layer specifies the basis model and the navigation model (cf. Section 2) using linear algebra (i.e., the formal approach we apply). In turn, the presentation layer deals with the visualization of business processes and related process information. It also provides different stencil sets enabling different process visualizations.

Distinguishing between navigation and presentation layer allows us to apply different visualizations in the context of the same navigation logic. In turn, this increases the flexibility of our framework as companies often prefer specific process visualizations [6]. Focus of this paper is on the *navigation layer*.

3.1   Running Example and Basic Issues

We first present a running example—an automotive requirements engineering process (see Use Case 1 in Section 1). The navigation space, which is shown in Figure 4, has been manually designed. Its schematic model, which is based on

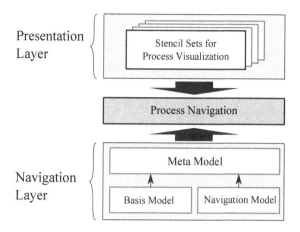

Fig. 3. Components of our process navigation framework

the three navigation dimensions introduced in Section 2, is shown in the center of Fig. 4. It assumes that the requirements engineer is currently working on activity "Create Component Profile" within the process step "General Specification". Assume further that the requirement engineer needs to know the activity succeeding the current one in order to find the right contact person for passing the specification document resulting from his work. For this purpose, he navigates from his current context, i.e., the default start state $(0,0,0)$ to state $(1,1,0)$ in which he then can gather the information needed.

Concerning the three dimensions of this simple example, we can define $g, s, v \in \{0,1\}$ instead of using natural numbers, i.e., every dimension is scaled in only two values. The overall number of possible navigation states is thus 2^3; note that in more complex navigation spaces, the number of navigation states increases exponentially with increasing number of navigation dimensions. In the following, NS_{total} is manually restricted by excluding two states: $(0,1,0)$ and $(0,1,1)$. These two states provide too many information items on the screen and would thus confuse the user. Think again of the Google Earth scenario, where all city names are shown in the semantic dimension, but the whole globe is shown in the geographic dimension at the same time. Considering (10a) and (10b), the basis model BM can be defined as shown in (10c):

$$NS_{total} = \{(0,0,0),(0,0,1),\dots,(1,1,1)\} \tag{10a}$$

$$NS_{forbidden} = \{(0,1,0),(0,1,1)\} \tag{10b}$$

$$BM = \{(0,0,0),(0,0,1),(1,0,0),(1,0,1),(1,1,0),(1,1,1)\} \tag{10c}$$

In our running example, we only allow for one-dimensional process interactions. Therefore we restrict Int_{total} by excluding all other possible process interactions $Int_{forbidden}$:

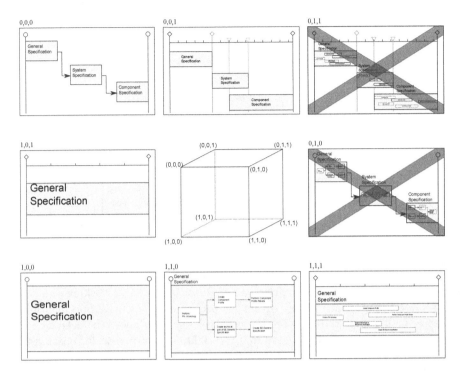

Fig. 4. Running example illustrating a navigation space with 8 navigation states

$$NM = \left\{ a \begin{pmatrix} 1 \\ 0 \\ 0 \end{pmatrix}, a \begin{pmatrix} 0 \\ 1 \\ 0 \end{pmatrix}, a \begin{pmatrix} 0 \\ 0 \\ 1 \end{pmatrix} \right\}, a \in \{1, -1\} \qquad (11)$$

Based on the notation of process navigation (9), we can now investigate user-driven navigation sequences. For each process interaction, we can calculate whether or not the requirement engineer leaves the BM (i.e., reaches a navigation state not being an element of BM). We assume that he applies navigation sequence $NavSeq$:

$$NavSeq = \left(i_1 = \begin{pmatrix} 1 \\ 0 \\ 0 \end{pmatrix}, i_2 = \begin{pmatrix} 0 \\ 1 \\ 0 \end{pmatrix} \right) \qquad (12)$$

$NavSeq$ consists of two process interactions. More precisely, i_1 corresponds to a geographical zooming without changing the level of information detail, whereas i_2 corresponds to an increase of the level of information detail. In the following, we apply both navigation interactions to our BM.

Step 1: We first calculate navigation state NS_1 resulting after the first process interaction of the requirement engineer, i.e., after adjusting the geographic dimension to zoom into the process step "General Specification" (cf. Fig. 4).

Therefore, we add the first vector i_1 to start state NS_0:

$$NS_0 + i_1 = NS_1 = (0,0,0)^T + (1,0,0)^T = (1,0,0)^T \tag{13}$$

As result, we obtain navigation state $(1,0,0) \in BM$. Hence, Step 1 constitutes a correct process interaction.

Step 2: From the newly obtained state NS_1 (i.e., the current start state) the requirement engineer now wants to increase the level of information detail, i.e., the value of the semantic dimension is increased to display the activities within the process step "General Specification". This process interaction i_2 is performed similar to Step 1:

$$NS_1 + i_2 = NS_2 = (1,0,0)^T + (0,1,0)^T = (1,1,0)^T \tag{14}$$

Finally, we check whether NS_2 is an element of BM. Since this is the case, $NavSeq$ can be constituted as allowed navigation sequence.

   If the user chooses another navigation sequence to reach the preferred end state $(1,1,0)$, the result may be different. For example, a navigation sequence may start with increasing the value of the semantic dimension, i.e., by applying process interaction $(0,1,0)$. The resulting state will then be $(0,1,0)$, which is not an element of BM and thus constitutes a forbidden state, i.e., the state to which the user must not navigate. By calculating allowed navigation possibilities in advance, i.e., before the user action takes place, we can guide the user in not taking a forbidden way through the navigation space.

3.2   Navigation Possibilities

Taking our running example (cf. Fig. 4), we further investigate possibilities to navigate from a given navigation state to other states. This becomes necessary to effectively support users moving within the navigation space. Think of a scenario in which a user is initially situated in navigation state $(0,0,0)$. As navigation spaces could become more complex than in our running example, the user does not necessarily know how the basis model BM looks like in detail, i.e., he does not know to which navigation state(s) he may navigate. To avoid incorrect navigation, like the one from $(0,0,0)$ to forbidden state $(0,1,0)$, it is important to give recommendations regarding allowed navigation options in a given state. Considering navigation states, for example, it is important to identify neighboring navigation states allowed.

   The *neighbor* characteristic describes the relation between two navigation states P_1 and P_2 that can be reached by applying exactly one single process interaction. Since we differentiate between one- and multi-dimensional process interactions, we also distinguish between one- and multi-dimensional neighbors.

One-Dimensional Neighbors. Two navigation states P_1 and P_2 constitute one-dimensional neighbors if a user can navigate from P_1 to P_2 by applying exactly one one-dimensional process interaction. In case only one-dimensional process interactions are allowed, the user may only navigate to one-dimensional neighbors of the current state:

$$P_1 \text{ is a one-dimensional neighbor of } P_2 \text{ iff}$$
$$P_1, P_2 \in BM \wedge \exists e \in Int_{oneDim} : P_1 + e = P_2 \tag{15}$$

Multi-dimensional Neighbors. Consider again our running example (cf. Fig. 4) and assume a user wants to navigate from $(0,0,0)$ to $(1,1,0)$. This could be accomplished by two consecutive one-dimensional process interactions. Generally, two states P_1 and P_2 are multi-dimensional neighbors, if P_2 is reachable from P_1 through a multi-dimensional process interaction:

$$P_1 \text{ is multi-dimensional neighbor of } P_2 \text{ iff}$$
$$P_1, P_2 \in BM \wedge \exists e \in Int_{multiDim} : P_1 + e = P_2 \tag{16}$$

Reachable Navigation States. A state P_2 is reachable from a state P_1 if there exists a navigation sequence that allows the user to navigate from P_1 to P_2. Thereby, the neighbor characteristics are applied in every process navigation step. As only pre-condition both P_1 and P_2 must be elements of BM:

$$P_1 \text{ is reachable from } P_2 \text{ iff}$$
$$P_1, P_2 \in BM \wedge \exists (n_1, \ldots, n_z) \text{ with } n_1, \ldots, n_z \in Int_{multiDim} \tag{17}$$
$$\wedge \ P_1 + \sum_{i=1}^{z} n_i = P_2 \wedge P_1 + \sum_{i=1}^{m} n_i \in BM \ \forall m$$

Knowing neighboring and reachable navigation states the navigation possibilities of a user can be determined. If a user is currently in a certain navigation state, we can guide further navigation interactions by recommending possible neighbors. This prevents users from a trial-and-error navigation.

3.3   Navigation Distance

Obviously, a navigation sequence applied by a user reflects the number of conducted process interactions. In turn, respective process interactions may require several user interactions (e.g., mouse clicks within an intranet portal). For the sake of simplicity assume that a user only applies one-dimensional process interactions. Then the number of user interactions corresponds to the number of mouse clicks. To decrease the latter (i.e., to enable a more efficient process navigation), the length of a navigation sequence required to navigate from a start state to a desired target state should be minimized. In the following, a method and metric to measure the length of navigation sequences are introduced.

As mentioned in Section 2, in general, we assume that the values of each navigation dimension correspond to any natural numbers. Hence, the distance between two arbitrary navigation states P_1 and P_2 can be easily calculated:

$$DIST(P_1, P_2) = \sqrt{(g_1 - g_2)^2 + (s_1 - s_2)^2 + (v_1 - v_2)^2} \tag{18}$$
$$\text{with } P_i = (g_i, s_i, v_i) \ ; \ i = 1, 2$$

Note that this metric can be applied to arbitrary states of the navigation space, i.e., these two states do not necessarily have to be one- or multi-dimensional neighbors. Furthermore, we can measure the overall walking distance of a user navigating within the navigation space. This distance corresponds to the sum of one- or multi-dimensional process interactions:

$$NAVDIST(NavSeq) = \sum_{i=1}^{n} \|a_i\| \text{ where } a_1, \ldots, a_n \in NavSeq \qquad (19)$$

3.4   Navigation Quality

To gain information about the quality of a chosen navigation sequence, we can measure its effectiveness , i.e., how quickly the user reaches his navigation goal when applying this navigation sequence. For this purpose, we consider the ratio of the distance between the start and end point of the navigation sequence on the one hand and its length on the other hand. Note that this metric does not only allow us to compare different navigation sequences, but it also enables better user assistence, e.g., based on recommendations about shorter navigation sequences:

$$Eff(P_1, P_2, NavSeq) = \frac{DIST(P_1, P_2)}{NAVDIST(NavSeq)} \qquad (20)$$

3.5   Discussion

When assisting users in searching for process information in complex business processes, we are facing various challenges. One challenge is to categorize available process information. Introducing different navigation dimensions simplifies this task.

Table 1. Table of requirements

ID	Requirement
R1	The dynamic adoption of different navigation paths must be enabled.
R2	Shortcuts and favorites are needed.
R3	Fast and lean calculation of new navigation situations is needed.
R4	User actions must be easily traceable.
R5	An expansion of the navigation space (up to n dimensions) must be supported.

Another challenge is to cope with the huge amount of process information and its classification as well as the formalization of process navigation. For the latter, several techniques may be applied. Table 1 lists main requirements on process navigation, we previously identified in two case studies [8]. Table 2 further shows that linear algebra, unlike other potential formalisms we can use for formalizing our navigation approach, fulfills all five requirements. Linear algebra is both generic enough to support the future expansion of our navigation approach (R5) and lean enough to allow a fast adaption to new navigation situations (R3). Therefore, linear algebra is most suitable in our context.

Table 2. Comparison of different formalization techniques. (+++: very good, ++: good, +:neutral)

ID	Finite state machines	Petri nets	State transition systems	Linear algebra	Predicate logic
R1	+++	+++	+++	+++	++
R2	++	++	++	+++	+
R3	++	++	+	+++	+
R4	+++	+++	+++	+++	++
R5	+	+	+	+++	+

4 Applying the Framework

We apply our navigation framework to a complex scenario comprising a large number of possible navigation states. Figure 5a shows a two-dimensional snippet of a three-dimensional navigation space. Black dots represent the basis model BM, i.e., the set of allowed navigation states. In turn, blank dots represent forbidden navigation states from set $NS_{forbidden}$. The horizontal dimension corresponds to the semantic and vertical one to the geographic dimension.

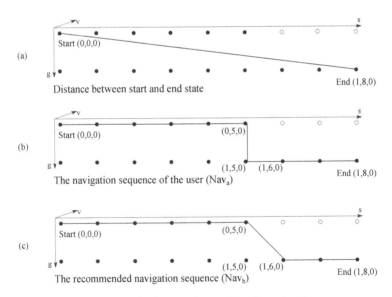

Fig. 5. Example of calculating the quality of navigation sequences

Assume that a user wants to navigate from start state $(0, 0, 0)$ to end state $(1, 8, 0)$. This corresponds to Use Case 2 from Section 1, where the project manager tries to identify project delays. Therefore, he needs detailed information about due dates, durations, and responsible persons (semantic dimension). Additionally, he requires an overview of all process steps of the project (geographic dimension).

First, we investigate the reachability of the end state from the start state. This way, we can check whether the needed information can be displayed at the desired geographic level, which shows all the process steps of the project.

When being in state $(0, 5, 0)$, a further increase of the information detail would result in an information overflow. The project manager then has to change the geographic level to zoom in, before he might be further allowed to increase the level of detail in the semantic dimension.

Second, we measure the distance between start and end state (cf. Fig. 5a):

$$DIST(Start, End) = \sqrt{8^2 + 1^2} \approx 8{,}06 \tag{21}$$

We now investigate the manager's navigation path while navigating within the navigation space, i.e., navigation sequence Nav_a from Fig. 5b. The manager applies nine one-dimensional process interactions to reach the end state. Hence, the distance is as follows:

$$DIST(Nav_a) = \sum_{i=0}^{n-1} DIST(P_{i+1}, P_i) = \sum_{0}^{8} 1 = 9 \tag{22}$$

Regarding our use case, the project manager might only be interested in adjusting the semantic dimension. The geographic dimension could be adjusted accordingly (from navigation state $(0, 5, 0)$ to state $(1, 6, 0)$) to avoid an overflow of the display with information. In this context, a multi-dimensional process interaction is applied automatically, reducing the user path by one interaction (cf. Fig. 5c). The distance of Nav_b can then be calculated as follows:

$$DIST(Nav_b) = \sum_{i=0}^{n-1} DIST(P_{i+1}, P_i) = 1+1+1+1+1+1+\sqrt{2}+1 \approx 8{,}41 \tag{23}$$

Using Eff, the following effectiveness ratios can be calculated for Nav_a and Nav_b respectively:

$$Eff(Start, End, Nav_a) = \frac{8{,}06}{9} \approx 89{,}55\% \tag{24a}$$

$$Eff(Start, End, Nav_b) = \frac{8{,}06}{8{,}41} \approx 95{,}79\% \tag{24b}$$

As can be seen in (24a) and (24b), suggesting navigation shortcuts leads to a more effective navigation path in Nav_b as indicated by the effectiveness ratios 95,79% and 89,55% respectively. This effect increases with the number of shortcuts. If typical navigation paths can be assigned to specific roles, further path suggestions could already be made before the user starts to navigate.

This example also indicates how our process navigation framework can be applied to Use Case 1. Again we use neighbors to measure distances and to calculate the effectiveness of navigation sequences. Doing so, more efficient navigation becomes possible by reducing unnecessary process interactions.

5   Related Work

According to Figure 6, related work stems from four areas:

First, *information retrieval* concerns information-seeking behavior of users [9]. We adopt this understanding in our process navigation framework. In this context, Belkin et al. [10] state that there are only little differences between information retrieval and information filtering. We apply this idea, by providing different navigation dimensions in our navigation framework.

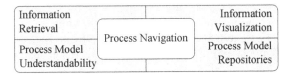

Fig. 6. Relevant areas of related work.

Second, *zoomable user interfaces* [11] have been developed to allow users to dynamically change views on information (*information visualization*). Specifically, they enable a decreasing fraction of an information space with an increasing magnification. Respective user interface concepts have been realized, for example, in *Squidy*, a zoomable design environment for natural user interfaces [12], in *ZEUS*, a zoomable explorative user interface for searching and presenting objects [13], and in *ZOIL*, a cross-platform user interface paradigm for personal information management [14]. Bederson also uses zooming techniques from JAZZ [15] and Pad++ [16] to develop intuitive user interfaces. Finally, Proviado applies aggregation and reduction techniques for creating views on large process models [17]. We adopt ideas from these approaches and extend them to ensure flexible process navigation.

Third, *process model repositories* [2,18] are discussed in literature. Current repositories, however, suffer from redundancies and complexity making changes costly and error-prone [19]. Recently, several approaches have been suggested to improve this situation [20,21].

Fourth, Mendling et al. [22] give insights into factors making process models better understandable. They investigate understandability as a proxy for the quality of process models, e.g., a relatively high number of arcs has a negative effect on a process model's understandability.

6   Summary and Outlook

Quickly finding the process information needed during process execution is crucial for knowledge workers. To support them in accomplishing this task, companies crave for new ways of delivering and visualizing processes together with associated process information. This paper has presented a framework for navigating in large process spaces and related process information on different levels of detail. Our framework allows achieving flexible navigation goals for users with different roles and different tasks. Specifically, we use linear algebra to formalize our framework and apply it to selected use cases. Our results show, how our process navigation framework facilitates information retrieval in complex processes and related process information.

Future research will address three topics. First, we will specify the presentation layer (cf. Fig. 3) in more detail, e.g., by defining and developing sophisticated concepts for process-oriented information visualization. Second, we will develop concepts for integrating the navigation layer and the presentation layer. Third, we will focus on the evaluation of the process navigation approach by performing user tests and surveys.

References

1. Müller, D., Herbst, J., Hammori, M., Reichert, M.: IT Support for Release Management Processes in the Automotive Industry. In: Dustdar, S., Fiadeiro, J.L., Sheth, A.P. (eds.) BPM 2006. LNCS, vol. 4102, pp. 368–377. Springer, Heidelberg (2006)
2. Weber, B., Reichert, M., Mendling, J., Reijers, H.A.: Refactoring large process model repositories. Computers in Industry 62(5), 467–486 (2011)
3. Fauvet, M.C., La Rosa, M., Sadegh, M., Alshareef, A., Dijkman, R.M., García-Bañuelos, L., Reijers, H.A., van der Aalst, W.M.P., Dumas, M., Mendling, J.: Managing Process Model Collections with AProMoRe. In: Maglio, P.P., Weske, M., Yang, J., Fantinato, M. (eds.) ICSOC 2010. LNCS, vol. 6470, pp. 699–701. Springer, Heidelberg (2010)
4. van Wijk, J.J., Nuij, W.A.A.: Smooth and efficient zooming and panning. In: IEEE Symp. on Inf. Visualization (INFOVIS), pp. 15–23. IEEE Comp. Society (2003)
5. Reichert, M., Bassil, S., Bobrik, R., Bauer, T.: The Proviado access control model for business process monitoring components. Enterprise Modelling and Information Systems Architectures (EMISA) 5(3), 64–88 (2010)
6. Bobrik, R., Bauer, T., Reichert, M.: Proviado – Personalized and Configurable Visualizations of Business Processes. In: Bauknecht, K., Pröll, B., Werthner, H. (eds.) EC-Web 2006. LNCS, vol. 4082, pp. 61–71. Springer, Heidelberg (2006)
7. Hipp, M., Mutschler, B., Reichert, M.: Navigating in Process Model Collections: A New Approach Inspired by Google Earth. In: Daniel, F., Barkaoui, K., Dustdar, S. (eds.) BPM Workshops 2011, Part II. LNBIP, vol. 100, pp. 87–98. Springer, Heidelberg (2012)
8. Hipp, M., Mutschler, B., Reichert, M.: On the context-aware, personalized delivery of process information: Viewpoints, problems, and requirements. In: Proc. 6th Int'l Conf. on Availability, Reliability and Security (ARES 2011), pp. 390–397. IEEE (2011)
9. Belkin, N.J., Croft, W.B.: Information filtering and information retrieval: two sides of the same coin? Communications of the ACM (CACM) 35(12), 29–38 (1992)
10. Belkin, N.J.: Interaction with texts: Information retrieval as information-seeking behavior. In: Proc. 1st Conf. on Information Retrieval 1993, pp. 55–66 (1993)

11. Reiterer, H., Büring, T.: Zooming techniques. In: Encyclopedia of Database Systems, pp. 3684–3689. Springer (2009)
12. König, W.A., Rädle, R., Reiterer, H.: Squidy: a zoomable design environment for natural user interfaces. In: Proc. 27th Int'l Conf. on Human Factors in Computing Systems (CHI 2009), pp. 4561–4566. ACM (2009)
13. Gundelsweiler, F., Memmel, T., Reiterer, H.: ZEUS – Zoomable Explorative User Interface for Searching and Object Presentation. In: Smith, M.J., Salvendy, G. (eds.) HCII 2007, Part I. LNCS, vol. 4557, pp. 288–297. Springer, Heidelberg (2007)
14. Zöllner, M., Jetter, H.C., Reiterer, H.: ZOIL: A design paradigm and software framework for post-WIMP distributed user interfaces. In: Distributed User Interfaces. HCI Series, pp. 87–94. Springer (2011)
15. Bederson, B.B., Meyer, J., Good, L.: Jazz: an extensible zoomable user interface graphics toolkit in java. In: Proc. 13th ACM Symposium on User Interface Software and Technology (UIST), pp. 171–180. ACM (2000)
16. Bederson, B.B., Hollan, J.D.: Pad++: A zooming graphical interface for exploring alternate interface physics. In: Proc. 7th ACM Symposium on User Interface Software and Technology (UIST), pp. 17–26. ACM (1994)
17. Reichert, M., Kolb, J., Bobrik, R., Bauer, T.: Enabling personalized visualization of large business processes through parameterizable views. In: Proc. 27th ACM Symposium On Applied Computing (SAC 2012), 9th Enterprise Engineering Track, pp. 1653–1660. ACM (2012)
18. Rinderle-Ma, S., Kabicher, S., Ly, L.T.: Activity-Oriented Clustering Techniques in Large Process and Compliance Rule Repositories. In: Daniel, F., Barkaoui, K., Dustdar, S. (eds.) BPM Workshops 2011, Part II. LNBIP, vol. 100, pp. 14–25. Springer, Heidelberg (2012)
19. Weber, B., Reichert, M.: Refactoring Process Models in Large Process Repositories. In: Bellahsène, Z., Léonard, M. (eds.) CAiSE 2008. LNCS, vol. 5074, pp. 124–139. Springer, Heidelberg (2008)
20. Choi, I., Kim, K., Jang, M.: An xml-based process repository and process query language for integrated process management. Knowledge and Process Management 14(4), 303–316 (2007)
21. Reuter, C., Dadam, P., Rudolph, S., Deiters, W., Trillsch, S.: Guarded Process Spaces (GPS): A Navigation System towards Creation and Dynamic Change of Healthcare Processes from the End-User's Perspective. In: Daniel, F., Barkaoui, K., Dustdar, S. (eds.) BPM Workshops 2011, Part II. LNBIP, vol. 100, pp. 237–248. Springer, Heidelberg (2012)
22. Mendling, J., Reijers, H.A., Cardoso, J.: What Makes Process Models Understandable? In: Alonso, G., Dadam, P., Rosemann, M. (eds.) BPM 2007. LNCS, vol. 4714, pp. 48–63. Springer, Heidelberg (2007)

Combining Information and Activities
in Business Processes

Giorgio Bruno

Politecnico di Torino, Torino, Italy
giorgio.bruno@polito.it

Abstract. This paper presents a notation, Chant, aimed at integrating the activi-ty-centric perspective and the data-centric one in the domain of business processes. Chant process models provide a network-oriented structure in which the tasks, the choice blocks and the flow handlers are the nodes and their inter-connections represent the information flows. The companion information models define the structure of the business entities: they place emphasis on mandatory attributes and associations so as to enable the readers to easily understand the effects brought about by the tasks.

Keywords: information systems, business processes, tasks, choices, informa-tion flows.

1   Introduction

The notion of PAIS (Process-Aware Information System) [1], which advocates a tigh-ter integration between the areas of information systems and business processes, has brought about several lines of research whose leitmotiv is the shift from the activity-centric perspective to the data-centric one.

While in the activity-centric perspective, whose standard representative is BPMN [2], the emphasis is placed on the tasks and their ordering provided by the control flow, what is important from the data-centric point of view is the discovery of the key business entities along with their life cycles.

Several approaches have been proposed to combine business entities (often called business artifacts) with their life cycles in higher-level business processes [3-10]. The most relevant issue is how to synchronize the states of the entities affected by the process, in a simple yet effective way. Most approaches keep the entity life cycles separate from the process models, for flexibility reasons, but they incur redundancy issues in that a number of states need to be repeated in the synchronization points.

In order to overcome these drawbacks, this paper proposes a notation, Chant (choice and task network), in which tasks, along with choice blocks and flow han-dlers, are nodes in a network and their interconnections represent information flows.

Chant is meant for conceptual models, i.e., models representing both static issues and dynamic ones, with the purposes of helping analysts understand the domain under consideration and of supporting communication between developers and users [11].

Chant models consist of two parts, the behavioral model and the information mod-el. The information model defines the structure (types, attributes and relationships) of

S.W. Liddle et al. (Eds.): DEXA 2012, Part II, LNCS 7447, pp. 481–488, 2012.

the business entities of the domain. Behavioral models encompass process models and role models. Role models include tasks that may be performed by any member of the role and are not part of processes.

A key issue for the integration of process models and information ones is the definition of the intended effects of the tasks. In Chant, the intended effects are expressed in a declarative manner either implicitly or explicitly. The implicit representation draws on mandatory attributes and relationships which are emphasized in the information models; the explicit effects are defined with post-conditions, while constraints are expressed with pre-conditions and invariants. The language used to express the conditions (pre-conditions, post-conditions and invariants) is a simplified version of the OCL language [12].

This paper is organized as follow. The next section introduces the Chant notation by presenting a simple role model and the companion information model. Section 3 presents the major features of the notation, then section 4 describes the related work, and section 5 provides the conclusion. The examples presented in the next sections refer to a process meant to handle the submission of papers to a conference.

2   Role Models and Generative Effects

Role models mainly include free tasks, i.e. tasks that may be performed by any member of the role when they want to. A simple role model with the companion information model is shown in Fig.1: task enterConferenceInfo enables a chairman to generate a new conference entity and to start a new instance of process HandleSubmissions. Tasks are depicted as rectangles with rounded corners and processes as rectangles with double borders.

Role model (Chairman)

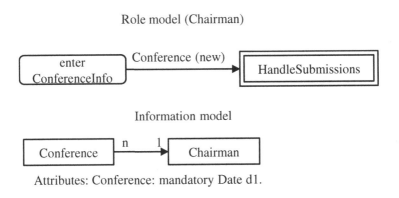

Attributes: Conference: mandatory Date d1.

Fig. 1. Example of role model

The effects of a task can be inferred from its local information flow, which consists of the input and output flows with their labels and qualifiers; the labels indicate the types of the entities the flows are made up of. If an output flow is qualified with the keyword new, then a generative effect is entailed. Therefore, the outcome of task enterConferenceInfo is the generation of a conference entity.

Complementary effects can be deduced from the conventions related to the information model, as follows.

Usually, when a new entity is generated, a number of attributes have to be initialized: they are called mandatory attributes and may be internal attributes or associative ones. Mandatory internal attributes are preceded by the "mandatory" qualifier.

In the Conference type, there is one mandatory attribute, d1, which indicates the deadline for the submission of papers to the conference.

Associative attributes implement associations and may refer to single entities or collections of entities depending on the cardinalities of the corresponding relationships. The cardinality of the relationship Conference-Chairman is many to one (n,1) and hence a conference is related to one chairman while a chairman may be related to several conferences. Such associations are represented by the associative attributes "chairman" in conference entities and the associative attributes "conferences" in chairman entities. The names of the associative attributes are based on the names of the corresponding entity types.

Associations may be mandatory on one side and may be optional on the other; in this case, the relationship is shown as an oriented link whose origin is the entity type for which the association is mandatory.

On the basis of these conventions, the execution of task enterConferenceInfo results in the generation of a new conference entity (in which the attribute d1 is initialized) and in the association of this entity with the chairman entity representing the performer of the task. The last result can be understood if the context of the task is taken into account: while the mandatory association Conference-Chairman calls for the connection to a chairman entity, the identification of the proper entity draws on the context of the task, which is provided by the process model. Since the task is meant to be performed by a chairman, the corresponding chairman entity is the item which makes the mandatory association fulfilled.

The output flow of task enterConferenceInfo is connected to process HandleSubmissions: the new conference entity is then passed to a newly generated instance of process HandleSubmissions. This process is illustrated in the next section.

3   Process Models

In this section the major features of Chant process models - i.e. the pre/post conditions of the tasks, the determination of the performers of the tasks, the representation of human choices and the flow handlers - are illustrated on the basis of process HandleSubmissions.

The process model and the companion information model are shown in Fig.2. The entity type written between parentheses after the process name is the process case. Then, each instance of process HandleSubmissions deals with a different conference entity (the case of the instance). The case entity determines the context of the tasks included in the process: it can be referred to in the pre (post)-conditions of the tasks with the type name written with the initial in lower case (e.g. conference).

The simplified requirements of the process are as follows; the implied tasks are written between parentheses.

Process HandleSubmissions enables authors to submit papers (task submitPaper) and reviewers to declare their interest in the conference (task join); these activities have to take place in the submission period, i.e., before deadline d1, where d1 is an attribute of the conference entities. After deadline d1, the chairman may assign papers to reviewers (task assignPaper) by issuing three requests for review for each paper, each request being directed to a reviewer. Reviewers may provide their reviews (task makeReview) or may reject the requests (task reject). In the second case, the chairman reassigns the paper to another reviewer (task reassignPaper). When three reviews are available, the chairman evaluates the paper (task evaluatePaper) and then a notification is automatically sent to the author (task sendEmail). Evaluating a paper results in setting the state to accepted or rejected.

Tasks may represent human activities or automated ones; the former are labeled with role names. All the tasks shown in Fig.2 are human tasks except for task sendE-mail which is an automated one.

The actual performers of the tasks may be generic actors or specific ones. A generic actor denotes any actor entitled to play the role needed for the task; on the contrary, a specific actor is an actor who bears a connection to the process case or to the input entities of the task. The actors performing task submitPaper and task join are generic actors in that any author (i.e. any user enrolled in this role) or any reviewer may perform them, respectively. On the contrary, the performer of task assignPaper is not a generic chairman but a specific one, i.e. the one associated with the conference which is the case of the process. When a specific performer is needed, the role label is enriched with the name of a performer rule written between parentheses; the rule indicates how to obtain the intended performer. The performer rule for the chairman's tasks is "conference.chairman": the chairman of the case entity is the intended performer. The performer rule for the reviewer's tasks is "request.reviewer" in that the intended performer is the one related to the input entities of the tasks.

A short description of the effects and constraints of the tasks is as follows.

Tasks submitPaper and join may be performed before deadline d1; this constraint is expressed by pre-condition "before conference.d1". The outcome of task submitPaper is the insertion of a new paper entity (inferred from the qualifier new) and its complementary effects are indicated by the mandatory attributes of the Paper type. There are two mandatory associations, one with a conference and the other with an author, and, on the basis of the context of the task, they may be actualized automatically: as a matter of fact, the conference is the case of the process instance and the author is the entity representing the performer of the task.

Task join has no output flow in that the effect is not the generation of a new entity, but the introduction of a membership association between the conference and the reviewer; the association is represented by the relationship Conference-Reviewer. The outcome is inferred from the post-condition "reviewer in conference.reviewers": it establishes that the performer entity "reviewer" will be an element of the collection of reviewers associated with the conference. A performer entity, i.e. the entity representing the performer of the task under consideration is indicated with the role name, e.g. reviewer;   the initial of the role name is written in lower case.

The effect of task AssignPaper is to generate three requests for reviews for each paper. The multiplicity 3 is indicated as a label, called weight, of the output flow. In general, the weights of the flows indicate how many input entities are taken from the input flows and how many output entities are added to the output flows.

Process HandleSubmissions (Conference)

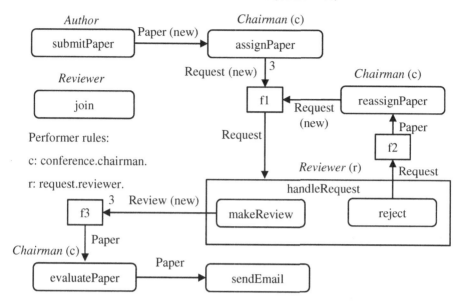

submitPaper: pre: before conference.d1.

join: pre: before conference.d1. post: reviewer in conference.reviewers.

assignPaper: pre: after conference.d1. evaluatePaper: post: state def.

Information model

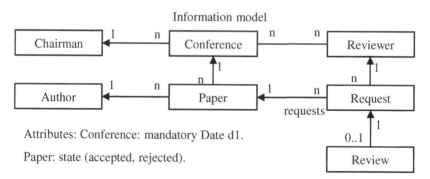

Attributes: Conference: mandatory Date d1.

Paper: state (accepted, rejected).

Fig. 2. Process HandleSubmissions

The weights are integer numbers shown close to the attachments of the flows to the tasks. If the weight is 1, it is omitted. The complementary effects of the generation of a request are two mandatory associations, one with a paper and the other with a reviewer. The context of the task brings about the actualization of only the first association in that the paper involved is the input paper. The second association cannot be actualized automatically because it is the result of a choice that has to be made during the execution of the task.

Since reviewers may provide their reviews or may reject the requests, two tasks are needed, i.e., makeReview and reject: they are the two options of a human choice. Chant represents human choices by means of rectangles including the options. The label of the rectangle (e.g. handleRequest) indicates the purpose of the choice. Task makeReview results in a new Review entity which is connected to the input request due to the mandatory relationship Review-Request.

If the request is rejected, the chairman has to make a new assignment and this is the purpose of task reassignPaper. This task produces a new Request entity and two associations as illustrated in the analysis of task assignPaper.

The process model includes three flow handlers, named f1, f2 and f3.

Flow handlers act as merging nodes, branching nodes or flow converters; f1 is a merging node while f2 and f3 are flow converters. The flow handler f1 merges its input flows into one output flow, in that the new requests for review, no matter where they come from, have to enter the choice block.

In a flow converter, the input flow and the output one have different types, but the types are interrelated through paths made up of mandatory relationships. The flow handler f2 has the purpose of converting the input requests for review into the associated papers; when it receives a request, it issues the paper associated.

The papers may be evaluated only when three reviews are available. The flow of reviews needs to be transformed into a flow of papers ready for evaluation. This is the purpose of the flow converter f3: it takes 3 reviews and issues one paper. Moreover, due to the path Review-Request-Paper, each review is indirectly associated with one paper and each paper is indirectly associated with a number of reviews: when 3 reviews are available for the same paper, the flow handler issues the paper.

Task evaluatePaper results in the acceptance or rejection of the input paper: the decision is written in the attribute state as indicated by the post-condition "state defined". Then the automatic task sendEmail notifies the author of the outcome.

4   Related Work

In the case-handling approach [3], a process is meant to take care of a specific entity type (e.g. an insurance claim), called the process case: the process evolution depends on the state of the case and not only on the tasks performed.

The BALSA framework [4] proposes four dimensions: (business) artifacts, life cycles, services and associations. The notion of artifact [5] encompasses not only the informational aspects pertaining to a business entity (e.g. a purchase order), but also its life cycle made up of the relevant stages of its evolution. The artifact-centric process model proposed in [6] draws on the BALSA framework and uses business rules to associate services with artifacts in a condition-action style.

If the starting point is the identification of the relevant business entities along with the definition of their life cycles, the job of the business processes is to coordinate such life cycles to achieve a certain goal. Such coordination takes advantage of the relationships that exist between the entity types, in particular of hierarchical ones. For example, hierarchical structures, such as those related to physical systems, are addressed by COREPRO [7], which provides specific means to achieve mutual synchronization between the states of compound objects and those of the components.

In PHILArmonicFlows [8], the life cycles of the entities are described by micro processes made up of states and transitions, while business processes are defined with macro processes consisting of macro steps and macro transitions. The approach presented in [9] is aimed at automatically generating a process model from the entity life cycles provided that the synchronization points are manually identified beforehand.

In Chant, the data flow is explicitly modeled and is integrated with the control flow. Process models handle the life cycles of several entity types and then the resulting model is more compact in that the states of the artifacts do not need to be repeated in the synchronization points. If separate life cycles are needed, an approach like the one illustrated in [10] can be used.

5   Conclusion and Future Work

This paper proposed a notation, Chant, whose purpose is to give participants as much information as possible on the tasks they have to perform and on the decisions they have to take, in terms of the business entities concerned. Chant process models provide a network-oriented perspective in which the tasks, the choice blocks and the flow handlers are the nodes and their interconnections represent the information flows. The companion information models define the structure (types, attributes and relationships) of the business entities: they place emphasis on mandatory attributes and associations so as to enable the readers to easily understand the effects brought about by the tasks.

Current work is dedicated to the implementation of Chant models. Two issues are under investigation; the structure of to-do lists [13] and the technology of the process engine. As to the process engine, two solutions are being compared: one is meant to automatically generate BPMN representations from Chant models so as to take advantage of existing open-software process engines, and the other is aimed at providing a specific execution environment based on ECA rules. The benefits of applying Complex Event Processing [14] technology to Chant models have been discussed in a previous paper [15].

Acknowledgments. The work presented in this paper has been partly supported by MIUR under the PRIN 2008 project "Documentation and processes".
The author wishes to thank the anonymous reviewers for their helpful comments.

References

1. Dumas, M., van der Aalst, W.M.P., ter Hofstede, A.H.M.: Process-Aware Information Systems: bridging people and software through process technology. Wiley, New York (2005)
2. BPMN, Business Process Model and Notation, V.2.0 (retrieved February 13, 2012), http://www.omg.org/spec/BPMN/
3. van der Aalst, W.M.P., Weske, M., Grünbauer, D.: Case handling: a new paradigm for business process support. Data & Knowledge Engineering 53, 129–162 (2005)

4. Hull, R.: Artifact-Centric Business Process Models: Brief Survey of Research Results and Challenges. In: Meersman, R., Tari, Z. (eds.) OTM 2008, Part II. LNCS, vol. 5332, pp. 1152–1163. Springer, Heidelberg (2008)
5. Nigam, A., Caswell, N.S.: Business artifacts: An approach to operational specification. IBM Systems Journal 42(3), 428–445 (2003)
6. Yongchareon, S., Liu, C.: A Process View Framework for Artifact-Centric Business Processes. In: Meersman, R., Dillon, T.S., Herrero, P. (eds.) OTM 2010, Part I. LNCS, vol. 6426, pp. 26–43. Springer, Heidelberg (2010)
7. Müller, D., Reichert, M., Herbst, J.: Data-Driven Modeling and Coordination of Large Process Structures. In: Meersman, R., Tari, Z. (eds.) OTM 2007, Part I. LNCS, vol. 4803, pp. 131–149. Springer, Heidelberg (2007)
8. Künzle, V., Reichert, M.: PHILharmonicFlows: towards a framework for object-aware process management. J. Softw. Maint. Evol.: Res. Pract. 23, 205–244 (2011)
9. Kumaran, S., Liu, R., Wu, F.Y.: On the Duality of Information-Centric and Activity-Centric Models of Business Processes. In: Bellahsène, Z., Léonard, M. (eds.) CAiSE 2008. LNCS, vol. 5074, pp. 32–47. Springer, Heidelberg (2008)
10. Küster, J.M., Ryndina, K., Gall, H.: Generation of Business Process Models for Object Life Cycle Compliance. In: Alonso, G., Dadam, P., Rosemann, M. (eds.) BPM 2007. LNCS, vol. 4714, pp. 165–181. Springer, Heidelberg (2007)
11. Wand, Y., Weber, R.: Research commentary: information systems and conceptual modeling - a research agenda. Information Systems Research 13(4), 363–376 (2002)
12. OCL, Object Constraint Language, V.2.3 (retrieved February 13, 2012), http://www.omg.org/spec/OCL/
13. Riss, U.V., Rickayzen, A., Maus, H., van der Aalst, W.M.P.: Challenges for business process and task management. Journal of Universal Knowledge Management (2), 77–100 (2005)
14. Luckham, D.: The power of events: an introduction to complex event processing in distributed enterprise systems. Pearson Education Inc., Boston (2002)
15. Bruno, G.: Emphasizing Events and Rules in Business Processes. In: Daniel, F., Barkaoui, K., Dustdar, S. (eds.) BPM Workshops 2011, Part I. LNBIP, vol. 99, pp. 395–406. Springer, Heidelberg (2012)

Opinion Extraction Applied to Criteria

Benjamin Duthil[1], François Trousset[1], Gérard Dray[1], Jacky Montmain[1],
and Pascal Poncelet[2]

[1] EMA-LGI2P, Parc Scientifique Georges Besse, 30035 Nîmes Cedex, France
`name.surname@mines-ales.fr`
[2] LIRMM, Université Montpellier 2, 161 Rue Ada, 34392 Montpellier, France
`name.surname@lirmm.fr`

Abstract. The success of Information technologies and associated services (*e.g.*, blogs, forums,...) eases the way to express massive opinion on various topics. Recently new techniques known as *opinion mining* have emerged. One of their main goals is to automatically extract a global trend from expressed opinions. While it is quite easy to get this overall assessment, a more detailed analysis will highlight that opinions are expressed on more specific topics: one will acclaim a movie for its soundtrack and another will criticize it for its scenario. Opinion mining approaches have little explored this multicriteria aspect. In this paper we propose an automatic extraction of text segments related to a set of criteria. The opinion expressed in each text segment is then automatically extracted. From a small set of opinion keywords, our approach automatically builds a training set of texts from the web. A lexicon reflecting the polarity of words is then extracted from this training corpus. This lexicon is then used to compute the polarity of extracted text segments. Experiments show the efficiency of our approach.

1 Introduction

Web technologies development have made numerous textual records available. The rapid increase of this mass of information requires efficient support system to ease the search of relevant information. Numerous tools are already designed in this way. For instance exhibiting customers opinion on a specific product, searching or automatically indexing documents are contemporary concerns. In particular, numerous tools have been developed for moviegoers to know the global trends of opinions on movies. However aggregated information found on the web does not always reflect the semantic richness provided by the critics. The aggregated score hides the divergence between the different criteria assessments. Thus this overall score will not reflect the semantic richness of the text. This observation is not specific to the domain of movies. This issue may be found in every domain when a sentiment analysis is done as in politics [1], e-commerce [2] or recommender systems [3].

Rather than allocating an overall score to the general opinion associated with a document, our approach propose a more detailed analysis by breaking down this overall assessment on a set a criteria. In a first step it extracts text segments

S.W. Liddle et al. (Eds.): DEXA 2012, Part II, LNCS 7447, pp. 489–496, 2012.

related to one criterion using the *Synopsis* method [4]. A second step identifies the polarity of each extracted segment relatively to the given criterion. This second step is the main contribution of this paper. Furthermore, we propose a method for automatic construction of the training corpus using minimal expertise (a few general opinion related words).

2   Description of the Opinion Mining Process

The way opinions are expressed may be quite different from one document to another and are often specific to the thematic the document deals with. Thus the vocabulary which is used depends on this thematic. Then a lexicon must be learned for each criterion for this thematic [5]. In the web context, creating annotated corpus for each criterion is very expensive and even discouraging. Indeed, considering the diversity of documents (e.g., blogs, forums, journalistic dispatches), numerous topics of interest (films, news, hi-tech ...), language levels may vary significantly from one medium to another. This considerably increases the number of significant words to learn before being able to get meaningful results. Moreover, the mass of data to process makes the manual task of annotation difficult and even impossible. All these facts highlight the interest of being able to automatically build training corpus with minimal human intervention. Based on statistical methods, our approach builds a lexicon of opinion descriptors for the selected thematic. This lexicon is then used to automatically extract polarity of each criterion relative excerpts of the document. Then, for each criteria, an overall assessment of the document is processed.

2.1   Acquiring the Training Corpus Automatically

To reflect the diversity of languages levels which are used to opinion expressed in various kinds of documents, the corpus is constructed by collecting documents from the web using a general web engine. To ensure the presence of positive and negative opinions in the texts constituting the training corpus, the approach focuses on texts from the thematic that contain at least one opinion word. These words are taken from two sets of given opinion words P (positive ones) and N (negative ones). Those words (called *seed words*) are widely used [6]:

$P = \{good, nice, excellent, positive, fortunate, correct, superior\}$;
$N = \{bad, nasty, poor, negative, unfortunate, wrong, inferior\}$;

By considering that a document containing at least one word from P (resp. N) but none from N (resp. P) contains positive (resp. negative) data, we can semantically attach documents to each seed word expressing positive (resp. negative) opinions. Documents are obtained by using a Web search engine and specifying the required thematic and seed word.

Thus, for each seed word g from P (resp. N) concerning the given thematic $\mathcal{T}$ (*movie*), K documents are collected. They express opinions related to g, with same polarity and in the context of $\mathcal{T}$. We thus get 14 corpus. Each of them contains K documents. Let us call S_g the corpus associated to the seed word g.

2.2 Learning Opinion Descriptors

The objective of the second phase consists in identifying the descriptors carrying opinion in collected documents. They are adjectives and specific "expressions". We call "expression" the concatenation of an adjective and all of the adverbs preceding it in the text. For example: "the ridiculously uneducated", "all bad", "very very good", "very nice".

In the documents collected in all the corpus, the approach will now search for adjectives and expressions which are correlated with the seed word associated with the document. Both adjectives and expressions which are carrying opinion are merged into the same concept called *descriptor*. The purpose of this learning phase is to enrich the original sets of seed words with opinion descriptors having same polarity.

To do that task, we consider the following assumption: the more a descriptor is correlated to a seed word (i.e. it is close to the seed word), the more it is likely to have the same polarity as this seed word. In opposite a "distant" descriptor (far from any seed word) is considered irrelevant to this seed word.

At the end of this process, each seed word g gets an associated set of correlated descriptors which is called its *class* C_g. In the opposite, the sets of descriptors that have been found "distant" from the seed word is called its *anti-class* AC_g. Thus the *class* is considered to have the polarity of its associated seed word and the *anti-class* to have opposite polarity. The basic idea carried by this discrimination is to identify relevant and irrelevant descriptors by studying their frequency in the *class* and the *anti-class*. If a descriptor is more common in the *class* than in the *anti-class* it will be considered relevant and with same polarity as the *class*. In the opposite, if a descriptor is more common in the *anti-class* than in the *class* it will be considered relevant with the polarity of the *anti-class*. Other descriptors having similar correlation to the *class* and to the *anti-class* are simply said irrelevant.

To compute the proximity of descriptors to seed words, we introduce the notion of window F of size sz. During this phase, windows are all centered on seed words g found in a document t belonging to S_g. Such a window is a set of words m and is defined as follows:

$$F(g, sz, t) = \{m \in t / d^t_{JJ}(g, m) \leq sz\} \tag{1}$$

$d^t_{JJ}(g, m)$ is the distance between the word m and the seed word g in text t. It is the number of adjectives (JJ) counted between m and g plus one.

The Learning Phase

The following will focus on operations done on the positive set of seed words P. Same operations are also applied on the negative one N.

Let $S_P = \bigcup_{g \in P} S_g$. Thus S_p is the corpus of text obtained from all positive seed words. Similarly we construct $C_P = \bigcup_{g \in P} C_g$, $AC_P = \bigcup_{g \in P} AC_g$ which are sets of descriptors. The learning phase is based on a discrimination technique *class/anti-class* as explained previously but is done on all corpus from S_P instead of doing it separately for each seed word $g \in P$. For any descriptor M a frequency $X_P(M)$ in the class C_P is computed as follows using windows of size 1 ($sz = 1$):

Let $\mathcal{O}(w, txt)$ be the set of occurrences of descriptor w (word, seed word or expression) in the text txt (a full document or any portion of text taken from any corpus). $X_P(M) = \sum_{g \in P} \sum_{t \in S_g} \sum_{\gamma \in \mathcal{O}(g,t)} |\mathcal{O}(M, F(\gamma, sz, t))|$

This score cumulates the frequencies of a descriptor M over all the windows in all the texts belonging to S_P.

The frequencies $\overline{X}(M)$ in the anti-classes AC_P are computed in a similar way using windows of size 2 ($sz = 2$). In this case, only descriptors outside of all windows are taken into account.

Let $\overline{F}(g, sz, t) = \{m \in t / d_{JJ}^t(g, m) > sz\}$. Thus we get $\overline{F}(g, sz, t) = t \setminus F(g, sz, t)$ (it is the complement of $F(g, sz, t)$ in the text t). Then $\overline{X}_P(M)$ may be computed: $\overline{X}_P(M) = \sum_{g \in P} \sum_{t \in T(g)} |\mathcal{O}(M, \bigcap_{\gamma \in \mathcal{O}(g,t)} \overline{F}(\gamma, sz, t))|$ The computed frequencies for $X_P(M)$ and $\overline{X}_P(M)$ already provides information about the polarity of the descriptors. These polarities are formally computed using the following discrimination function $Sc_P(M)$: $Sc_P(M) = \frac{(X_P(M) - \overline{X}_P(M))^3}{(X_P(M) + \overline{X}_P(M))^2}$. Positive values means that polarity of the descriptor is similar to the one of germs $g \in P$. Negative values are discarded because they do not actually mean that the descriptor are of opposite polarity but only that they do not have the same polarity: in fact they may be neutral.

The final score $Sc(M)$ of a descriptor M is finally computed from $Sc_N(M)$ and $Sc_P(M)$ using a similar operation: $Sc(M) = \frac{(Sc_P(M) - Sc_N(M))^3}{(Sc_P(M) + Sc_N(M))^2}$. (undefined values of either $Sc_P(M)$ or $Sc_N(M)$ are taken to be zero). Thus descriptors in C_p will get positive values and those in C_n will get negative ones.

2.3   Opinion Mining

Once the lexicon is built, opinion can be automatically extracted from a text or an excerpt of text dealing with thematic $\mathcal{T}$. A window is again introduced. But this time, instead of centering the windows on seed words and propagating the polarity of the seed word to the descriptor, we reverse the process: windows are centered on every adjective and their polarity is inferred from the ones of all the descriptors it contains. This is done by computing a score for each window f: $WSc(f) = \sum_{M \in f} Sc(M)$

The polarity of a text t is then identified by studying the sign of the score of the text: $Score(t) = \frac{\sum_{f \in t} WSc(f)}{|\{f \in t\}|}$

The polarity of the text t is given by the sign of its score: if $Score(t) < 0$ then t is negative and if $Score(t) > 0$ the t is positive.

3   Experiments

We have chosen to validate our approach against the test corpus proposed by [7] designed on the thematic of cinema. This corpus is a compilation of film critics coming from the *Internet Movie DataBase* [1]. Each text has got a human expertize and is labeled positive or negative. The process of validation consists in extracting parts of these texts relative to criteria *actor* and *scenario* (using *Synopsis*) and then identifying their polarity with our approach.

Our results are then compared to those obtained using the well known *SenticNet* [8] resource for opinion mining using common indicators as precision, recall and F1Score.

Validation of the Approach for Text Classification

In a first step we compare our opinion mining approach with *SenticNet* one on the whole text without using *Synopsis*. We just learn descriptors relative to the thematic of cinema and compute a global opinion on the whole texts. The results are given in table 1.

This highlights that our approach is efficient in a context of document classification: its *FScore* is better than the one obtained using *SenticNet*.

Table 1. Text classification results obtained on the corpus [7]

	Our approach		SenticNet	
	positive	*negative*	*positive*	*negative*
FScore	**0.73**	**0.69**	0.68	0.68
precision	**0.68**	**0.75**	0.54	0.74
recall	0.79	**0.63**	0.91	0.25

Table 2. Classification results for criteria *actor* and *scenario* on corpus [7]

	Criterion *Actor*					Criterion *Scenario*			
	Our approach		SenticNet			Our approach		SenticNet	
	positive	*negative*	*positive*	*negative*		*positive*	*negative*	*positive*	*negative*
FScore	**0.92**	**0.92**	0.70	0.70	FScore	**0.83**	**0.80**	0.69	0.69
precision	**0.90**	**0.95**	0.60	0.74	precision	**0.76**	**0.90**	0.55	0.74
recall	**0.95**	**0.90**	0.87	0.38	recall	**0.92**	**0.71**	0.91	0.26

We may notice some weakness of our approach concerning the detection of positive texts. But this is widely compensated by the one of negative texts for which we obtain a recall of 0.63 which is more than twice the one obtained by *SenticNet* (0.25). These results may indicate that non specific opinion vocabulary may be sufficient to get a global evaluation of a document. We can also notice that in this text classification our approach gives similar results as another global approach. That means that the vocabulary which is automatically learned is relevant for this kind of task and that the sole learning of the adjectives and expression appears sufficient. Furthermore we must point out that our approach is completely automated contrary to the one of *SenticNet* which is fully supervised. On our side we just need a minimal expertise to tell the system what means positive and negative through a small set of seed words.

---

[1] http://reviews.imdb.com/Reviews

Validation upon Criteria The second step of validation consists in evaluating the opinion mining task on chosen criteria excerpts. To do that we have chosen the two concepts *actor* and *scenario*. Once again we use the annotated corpus proposed by [7]. First we extract all the segments of text belonging to any of the two criteria *actor* or *scenario*. This is done by using *Synopsis* [4]. We then extract all the opinions of excerpts related to a criterion in a text and aggregate them to assess the global polarity of the text relatively to each criterion. We simply uses an average as aggregation operator. This is done similarly using the lexicon learned with our approach and the one from *SenticNet*. Results are compared in table 2.

Analysis of the results shows that opinions formulated on those critics highly depend on criteria. We can notice that criterion *actor* is more in line with the overall opinion expressed in the critics than the *scenario* one. This is highlighted by the lower score obtained for criterion *scenario* than for *actor*. In a multicriteria approach this might correspond to allocate a lower weight to criterion *scenario* than to *actor* one and to check the overall score.

Results obtained in table 1 represent an average of all the opinions expressed in the whole text. By comparing them to the ones obtained in table 2 we can deduce the well known fact that human mind uses a complex preference aggregator from various criteria rather than a simple average to identify the opinion that rises from the text.

We may notice that the *SenticNet* approach gives lower results than our approach in all cases. It highlights as expected that a vocabulary of opinion specific to the studied thematic gives better results when applied to criterion of this thematic rather than using a universal vocabulary as *SenticNet*. The lower results obtained in the previous experiment (table 1) can be explained as follows: opinions from thematic specific descriptors were covered by other common language opinion words in the first experiment. This significantly decreases the performance of the system. Finally, using our method, these results are very easy to reproduce on any thematic and criteria. As both thematic and opinion extraction methods used in this paper are unsupervised, the required human expertise is minimal. It only consists in giving a very small set of words to *Synopsis* [4] to specify what the criteria are. In the opinion mining task once again we only need a few words to identify the thematic and the polarities This is not the case for approaches like *SenticNet* which need human expertise for all of the descriptors.

4   State of the Art

This work focuses on the detection of opinions and feelings in textual data. Today opinion mining concerns are also shared both with natural language processing and information retrieval approaches. It uses lot of techniques from both of these disciplines as text analysis, machine learning, semantic analysis, text mining...

Mainly we can identify two kinds of approaches designed to extract and classify opinions. The binary classification ones that identify two classes of documents (positive and negative ones) and the multi-classes ones that offer several degrees of opinion (example: strongly positive, positive, neutral, negative, strongly negative). The major part of the works has focused on binary classification of documents especially when they are dedicated to recommender systems or point of view confrontation [9]. Another opinion mining investigation field concerns sentiment summary [10]. [11] is interested in opinion detection with arguments (*Opinion Mining Reason*) which are subjective sentences extracted from the document that consolidate the opinion assessment.

Other works focus on determining the intensity of expressed opinions. This may be achieved by using preselected sets of seed words and/or linguistic heuristics. Some of them have highlighted that restricting the analysis to sole adjectives improves the performance [12] but is not sufficient: adverbs, some nouns and verbs shall be considered too.

Automated text annotation is mainly done in two ways: either based on training corpus or on dictionaries. The first ones mostly rely on syntactic analysis or on word co-occurrences. The second ones take advantage of the semantics orientation of words given by dictionaries like *Word-Net* (http://wordnet.princeton.edu/) to deduce the polarity of sentences and texts [9].

Even if the results obtained by all of these approaches are very close to ours, they mainly differ at a very important point. In fact all of them need a huge human expertise to be applied (dictionaries, human annotated corpus, language semantics...). In the opposite we think that human expertise is a major constraint that interferes in the objective of having wide used, very flexible and highly adaptive opinion mining methods.

5 Conclusion

This paper presents a novel approach to automatically extract opinions from texts. The orientation given to this work was to minimize the human expertise to be provided to obtain relevant results. Furthermore we relate opinion extraction to multicriteria analysis. Following this guideline, our approach automatically builds the training corpus from which it learns the polarity of descriptors and automatically builds its lexicon lately used for opinion extraction. We have shown that the obtained lexicon pertinence is similar to human annotated ones like *SenticNet* when used in a general context. We have also demonstrated that the descriptors may bring different opinions in different contexts and that our criteria based method is able to automatically construct such a context oriented lexicon. Using such a lexicon in proper context significantly enhances the results. This work illustrates the complexity of human mind in the process of extracting the opinion carried by a text or a sentence. It also shows that this process is not based on an overall assessment but rather on local opinions expressed in terms of criteria which are lately aggregated. This consideration complicates cognitive automation of sentiment analysis task in the way that opinion mining should be combined with thematic extraction to enhance the results.

Our method has successfully been tested on other thematics like restaurant assessment and further experimentations are still in progress. The main task is to construct human annotated test corpus to validate our approach. This is a very expensive task that still consolidates our point of view which tends to leave human outside the learning process as much as possible.

References

1. Thomas, M., Pang, B., Lee, L.: Get out the vote: Determining support or opposition from congressional floor-debate transcripts. In: Proceedings of EMNLP, pp. 327–335 (2006)
2. Castro-Schez, J.J., Miguel, R., Vallejo, D., López-López, L.M.: A highly adaptive recommender system based on fuzzy logic for B2C e-commerce portals. Expert Systems with Applications 38(3), 2441–2454 (2011)
3. Garcia, I., Sebastia, L., Onaindia, E.: On the design of individual and group recommender systems for tourism. Expert Systems with Applications 38(6), 7683–7692 (2011)
4. Duthil, B., Trousset, F., Roche, M., Dray, G., Plantié, M., Montmain, J., Poncelet, P.: Towards an Automatic Characterization of Criteria. In: Hameurlain, A., Liddle, S.W., Schewe, K.-D., Zhou, X. (eds.) DEXA 2011, Part I. LNCS, vol. 6860, pp. 457–465. Springer, Heidelberg (2011)
5. Harb, A., Plantié, M., Dray, G., Roche, M., Trousset, F., Poncelet, P.: Web opinion mining: how to extract opinions from blogs? In: International Conference on Soft Computing as Transdisciplinary Science and Technology (2008)
6. Turney, P.: Thumbs up or thumbs down? semantic orientation applied to unsupervised classification of reviews. In: Proceedings of 40th Meeting of the Association for Computational Linguistics, pp. 417–424 (2002)
7. Pang, B., Lee, L., Vaithyanathan, S.: Thumbs up? sentiment classification using machine learning techniques. In: Proceedings of the 2002 Conference on Empirical Methods in Natural Language Processing, EMNLP (2002)
8. Cambria, E., Speer, R., Havasi, C., Hussain, A.: Senticnet: A publicly available semantic resource for opinion mining. Artificial Intelligence, 14–18 (2010)
9. Xu, K., Liao, S.S., Li, J., Song, Y.: Mining comparative opinions from customer reviews for Competitive Intelligence. Decision Support Systems 50(4), 743–754 (2011)
10. Ku, L.W., Lee, L.Y., Wu, T.H., Chen, H.H.: Major topic detection and its application to opinion summarization. In: Proceedings of the 28th Annual International ACM SIGIR Conference on Research and Development in Information Retrieval, SIGIR 2005, pp. 627–628. ACM, New York (2005)
11. Gopal, R., Marsden, J.R., Vanthienen, J.: Information mining? Reflections on recent advancements and the road ahead in data, text, and media mining. Decision Support Systems (2011)
12. Andreevskaia, A., Bergler, S.: Mining wordnet for fuzzy sentiment: Sentiment tag extraction from wordnet glosses. In: Proceedings EACL 2006, Trento, Italy (2006)

SOCIOPATH: Bridging the Gap between Digital and Social Worlds

Nagham Alhadad[1], Philippe Lamarre[2], Yann Busnel[1],
Patricia Serrano-Alvarado[1], Marco Biazzini[1], and Christophe Sibertin-Blanc[3]

[1] LINA/Université de Nantes
2, rue de la Houssinière
44322 Nantes, France
Name.LastName@univ-nantes.fr
[2] Liris/Université de Lyon
37, Rue du Repos
69007 Lyon, France
Philippe.Lamarre@liris.cnrs.fr
[3] IRIT/Université de Toulouse 1
2, rue du Doyen-Gabriel-Marty
31042 Toulouse, France
Christophe.Sibertin-Blanc@univ-tlse1.fr

Abstract. Everyday, people use more and more digital resources (data, application systems, Internet, *etc.*) for all aspects of their life, like financial management, private exchanges, collaborative work, *etc*. This leads to non-negligible dependences on the digital distributed resources that reveal strong reliance at the social level. Users are often not aware of their real autonomy regarding the management of their digital resources. People underestimate social dependences generated by the system they use and the resulting potential risks. We argue that it is necessary to be aware of some key aspects of system's architectures to be able to know dependences. This work proposes SOCIOPATH, a generic meta-model to derive dependences generated by system's architectures. It focuses on relations, like access, control, ownership among different entities of the system (digital resources, hardware, persons, *etc.*). Enriched with deduction rules and definitions, SOCIOPATH reveals the dependence of a person on each entity in the system. This meta-model can be useful to evaluate a system, as a modeling tool that bridges the gap between the digital and the social worlds.

1 Introduction

Nowadays, the most widespread architectures belong to the domain of distributed systems [1]. Most of participants' activities on these systems concern their data (sharing and editing documents, publishing photos, purchasing online, *etc.*) [2]. Using these systems implies some relationships with a lot of people. These people may be partly unknown, but one depends on them in several ways:

- Which person(s)/resource(s) a user depends on to perform an activity?
- Whom can prevent a user from performing an activity?
- Which persons a user is able to avoid to perform an activity?
- Who is possibly able to access user's data?

S.W. Liddle et al. (Eds.): DEXA 2012, Part II, LNCS 7447, pp. 497–505, 2012.

Some of these questions raise several issues as someone may be able to grab information about who the user is, what the user does, and so forth. That directly leads to privacy [3], trust [4] and security issues [5].

The analysis of such systems is usually limited to technical aspects as latency, QoS, functional performance, failure management [6], *etc.* The aforementioned questions give some orthogonal but complementary criteria to the classical approach. Currently, people underestimate social dependences [7,8,9] generated by the systems they use and the resulting potential risks. We argue that to be able to know dependences, one should be aware of some key aspects of systems.

This paper proposes the SOCIOPATH meta-model. This approach is based on notions coming from many fields, ranging from computer science to sociology. SOCIOPATH is a generic meta-model that considers two worlds: the social world and the digital world. SOCIOPATH allows us to draw a representation of a system that identifies persons, hardware, software and the ways they are related. Enriched with deduction rules, SOCIOPATH analyzes the relations in the digital world, to deduce the relations of dependences in the social world, given an activity concerning some data. As a modeling tool that bridges the gap between the digital and the social world, SOCIOPATH can be useful in the evaluation process of a system with respect to security, privacy, risk and trust requirements.

The paper is organized as follows. Section 2 introduces the SOCIOPATH meta-model and gives a simple example of its use. Section 3 defines the way to compute the user's digital and social dependences. Section 4 presents a brief overview of related works. Finally, Section 5 concludes and points out our ongoing work.

2   SOCIOPATH Meta-model

The SOCIOPATH meta-model describes the architecture of a system in terms of the components people use to access digital resources, so that chains of dependences can be identified. It distinguishes the *social world*, where humans or organizations own physical resources and data, and the *digital world*, where instances of data are stored and processes run. Figure 1 shows the graphical representation of SOCIOPATH, which we analyze in the following sections.

2.1   The Social World

The social world includes persons (users, enterprises, companies, *etc.*), physical resources, data and relations among them.

- *Data* represents an abstract notion that exists in the real life, and does not necessarily imply a physical instance (*e.g.*, address, age, software design, *etc.*);
- *Physical Resource* represents any hardware device (*e.g.*, PC, USB device, *etc.*);
- *Person* represents a generic notion that defines an Individual like Alice or a Legal Entity like Microsoft.

2.2   The Digital World

The digital world has nodes that are defined as follows:

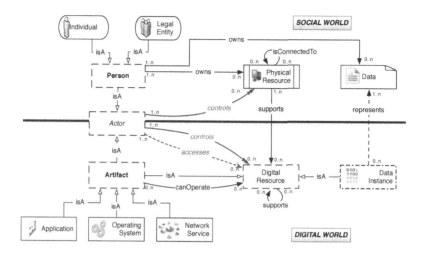

Fig. 1. Graphical view of SOCIOPATH as a UML class diagram.

- *Data Instance* represents a digital representation of *Data*;
- *Artifact* represents an abstract notion that describes a "running software". This can be an *Application*, an *Operating System* or a *Network Service*;
- *Digital Resource* represents an *Artifact* or a *Data Instance*;
- *Actor* represents a *Person* in the social world or an *Artifact* in the digital world.

2.3   The Relations in SOCIOPATH

Several relations are drawn in SOCIOPATH. In this section, we briefly describe them.

- *owns*: this means ownership. This relation only exists in the social world;
- *isConnectedTo*: this means that two nodes are physically connected, through a network for instance. This symmetric relation exists only in the social world;
- *canOperate*: this means that an artifact is able to process, communicate or interact with a target digital resource;
- *accesses*: this means that an *Actor* can access a *Digital Resource*;
- *controls*: this means that an *Actor* controls a *Digital* or a *Physical Resource*. There exists different kinds of control relations. A legal entity, who provides a resource, controls its functionalities. The persons who use this resource have some kind of control on it;
- *supports*: this means that the target node could never exist without the source node;
- *represents*: this is a relation between data and their instances in the digital world.

Persons own some data in the social world. *Data* have a concrete existence in the digital world if they are represented by some *Data Instance* and supported by some *Physical Resource*. As an *Actor* in the digital world, a *Person* can access and control *Data Instances* representing her (and others') *Data*. This may be done through different resources, thus implying some dependences on other persons. Moreover, we consider that a person *provides* an artifact (*cf.* the rightmost part of Figure 2) if this person owns data represented by a data instance which supports the artifact. Applying SOCIOPATH makes

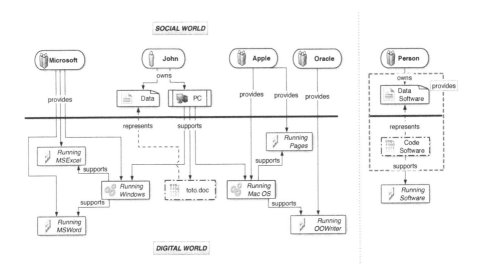

Fig. 2. Use case example: a document accessed by 2 different operating systems

Table 1. Glossary of notations

Basic type of instance	The set of all instances		A subset of instances		One instance	
	Notation	Remark	Notation	Remark	Notation	Remark
Person	$\mathbb{P}$	$\{P : person(P)\}$	$\mathcal{P}$	$\mathcal{P} \subset \mathbb{P}$	P	$P \in \mathbb{P}$
Actors	$\mathbb{A}$	$\{A : actor(A)\}$	$\mathcal{A}$	$\mathcal{A} \subset \mathbb{A}$	A	$A \in \mathbb{A}$
Artifact	$\mathbb{F}$	$\{F : artifact(F)\}$	$\mathcal{F}$	$\mathcal{F} \subset \mathbb{F}$	F	$F \in \mathbb{F}$
Digital resource	$\mathbb{DR}$	$\{DR : resource(DR)\}$	$\mathcal{DR}$	$\mathcal{DR} \subset \mathbb{DR}$	DR	$DR \in \mathbb{DR}$
Physical resource	$\mathbb{PR}$	$\{PR : phyresource(PR)\}$	$\mathcal{PR}$	$\mathcal{PR} \subset \mathbb{PR}$	PR	$PR \in \mathbb{PR}$
Data	$\mathbb{D}$	$\{D : data(D)\}$	$\mathcal{D}$	$\mathcal{D} \subset \mathbb{D}$	D	$D \in \mathbb{D}$
Data instance	$\mathbb{DI}$	$\{DI : dataInstance(DI)\}$	$\mathcal{DI}$	$\mathcal{DI} \subset \mathbb{DI}$	DI	$DI \in \mathbb{DI}$
Operating System	$\mathbb{O}$	$\{OS : operatingSystem(OS)\}$	$\mathcal{O}$	$\mathcal{O} \subset \mathbb{O}$	OS	$OS \in \mathbb{O}$
Path	Γ	$\{\sigma : path(\sigma)\}$	Υ	$\Upsilon \subset \Gamma$	σ	$\sigma \in \Gamma$
Activity	$\mathbb{W}$	—	—	—	ω	$\omega \in \mathbb{W}$
Set of activity restrictions	$\mathbb{S}$	$\{\mathcal{S} = \mathscr{P}(\mathbb{F}^{\mathbb{N}})\}$	—	—	$\mathcal{S}$	$\mathcal{S} \in \mathbb{S}$

possible non-trivial deductions about relations among nodes. For instance, an actor may be able to access digital resources supported by different physical resources connected to each other (*e.g.*, a user can access processes running in different hosts).

Figure 2 shows a basic SOCIOPATH model[1] of a use-case on a unique PC. In the social world, a user John owns some Data and a PC. There are also legal entities as: Microsoft, provider of Windows, Word (MSWord) and Excel (MSExcel); Apple, provider of MacOS and Pages; and Oracle, provider of Open Office Writer (OOWriter). In the digital world, two operating systems exist on John's PC: Windows and MacOS. On Windows, two applications are available: MSWord and MSExcel. On MacOS are installed OOWriter and Pages. John's Data are represented in the digital world by the document toto.doc. We use this example to illustrate some deductions in the following. Table 1 summarizes the notations we use.

---

[1] We consider that a model conforms to a meta-model.

Table 2. Deduced access and control relations

Rule	Formal definition
Rule 1	$\forall F \in \mathbb{F}, \forall DR \in \mathbb{DR},$ $\forall PR1, PR2 \in \mathbb{PR} :$ $\bigwedge \left\{ \begin{array}{l} canOperate(F, DR) \\ supports(PR1, F) \\ \bigvee \left\{ \begin{array}{l} supports(PR1, DR) \\ \bigwedge \left\{ \begin{array}{l} supports(PR2, DR) \\ isConnectedTo(PR1, PR2) \end{array} \right. \end{array} \right. \end{array} \right. \Rightarrow accesses(F, DR)$
Rule 2	$\forall P \in \mathbb{P}, \forall PR \in \mathbb{PR}, \forall OS \in \mathbb{O} : \bigwedge \left\{ \begin{array}{l} owns(P, PR) \\ supports(PR, OS) \end{array} \right. \Rightarrow \bigwedge \left\{ \begin{array}{l} accesses(P, OS) \\ controls(P, OS) \end{array} \right.$
Rule 3	$\forall F \in \mathbb{F}, \forall OS \in \mathbb{O} : \bigwedge \left\{ \begin{array}{l} supports(OS, F) \\ canOperate(OS, F) \end{array} \right. \Rightarrow controls(OS, F)$
Rule 4	$\exists P \in \mathbb{P}, \exists D \in \mathbb{D}, \exists DI \in \mathbb{DI}, \exists F \in \mathbb{F} : \bigwedge \left\{ \begin{array}{l} owns(P, D) \\ represents(DI, D) \\ supports(DI, F) \end{array} \right. \Rightarrow controls(P, F)$
Rule 5	$\forall A \in \mathbb{A}, \forall F \in \mathbb{F}, \forall DR \in \mathbb{DR} : \bigwedge \left\{ \begin{array}{l} accesses(A, F) \\ accesses(F, DR) \end{array} \right. \Rightarrow accesses(A, DR)$
Rule 6	$\forall A \in \mathbb{A}, \forall F_1, F_2 \in \mathbb{F} : \bigwedge \left\{ \begin{array}{l} controls(A, F_1) \\ controls(F_1, F_2) \end{array} \right. \Rightarrow controls(A, F_2)$
Rule 7	$\exists PR1, PR2 \in \mathbb{PR},$ $\exists OS1, OS2 \in \mathbb{O} :$ $\bigwedge \left\{ \begin{array}{l} isConnectedTo(PR1, PR2) \\ supports(PR1, OS1) \\ supports(PR2, OS2) \end{array} \right. \Rightarrow accesses(OS1, OS2)$

3 Using SOCIOPATH to Derive Dependences for an Activity

SOCIOPATH features a set of rules we can apply to underline and discover chains of *accesses* and *controls* relations. Table 2 shows the formal definitions of some rules that we can deduce from the meta-model basic definitions. More details are given in [14]. In this work, we want to use these relations to better understand the "social and digital dependences" among entities in the model. Thus, informally, the sets of *digital* dependences of a person are composed by the artifacts a user passes through to reach a particular element. The sets of *social* dependences are composed by the persons who control these artifacts. In the following, all those concepts are formally defined. The examples refer to Figure 2.

3.1 Activities and Paths

A user follows a path to perform an activity in a system. Some restrictions may be given to the ways the person might do their activity. For instance, if a person wants to read a .doc document, she must use an artifact that can "understand" this type of document, so activity is defined as follows.

Definition 1 (Activity)
We define an activity ω as a triple $(P, D, \mathcal{S})$, where P is a person, D is data and $\mathcal{S}$ is a set of ordered multisets of $\mathbb{F}$ in a model, so an activity ω is an element of $\mathbb{P} \times \mathbb{D} \times \mathbb{S}$. The sets in the $\mathbb{S}$ component of an activity are alternative sets of artifacts that are necessary for the person to perform her activity. For instance, the activity "John edits toto.doc" is defined as ω= (John, Data, {{MSWord}, {Pages}, {OOWriter}}).

We call *paths* the lists of actors and digital resources that describe the ways an actor may access a resource. A person may perform an activity in different ways and using different intermediate digital resources. Each possibility can be described by a path.

Definition 2 (Activity path, or ω-path)
A path σ for an activity $\omega = (P, D, \mathcal{S}) \in \mathbb{P} \times \mathbb{D} \times \mathbb{S}$ is a list of actors and digital resources such that:

- *$\sigma[1] = P$;*
- *$\sigma[|\sigma|] = D$;*
- *$represents(\sigma[|\sigma| - 1], \sigma[|\sigma|])$;*
- *$\forall i \in [2 : |\sigma| - 1], artifact(\sigma[i]) \wedge accesses(\sigma[i - 1], \sigma[i])$;*
- *$\exists s \in \mathcal{S}, s \subseteq \sigma$;*

where $\sigma[i]$, denotes the ith element of σ, and $|\sigma|$ the length of σ.

Notation: *Assuming that there is no ambiguity on the model under consideration, the set of ω-paths where $\omega = (P, D, \mathcal{S})$ is noted Υ^ω and the set of all the paths in the model is noted Υ.*

For example, concerning the ω-paths for the activity $\omega =$ "John edits toto.doc" in Figure 2, we have:

$$\left\{ \begin{array}{c} \{\texttt{John, Windows, MSWord, Windows, MSExcel, Windows, toto.doc}\} \\ \{\texttt{John, MacOS, OOWriter, MacOS, toto.doc}\} \end{array} \right\}.$$

In the first ω-path presented above, `MSExcel` is not mandatory to edit `toto.doc`. It appears in the ω-path because of the relation *accesses* between it and the artifact `Windows`. In order to exclude all the unnecessary elements from the ω-path, so we define the minimal paths as follows.

Definition 3 (Minimal path)
Let Υ^ω be a set of paths for an activity ω.
A path $\sigma \in \Upsilon^\omega$ is said to be minimal in Υ^ω iff there exists no path σ' such that:

- *$\sigma[1] = \sigma'[1]$ and ; $\sigma[|\sigma|] = \sigma'[|\sigma'|]$;*
- *$\forall i \in [2 : |\sigma'|], \exists j \in [2 : |\sigma|], \sigma'[i] = \sigma[j]$.*

Notation: *The set of minimal paths enabling an activity $\omega = (P, D, \mathcal{S})$ is noted $\widehat{\Upsilon^\omega}$. For sake of simplicity, we name this set the ω-minimal paths.*

For instance, for the activity $\omega =$ "John edits `toto.doc`", the set of the ω-minimal paths are:

$$\widehat{\Upsilon^\omega} = \left\{ \begin{array}{c} \{\texttt{John, Windows, MSWord, Windows, toto.doc}\} \\ \{\texttt{John, MacOS, OOWriter, MacOS, toto.doc}\} \\ \{\texttt{John, MacOS, Pages, MacOS, toto.doc}\} \end{array} \right\}.$$

Notation: Let say $F \in \sigma$ iff $\exists i$ such that $\sigma[i] = F$, and $s \subseteq \sigma$ iff $\forall F \in s, F \in \sigma$.

3.2   Dependence

We can now introduce the definitions of digital dependences (Definition 4 and 5) and social dependences (Definition 6 and 7). We say that a person depends on a set of artifacts for an activity ω if each element of this set belongs to one or more paths in the set of the ω-minimal paths.

Definition 4 (Person's dependence on a set of artifacts for an activity)
Let $\omega = (P, D, \mathcal{S})$ be an activity, $\mathcal{F}$ be a set of artifacts and $\widehat{\Upsilon^\omega}$ be the set of ω-minimal paths. P depends on $\mathcal{F}$ for an activity ω iff $\forall F \in \mathcal{F}, \exists \sigma \in \widehat{\Upsilon^\omega} : F \in \sigma$.

For instance, one of the sets on which John depends for the activity "John edits `toto.doc`" is {`MacOS, MSWord`}.

A person does not depend in the same way on all the sets of artifacts. Some sets may be avoidable *i.e.*, the activity can be executed without them. Some sets are unavoidable *i.e.*, the activity cannot be performed without them. To distinguish the way a person depends on artifacts, we define the degree of a person's dependence on a set of artifacts for an activity as the ratio of the ω-minimal paths that contain these artifacts.

Definition 5 (Degree of person dependence on a set of artifacts for an activity)
Let $\omega = (P, D, S)$ be an activity, $\mathcal{F}$ be a set of artifacts and $\widehat{\Upsilon^\omega}$ be the set of ω-minimal paths and $|\widehat{\Upsilon^\omega}|$ is the number of the minimal ω-paths. The degree of dependence of P on $\mathcal{F}$, denoted $d_{\mathcal{F}}^\omega$, is:

$$d_{\mathcal{F}}^\omega = \frac{|\{\sigma : \sigma \in \widehat{\Upsilon^\omega} \wedge \exists F \in \mathcal{F}, F \in \sigma\}|}{|\widehat{\Upsilon^\omega}|}$$

For instance, the degree of dependence of John on the set {`MacOS, MSWord`} for the activity "John edits `toto.doc`" is equal to one, while the degree of dependence of John on the set {`Pages, OOWriter`} is equal to 2/3.

From digital dependences we can deduce social dependences. A person depends on a set of persons for an activity if the persons of this set control some of the artifacts the person depends on.

Definition 6 (Person's dependence on a set of persons for an activity)
Let $\omega = (P, D, S)$ be an activity, and $\mathcal{P}$ be a set of persons.

$$P \text{ depends on } \mathcal{P} \text{ for } \omega \text{ iff } \wedge \begin{cases} \exists \mathcal{F} \subset \mathbb{F} : P \text{ depends on } \mathcal{F} \text{ for } \omega \\ \forall F \in \mathcal{F}, \exists P' \in \mathcal{P} : controls(P', F) \end{cases}$$

For instance, one of the sets John depends on for the activity "John edits `toto.doc`" is {`Oracle, Apple`}.

The degree of a person's dependence on a set of persons for an activity is given by the ratio of the ω-minimal paths that contain artifacts controlled by this set of persons.

Definition 7 (Degree of person's dependence on a set of persons for an activity)
Let $\omega = (P, D, S)$ be an activity, $\mathcal{P}$ be a set of persons and $\widehat{\Upsilon^\omega}$ be the ω-minimal paths. The degree of dependence of P on $\mathcal{P}$, noted $d_{\mathcal{P}}^\omega$ is:

$$d_{\mathcal{P}}^\omega = \frac{|\{\sigma : \sigma \in \widehat{\Upsilon^\omega} \wedge \exists P' \in \mathcal{P}, \exists F \in \sigma, controls(P', F)\}|}{|\widehat{\Upsilon^\omega}|}$$

For instance, the degree of dependence of John on the set {`Oracle, Apple`} for the activity "John edits the `toto.doc`" is equal to 2/3. We recall that `Oracle` *controls* `OOWriter` and `Apple` *controls* `MacOS`.

4   Related Work

Frameworks and tools modeling IT (Information Technology) systems are widely used in the context of Enterprise Architecture Management (EAM) [11]. EAM aims at giving a structured description of large IT systems in terms of their business, application, information and technical layers, with the additional goal of understanding how existing architectures and/or applications should be changed to improve business or strategic goals. SOCIOPATH rather focuses on dependence relationships, although converging with these frameworks in some aspects. For instance, RM-ODP (Reference Model of Open Distributed Processing) [12], is a generic set of standards and tools to create and manage aspect-oriented models of systems. RM-ODP analyzes and decomposes the systems in detail, mainly focusing on standard compliance. Aiming at different goals, SOCIOPATH gives a simpler overview of a system that is meant to inform the users about the relations that are implied by the system architecture, without exposing technical details. TOGAF (The Open Group Architecture Framework) [13], an approach for designing, planning, implementation, and governance of an enterprise information architecture, converges with SOCIOPATH in some of the concepts used to model the technical architecture of the enterprise. Unlike SOCIOPATH, TOGAF focuses on several aspects of the software engineering process, while describing the hardware, software and network infrastructures.

None of these works considers dependence relations among users in the social world. SOCIOPATH aims at improving awareness and exposing the relations of dependence generated on the social world from the digital world. More details about the related works are available in [14]

5   Ongoing Work and Conclusion

SOCIOPATH allows to deduce the degree of dependence of a person on the entities of a system architecture for an activity. With SOCIOPATH it is possible to know, for instance, who can prevent a person to perform an activity or whom a person can avoid while performing an activity.

Our ongoing work concerns several aspects. Currently SOCIOPATH does not distinguish the different kinds of access and control of an actor to a digital resource. In order to consider intentions and expectations of users regarding digital resources, SO-CIOPATH can be enriched with access and control typologies, to define different kinds of dependences. Moreover, no difference is made between what persons can do and what they are allowed to do according to the law, the moral rules *etc*. We aim at distinguishing between dependences related to the system's architecture, and dependences related to social commitments. Finally, the results of SOCIOPATH can be related with the notion of trust of a person toward the persons she depends on. Indeed, if a user does not trust some person, she will be concerned about architecture-induced dependences on this person, whereas if she trusts a person, she will only be concerned about commitments-related dependences.

References

1. Coulouris, G., Dollimore, J., Kindberg: Distributed Systems: Concepts and Design, 2nd edn. Addison Wesley Publishing Company, Harlow (1994)
2. Weaver, A.C., Morrison, B.B.: Social Networking. Computer 41, 97–100 (2008)
3. Westin, A.F.: Privacy and Freedom, 6th edn. Atheneum, New York (1970)
4. Marti, S., Garcia-Molina, H.: Taxonomy of Trust: Categorizing P2P Reputation Systems. Computer Network Journal 50, 472–484 (2006)
5. Lampson, B., Abadi, M., Burrows, M., Wobber, E.: Authentication in Distributed Systems: Theory and Practice. ACM Transactions on Computer Systems 10, 265–310 (1992)

6. Aurrecoechea, C., Campbell, A.T., Hauw, L.: A Survey of QoS Architectures. Multimedia Systems 6, 138–151 (1996)
7. Emerson, R.M.: Power-Dependence Relations. Amer. Sociological Review 27, 31–41 (1962)
8. Molm, L.: Structure, Action, and Outcomes: The Dynamics of Power in Social Exchange. American Sociological Association Edition (1990)
9. Blau, P.: Exchange and Power in Social Life. John Wiley and Sons (1964)
10. Alhadad, N., Lamarre, P., Busnel, Y., Serrano-Alvarado, P., Biazzini, M.: SocioPath: In Whom You Trust? Journes Bases de Donnes Avances (2011)
11. Goethals, F., Lemahieu, W., Snoeckand, M., Vandenbulcke, J.: An Overview of Enterprise Architecture Framework Deliverables. In: Banda, R.K.J. (ed.) Enterprise Architecture-An Introduction. ICFAI University Press (2006)
12. Farooqui, K., Logrippo, L., de Meer, J.: The ISO Reference Model for Open Distributed Processing: An Introduction. Computer Networks and ISDN Systems 27, 1215–1229 (1995)
13. TOGAF: Welcome to TOGAF - The Open Group Architectural Framework (2002), http://pubs.opengroup.org/architecture/togaf8-doc/arch
14. Alhadad, N., Lamarre, P., Busnel, Y., Serrano-Alvarado, P., Biazzini, M., Sibertin-Blanc, C.: SOCIOPATH: In Whom You Trust? Technical report, LINA – CNRS: UMR6241 (2011)

Detecting Privacy Violations in Multiple Views Publishing

Deming Dou and Stéphane Coulondre

Université de Lyon, CNRS
INSA-Lyon, LIRIS, UMR5205, 69621, Lyon, France
{deming.dou,stephane.coulondre}@liris.cnrs.fr

Abstract. We present a sound data-value-dependent method of detecting privacy violations in the context of multiple views publishing. We assume that privacy violation takes the form of linkages, that is, identifier-privacy value pair appearing in the same data record. At first, we perform a theoretical study of the following security problem: given a set of views to be published, if linking of two views does not violate privacy, how about three or more of them? And how many potential leaking channels are there? Then we propose a pre-processing algorithm of views which can turn multi-view violation detection problem into the single view case. Next, we build a benchmark with publicly available data set, Adult Database, at the UC Irvine Machine Learning Repository, and identity data set generated using a coherent database generator called Fake Name Generator on the internet. Finally, we conduct some experiments via Cayuga complex event processing system, the results demonstrate that our approach is practical, and well-suited to efficient privacy-violation detection.

Keywords: Privacy violation, Multi-view publishing, Pre-processing algorithm, Cayuga system.

1 Introduction

Recently, privacy-preserving data publishing has been proposed whereby the data owner prevents linking some identity to a specific data record and sensitive information in the released data while, at the same time, preserves useful information for data mining purpose. Data publishing exports a set of selection-projection views instead of the whole proprietary table, data users can only access data by submitting queries against these views. Even though each published view is secure after de-identification, while an adversary can re-identify an identity by linking two or more views on their common attributes. Consider a private table $Patient(SSN, Name, ZIP, DOB, Sex, Job, Disease)$ in Table 1, SSN and Name are explicit identifiers (EIs) which can explicitly identify record holders, Disease is one of the sensitive attributes (SAs) which contain some privacy information. Suppose that the data owner releases views $v_1 = \Pi_{SSN,DOB,Job,Disease}(Patient)$, $v_2 = \Pi_{Name,ZIP,Job,Sex}(Patient)$ to a data recipient, and we define privacy violations by $\Pi_{SSN,Disease}(Patient)$, $\Pi_{Name,Disease}(Patient)$ and $\Pi_{SSN,Name,Disease}(Patient)$ which means that an EI and a SA cannot appear at the same time. Obviously, each of them does not violate privacy, however, by joining v_1 and v_2 using $v_1.Job = v_2.Job$, an adversary can uniquely re-identify that Dan is a Flu patient which violates Dan's privacy.

S.W. Liddle et al. (Eds.): DEXA 2012, Part II, LNCS 7447, pp. 506–513, 2012.
ⓒ Springer-Verlag Berlin Heidelberg 2012

Table 1. Private table P

SSN	Name	ZIP	DOB	Sex	Job	Disease
387-399	Alice	47677	09/09/80	F	Manager	OC
387-200	Bob	47602	24/05/87	F	Engineer	OC
387-486	Carol	47678	08/11/82	M	Engineer	PC
387-756	Dan	47905	27/08/66	M	Dancer	Flu
387-665	Ellen	47909	04/10/57	F	Engineer	HD
387-588	Jack	47906	10/10/62	M	Manager	HD

The privacy violation presented here is not schema-dependent but data-value-dependent, for example, if we release another view $v_2' = \prod_{Name,ZIP,Job,Sex}(\sigma_{Name!='Dan'}Patient)$, v_2' has the same schema with v_2, but the joining of v_1 and v_2' does not violate any privacy.

Now, we further suppose that the data holder releases $v_3 = \prod_{SSN,DOB,Sex}(Patient)$, $v_4 = \prod_{ZIP,DOB,Job}(Patient)$ and $v_5 = \prod_{ZIP,Job,Disease}(Patient)$ to another data recipient, Each of these views does not violate privacy, and neither does their pairwise joining. However, if an adversary performs at first a joining of v_3, v_4 using $v_3.SSN = v_4.SSN$ and $v_3.DOB = v_4.DOB$ which generates a mediated view $v_{34}(SSN,ZIP,DOB,Sex,Job)$, and eventually the adversary can also uniquely deduce that Dan is a Flu patient by joining v_{34} and v_5 using $v_{34}.ZIP = v_5.ZIP$ and $v_{34}.Job = v_5.Job$. Furthermore, if these two data recipients collude with each other behind the scene, the situation will inevitably become more complicated, they have 5 views in all, v_1, v_2, v_3, v_4 and v_5, and there will be much more violation channels and the risk increases. The challenge is how to detect all these potential violation channels when the number of published views is in dozens. To the best of our knowledge, no previous work takes this problem into account, and this paper is the first to present a formal analysis of privacy violation in multiple views publishing, and the first to propose an algorithm of multi-view pre-processing.

2   Related Work

Since Sweeney [8] introduced the concept of *linking attack* and the concept of *k-anonymity* in 1998, most of previous works have addressed the case of privacy protection using anonymization-based methods, either over traditional data tables, such as *l*-diversity [4], *t*-closeness [6], or over data streams, such as SWAF [10], SKY [5], CASTLE [1] and so on. We refer readers to [3] for a survey of different privacy preserving approaches with respect to privacy-preserving data publishing. Miklau and Suciu [7] performed a theoretical study of the query-view perfect-security problem to analyze the privacy violation between published relational views and queries, while the security standard is so strong that it has some practical limitations. They also presented a spectrum of information disclosure (privacy violation), total, partial and minute among which we aim to detect the first two types. In 2009, Vincent et al. [9] presented a method of detecting privacy violations using disjoint queries in database publishing, with the aim of overcoming the limitations of the perfect-security approach and provided an alternative in applications where a less stringent, but computationally tractable, method of detecting privacy violations is required. Yao et al. [11] assume that privacy violation takes the form of linkages, that is, pairs of values appearing in the same data record, the assumption of privacy violation in this paper is similar to Yao's.

3   Contribution

Similar with steps that compose the KDD (Knowledge Discovery in Databases) process, we proposed a privacy violation detection method consisting of three steps, multi-view pre-processing, data merging and event processing. In this section, we will at first give an formal analysis of the problem of detecting privacy violations in the context of multiple views publishing, then we will introduce all the steps mentioned above.

3.1   Problem Formulation and Definitions

We assume an instance I of an universal relation R with schema D, where D contains EIs, SAs and other attributes which do not fall into these two categories, written as OAs, and we use R^D to denote the set of all possible relations on D. Data are being published under the form of a being set which is a pair (q, r), where q is a list of selection-projection queries $(q_1, q_2, \cdots, q_n)$ on I, and r is a list of relations $(r_1, r_2, \cdots, r_n)$ such that $\exists G \in R^D, r_i = q_i(G)$ for each $i = 1, 2, \ldots, n$. If q is given, we can abbreviate (q, r) to V which contains n views, and denote (q_i, r_i) by v_i where v_i is in V.

Definition 1 (privacy violation). *Given a direct published or a mediated view v_i of a underlying data table, $\alpha \subseteq EIs$ and $\beta \subseteq SAs$, and a is a constant value of α and b of β, if the result of $\prod_{\alpha, \beta}(v_i)$ is a finite set of data records, a privacy violation occurs if and only if $\mid \sigma_{\alpha=a, \beta=b}(\prod \alpha, \beta(v_i)) \mid = 1$.*

Problem Definition. Given a set of published views $v_1, v_2, \cdots, v_n$ of the same underlying data table, detect any total and partial privacy violation which could be deduced from one single view v_i or the joins $\bowtie (v_i, v_j, \cdots, v_k)$, where $2 \le k - i \le n$.

3.2   Multi-view Pre-processing

In this paper, we regard the views as nodes of a graph with an edge between two nodes representing a natural join of them. We model our approach as a simple undirected graph $G(V, E)$ with a finite non-empty view set V and a set of symmetric view pairs E. For each ordered view pair $(w, u) \in E$, the pair (u, w) also belongs to E. G_i owns some part of G, denoted as $G(V_i, E_i)$ where $\bigcup_i V_i = V$, $\bigcup_i E_i \subseteq E$. The natural join of each view pair in E is an edge, E itself is also called the *edge set* of G.

As we mentioned above, two views having common attributes do not always violate privacy, for graph G, that can be saying as inexistence of some edges of E. Take v_1, v_2, v_3 as an example, the sub-graph G_1 can be defined by the view set $V_1 = \{v_3, v_4, v_5\}$ and $E_1 = \{\{(v_3, v_4), (v_4, v_3)\}, \{(v_3, v_5), (v_5, v_3)\}, \{(v_4, v_5), (v_5, v_4)\}\}$ among which the view pairs $\{(v_3, v_4), (v_4, v_3)\}$ generate edge v_{34} and $\{(v_4, v_5), (v_5, v_4)\}$ generate edge v_{45}, while the view pairs $\{(v_3, v_5), (v_5, v_3)\}$ generate no edge.

Definition 2 (an edge). *Given view pairs* $\{(v_a, v_b), (v_b, v_a)\}$, $v_a \in G_i$, $v_b \in G_j$, *an edge exists if and only if* $\exists t : t \in v_a \bowtie v_b$, *t is a data record. If* $i = j$, *the edge is called an "inner-edge", otherwise we call it an "inter-edge".*

In our approach, we define the published views $v_1, v_2, \cdots, v_n$ as V_1 in level G_1, then $|V_1| = n$, and $|E_1| \le C_2^n$, E_1 can be denoted as following:

$$E_1 = \{\{(v_1, v_2), (v_2, v_1)\}, \{(v_1, v_3), (v_3, v_1)\}, \cdots, \{(v_a, v_b), (v_b, v_a)\}\}, 1 \le a, b \le n$$

According to Definition 2, we know that the edge number of G_1 is no larger than C_2^n. It is possible that there is no edge in level G_1.

Proposition 1. *Given* $v_a \in V_i$, $v_b \in V_j$, $i \ne j$, *an "inter-edge" equals to a view* v_c *in* V_m *or* v_d *in* V_j *where* $m > j > i$.

Proof. By the commutative property, distributive property and associative property of binary operations in relational algebra.

As shown in Fig.1, for G_1, we assume that all view pairs in E_1 generate m edges, next we take these m edges as the view set of G_2, by executing natural join of each view pair in E_2, we get k edges, and then we take these k edges as the view set of G_3, the rest can be deduced by analogy. Corresponding to these operations, we propose a multi-view pre-processing algorithm, as shown in Algorithm 1, in this algorithm, we take n views $v_1, v_2, \ldots, v_n$ and their respective schemes $r_1, r_2, \ldots, r_n$ as original inputs, at first we get the universal relation R by the union loop $R = R \cup r_i$, we assume that we know the number of views in the first level G_1 is $L_1 = n$, and after pre-processing of these n views, we get L_2 views in second level $\acute{G}_2$ and their corresponding schemes, the rest operations can be done in the same manner until there is no more view in next level. Finally, we collect the views in each level and get $n + l$ views $v_1, v_2, \ldots, v_n, v_{n+1}, v_{n+2}, \ldots, v_{n+l}$.

One advantages to perform this multi-view pre-processing operation is that we can transform partial information disclosure detection problem to total information detection case. Take v_3, v_4 and v_5 as an example, if they are the input views in our algorithm, after being processed using our multi-view pre-processing algorithm, we get three more

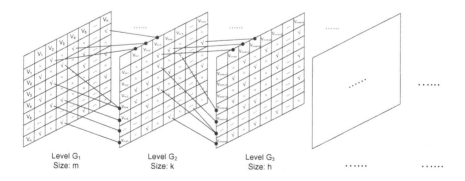

Fig. 1. Schematic diagram for multi-view pre-processing

Algorithm 1. Pre-processing Algorithm of Multiple Views

Input: n views: $v_1, v_2, \ldots, v_n$ and corresponding schemes $r_1, r_2, \ldots, r_n$.

Output: $(n+l)$ views: $v_1, v_2, \ldots, v_n, v_{n+1}, v_{n+2}, \ldots, v_{n+l}$.

1: Let $R = r_1, s = 1, i = 1, l = 0$
2: **for** $i \leftarrow 2 \ldots n$ **do**
3:     $R = R \cup r_i$ // R is a universal relation
4: **end for**
5: Let L_i be the number of views in level G_i, $L_1 = n$.
6: **repeat**
7: **for** *each level G_i in G* **do**
8:     **for** $j \leftarrow (l+1) \ldots L_i - 1$ **do**
9:         **for** $k \leftarrow (j+1) \ldots L_i$ **do**
10:             **if** $r_j \neq R$ and $r_k \neq R$ **then**
11:                 **if** $r_j \cap r_k \neq \emptyset, r_j \subset (r_j \cup r_j), r_k \subset (r_j \cup r_k)$ **then**
12:                     $v = v_j \bowtie v_k$ //natural join of two views
13:                     **if** $v \neq \emptyset$ **then**
14:                         $v_{L_i+s} = v_j \bowtie v_k$ //generate views for next level G_{i+1}
15:                         $r_{L_i+s} = r_j \cup r_k$ //generate corresponding relations
16:                         $s = s + 1$
17:                     **end if**
18:                 **end if**
19:             **end if**
20:         **end for**
21:         $L_{i+1} = L_i + s$ //number of views in next level
22:         $l = l + s$ //total number of views in all levels
23:         $i = i + 1$ //total number of views in all levels
24:     **end for**
25: **end for**
26: **until** $s = 0$ //no views in next level

views v_{34}, v_{45} and v_{345}. At this time, v_3, v_4, v_5, v_{34}, v_{45} and v_{345} are the final inputs to data merging step, what we need to do is to check if every tuple of these final views violate specific privacy.

3.3   Data Merging

We use Cayuga [2] as our event processing engine, however Cayuga can only read stream data from disk files, such as a text file, or TCP sockets, this implies that we must merge our multiple pre-processed views into a single file and then store it in a disk. In our system, we adopt an universal relational assumption, however, the original published views and the views generated from the pre-processing algorithm have different schemes, we must add to different views different additional attributes and set their values as *Null*. Take v_1 and v_2 as an example, the universal relation is {Name, ZIP, Job, DOB, Sex, Disease}, v_1 does not have attributes {Name, Sex}, while v_1 does not have {DOB, Disease}, so we merge them like follows in Table 2:

Table 2. Merged data of v_1 and v_2

Name	ZIP	DOB	Job	Sex	Disease
	47677	09/09/80	Manager		OC
	47602	24/05/87	Engineer		OC
	......	......	......		......
Alice	47677		Manager	F	
Bob	47602		Engineer	M	
......	......			......	

3.4 Event Processing

After data merging, we now investigate the details of event processing step which is Cayuga-based. Cayuga system is an expressive and scalable Complex Event Processing (CEP) system developed at the Cornell Database Group, it provides a SQL-like query language, Cayuga Event Language (CEL), for expressing complex event patterns, instead of operating on tables, Cayuga system performs continuous queries on data streams with multi-query optimization, it can serve as an event engine in a larger software environment without being obtrusive where the end user/application interacts with Cayuga by submitting queries written in CEL to it and receiving query result streams from it. For further information, we refer readers to Cayuga's official website http://www.cs.cornell.edu/bigreddata/cayuga/. The benefit of using Cayuga system other than the other DBMS is that it has an high-speed query processing engine which can handles simultaneous events for large-scale data-sets and data streams. What the data owner should do is to launch specific continuous queries over the final merged data.

4 Case Study

4.1 Experimental Settings

Dataset Benchmark. We adopt the publicly available Adult Database at the UCI Machine Learning Repository with 15 attributes, if records with missing values and unknown values are removed, there are 45,222 instances. In order to build a representative benchmark *BP*, we get another identity data set generated using a coherent database generator called Fake Name Generator on the internet which contains 30 attributes. We integrate these two datasets and keep the same number of instance, 45,222 tuples, and only retain some of their attributes. Then we add a column with sensitive values called "Health Condition" consisting of HIV, Cancer, Phthisis, Hepatitis, Obesity, Asthma, Flu, Indigestion to the extracted data and randomly assign one sensitive value to each record. The random technique works in the following way. First, assign a number to each sensitive attribute, i.e., 1: HIV, 2: Cancer, 3: Phthisis, 4: Hepatitis, 5: Obesity, 6: Asthma, 7: Flu, 8: Indigestion. Second, for each tuple (record), generate a random number from 1-8. Then, assign the corresponding sensitive attribute value to the tuple. For example, for the first tuple in the data set, if the random number is 5, then this record has the sensitive value "Obesity". "NationalId" and "Givenname" are chosen as the *EIs*, and "Health Condition" and "Password" are *SAs*.

```
SELECT *
FROM FILTER {count = 1}
(FILTER{NationalId != '' and Password != ''}
(SELECT *, 1 as count FROM BP)
FOLD {NationalId != '' and Password != '',,S.count + 1 as count} S)

SELECT *
FROM FILTER {count = 1}
(FILTER{NationalId != '' and Health Condition != ''}
(SELECT *, 1 as count FROM BP)
FOLD {NationalId != '' and Health Condition != '',,S.count + 1 as count} S)
............................
```

Fig. 2. Cayuga continuous queries

Table 3. Six views to be published

Views to be published
$v_1 = \Pi_{NationalId,Birthday,Gender}(BP)$
$v_2 = \Pi_{Givenname,Zipcode,Education}(BP)$
$v_3 = \Pi_{Zipcode,Company,CCType}(BP)$
$v_4 = \Pi_{Gender,CCType,Password}(BP)$
$v_5 = \Pi_{Birthday,Zipcode,Health\ Condition}(BP)$
$v_6 = \Pi_{Givenname,CCType,Birthday}(BP)$

Views and Continuous Queries. In our experiments, we try to publish six views $v_1, v_2, v_3, v_4, v_5, v_6$ on our benchmark BP, as shown in Table 3. In our benchmark BP, we assume that there are two EIs, "NationalId" and "Givenname", and two SAs "Health Condition" and "Password", therefore the number of continuous queries we should launch is $C_1^2 \times C_1^2 = 4$, as shown in Fig.2, the continuous queries are written in the structured Cayuga Event Language (CEL) and will be executed in the query engine of Cayuga system.

4.2 Detecting Results

For the six published views, the universal relation contains 10 attributes, they are NationalId, Givenname, Birthday, Gender, Zipcode, Education, Company, CCType, Password, and Health Condition. After multi-view pre-processing, the second level has 13 views, and every two of them have common attributes, meanwhile their natural joins have more than one data records, therefore in the third level we have $C_2^{13} = 78$ views, while for the fourth level and fifth level, the number of view is not a combination of the previous levels.

After being processed via all steps of multi-view pre-processing, data merging, and event processing using Cayuga system, we successfully get more than 45222 records which violate privacy, because privacy violations occur in some levels of G, as shown in Table 4.

Table 4. Detection outputs of privacy violation

NationalId	Givenname	Birthday	Gender	⋯	Password	Health Condition
170-40-9312	John	6/5/1988	male	⋯	be6Poowooph	Asthma
508-98-9112	Steven	6/30/1983	male	⋯	acoTo5ah	Asthma
130-84-3154	Patrick	2/2/1941	male	⋯	paiS0roo3	Flu
......	......	......	......	⋯	......	......

5   Conclusion and Future Work

We present a formal analysis of privacy violation in the context of multi-view publishing: given a set of published views, if linking of two views does not violate privacy, how about linking of multiple views? And how many potential violation channels are there? We propose a pre-processing algorithm of multiple views which can transform partial violation problem to the total case. Then, we build a benchmark with Adult Database at UCI Machine Learning Repository, and an identity data set generated using Fake Name Generator on the internet. Finally, we conduct some experiments via Cayuga system, the results demonstrate that our approach is practical, and well-suited to efficient privacy-violation detection. However, we did not conduct the performance evaluation of the proposed algorithm given that the number and the size of the input views can be very large, and we did not take the timestamps of views into account, they are useful if we want to detect privacy violation within a specific time interval. Therefore, future work includes performance evaluation, detecting minute privacy violation, algorithm optimization and develop a Cayuga-based privacy violation detector.

References

1. Cao, J., Carminati, B., Ferrari, E., Tan, K.-L.: Castle: A delay-constrained scheme for ks-anonymizing data streams. In: ICDE, pp. 1376–1378 (2008)
2. Demers, A., Gehrke, J., Cayuga, B.P.: A general purpose event monitoring system. In: CIDR, pp. 412–422 (2007)
3. Fung, B.C.M., Wang, K., Chen, R., Yu, P.S.: Privacy-preserving data publishing: A survey of recent developments. ACM Comput. Surv. 42(4), 1–53 (2010)
4. Kifer, D., Gehrke, J.: l-diversity: Privacy beyond k-anonymity. In: ICDE 2006: Proceedings of the 22nd International Conference on Data Engineering, p. 24 (2006)
5. Li, J., Ooi, B.C., Wang, W.: Anonymizing streaming data for privacy protection. In: ICDE 2008: Proceedings of the 2008 IEEE 24th International Conference on Data Engineering, pp. 1367–1369. IEEE Computer Society, Washington, DC (2008)
6. Li, N., Li, T., Venkatasubramanian, S.: t-closeness: Privacy beyond k-anonymity and l-diversity. In: ICDE (2007)
7. Miklau, G., Suciu, D.: A formal analysis of information disclosure in data exchange. In: SIGMOD 2004: Proceedings of the 2004 ACM SIGMOD International Conference on Management of Data, pp. 575–586. ACM, New York (2004)
8. Samarati, P., Sweeney, L.: Generalizing data to provide anonymity when disclosing information. Technical report (March 1998)
9. Vincent, M.W., Mohania, M., Iwaihara, M.: Detecting privacy violations in database publishing using disjoint queries. In: EDBT 2009: Proceedings of the 12th International Conference on Extending Database Technology, pp. 252–262. ACM, New York (2009)
10. Wang, W., Li, J., Ai, C., Li, Y.: Privacy protection on sliding window of data streams. In: COLCOM 2007: Proceedings of the 2007 International Conference on Collaborative Computing: Networking, Applications and Worksharing, pp. 213–221. IEEE Computer Society, Washington, DC (2007)
11. Yao, C., Wang, X.S., Jajodia, S.: Checking for k-anonymity violation by views. In: VLDB 2005: Proceedings of the 31st International Conference on Very Large Data Bases, pp. 910–921. VLDB Endowment (2005)

Anomaly Discovery and Resolution in MySQL Access Control Policies

Mohamed Shehab, Saeed Al-Haj, Salil Bhagurkar, and Ehab Al-Shaer

Department of Software and Information Systems
University of North Carolina at Charlotte
Charlotte, NC, USA
{mshehab,salhaj,sbhagurk,ealshaer}@uncc.edu

Abstract. Managing hierarchical and fine grained DBMS policies for a large number of users is a challenging task and it increases the probability of introducing misconfigurations and anomalies. In this paper, we present a formal approach to discover anomalies in database policies using Binary Decision Diagrams (BDDs) which allow finer grain analysis and scalability. We present and formalize intra-table and inter-table redundancy anomalies using the popular MySQL database server as a case study. We also provide a mechanism for improving the performance of policy evaluation by upgrading rules from one grant table to another grant table. We implemented our proposed approach as a tool called *MySQLChecker*. The experimental results show the efficiency of *MySQLChecker* in finding and resolving policy anomalies.

Keywords: Policy, Access Control, Policy Analysis, Anomaly Detection.

1 Introduction

Large DBMSs used in applications such as banks, universities and hospitals can contain a large number of rules and permissions which have complex relations. Managing such complex policies is a highly error-prone manual process. Several surveys and studies have identified human errors and misconfigurations as one of the top causes for DBMS security threats [3]. This is further magnified by the adoption of fine grain access control that allows policies to be written at the database, table and row levels. In addition, once policy misconfigurations are introduced they are hard to detect.

There has been recent focus on policy anomaly detection, especially in firewall policy verification [2,8]. The approaches proposed are not directly applicable to database policies due to several reasons. First, firewall policies are based on first-match rule semantics, in database policies other semantics such as most specific are adopted. Second, the firewall polices are flat, on the other hand database policies are hierarchical and are applied at different granularity levels. Third, firewall policies are managed in a single access control list, while database policies are managed by multiple access control lists. Therefore, new mechanisms for policy analysis and anomaly detection are required database policies.

S.W. Liddle et al. (Eds.): DEXA 2012, Part II, LNCS 7447, pp. 514–522, 2012.

In this paper, we use MySQL as a case study due to its wide adoption. MySQL provides a descriptive fine grain policy language which provides an expressive policy language. First we start by introducing the details of group-based policies used in MySQL. We define and formalize the set of possible access violations. To aid the administrator in detecting and avoiding these violations we propose a framework to detect and resolve the detected violations. To the best of our knowledge, this is the first time such a policy violation detection and resolution mechanism for DBMS policies has been proposed. We implemented our proposed anomaly detection and resolution approaches as a tool (*MySQLChecker*). The efficiency and scalability of *MySQLChecker* was demonstrated through our experimental evaluations. The main contributions of the paper are:

- We formulate and model database access control policies using BDDs.
- We classified the set of possible database policy anomalies and proposed algorithms to detect and resolve the detected anomalies using BDDs.
- With a proof of concept implementation of our proposed anomaly detection and resolution, we have conducted experiments using synthetic policies.

The rest of the paper is organized as follows: In Section 2, we provide a brief background of database policies, access control in the MySQL framework and Binary Decision Diagrams. In Section 3, we define the formal representation of database grant rules, tables, and policies using BDDs. In Section 4, we define the set of possible anomalies in the MySQL context, and describe how the BDDs are used to detect and resolve these anomalies. Our implementation and experimental results are described in Section 5. Finally, we wrap up the paper with related work and conclusions.

2   Preliminaries

In this section we present the preliminaries related to MySQL access control, and Binary Decision Diagrams.

2.1   Database Policies

Most DBMS use access control lists to specify and maintain access control policies. For example, in MySQL access control lists (grant tables) are stored in multiple tables within the DBMS itself. The policy adopted is based on a closed world policy model, where access is allowed only if an explicit positive authorization is specified, otherwise access is denied. The grant tables include *user*, *db*, *tables_priv*, *columns_priv* and *procs_priv* which are used to store privileges at global, database, table, column and procedure levels respectively. Database privileges include several permissions, for example SELECT, INSERT, etc.

An access rule R can be represented as a tuple that indicates the permissions granted to a user over a given object. A rule is defined as $R = \{user, host, db, table, column, privs\}$, where $R[user]$, $R[host]$, $R[db]$, $R[table]$, $R[column]$, $R[privs]$ refers to the rule's user, host, database, table column names and set of privileges respectively. To refer to a group of hosts, IP address and domain wildcards are allowed in $R[host]$. For example, *192.168.1.%* refers to all hosts in the *192.168.1* class C network, and *%.abc.com* refers to any host in the *abc.com* domain. Database names can also include wild characters.

Access control in MySQL involves both *Connection* and *Request Verification*. Connection verification verifies if a user connecting from a specific host is allowed to login to the database. If access is granted, the request verification stage verifies if the user has access to the objects requested. Request verification checks user access from higher to lower granularity levels, for example database level, then table level, and then column level.

2.2 Binary Decision Diagrams (BDDs)

Binary Decision Diagrams (BDDs) are a type of symbolic model checking. A BDD is a directed acyclic graph [4, 5] used to represent boolean functions. The graph has a root and set of terminal and non-terminal nodes. Each non-terminal node represents a binary variable. Non-terminal nodes have two edges at most, high and low. High edge represents the true assignment for that variable while low edge represents the false assignment. Terminal nodes are two nodes representing the values *true* and *false*. BDDs were used efficiently in anomaly discover in access control list in Firewalls [2] and XACML [7] policies. Utilizing BDD operations such as and, or and negation can be used to efficiently implement set operations on Boolean expressions such as intersection and union.

3   Formal Representation

In this section, we will present the formal representation for the MySQL policy evaluation process using BDDs to encode the MySQL grant tables and rules.

3.1 Rules Modeling

A rule has two parts, condition and privilege vector. The matching request has to match rule conditions in order to grant a set of privileges. A rule can be represented formally as $R_i : C_i \rightsquigarrow PV_i$, where C_i are the conditions for the ith rule that must be satisfied in order to grant privileges Vector PV_i. The condition C_i can be represented as a Boolean expression of the filtering fields, f_1, f_2, ..., f_k as $C_i = f_1 \wedge f_2 \wedge \ldots \wedge f_k$. The fields can be *host IP*, *user name*, *database name*, *table name* and *column name*.

Each grant table has different set of fields to represent the matching condition for rules in that table. For example, *users* grant table uses host IP and user name to match rules in the table. The *db* grant table uses host name, user name, and database name for condition matching. Privileges vector is the decision vector for each rule that specifies what are the privileges that will be granted when the rule is triggered. Formally as $PV = P_1 \wedge P_2 \wedge \ldots \wedge P_m$, where m is the total number of permissions in the system, P_i is a Boolean variable representing the ith permission. When a permission is granted, its corresponding Boolean variable is set to TRUE. For example, in Figure 1, the permission vector for *bob@152.150.10.2* is $|1, 0, 0|$ which means allowing SELECT and denying INSERT and UPDATE.

Rule	User	Host	DB name	Select	Insert	Update
r_1	alice	localhost	%	Y	Y	Y
r_2	bob	152.150.10.2	Emp	Y	N	N

Fig. 1. *db* grant table example

3.2   Grant Tables Modeling

To evaluate a request in MySQL policy, there has to be a rule in one of the grant tables that allows this request. A grant table is modeled as a Boolean formula using BDDs. Each grant table is a sequence of filtering rules, $R_1, R_2, ..., R_n$. The rules are checked in the first match semantic when a request is matched.

In the first match semantic, policy evaluation starts from the first rule, then the second rule and so on until a matching rule is found. For example, when the third rule is matched, the first and the second rule are not matched. The formal rule ordering in this is represented as $R_1 \vee (\neg R_1 \wedge R_2) \vee (\neg R_1 \wedge \neg R_2 \wedge R_3)$. The previous formula shows the rules ordering only, it did not show the how rules conditions and permissions are encoded in the grant table. Formally, each grant table will be constructed as $X_{BDD} = \bigwedge_{i=1}^{m} X_{BDD}^i$, where X_{BDD} is the BDD representation for any of the grant tables, and X_{BDD}^i is the BDD representation for the ith permission in X grant table.

Each permission in the permission vector is represented by a BDD. The kth permission BDD is $X_{BDD}^k = \bigvee_{i=1}^{n} \neg C_1 \wedge \neg C_2 \ldots \neg C_{i-1} \wedge C_i \wedge P_k$, where n is the total number of the rules in the X grant table and k is the kth permission in the permission vector PV. Each of the grant tables will be encoded as a BDD. Where $U_{BDD}, D_{BDD}, T_{BDD}$ and C_{BDD} represent the *users*, *db*, *tables_priv*, and *columns_priv* grant tables respectively.

3.3   Policy Modeling

In order to determine if a request is to be granted or not, the *users* grant table is checked first. If a matching rule is found in the table, the associated privileges are granted. Otherwise, *db* grant table is checked for a matching rule. MySQL policy checks grant tables in the following order: *users*, *db*, *tables_priv* and *columns_priv*. Formally, the BDD first match semantic for MySQL policy (G_{MySQL}) is represented as:

$$G_{MySQL} = U_{BDD} \bigvee (\neg U_{BDD} \wedge D_{BDD}) \bigvee (\neg U_{BDD} \wedge \neg D_{BDD} \wedge T_{BDD})$$
$$\bigvee (\neg U_{BDD} \wedge \neg D_{BDD} \wedge \neg T_{BDD} \wedge C_{BDD})$$

Given a request Q, the decision whether to grant this request or not is evaluated by intersecting G_{MySQL} and Q. If the resultant Boolean expression is FALSE, then the request is denied. Otherwise, it is granted. Formally this operation can be represented as $Q \wedge G_{MySQL} \Leftrightarrow action, action \in \{TRUE \mid FALSE\}$.

4   MySQL Policy Anomalies

The hierarchal relation between grant tables and the first match semantic in evaluating requests can introduce different types of anomalies. There two types of anomalies in MySQL policies namely intra-table and inter-table redundancy.

4.1   Intra-table Redundancy

Intra-table redundancy is the redundancy within the same grant table. Subset and superset relationships are the primary causes for rule redundancy. Two rules are intra-table redundant if they grant the same privileges for the same conditions. Rule redundancy can occur in all grant tables. The following intra-table redundancy definition applies for any grant table.

Definition 1. *Given a grant table X_{BDD}, a rule R_i is intra-table redundant to rule R_j for $i < j$ if:*

$$(C_i \subseteq C_j) \bigwedge (PV_i = PV_j) \bigwedge (\nexists R_k (i < k < j \wedge \ C_i \subseteq C_k \wedge PV_i \neq PV_k)).$$

Definition 1 covers intra-table redundancy case in which the preceding rule R_i is a subset from the superset rule R_j. Unlike firewall policies, a superset rule cannot appear before a subset rule because of the pre-sorting process. A rule is intra-table redundant if there is a rule with the same privileges vector follows the redundant rule. Definition 1 excludes the case of exceptions. Exceptions are not considered as redundant rules because they intended to perform different action on a subset from the later rule. We demonstrate this condition through the example in Figure 2(a). Note that rules R_2 and R_3 are not intra-redundant because they have different permission vectors. In the case of R_2 and R_5, even both rules have the same privilege vector and subset relationship between conditions, they are not considered redundant because there is another rule, R_3 that has different permission vector. When deleting rule R_2, bob@152.150.10.1 will not be able *insert* because the *insert* privilege has been revoked in rule R_3. Rules R_1 and R_4 are redundant. Deleting R_1 will not affect the policy semantics.

4.2   Inter-table Redundancy

Inter-table redundancy appears between two rules in different grant tables. Considering all grant tables, we have *six* inter-table redundancy cases between all grant tables. Inter-table redundancy is *partial* or *complete*. Complete inter-table redundancy occurs when two rules in different grant tables have the same privileges vector for some common conditions. While the partial inter-table redundancy when some privileges are similar for some common conditions.

Definition 2. *Given two grant tables X and Y, $X < Y$, having the BDD representation X_{BDD} and Y_{BDD} respectively, a rule $R_i \in Y_{BDD}$ is completely inter-table redundant by $R_j \in X_{BDD}$ if: $(X_{BDD} \cap Y_{BDD} \neq \phi) \wedge (C_i \subseteq C_j)$.*

Rule	User	Host	Select	Insert	Update
R_1	alice	152.150.40.55	Y	N	N
R_2	bob	152.150.10.1	Y	Y	N
R_3	bob	152.150.10.%	Y	N	N
R_4	alice	152.150.40.%	Y	N	N
R_5	bob	152.150.%.%	Y	Y	N

(a) *users* grant table example

Rule	User	Host	DB name	Table	Select	Insert	Update
R_1	bob	152.150.10.5	Emp	manager	Y	Y	N
R_2	bob	152.150.10.%	Emp	human_resources	Y	N	N
R_3	bob	152.150.%.%	Acc	human_resources	Y	N	Y
R_4	alice	169.12.%.%	Emp	manager	Y	Y	Y

(b) *tables_priv* grant table example

Rule	User	Host	DB name	Table	Column	Select	Insert	Update
R_1	bob	152.150.10.%	Emp	manager	name	Y	Y	N
R_2	bob	152.150.10.%	Emp	human_resources	id	Y	Y	Y
R_3	bob	152.150.%.%	Acc	human_resources	salary	Y	N	N
R_4	alice	169.12.25.%	Emp	manager	id	Y	Y	Y

(c) *columns_priv* grant table example

Fig. 2. Intra and Inter-Table Redundancy Example

To find complete inter-table redundancy, grant tables BDDs are compared together. Complete inter-table redundancy requires two conditions: 1) there is an overlap between grant table BDDs and 2) the superset rule appears in the upper level grant table.

Definition 3. *Given two grant tables X and Y, $X<Y$, having the BDD representation X_{BDD} and Y_{BDD} respectively, a rule $R_i \in Y_{BDD}$ is partially inter-table redundant by $R_j \in X_{BDD}$ if: $\exists m \exists k (X_{BDD}^m \cap \neg Y_{BDD}^m \neq \phi) \wedge (X_{BDD}^k \cap Y_{BDD}^k \neq \phi)$.*

Where X_{BDD}^m is the BDD representation for the mth permission in X grant table. In partial inter-table redundancy, the privileges vectors are not the same. There are some permissions allowed in the upper level table and denied in the lower level table. Partial inter-table redundancy does not hold if a permission is denied in the upper level table and allowed in the lower level table. Partial inter-table redundancy requires at least one permission to be allowed in both rules. This is necessary to eliminate the case in which there is no rule exists in the lower level grant table. Figures 2(b)-2(c) provide inter-table redundancy examples. Rules R_1 in *tables_priv* and R_1 in *columns_priv* are not redundant because the upper table rule is not a superset rule. Rules R_2 in *tables_priv* and R_2 in *columns_priv* are not complete inter-table redundant because privileges vectors are not the same, also they are not partial inter-table redundant because the upper table rule denies the *insert* privilege while the lower table rule allows it. Rules R_3 in *tables_priv* and R_3 in *columns_priv* are partially inter-table redundant. Rules R_4 in *tables_priv* and R_4 in *columns_priv* are complete inter-table redundant.

4.3 Violations and Safety

A policy operation is safe if it does not cause the new policy to allow (deny) requests that were previously denied (allowed) by the original policy. Mainly safety is focused on maintaining the allow and deny space of the original policy.

Definition 4. *Policy P_A is allow (deny) safe w.r.t policy P_B iff every request allowed (denied) by P_A is also allowed (denied) by P_B. An operation OP that transforms P_A to P_B is safe iff P_A and P_B are both deny and allow safe.*

In what follows we will investigate the safety of the operations involving the removal of rules identified as redundant.

Proof. Let the original policy P_A includes a rule R_s is identified as a redundant violation. Let the policy P_B be the policy generated after removing rule R_s from P_A. Let P_A be a policy with three rules: R_1, R_2 and R_3. Assume rule R_1 is an intra-table redundant with R_3. Using the BDD modeling described earlier, the formal representation for P_A is $R_1 \vee (\neg R_1 \wedge R_2) \vee (\neg R_1 \wedge \neg R_2 \wedge R_3)$. Redundancy check requires $R_1 \subset R_3$. When $R_1 \subset R_3$, then $R_1 \cap R_3 \Rightarrow R_1$ and $R_1 \cup R_3 \Rightarrow R_3$. To simplify the formula, we can apply demorgan's law. The final result will be similar to P_B that has only two rules R_2 and R_3. The formal representation for P_B is $R_2 \vee (\neg R_2 \wedge R_3)$. After simplifying P_A and P_B, the same Boolean expression is reached. Therefore, removing an intra-table redundant rule will not change the matching semantic for MySQL final policy. For the sake of simplicity we used 3 rules, the proof can be easily extend to any number of rules.

In case of inter-table redundancy anomaly, the redundant rule is located in the lower level grant table. Let the original policy P_A includes a rule R_s is identified as an inter-table redundant violation. Let the policy P_B be the policy generated after removing rule R_s from P_A. Let the subset rule, R_s, be in Y_{BDD} grant table and the superset rule be in X_{BDD}. Before removing the redundant rule, the intersection of grant table BDDs is: $X_{BDD} \cap Y_{BDD} \neq \phi$. After removing the subset rule R_s from Y_{BDD} the intersection of the two grant table BDDs is ϕ and $R_s \cap X_{BDD} = R_s$, because of the presence of a superset rule that covers R_s in X_{BDD}. Therefore, any request will be matched in the upper grant table after removing the inter-table redundant rule from the lower grant table.   □

4.4   Algorithms

This section defines algorithms that are used to extract the list of intra/inter table redundant rule conditions in MySQL based on the BDD representation of the policy. Algorithm 1 detects intra-table redundant rule by comparing each rule condition with its possible supersets succeeding it. Algorithm 2 detects complete inter-table redundant rules in the *db* grant table based on U_{BDD} and D_{BDD}.

5   Implementation and Evaluation

We developed and tested our framework against synthetic policies to show the scalability of the framework. It is difficult to get a large number of real-life MySQL policies as these policies are often regarded as confidential. We developed a policy generator engine that generates synthetic policies, with a specific number of rules at each level and probabilities of intra-table (P_{intra}) and inter-table (P_{inter}) redundancy. The experiments were performed on Mac OS X 10.5.5 with

Algorithm 1. IntraRedRules

```
Input: L_u (Sorted list of user rule BDDs)
Output: L_r
1  for i = 0 to |L_u| - 2 do
2      for j = (i + 1) to |L_u| - 2 do
3          P = L_u[i];
4          N = L_u[j];
5          I = N ∩ P;
6          if P == I and PV_i == PV_j then
7              L_r ∪ {(i, j)};
8          end
9          if P == I and PV_i != PV_j then
10             break;
11         end
12     end
13 end
14 return L_r
```

Algorithm 2. InterRedRules

```
Input: U_BDD, D_BDD
Output: L_s
1  I = U_BDD ∩ D_BDD;
2  L_s = φ;
3  while (I) has satisfying assignments do
4      b = One satisfying assignment of (I);
5      L_s = L_s ∪ info(b);
6      I = I ∩ (¬b);
7  end
8  return L_s;
```

Fig. 3. Algorithms

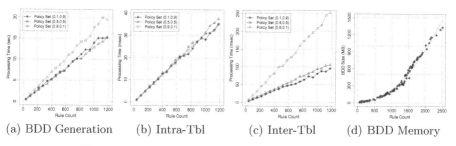

(a) BDD Generation   (b) Intra-Tbl   (c) Inter-Tbl   (d) BDD Memory

Fig. 4. User Study Results and Algorithms Performance

4GB RAM and a 2.4GHz Dual Core, using the BuDDy library v2.4 and MySQL client library v5.1.

The synthetic policies were analyzed by our framework and the BDD generation, intra-table and inter-table redundancy average processing times were recorded. Figure 4(a), shows the initial processing time required to build the BDD for different policy sets where (P_{inter}, P_{intra}) are the inter-table and intra-table redundancy probabilities respectively. The intra-table and inter-table redundancy processing times are reported in Figures 4(b) and 4(c) respectively. Note, that the BDD generation, intra-table and inter-table processing times are linear with respect to the number of policy rules. The discovery and resolution of the inter-table redundancy depends on the percentage of inter-table introduced in the synthetic policy, for a policy containing 1200 policy rules it takes around 90ms and 250ms for P_{inter} values of 0.1 and 0.9 respectively, refer to Figure 4(c). In addition, we recorded the memory requirements for storing the BDD generated, the memory required is polynomial (degree 2) w.r.t the number of policy rules, the regression estimate $(R^2 = 0.983)$ is plotted in Figure 4(d).

6   Related Work

BDDs were utilized to resolve anomalies in access control lists [2, 6, 7]. The work presented by Al-Shaer et al. in [2,6] used BDDs to discover and resolve anomalies in network devices such as firewalls, IPSecs, etc. The work introduced enter-policy and intra-policy anomalies. Redundancy, shadowing, correlation and exception were resolved in this work. Web access control list anomalies were studies by Hu *et al.* in [7], where XACML policies were modeled using BDDs, which was used to discover and resolve conflicts and redundancy in both XACML policy and policy set levels.

Role based access control, which has made significant simplifications in the management of security policies. Roles represent functional roles in an enterprise and individual users acquire authorizations through their assigned roles. Research related to RBAC policy verification [1,9] has focused on verifying RBAC implementation, Separation of Duty and role hierarchy constraints.

7 Conclusion

In this paper, we presented a formal approach to model and define anomalies in MySQL policies. We utilized Binary Decision Diagrams (BDDs) to encode MySQL policy and grant tables. We presented and formalized intra-table and inter-table redundancy anomalies. In addition, we provided a mechanism for improving the performance of policy evaluation by upgrading rules from one grant table to another grant table. We implemented our proposed approach as a tool called *MySQLChecker*. The experimental evaluation conducted on the *MySQLChecker* shows the efficiency and scalability of finding and resolving the presented policy anomalies.

References

1. Ahn, G.-J., Hu, H.: Towards realizing a formal RBAC model in real systems. In: Proceedings of the 12th ACM Symposium on Access Control Models and Technologies, SACMAT 2007, pp. 215–224. ACM, New York (2007)
2. Al-Shaer, E.S., Hamed, H.H.: Discovery of policy anomalies in distributed firewalls. In: INFOCOM 2004, vol. 4 (March 2004)
3. Application Security Inc. Database security tips for 2012 (2011), http://www.appsecinc.com/santa-breach/Database_Security_Tips_2012.pdf
4. Brace, K.S., Rudell, R.L., Bryant, R.E.: Efficient implementation of a BDD package. In: Proceedings of the 27th ACM/IEEE Design Automation Conference, DAC 1990, pp. 40–45. ACM, New York (1990)
5. Bryant, R.E.: Graph-based algorithms for boolean function manipulation. IEEE Trans. Comput. 35, 677–691 (1986)
6. Hamed, H.H., Al-Shaer, E.S., Marrero, W.: Modeling and verification of IPSec and VPN security policies. In: 13th IEEE International Conference on Network Protocols, ICNP 2005, pp. 259–278 (2005)
7. Hu, H., Ahn, G.-J., Kulkarni, K.: Anomaly discovery and resolution in web access control policies. In: Proceedings of the 16th ACM Symposium on Access Control Models and Technologies, SACMAT 2011, pp. 165–174. ACM, New York (2011)
8. Lupu, E.C., Sloman, M.: Conflicts in policy-based distributed systems management. IEEE Trans. Softw. Eng. 25, 852–869 (1999)
9. Shafiq, B., Masood, A., Joshi, J., Ghafoor, A.: A role-based access control policy verification framework for real-time systems. In: WORDS 2005 (February 2005)

Author Index